# Higher Franz-Reidemeister Torsion

AMS/IP

# Studies in Advanced Mathematics

Volume 31

# Higher Franz-Reidemeister Torsion

Kiyoshi Igusa

American Mathematical Society • International Press

Shing-Tung Yau, General Editor

2000 *Mathematics Subject Classification.* Primary 19D10; Secondary 55R40, 57R45, 19F27.

---

**Library of Congress Cataloging-in-Publication Data**

Igusa, Kiyoshi, 1949–
Higher Franz-Reidemeister torsion / Kiyoshi Igusa.
p. cm. — (AMS/IP studies in advanced mathematics, ISSN 1089-3288 ; v. 31)
Includes bibliographical references and index.
ISBN 0-8218-3170-4 (alk. paper)
1. Reidemeister torsion. I. Title. II. Series.

QA613.4 .I38 2002
514′.3—dc21 2002027975

---



∞ The paper used in this book is acid-free and falls within the guidelines
established to ensure permanence and durability.
Visit the AMS home page at http://www.ams.org/
Visit the International Press home page at URL: http://www.intlpress.com/

10 9 8 7 6 5 4 3 2 1 07 06 05 04 03 02

To the three men who taught me how to do mathematics:
Jun-Ichi Igusa, Allen Hatcher and Friedhelm Waldhausen.

# Contents

# List of Figures

# Introduction

The purpose of this book is to develop the theory of higher Franz-Reidemeister torsion (FR-torsion) from its foundations up to the current results. We give several descriptions of the higher FR-torsion, show that they are equivalent, use them to define the higher FR-torsion in a very general setting and also show that it can now be easily computed in many cases. This invariant is not the same as higher Whitehead torsion although they are closely related. Also, higher FR-torsion is conceptually very different from the higher analytic torsion of Bismut and Lott although the numerical results appear to agree in many but not all cases. One of the big challenges is to find some theoretical basis for the similarity in the results of the two theories.

## Classic Whitehead torsion, Reidemeister torsion and Ray-Singer analytic torsion

There are three classic types of torsion: Whitehead torsion, Reidemeister torsion and Ray-Singer analytic torsion.

*Whitehead torsion* is an invariant associated to a relative cell complex $(X, A)$ having a finite number of cells:

$$X = A \cup e_1 \cup e_2 \cup \cdots \cup e_n$$

so that $A$ is a deformation retract of $X$. By taking the mapping cylinder $(Z(f), X)$ one can also associate a Whitehead torsion invariant to a homotopy equivalence $f : X \to Y$ between finite cell complexes. If the torsion of $f$ is zero then $f$ is called a *simple homotopy equivalence*.

By Morse theory one can associate to any compact smooth (i.e. $C^\infty$) manifold a finite cell complex which is well defined up to simple homotopy equivalence. Consequently, a homotopy equivalence between compact smooth manifolds has a well defined Whitehead torsion invariant. Similarly, a compact smooth manifold pair $(M, A)$ (where $A$ is a compact smooth submanifold of $M$) has a Whitehead torsion invariant if $A$ is a deformation retract of $M$, for example if $M$ is a smooth $h$-cobordism of $A$.

The Whitehead torsion invariant lies in the Whitehead group $Wh_1(\pi)$ which is an abelian quotient of the infinite general linear group of $\mathbb{Z}[\pi]$ where $\pi$ is the fundamental group of the manifold. Typically, Whitehead torsion is a *complete invariant* in the sense that it is usually zero if and only if the objects are topologically equivalent. For example, the $s$-cobordism theorem says that, in the stable range $\dim M \geq 6$, an $h$-cobordism $(M, \partial_0 M)$ is diffeomorphic to a product $(\partial_0 M \times I, \partial_0 M)$ if and only if the Whitehead torsion of $(M, \partial_0 M)$ is zero. Therefore a computation of the Whitehead group gives a classification of $h$-cobordisms up to diffeomorphism.

*Reidemeister torsion* is a real number associated to any finite relative cell complex $(X, A)$ which is acyclic with respect to a Hermitian locally trivial coefficient sheaf $\mathcal{F}$. (This means that $\mathcal{F}$ has stalks $\mathbb{C}^n$ which are twisted by a unitary representation $\rho : \pi_1 X \to U(n)$ of the fundamental group of $X$.) By Morse theory this also applies to compact smooth manifold pairs $(M, A)$. Typically, we take $A$ to be empty. For example, lens spaces have Reidemeister torsion with respect to several different unitary representations of their fundamental group $\pi_1 L = \mathbb{Z}/p$.

Although Reidemeister torsion is defined in more general situations than Whitehead torsion (since any homotopy equivalence is a homology equivalence), it requires a unitary representation of the fundamental group. When they are both defined the Reidemeister torsion is a function of the Whitehead torsion: for each unitary representation $\rho$ of $\pi$, the Reidemeister torsion is the real part of the log of the determinant of the complex matrix induced by $\rho$ from any matrix representing the Whitehead torsion.

The acyclicity condition is not always necessary to define Reidemeister torsion. It is sufficient to have a volume form on the homology $H_*(M, A; \mathcal{F})$.

In the 1930's Franz and DeRham classified lens spaces up to homeomorphism using Reidemeister torsion. (See [Mil66].) For this reason it is sometimes called *Franz-Reidemeister-DeRham torsion.*

In 1971, Ray and Singer [RS71] defined *analytic torsion* (also called *Ray-Singer analytic torsion*) and conjectured that it was the same as FR-torsion. This was proved by Cheeger [Che79] and Müller [Mül78]. (See also [BZ92].)

## Higher Whitehead torsion

The history of higher Whitehead torsion begins in 1969 when Jean Cerf showed that pseudoisotopy implies isotopy for smooth simply connected manifolds of dimension $\geq 5$ [Cer70]. In other words, the pseudoisotopy space $\mathcal{C}(M)$ is connected. Allen Hatcher and John (Jack) Wagoner realized that the algebraic part of Cerf's argument was essentially equivalent to Matsumoto's theorem ([Mat69], also [Mil71], pp 81-92) which says that the algebraic $K$-group $K_2\mathbb{Z}$ has only two elements. In [HW73] Hatcher and Wagoner defined $Wh_2(\pi)$ as a quotient of the algebraic $K$-group $K_2\mathbb{Z}[\pi]$ (so that, e.g., $Wh_2(1) = 0$) and obtained a surjective mapping

$$\xi_0 : \pi_0 \mathcal{C}(M) \twoheadrightarrow Wh_2(\pi_1(M)).$$

In [Igu79], I extended this to define a natural mapping

$$\xi_1 : \pi_1 \mathcal{C}(M) \to Wh_3(\pi_1(M))$$

where $Wh_3(\pi_1(M))$ is a quotient of $K_3\mathbb{Z}[\pi_1(M)]$ and in [Igu84] I gave a formula for the cokernel of $\xi_1$.

The higher homotopy groups of $\mathcal{C}(M)$ map to higher Whitehead torsion groups of $\pi_1(M)$. However, to obtain a complete invariant we need a fancy version of these higher Whitehead torsion groups which incorporates the higher homotopy groups of $M$ as well as $\pi_1(M)$. Friedhelm Waldhausen accomplished this around 1976 ([Wal85], [Wal82], [Wal87b]) when he constructed a $K$-theory space of $M$ which he called $A(M)$. Rationally, this is the $K$-theory of the integer group ring of the entire loop group of $M$ which incorporates all of the higher homotopy groups of $M$. (Integrally, it is the algebraic $K$-theory of the $A_\infty$-ring space $\Omega^\infty S^\infty(\Omega M_+)$ in the language of Peter May [May78].) Waldhausen also constructed the smooth Whitehead space of $M$ which he called $Wh^{diff}(M)$ and proved the following.

- $\Omega Wh^{diff}(M) \simeq B\mathcal{P}(M)$ where $\mathcal{P}(M) = \lim_{\rightarrow} \mathcal{C}(M \times I^n)$. In particular, there is a natural mapping $B\mathcal{C}(M) \to \Omega Wh^{diff}(M)$.
- $A(M) \simeq Wh^{diff}(M) \times \Omega^\infty S^\infty(M_+)$.

Composing these we get a mapping

$$B\mathcal{C}(M) \to \Omega A(M). \tag{0.1}$$

By naturality of Waldhausen $K$-theory there is also a mapping

$$A(M) \to \mathbb{Z} \times BGL(\mathbb{Z}[\pi_1 M])^+. \tag{0.2}$$

(Recall that $K_i(R) = \pi_i(BGL(R)^+)$ for $i \geq 1$ and any ring $R$.) It was shown in [Igu88] that the mapping $\mathcal{C}(M) \to \mathcal{P}(M)$ is a homotopy equivalences in a stable range. The map (0.2) is a rational homotopy equivalence in the range in which the higher homotopy groups of $M$ are zero. Consequently, for example, in the special case when $M$ is a disk, Waldhausen's theorem says that, in the stable range:

$$\pi_{4k-1}(Diff(D^n)/O(n)) \otimes \mathbb{Q} \cong (K_{4k+1}\mathbb{Z}) \otimes \mathbb{Q}.$$

By the work of Armand Borel ([Bor74b], [Bor74a]) we know the rank of the higher $K$-theory of any number field, or ring of integers of any number field, say $F$. For the case $F = \mathbb{Z}$, $K_{4k+1}\mathbb{Z}$ has rank 1 for all $k \geq 1$. Borel detected all of the rational higher $K$-theory of $F$ by composing the maps $K_{2k+1}F \to K_{2k+1}\mathbb{C}$ induced by embeddings $\rho : F \to \mathbb{C}$ with a fixed mapping $K_{2k+1}\mathbb{C} \to \mathbb{R}$ called the *Borel regulator map*.

It will turn out that the higher FR-torsion, in the case of pseudoisotopies, is equivalent to the composition of the two known mappings:

1. Waldhausen's map:

$$\pi_{i-1}\mathcal{C}(M) \to \pi_{i+1}A(M) \to K_{i+1}\mathbb{Z}[\pi_1 M]$$

2. The Borel regulator map:

$$K_{2k+1}\mathbb{Z}[\pi_1 M] = \pi_{2k+1}BGL(\mathbb{Z}[\pi_1 M])^+ \xrightarrow{Hurewicz}$$
$$H_{2k+1}(GL(\mathbb{Z}[\pi_1 M])) \xrightarrow{\rho_*} H_{2k+1}(GL(\mathbb{C})) \xrightarrow{\cap c_k} \mathbb{R}$$

where $c_k \in H_c^{2k+1}(GL(\mathbb{C}); \mathbb{R})$ is a fixed continuous cohomology class.

By the work of Alexander Beĭlinson [Beĭ85] (see also [Esn89], [JLDZ00], [Ram88], [Rap88]), it was also already known that the value of the Borel regulator on a generator of $K_{2k+1}\mathbb{Q}(\zeta)$ (generator by the Galois action) for $\zeta$ an $n$-th root of unity is a rational multiple of the real part of

$$\frac{1}{i^k}\sum_{n=1}^{\infty}\frac{\zeta^n}{n^{k+1}}.$$

Thus, the higher FR-torsion for pseudoisotopies was already "known." However, pseudoisotopies are a very special case, whereas the higher FR-torsion is defined in a very general setting (for smooth compact manifold bundles over all simply connected spaces and more generally when the fundamental group of the base acts trivially on the homology of the fiber). Also, despite the wealth of knowledge we have about pseudoisotopies, Waldhausen $K$-theory, algebraic $K$-theory and the homotopy type of diffeomorphism spaces, the higher FR-torsion for even the simplest case of linear circle bundles was not known until very recently ([IK93] and Chapter VII below). In this book we will, for the most part, avoid the pseudoisotopy case and concentrate on the other cases. One exception will be Hatcher's construction in the pseudoisotopy of disks (§6.4).

# Higher analytic torsion

In 1995, Jean-Michel Bismut and John Lott constructed higher analytic torsion forms

$$\mathcal{T}_k(E;\mathcal{F}) \in \Omega^{2k}(B)$$

for smooth bundles $p : E \to B$ with Hermitian locally trivial coefficient systems $\mathcal{F}$ and certain other additional structures. In certain cases these forms are closed giving cohomology classes

$$\tau_k^{BL}(E;\mathcal{F}) \in H^{2k}(B;\mathbb{R}),$$

for example, when the fiber $M$ is odd dimensional and acyclic (with coefficients in $\mathcal{F}$). This cohomology class is topological (defined for bundles with structure group $Diff(M)$ and independent of the horizontal distribution and vertical metric) in many cases and has been explicitly computed in the case of circle bundles by Bismut and Lott [BL95], odd dimensional sphere bundles by Bunke [Bun] and in other cases by Bismut and Goette [BG00].

# Higher FR-torsion

As noted earlier, higher FR-torsion invariants in the case of pseudoisotopies can be obtained as a function of higher Whitehead torsion invariants using the Waldhausen $K$-theory space $A(X)$ and its relationship to the algebraic $K$-theory of discrete rings. It was John Wagoner [Wag78] who first proposed that higher FR-torsion might be defined more generally. John Klein [Kle89] was the first to accomplish this. Klein used a variation $A^\rho(X)$ of Waldhausen's $A(X)$ which replaced homotopy equivalences with $\rho$-homology equivalences where $\rho$ is a complex representation of $\pi_1 X$. Later Klein and I developed two other methods to describe the higher FR-torsion. (We take the Platonic point of view that the higher FR-torsion already exists and we are looking for ways to describe and compute it.) The first was the space of acyclic filtered chain complexes $FC^h(X, \xi)$ which we used to define higher FR-torsion in a general setting in ([Igu91], [IK95]) and the second was the computation in the special case of circle bundles over $S^2$ using "pictures" and the Bloch-Wigner dilogarithm [IK93]. In the meantime Dwyer, Weiss and Williams obtained a different topological construction for higher FR-torsion [DWW]. These different topological constructions are probably equivalent.

In this book we develop what is essentially an algebraic refinement of the filtered chain complex approach to higher FR-torsion.

# Contents of this book

In this book we show how higher Franz-Reidemeister torsion can be defined and how it can be computed. Higher FR-torsion is defined for all smooth manifold bundles $p : E \to B$ with compact manifold fiber $M$ so that $\pi_1(B)$ acts trivially on the homology groups of the fiber $H_*(M, \mathcal{F})$ where $\mathcal{F}$ is a Hermitian coefficient system for $E$. They are cohomology classes:

$$\tau_k(E; \mathcal{F}) \in H^{2k}(B; \mathbb{R}).$$

The construction of these higher FR-torsion classes is analogous to the pseudoisotopy case explained above. First, we use Morse theory in the form of the framed function theorem ([Igu87], explained in §4.6 below) to go from the smooth bundle $E \to B$ to a mapping from $B$ into a Whitehead space. Then we compose with the Borel regulator map on Volodin $K$-theory.

## Mapping to the Whitehead space

The framed function theorem tells us that there exists a smooth map

$$f : E \to \mathbb{R}$$

which is a "framed" generalized Morse function on each fiber provided that $\dim M \geq \dim B$. (When this is not satisfied we stabilize $E$ by replacing it with a product $E \times D^n$.) Taking the resulting family of cellular chain complexes $\{C(M_\sigma)\}$ for all generic small simplices $\sigma$ in $B$ we obtain a functor $\xi_E$ from the category of generic small smooth simplices in the base $B$ to the (simplicial) Whitehead category $\mathcal{W}h_\bullet(\mathbb{Z}[\pi_1 E], \pi_1 E)$ of based free chain complexes over the ring $\mathbb{Z}[\pi_1 E]$:

$$\xi_E : \operatorname{simp} B \to \mathcal{W}h_\bullet(\mathbb{Z}[\pi_1 E], \pi_1 E). \tag{0.3}$$

Furthermore, for every unitary representation $\rho : \pi_1 E \to U(n)$ we obtain a simplicial functor

$$\rho_* : \mathcal{W}h_\bullet(\mathbb{Z}[\pi_1 E], \pi_1 E) \to \mathcal{W}h_\bullet(M_n(\mathbb{C}), U(n)).$$

In the special case when the fiber $M$ is acyclic with respect to the representation $\rho$ the composition $\rho_* \circ \xi_E$ lifts to the simplicial full subcategory $\mathcal{W}h^h_\bullet(M_n(\mathbb{C}), U(n))$ of acyclic based free chain complexes over the ring $M_n(\mathbb{C})$.

By a theorem of John Klein and myself ([IK95], also Theorem 0.0.2 below) we know that there are universal classes for higher FR-torsion:

$$\tau_k \in H^{2k}(\mathcal{W}h^h_\bullet(M_n(\mathbb{C}), U(n)); \mathbb{R}) \tag{0.4}$$

which we can now pull back to give higher FR-torsion classes for $E$ (since $B \simeq |\operatorname{simp} B|$):

$$\tau_k(E, \rho) = \xi_E^* \rho^*(\tau_k) \in H^{2k}(B; \mathbb{R}).$$

This was the procedure which was outlined in [Igu91]. It was also announced that higher FR-torsion could be defined in the case when the homology of the fiber injects into the homology of the total space. In this book this is strengthened to the following. (See Corollary 3.6.8.)

**Theorem 0.0.1.** *Let $\rho : E \to B$ be a smooth bundle with compact manifold fiber $M$ and suppose that $G$ is a subgroup of the group of units of a ring $R$ and $\rho : \pi_1 E \to G$ is a homomorphism with the property that the fiberwise homology $H_*(M; R)$ of $M$ with twisted $R$ coefficients is a projective $R$-module with trivial action of $\pi_1 B$. Then the natural mapping*

$$B \to \mathcal{W}h_\bullet(R, G)$$

*lifts canonically to $\mathcal{W}h^h_\bullet(R, G)$. In particular when $(R, G) = (M_n(\mathbb{C}), U(n))$ we obtain well-defined higher FR-torsion invariants for $E$.*

Since $M_n(\mathbb{C})$ is semisimple, all modules are projective and the only condition for the existence of higher FR-torsion is the trivial action of $\pi_1 B$ as we claimed earlier.

In the special case when the coefficient ring $R$ is a field we have a further extension of this result to the case when $\pi_1 B$ acts on $H_*(M; R)$ by upper triangular transformations with respect to some basis. (See Theorem 3.5.11, Corollary 3.6.12.)

## Computation of higher FR-torsion

In order to explicitly compute this higher FR-torsion class we need an explicit formula for the universal class $\tau_k$. In [IK93] John Klein and I used the Block-Wigner dilogarithm to find a formula for $\tau_1$ and compute it in the case of oriented linear circle bundles over $S^2$.

In the present work I use a rational multiple of the Kamber-Tondeur-Dupont formula [Dup76] for the continuous cohomology class

$$c_k \in H_c^{2k+1}(BGL(\mathbb{C});\mathbb{R})$$

which gives the Borel regulator map $K_{2k+1}\mathbb{C} \to \mathbb{R}$. I use this to compute the higher FR-torsion invariants for oriented linear circle bundles, oriented linear sphere bundles, lens-space bundles, nonlinear disk bundles (Hatcher's example), the mapping class group and the Torelli group.

The first step is to convert this $(2k+1)$-cocycle for $GL(\mathbb{C})$ into a $2k$-cocycle for the Volodin space $V(\mathbb{C}) \simeq \Omega BGL(\mathbb{C})^+$. This is carried out in Chapter I. The Volodin space of a ring $R$ is a space of invertible matrices over $R$ which locally vary by elementary column operations. In Chapter II we embed this space into the larger space of invertible matrices over $R$ which locally vary by elementary row and column operations and by $G$-monomial basis change. We call this $\mathcal{W}_\bullet(R,G)$ and we extend the Volodin $K$-theory cohomology class to a class

$$\tau_k \in H^{2k}(\mathcal{W}_\bullet(M_n(\mathbb{C}),U(n));\mathbb{R}). \tag{0.5}$$

This allows us to define and compute the higher FR-torsion for all oriented circle bundles. The computation is reduced to one special case using transfer arguments in Chapter II and this special case is worked out in detail in Chapter VII.

When the cellular chain complex of the fiber is both acyclic and in two indices, it is given by a single invertible matrix and therefore we get a mapping from $B$ into some $\mathcal{W}_\bullet(R,G)$. In order to define the higher FR-torsion classes more generally we need a chain complex version of $\mathcal{W}_\bullet(R,G)$. In Chapter III we construct the simplicial category $\mathcal{W}h_\bullet(R,G)$ of based free $R$-complexes which locally vary by $G$-monomial and elementary basis change in each degree. Since invertible matrices are the same as acyclic based free chain complexes which are nonzero in only two consecutive degrees, $\mathcal{W}_\bullet(R,G)$ is in some sense a full subcategory of $\mathcal{W}h_\bullet^h(R,G)$ where $(\ )^h$ means we take only acyclic chain complexes. The two-index theorem (proved in §5.9) says that

$$\mathcal{W}_\bullet(R,G) \simeq \mathcal{W}h_\bullet^h(R,G). \tag{0.6}$$

In particular they have the same cohomology for any $(R,G)$ so the cohomology classes $\tau_k$ in (0.5) uniquely determined the universal classes in (0.4).

Chapter III also contains the basic result by which the higher FR-torsion can be extended to the non-acyclic case when one of the following conditions holds (Theorem 3.6.7).

1. The homology $H_*(M;R)$ of the fiber $M$ is a projective $R$-module with trivial $\pi_1 B$-action or

2. $R$ is a field and $H_*(M;R)$ has a filtration by $R[\pi_1 B]$-submodules so that $\pi_1 B$ acts trivially on the associated graded module.

Chapters IV and V explains the relationship between $\mathcal{W}h_\bullet(R,G)$ and the space of filtered chain complexes. The result is:

$$|\mathcal{W}h^h_\bullet(R,G)| \simeq |FC^h(BG,R)| \tag{0.7}$$

and the same with "$h$" deleted. We need this result for the following reason. When we use the framed function theorem we get a canonical family of oriented fiberwise generalized Morse functions (GMF's) on the fibers. This gives us, in a canonical way, a mapping

$$\operatorname{simp} B \to FC(B\pi_1 E, \mathbb{Z}[\pi_1 E])$$

by sending each simplex to the total singular complex of its inverse image in $E$ together with a filtration given by the fiberwise GMF. The homotopy equivalence (0.7) tells us that we get a mapping $\xi_E$ as in (0.3) which is well defined up to homotopy.

In Chapter V, we prove the two-index theorem (0.6) and explain the main theorem of [IK95]:

**Theorem 0.0.2.** *There is a homotopy fiber sequence*

$$|FC^h(X_\bullet,R)| \to \Omega^\infty S^\infty(|X_\bullet|_+) \to \overline{\mathbb{Z}} \times BGL(R)^+$$

*where $\overline{\mathbb{Z}}$ is the image of $K_0\mathbb{Z} = \mathbb{Z}$ in $K_0R$.*

This theorem is important for several reasons. First, it says that the space of filtered chain complexes (and also $\mathcal{W}_\bullet(R,G)$, $\mathcal{W}h^h(R,G)$ by (0.6) and (0.7)) has the "correct" homotopy type and therefore we have the correct definition of higher FR-torsion. Secondly, this implies that the involution of reversing the sign of the fiberwise GMF changes the sign of the higher FR-torsion in the expected way. This is proved in §5.8.

In Section 5.7 we use the two-index theorem to make calculations of higher FR-torsion for linear and sphere bundles and their associated lens space bundles. These agree with Ulrich Bunke's computations when the dimension of the fiber is odd. For even dimensional fibers the computations do not agree. However, the discrepancy seems to vanish if we replace the higher FR-torsion defined by a framed function $f$ with the average of the higher FR-torsions defined by $f$ and $-f$. (This is an application of Theorem 0.0.2 explained in Section 5.8.)

In Chapter VI we derive a general rule which we call the "framing principle" by which we may determine the higher FR-torsion invariant of a bundle $E \to B$ from an oriented fiberwise GMF on $E$ which is not framed. Section 6.1 gives the precise statement of the framing principle and proves it as an easy consequence of the higher FR-torsion computation for linear sphere bundles. Sections 6.2 and 6.3 contain a review of pseudoisotopy theory and results and methods of Jean Cerf. In §6.4 we give Hatcher's construction for a rationally nontrivial element of $\pi_{4k-1}(Diff(D^n)/O(n))$. The well-known argument ([FH78]) for the existence of such exotic elements is explained in §6.5. Hatcher's example in particular was shown to be equivalent to the known exotic classes by Marcel

Bökstedt using homotopy theoretic methods ([Bök84]). We give another proof of this by computing the higher FR-torsion invariant for Hatcher's example using the framing principle. Finally, in §6.6 we give a product formula for higher FR-torsion and prove it for fiber products of oriented linear sphere bundles and in two other cases.

In Chapter VIII we look at the moduli space of curves and the Torelli group. It is a result of Richard Hain, Robert Penner and myself, based on work of John Klein [Kle93], that the higher FR-torsion $\tau_{2k}$ of the mapping class group is a nonzero multiple of the Miller-Morita-Mumford classes $\kappa_{2k}$ plus a linear combination of products of lower terms. By restriction the same is true for the Torelli group. Using the framing principle we refine this to obtain an explicit computation (Theorem 8.5.19):

$$\tau_{2k}(M_g^s) = (-1)^k \zeta(2k+1) \frac{\kappa_{2k}}{(2k)!}.$$

Combinatorial constructions on the category of graphs which are needed to fill in the details of this argument are given. We define a structure called a "framed graph" which is the additional structure on the graph needed to construct a framed function on a tubular neighborhood of the graph when it is immersed in $R^n$. Since the category of connected graphs with $H_1 = \mathbb{Z}^n$ is the classifying space of the outer automorphism group of the free group on $n$ letters by Culler and Vogtmann [CV86] , we obtain higher FR-torsion classes

$$\tau_{2k}(Out^h(F_n)) \in H^{4k}(Out^h(F_n); \mathbb{R})$$

which restrict to the higher FR-torsion classes on the Torelli group (Corollary 8.5.17).

The mapping $BOut^h(F_n) \to \mathcal{W}h_\bullet^h(\mathbb{Z}, 1)$ used to define the higher FR-torsion on $Out^h(F_n)$ fits into the following commuting diagram where the rows are homotopy fiber sequences.

$$\begin{array}{ccccc} BOut^h(F_n) & \longrightarrow & BOut(F_n) & \longrightarrow & BGL_n(\mathbb{Z}) \\ \downarrow & & \downarrow & & \downarrow \\ |\mathcal{W}h_\bullet^h(\mathbb{Z},1)| & \longrightarrow & |\mathcal{W}h_\bullet(\mathbb{Z},1)| & \longrightarrow & \mathbb{Z} \times BGL(\mathbb{Z})^+ \end{array}$$

Since $|\mathcal{W}h_\bullet(\mathbb{Z},1)| \simeq QS^0$ is rationally trivial, this implies that the mapping $Out(F_n) \to GL(\mathbb{Z})$ is trivial in rational homology. (See Theorem 8.5.3.)

The logical dependency of the chapters is given as follows. However, from Chapter VII only Theorem 7.10.2 is needed. The proof and the rest of Chapter VII can be skipped.

$$\begin{array}{ccccccccc} & & Chap\ I & \longrightarrow & Chap\ II & \longrightarrow & Chap\ VII & & \\ & & & & & & \downarrow & & \\ Chap\ III & \longrightarrow & Chap\ IV & \longrightarrow & & & Chap\ V & \longrightarrow & Chap\ VI \longrightarrow Chap\ VIII \end{array}$$

I am grateful for the help and encouragement I received from many people during the many years that I worked on this project: Mel Rothenberg, Donggeng Gong, Igor Najfeld, Richard Palais, John Klein, Tom Goodwillie, Robert Penner, Richard Hain and Sebastian Goette. I also owe a great debt to Jean Cerf whose pioneering works on diffeomorphism groups ([Cer68], [Cer70]) inspired me to study pseudoisotopy theory. Also, both John Klein and I were motivated by John Wagoner's original article on higher torsion [Wag78]. I should thank the National Science Foundation for their generous support. I would also like to thank my wife Gordana Todorov and her mother Rakila Todorov for their support and understanding while I concentrated all my efforts on finishing this book. Also Gordana deserves credit for revising this introduction so that it is readable.

November 2001, Kiyoshi Igusa

# Chapter 1

# Cocycles in Volodin K-Theory

## Introduction

The Borel regulator map in algebraic $K$-theory is a homomorphism

$$K_{2k+1}(\mathbb{C}) = \pi_{2k+1}(BGL(\mathbb{C})^+) \to H_{2k+1}(BGL(\mathbb{C})) \xrightarrow{\cap[I_{2k+1}]} \mathbb{R} \tag{1.1}$$

given by a continuous cohomology class $[I_{2k+1}]$ on $GL(\mathcal{C})$ which is represented by a continuous $(2k+1)$-cocycle

$$I_{2k+1} : GL(\mathbb{C})^{2k+2} \to \mathbb{R}.$$

To compute the higher Reidemeister torsion, however, we need a formula for the Borel regulator on the level of Volodin $K$-theory:

$$K_{2k+1}(\mathbb{C}) \cong K^V_{2k+1}(\mathbb{C}) = \pi_{2k}(V(\mathbb{C})) \to H_{2k}(V(\mathbb{C})) \xrightarrow{\cap[D_{2k}]} \mathbb{R}. \tag{1.2}$$

For this we need a $2k$-cocycle $D_{2k}$ for the Volodin space $V(\mathbb{C})$. In this first chapter we obtain a formula for $D_{2k}$.

§1.1 reviews the definition and some properties of the Volodin description of $K$-theory. In §1.2 we modify the continuous cocycle $I_{2k+1}$ to obtain a $2k$-cochain $F_{2k}$ which is (the correct) cocycle if $k = 0$ but is not closed if $k \geq 1$.

In §1.3 we use cyclic homology to obtain a formula for the "correction term" $G_{2k}$ which makes $F_{2k}$ into a cocycle

$$D_{2k} = F_{2k} - G_{2k}.$$

In §1.4 we prove that the homomorphism (1.2) is equal to the homomorphism (1.1) up to sign.

§1.7 contains many computations of $F_{2k}$ and $G_{2k}$. The results are very messy but we extract an important simple conclusion: $G_{2k}$ is polynomial in the real and imaginary parts of the entries of the comparison matrices $(g^i)^{-1}g^j$.

I am indebted to Mel Rothenberg and Donggeng Gong for their help and hospitality during the development of some of the details of this chapter and also to Igor Najfeld for showing me his matrix computation techniques.

## 1.1 Volodin $K$-theory

In this section we will review the definition and main properties of the Volodin space $V_n(R)$ whose homotopy groups give the unstable Volodin $K$-theory of $R$:

$$K_{i,n}^V(R) = \pi_{i-1}(V_n(R)).$$

Given an associative ring $R$ with 1 we have the following sets of $n \times n$ matrices.

$M_n(R)$ is the ring of $n \times n$ matrices with coefficients in $R$.

$GL_n(R) = M_n(R)^\times$ is the group of invertible elements in $M_n(R)$.

$E_n(R)$ is the subgroup of $GL_n(R)$ generated by the elementary matrices $E_{ij}(r)$, $r \in R$, $1 \le i, j \le n$, $i \ne j$.

Let $GL(R) = GL_\infty(R), E(R) = E_\infty(R)$ be the limits of $GL_n(R), E_n(R)$ with respect to the stabilization map given by appending 1's to the lower right hand corner of all matrices. Then $E(R)$ is the commutator subgroup of $GL(R)$ and the first $K$-theory of $R$ can be defined by:

$K_1(R) = GL(R)/E(R) = GL(R)/GL(R)'$

$K_{1,n}(R) = GL_n(R)/E_n(R)$ is the set of right cosets $gE_n(R)$, $g \in GL_n(R)$.

In other words, an invertible $n \times n$ matrix represents the trivial element of $K_{1,n}(R)$ if it can be reduced to the identity matrix by elementary column operations. If $R$ is commutative then a necessary condition is that the determinant of the matrix be equal to 1.

The *Steinberg group* $St_n(R)$ is defined to be the group generated by the symbols $x_{ij}(r)$ where $r \in R$, $1 \le i, j \le n$, $i \ne j$ modulo the *Steinberg relations*:

1. $x_{ij}(r)x_{ij}(s) = x_{ij}(r+s)$
2. $[x_{ij}(r), x_{jk}(s)] = x_{ik}(rs)$ for $i, j, k$ distinct.
3. $[x_{ij}(r), x_{k\ell}(s)] = 1$ if $j \ne k, i \ne \ell$.

where $[a, b] = aba^{-1}b^{-1}$. Then we have an epimorphism

$$\phi : St_n(R) \to E_n(R)$$

given by $\phi(x_{ij}(r)) = E_{ij}(r)$. When $n \ge 3$, both $St_n(R)$ and $E_n(R)$ are (obviously) *perfect groups* in the sense that they are equal to their commutator subgroups.

Letting $St(R) = \lim_{\to} St_n(R)$ we can define the second K-group of $R$ by:

$K_2(R) = \ker(\phi : St(R) \to E(R))$

$K_{2,n}(R) = \ker(\phi : St_n(R) \to E_n(R))$.

In other words, an element of $K_2(R)$ is represented by a sequence of elementary operations whose composition is the identity operation.

To define the higher $K$-groups we need to look at the *$\sigma$-upper triangular matrix group* $T_n^\sigma(R)$ which is defined to be the subgroup of $E_n(R)$ generated by all elementary matrices $E_{ij}(r)$ where $i < j$ in the partial ordering $\sigma$ of the set $\{1, \cdots, n\}$. Since this is a nilpotent group with all commutators defined by the Steinberg relations, $T_n^\sigma(R)$ is isomorphic to the subgroup of $St_n(R)$ generated by the elements $x_{ij}(r)$ where $i <_\sigma j$. Consequently we have:

**Lemma 1.1.1.** *$E_n(R)$ is the subgroup of $GL_n(R)$ generated by $\bigcup_\sigma T_n^\sigma(R)$ and $St_n(R)$ is the direct limit* $\lim_\to T_n^\sigma(R)$ *in the category of groups of the diagram consisting of all $T_n^\sigma(R)$ with all inclusion maps between them.*

(In other words, for any group $G$ and any system of homomorphisms $\{f_\sigma : T_n^\sigma(R) \to G\}$ so that $f_\sigma | T_n^\tau(R) = f_\tau$ for any $\tau \subseteq \sigma$, there exists a unique homomorphism $f : St_n(R) \to G$ so that $f|T_n^\sigma(R) = f_\sigma$.)

Let $EGL_n(R)$ denote the contractible simplicial set which in degree $k$ is

$$EGL_n(R)[k] = GL_n(R)^{k+1}$$

with simplicial operators given by $a^*(g_0, \cdots, g_k) = (g_{a(0)}, \cdots, g_{a(j)})$ for $a : [j] \to [k]$. (Recall that a *simplicial set* is a contravariant functor $X_\bullet : \Delta \to Sets$ where $\Delta$ has objects $[k] = \{0, 1, 2, \cdots, k\}$ for $k \geq 0$ and morphisms all set mappings $a : [j] \to [k]$ where $a(i+1) \geq a(i)$ for every $i < j$.) $GL_n(R)$ acts freely on the right on $EGL_n(R)$ and the realization of the quotient simplicial set is the classifying space of the discrete group $GL_n(R)$:

$$BGL_n(R) = |EGL_n(R)/GL_n(R)|.$$

(Recall that the *(geometric) realization* of a simplicial set $X_\bullet$ is the identification space:

$$|X_\bullet| = \coprod_{k \geq 0} X_k \times \Delta^k / \sim$$

with the quotient topology where the identifications are given by $(a^*(x), t) \sim (x, a_*(t))$ for all $(x,t) \in X_k \times \Delta^j$ and $a : [j] \to [k]$ where $a_* : \Delta^j \to \Delta^k$ is given in barycentric coordinates by $a_*(t)_i = \sum_{a(s)=i} t_s$.)

**Definition 1.1.2.** The *Volodin space* $V_n(R)$ is defined to be the simplicial subset of $EGL_n(R)$ which in degree $k$ is the set of all $(k+1)$-tuples of matrices $(g_0, g_1, \cdots, g_k)$ so that for some partial ordering $\sigma$ of $1, 2, \cdots, n$ the matrices $t_{ij} = g_i^{-1} g_j$ are $\sigma$-upper triangular for all $i, j$ (so that $g_j = g_i t_{ij}$).

Let $V(R) = V_\infty(R) = \lim_\to V_n(R)$. Then the *Volodin $K$-theory* of $R$ can be defined by

$$K_i^V(R) = \pi_{i-1}|V(R)|$$

$$K_{i,n}^V(R) = \pi_{i-1}|V_n(R)|$$

It follows from Lemma 1.1.1 that this agrees with the definitions above for $i = 1, 2$:

**Proposition 1.1.3.** *$K_i^V(R) \cong K_i(R)$ and $K_{i,n}^V(R) \cong K_{i,n}(R)$ for $i = 1, 2$.*

It is a theorem of Wagoner and Vasserstein best explained by Suslin [Sus81], [Sus82] that the higher Volodin $K$-theory of $R$ also gives the usual algebraic $K$-theory groups of $R$ as defined by Quillen. But, we need to know a little bit more about this isomorphism.

Since $GL_n(R)$ acts freely on the left on $V_n(R)$ the quotient of the realization $GL_n(R)\backslash|V_n(R)|$ (which is naturally homeomorphic to the realization of the quotient $GL_n(R)\backslash V_n(R)$) is homotopy equivalent to the *balanced product*

$$X_n(R) = |EGL_n(R)| \times_{GL_n(R)} |V_n(R)|.$$

(Recall that this is by definition the quotient space of $|EGL_n(R)| \times |V_n(R)|$ modulo the equivalence relation

$$(xg, ((g_0, \cdots, g_j), t)) \sim (x, ((gg_0, \cdots, gg_j), t))$$

where $((g_0, \cdots, g_j), t) \in g \times \Delta^j \subseteq |V_n(R)|$ represents the point with barycentric coordinates $t$ in the geometric $j$-simplex $g \times \Delta^j$ with vertices $g_0, \cdots, g_j$.)

This space fits into a fibration sequence:

$$|V_n(R)| \to X_n(R) \to BGL_n(R).$$

**Theorem 1.1.4.** *(See [Sus82].) a) $X(R) = |EGL(R) \times_{GL(R)} V(R)|$ is an acyclic space with perfect fundamental group $\pi_1 X(R) = St(R)$ the Steinberg group of $R$, and furthermore*

*b) $X_n(R)$ is $(n-1)/2$- acyclic, i.e., $\overline{H}_i(X_n(R); \mathbb{Z}) = 0$ for $n > 2i$.*

*c) The Quillen plus construction gives a homotopy fiber square as follows where $X(R)^+$ is contractible:*

$$\begin{array}{ccc} X(R) & \longrightarrow & X(R)^+ \\ \downarrow & & \downarrow \\ BGL(R) & \longrightarrow & BGL(R)^+ \end{array}$$

*d) Consequently, $|V(R)| \simeq \Omega BGL(R)^+$.*

*e) If $A$ is a finite dimensional algebra over a field then $K_{i,n}^V(A) \cong K_i(A)$ for $n > i$ and $\pi_i(BGL_n(A)^+) \cong K_i(A)$ for $n \geq 2i$.*

**Remark 1.1.5.** a) Suslin's space $X_n(R)$ is the quotient $St_n(R)\backslash\widetilde{V}_n^0(R)$ where $\widetilde{V}_n^0(R)$ is the universal covering of the identity component $V_n^0(R)$ of $V_n(R)$. For $n \geq 3$ this is the same as $E_n(R)\backslash V_n^0(R) = GL_n(R)\backslash V_n(R)$ which is homotopy equivalent to $X_n(R)$ and, for $n \leq 2$, (b) is obvious since $GL_n(R)\backslash V_n(R)$ has only one vertex.

b) Suslin shows in [Sus82] (building on van der Kallen's work [vdK80]) that $K^V_{i,n}(R) \cong K_i(R)$ for $n \geq i + sr(R)$ and $\pi_i(BGL_n(R) \cong K_i(R)$ for $n \geq \max(i + sr(R), 2i+1)$ where $sr(R)$ is the "stable rank" of $R$ in [Sus82] which is equal to one plus the "stable dimension" $d = sdim(R)$ in [vdK80], so that the "stable range condition" $\mathrm{R}_{sr(R)+1}(R) = \mathrm{R}_{d+2}(R)$ of [Bas68] holds. (e) is a special case of Suslin's stability result since $d = 0$ for any finite dimensional algebra over any field. (See [Bas68], p.239.)

### 1.1.1 The smooth Volodin space of $\mathbb{C}$

Now suppose that $R = \mathbb{C}$, or more generally, let $R$ be any finite dimensional (f.dim) algebra over the real numbers $\mathbb{R}$. Then $GL_n(\mathbb{C}) \subseteq M_n(\mathbb{C}) \cong \mathbb{C}^{n^2}$ has a smooth ($C^\infty$) structure and we can use this to define a smooth version of the Volodin space $V_n(\mathbb{C})$.

**Definition 1.1.6.** Let $V_n^{diff}(\mathbb{C})$ denote the simplicial set whose $k$ simplices are smooth mappings $\gamma : \Delta^k \to GL_n(\mathbb{C})$ with image $\gamma(\Delta^k)$ contained in some right coset of some $\sigma$-upper triangular matrix group $T_n^\sigma(\mathbb{C})$. Let $V_n^{diff}(A)$ be defined analogously for any f.dim algebra $A$.

There is a *simplicial forgetful map*

$$F : V_n^{diff}(\mathbb{C}) \to V_n(\mathbb{C})$$

given by $F(\gamma) = (g_0, \cdots, g_k)$ where $g_i = \gamma(v_i)$ is the value of $\gamma$ on the $i$-th vertex $v_i$ of $\Delta^k$. (Recall that a *simplicial map* $X_\bullet \to Y_\bullet$ is a natural transformation of contravariant functors on $\Delta$.)

Since the cosets $gT_n^\sigma(\mathbb{C})$ are convex in the $\mathbb{C}$-linear space of all complex matrices, there is right inverse

$$\lambda : V_n(\mathbb{C}) \to V_n^{diff}(\mathbb{C})$$

for $F$ given by *linear interpolation* :

$$\lambda(g_0, \cdots, g_k)(t) = \sum_{i=0}^{k} t_i g_i.$$

The simplicial mapping $\lambda$ is also a homotopy left inverse for $F$. This follows easily from the fact that $\lambda F(\gamma)(\Delta^k)$ lies in any coset $gT_n^\sigma(\mathbb{C})$ containing the image of $\gamma$. Since this is the first example of a simplicial homotopy in this book we will work it out.

First, we recall that a *simplicial homotopy* $h : F \simeq G$ between simplicial maps $F, G : X_\bullet \to Y_\bullet$ is a simplicial map $h : X_\bullet \times I \to Y_\bullet$ where $I$ is the simplicial set $I_k = \mathrm{Hom}_\Delta([k], [1])$, elements being given by $(k+1)$-tuples $(0, \cdots, 0, 1, \cdots, 1)$, the product of simplicial sets is given by $(X \times Y)_k = X_k \times Y_k$ and $F, G$ are related to $h$ by $F(x) = h(x, (0, \cdots, 0))$ and $G(x) = h(x, (1, \cdots, 1))$ for all $x \in X_k$. It is well-known that geometric realization commutes with product and that $|I|$ is homeomorphic to the closed interval $\Delta^1$. Thus a simplicial homotopy induces a homotopy of continuous mappings $|h| : |F| \simeq |G|$. (See [May67], [Seg74].)

With this notation the desired simplicial homotopy $h : \lambda \circ F \simeq id$ is given by

$$h(\gamma, (\overbrace{0, \cdots, 0}^{j+1}, \overbrace{1, \cdots, 1}^{k-j}))(t) = \sum_{i=0}^{j} t_i \gamma(v_i) + s\gamma\left(\sum_{i=j+1}^{k} \frac{t_i}{s} v_i\right)$$

where $s = t_{j+1} + \cdots + t_k$. Thus we have the following.

**Theorem 1.1.7.** *$F, \lambda$ induce (inverse) homotopy equivalences on realizations:*

$$|F| : |V_n^{diff}(\mathbb{C})| \simeq |V_n(\mathbb{C})|,$$

$$|\lambda| : |V_n(\mathbb{C})| \simeq |V_n^{diff}(\mathbb{C})|$$

*for all $n \geq 1$. Similarly, $V_n^{diff}(A) \simeq V_n(A)$ for any f.dim $\mathbb{R}$-algebra $A$.*

We also note that $F, \lambda$ commute with transfer.

**Definition 1.1.8 (Transfer for Volodin space).** For any ring $R$ the *simplicial transfer map*

$$\mathrm{N}_\bullet : V_n(M_m(R)) \to V_{nm}(R)$$

is the simplicial map induced by the natural isomorphism

$$\mathrm{N} : GL_n(M_m(R)) \xrightarrow{\cong} GL_{nm}(R).$$

Since N maps $T_n(M_m(R))$ into $T_{nm}(R)$, the simplicial map $\mathrm{N}_\bullet$ can be defined by $\mathrm{N}_p(g_0, \cdots, g_p) = (\mathrm{N}(g_0), \cdots, \mathrm{N}(g_p))$. For any f.dim $\mathbb{R}$-algebra $A$, the *smooth transfer*

$$\mathrm{N}_\bullet^{diff} : V_n^{diff}(M_m(A)) \to V_{nm}^{diff}(A)$$

can be given by $\mathrm{N}_p^{diff}(\gamma) = \mathrm{N} \circ \gamma$ since N is smooth for $R = A$.

We have the following easy observation.

**Proposition 1.1.9.** *The following diagram commutes for any f.dim algebra $A$ over $\mathbb{R}$.*

$$\begin{array}{ccccc}
V_n(M_m(A)) & \xrightarrow{\lambda} & V_n^{diff}(M_m(A)) & \xrightarrow{F} & V_n^{diff}(M_m(A)) \\
\downarrow{\scriptstyle \mathrm{N}_\bullet} & & \downarrow{\scriptstyle \mathrm{N}_\bullet^{diff}} & & \downarrow{\scriptstyle \mathrm{N}_\bullet} \\
V_{nm}(A) & \xrightarrow{\lambda} & V_{nm}^{diff}(A) & \xrightarrow{F} & V_{nm}^{diff}(A)
\end{array}$$

We will view $V_n(\mathbb{C})$ as a simplicial subset of $V_n^{diff}(\mathbb{C})$ using the linear interpolation map $\lambda$ as the embedding with $F$ being a deformation retraction of $V_n^{diff}(\mathbb{C})$ to $V_n(\mathbb{C})$. If we stabilize then $V(\mathbb{C})$ will be a deformation retract of the direct limit

$$V^{diff}(\mathbb{C}) = \lim_{\rightarrow} V_n^{diff}(\mathbb{C}).$$

One consequence of Theorem 1.1.7 is that $X_n(\mathbb{C}) = |EGL_n(\mathbb{C})| \times_{GL_n} |V_n(\mathbb{C})|$ is homotopy equivalent to its smooth version:

$$X_n^{diff}(\mathbb{C}) = |EGL_n(\mathbb{C})| \times_{GL_n} |V_n^{diff}(\mathbb{C})|$$

and similarly for any f.dim $\mathbb{R}$-algebra $A$. Therefore, by Suslin (Theorem 1.1.4(b) above), we have:

**Corollary 1.1.10.** *$X_n^{diff}(\mathbb{C})$ and $X_n^{diff}(A)$ are $k$-acyclic for $n \geq 2k+1$.*

## 1.2 First formula for the cocycle

We continue with the special case $R = \mathbb{C}$. Then $K_1(\mathbb{C})$ is easily seen to be equal to $\mathbb{C}^\times$. Also, $K_{1,n}(\mathbb{C}) = \mathbb{C}^\times$ for all $n > 0$. The *Borel regulator map* on $K_1(\mathbb{C})$ is defined to be the additive homomorphism

$$D_0 : K_{1,n}(\mathbb{C}) = \pi_0|V_n(\mathbb{C})| \to \mathbb{R}$$

given by the formula

$$D_0(g) = \log|\det(g)| = \Re \log\det(g) = -\Re\left(\sum_{s=1}^{\infty} \frac{(1-\det g)^s}{s}\right) \tag{1.3}$$

We view $D_0$ as a 0-cocycle on the smooth Volodin space $V_n^{diff}(\mathbb{C})$. We will generalize these to $2k$-cocycles $D_{2k}$ on $V_n^{diff}(\mathbb{C})$ and by restriction we will also obtain cocycles on the usual Volodin space $V_n(\mathbb{C})$. These cocycles will be compatible with stabilization so they will be the restrictions to $V_n^{diff}(\mathbb{C})$ and $V_n(\mathbb{C})$ of cocycles on $V^{diff}(\mathbb{C})$ and $V(\mathbb{C})$.

Let $H_n$ be the space of all positive definite $n \times n$ Hermitian matrices. For each $k \geq 0$ let $c_{2k+1}$ be the $(2k+1)$-form on $H_n$ given by

$$c_{2k+1} = \frac{1}{(2k+1)!2i^k} \operatorname{Tr}((h^{-1}dh)^{2k+1})$$

where $i = \sqrt{-1}$.

This is a closed $(2k+1)$-form and thus exact since $H_n$ is contractible. To see this note that $d\operatorname{Tr} = \operatorname{Tr} d$ and $d(h^{-1}) = -h^{-1}dhh^{-1}$. Thus, up to a constant, $dc_{2k+1} = \operatorname{Tr}((h^{-1}dh)^{2k+2}) = 0$ since $\operatorname{Tr}(\omega^{2k+2}) = -\operatorname{Tr}(\omega^{2k+2}) = 0$ for any matrix valued 1-form $\omega$.

The $(2k+1)$-form $c_{2k+1}$ is also invariant under the left action of $GL_n(\mathbb{C})$ on $H_n$ given by $ghg_*$ where $^*$ denotes complex conjugate transpose. This follows from the fact that

$$(ghg^*)^{-1}d(ghg^*) = (g^*)^{-1}h^{-1}dhg^*$$

is a conjugate of $h^{-1}dh$ and thus $\operatorname{Tr}(((ghg^*)^{-1}d(ghg^*))^{2k+1}) = \operatorname{Tr}((h^{-1}dh)^{2k+1})$.

The constant coefficient of $c_{2k+1}$ can be explained as follows. First of all the complex conjugate of $\operatorname{Tr}((h^{-1}dh)^{2k+1})$ is given by $(-1)^k \operatorname{Tr}((h^{-1}dh)^{2k+1})$

so this term is real for $k$ even and purely imaginary for $k$ odd. When we divide by $i^k$ we make this a real number. If $h^{-1}dh$ is written as a sum of 1-forms it will have $2k+1$ terms and we see that $(h^{-1}dh)^{2k+1}$ contains $(2k+1)!$ nonzero terms, one for nonzero product of these $(2k+1)$ 1-forms. Thus division by $(2k+1)!$ gives the "average" of these.

The last factor $\frac{1}{2}$ compensates for the fact that each 1-form $h^{-1}dh$ is too big by a factor of 2 since, if $h = gg^*$, then $h^{-1}dh = g^{-*}gdgg^* + g^{-*}dg^*$ where we write $g^{-*}$ for $(g^*)^{-1}$. However, $(h^{-1}dh)^{2k+1}$ is also only too big by a factor of 2 since, as we will see later, only the two terms $\operatorname{Tr}\left((g^{-1}dg)^{2k+1}\right)$ and $\operatorname{Tr}\left((g^{-*}dg^*)^{2k+1}\right)$ contribute.

If $g_0, \cdots, g_{2k+1} \in GL_n(\mathbb{C})$, let $\Delta(g_0, \cdots, g_{2k+1})$ denote the linear simplex in $H_n$ spanned by $g_i^{-1}g_i^{-*}$. (We consider $H_n$ as a convex open subset of $\mathbb{R}^{n^2}$.) Let $I_{2k+1}(g_0, \cdots, g_{2k+1})$ be the integral of $c_{2k+1}$ over this simplex. Since $c_{2k+1}$ is $GL_n(\mathbb{C})$-invariant and the action of $GL_n(\mathbb{C})$ on $H_n$ is linear (and thus takes linear simplices to linear simplices) the cocycle $I_{2k+1}$ is right $GL_n(\mathbb{C})$-invariant and thus determines an element of the continuous cohomology group

$$[I_{2k+1}] \in H_c^{2k+1}(GL_n(\mathbb{C}); \mathbb{R}).$$

Furthermore this cohomology class is the same as the Kamber-Tondeur class [KT74] as given in [Dup76] divided by $2i^k(2k+1)!$. (Dupont uses geodesic simplices instead of linear simplices but these give cohomologous cocycles.)

Let $F_{2k}$ be the $2k$-cochain on $V_n^{diff}(\mathbb{C})$ given by

$$F_{2k}(\gamma) = \int_{\Delta^{2k}\times I} \phi^* c_{2k+1} \tag{1.4}$$

where $\phi : \Delta^{2k} \times I \to H_n$ is the map given by $\phi(t,u) = (\gamma(t)\gamma(t)^*)^u$. Note that $h^u, u \in [0,1]$, defines a smooth contraction of $H_n$ but there are other such contractions such as $(1-u)I_n + uh$. Any of these can be used to define $\phi(t,u)$ with cohomologous results.

Take the special case $k = 0$. Then we are looking at a single complex matrix $g$ and $\phi(u) = (gg^*)^u = h^u$. The Hermitian matrix $h$ is diagonalizable: $h = UDU^*$ where $U$ is unitary and $D$ is diagonal with real diagonal entries $|\lambda_i|^2$ where $\lambda_i$ are the eigenvalues of $g$. Then $dh^u = U(dD^u)U^*$ where $dD^u$ is diagonal with diagonal entries $d|\lambda_i|^{2u} = 2|\lambda_i|^2 \log|\lambda_i| du$. Consequently,

$$\phi^* c_1(g) = \frac{1}{2}\operatorname{Tr}(h^{-u}dh^u) = \log|\det g| du$$

and

$$F_0(g) = \int_0^1 \phi^* c_1(g) = \log|\det g| = D_0(g),$$

i.e., $F_0 = D_0$.

For $k \geq 1$ the cochain $F_{2k}$ is not a cocycle on $V_n^{diff}(\mathbb{C})$. However, when we subtract a "correction term" $G_{2k}$ we will obtain the following.

**Theorem 1.2.1.** *a) $D_{2k} = F_{2k} - G_{2k}$ is a $2k$-cocycle on $V_n^{diff}(\mathbb{C})$ and thus by restriction also gives a $2k$-cocycle on $V_n(\mathbb{C})$.*

*b) For $n \geq 2k+2$, evaluation by the cohomology class $D_{2k}$ gives a map*

$$K_{2k+1}\mathbb{C} \cong \pi_{2k}V_n(\mathbb{C}) \xrightarrow{Hurewicz} H_{2k}V_n(\mathbb{C}) \xrightarrow{\cap[D_{2k}]} \mathbb{R}$$

*which, for $n \geq 4k+3$, is the same (up to sign) as the Borel regulator map:*

$$K_{2k+1}\mathbb{C} \cong \pi_{2k+1}BGL_n(\mathbb{C})^+ \xrightarrow{Hurewicz} H_{2k+1}(BGL_n(\mathbb{C})) \xrightarrow{\cap[I_{2k+1}]} \mathbb{R}$$

*given by cap product with the continuous cohomology class $[I_{2k+1}]$:*

$$I_{2k+1}(g_0, \cdots, g_{2k+1}) = \int_{\Delta^{2k+1}} \Psi^* c_{2k+1}$$

*where $\Psi : \Delta^{2k+1} \to H_n$ is the unique linear map sending the i-th vertex of $\Delta^{2k+1}$ to $g_i^{-1}(g_i^*)^{-1}$.*

The proof of this theorem occupies the next two sections.

# 1.3 The correction term

## Introduction

Let $\gamma$ be a $(2k+1)$-simplex in $V_n^{diff}(\mathbb{C})$, i.e., $\gamma$ is a smooth mapping $\gamma : \Delta^{2k+1} \to GL_n(\mathbb{C})$ with image in some unspecified right coset $g_0 T_n^\sigma(\mathbb{C})$ of some $\sigma$-upper triangular matrix group $T_n^\sigma(\mathbb{C})$. Then the value of the coboundary $dF_{2k}$ of $F_{2k}$ on $\gamma$ is the integral of the $(2k+1)$-form $c_{2k+1}$ over all but one side of a $(2k+2)$-simplex in $H_n$ (with reversed orientation). Since $c_{2k+1}$ is closed this is equal to the integral of $c_{2k+1}$ on the remaining side. Thus:

$$dF_{2k}(\gamma) = \int_{\Delta^{2k+1}} \phi^*(c_{2k+1})$$

where $\phi : \Delta^{2k+1} \to H_n$ is given in barycentric coordinates by $\phi(t) = \gamma(t)\gamma(t)^*$. The smooth mapping $\phi$ is the composition of

$$\gamma : \Delta^{2k+1} \to g_0 T_n^\sigma(\mathbb{C})$$

with the map

$$\psi : g_0 T_n^\sigma(\mathbb{C}) \to H_n$$

given by $\psi(g) = gg^*$.

We note that since $c_{2k+1}$ is invariant under the action of $GL_n(\mathbb{C})$ the induced form $\psi^*(c_{2k+1})$ is invariant under left translation. Thus, if $\lambda : g_0 T_n^\sigma(\mathbb{C}) \to T_n^\sigma(\mathbb{C})$ is given by left multiplication by $g_0^{-1}$, we have $\lambda^*\psi^*(c_{2k+1}) = \psi^*(c_{2k+1})$. Therefore $\phi^*(c_{2k+1}) = \gamma_0^*\psi^*(c_{2k+1})$ where

$$\gamma_0 : \Delta^{2k+1} \to T_n^\sigma(\mathbb{C})$$

is the smooth map given by $\gamma_0(t) = g_0^{-1}\gamma(t)$.

In this section we will construct a left invariant $2k$-form $b_{2k}$ on the Lie group $T_n^\sigma(\mathbb{C})$ so that $db_{2k} = (2k+1)!2i^k\psi^*(c_{2k+1})$. By Stokes' theorem we will then have

$$\int_{\Delta^{2k+1}} \phi^*(c_{2k+1}) = \frac{1}{(2k+1)!2i^k}\int_{\partial\Delta^{2k+1}} \gamma_0^*(b_{2k}) \tag{1.5}$$

Consequently a $2k$-cocycle $D_{2k}$ on the Volodin space $V_n^{diff}(\mathbb{C})$ can be given by

$$D_{2k}(\gamma) = F_{2k}(g\gamma) - \frac{1}{(2k+1)!2i^k}\int_{\Delta^{2k}} \gamma_0^*(b_{2k}). \tag{1.6}$$

Since $b_{2k}$ is left invariant, $\gamma_0^*(b_{2k})$ is independent of the choice of the coset representative $g_0$.

The differential form $\psi^*(c_{2k+1})$ on $T_n^\sigma(\mathbb{C})$ is given by $\psi^*(c_{2k+1}) = \operatorname{Tr}(w^{2k+1})$ where $w$ is the matrix valued 1-form on $T_n^\sigma(\mathbb{C})$ given by

$$w = (EE^*)^{-1}d(EE^*) = (EE^*)^{-1}dEE^* + (E^*)^{-1}dE^*$$

where $E$ is the identity mapping on $T_n^\sigma(\mathbb{C})$. If we conjugate by $E^*$ this becomes left invariant: $E^*w(E^*)^{-1} = E^{-1}dE + dE^*(E^*)^{-1}$. The existence of the desired $2k$-form $b_{2k}$ on $T_n^\sigma(\mathbb{C})$ will follow from the following lemma.

**Lemma 1.3.1.** *Let $x = E^{-1}dE$ and $y = dE^*(E^*)^{-1}$. Then*

*(a) $x, y$ are both left invariant matrix 1-forms on $T_n^\sigma(\mathbb{C})$.*

*(b) $dx = -x^2, dy = y^2$*

*(c) $\operatorname{Tr}(x^{2k+1}) = \operatorname{Tr}(y^{2k+1}) = 0$*

*(d) $x + y$ is a conjugate of a matrix 1-form $w$ which also has the property $dw = -w^2$.*

*Proof.* We comment only on (c). Since $E$ is $\sigma$-upper triangular, $dE$ is strictly $\sigma$-upper triangular. Consequently $x^{2k+1}$ is strictly $\sigma$-upper triangular with zero trace. Similarly $y$ and its powers are strictly $\sigma$-lower triangular. □

### 1.3.1 Cyclic homology

Now we will use the formal properties of $x$ and $y$ as given in Lemma 1.3.1 to construct the $2k$-form $b_{2k}$. To avoid confusion we start from scratch.

Let $A = \mathbb{C}^2$ be the complex semisimple commutative algebra with 2 orthogonal idempotents $u, v$. (We could take $A$ to be the ring of diagonal $2 \times 2$ complex matrices and $u, v$ to be the diagonal matrices $u = diag(1,0), v = diag(0,1)$.) For each $j \geq 0$ let $C_j$ be the quotient of the $(j+1)$-fold tensor product

$$C_j = A^{\otimes(j+1)}/(t-1) = (A \otimes A \otimes \cdots \otimes A)/(t-1)$$

where the cyclic operator $t$ is given by

$$t(a_0, a_1, \cdots, a_j) = (-1)^j (a_j, a_0, \cdots, a_{j-1})$$

where tensor products are represented by commas. Let $b : C_j \to C_{j-1}$ be given by

$$b(a_0, \cdots, a_j) = \sum_{0 \le i \le j} (-1)^{ij} (a_i a_{i+1}, a_{i+2}, \cdots, a_{i-1})$$

Then $b^2 = 0$ and $H_j(C_*, b) \cong HC_j(A)$ is, by definition, the $j$-th cyclic homology group of $A$. (Over fields of nonzero characteristic the definition is different. See [Lod97].)

As a vector space $C_j$ is generated by elements of the form $(a_0, \cdots, a_j)$ where each $a_i$ is either $u$ or $v$. Such elements will be called *monomials*. The *weight* $w$ of a monomial is defined to be the number of indices $i$ so that $a_i \ne a_{i+1}$ where $a_{j+1} = a_0$. Note that $w$ is always even and nonnegative and that the boundary operator $b$ preserves weight. Thus

$$(C_*, b) \cong \bigoplus_{w \ge 0} (C_*^w, b)$$

where $C_*^w$ is the weight $w$ subcomplex of $C_*$.

It is well-known that $HC_j(A) \cong C_j^0$, or equivalently, $C_*^w$ is contractible for all $w > 0$. This means that $HC_j(A) = 0$ for odd $j$ and $HC_j(A)$ is 2-dimensional for even $j$ with basis given by the two monomials of weight zero $(u, u, \cdots, u)$ and $(v, v, \cdots, v)$. We will need to go through the details of this well-know computation.

A chain contraction of the weight $w \ne 0$ complex $C_*^w$ is given by $\frac{1}{w} d$ where

$$d(a_0, \cdots, a_j) = \sum_{0 \le i \le j} (-1)^i (a_0, \cdots, da_i, \cdots, a_j)$$

where $d : A \to A \otimes A$ is the diagonal map given by $du = u \otimes u, dv = v \otimes v$. If we denote tensor powers by exponents then $d$ can be given by

$$d(u^a v^b u^c \cdots) = (du^a) v^b u^c \cdots + (-1)^a u^a (dv^b) u^c \cdots + \cdots$$

where $du^a = 0$ for even $a$ and $du^a = u^{a+1}$ for odd $a$. An easy computation shows that $bd + db$ is multiplication by $w$. (Use the fact that $bu^a = 0$ for odd $a$ and $bu^a = u^{a-1}$ for even $a$.)

Now let $\phi$ be the linear operator on $A$ taking values in matrix valued 1-forms on $T_n^\sigma(\mathbb{C})$ given by $\phi(u) = -x = -E^{-1}dE$ and $\phi(v) = y = dE^*(E^*)^{-1}$. (Thus $\phi(ru + sv) = -rx + sy$.) Extend this to tensor powers of $A$ by matrix multiplication:

$$\phi(a_0, \cdots, a_j) = \phi(a_0)\phi(a_1) \cdots \phi(a_j)$$

Since $\operatorname{Tr} \phi(a_0, \cdots, a_j) = (-1)^j \operatorname{Tr} \phi(a_j, a_0, \cdots, a_{j-1})$ we get a well-defined linear map

$$\Phi : C_j \to \Omega^{j+1}(T_n^\sigma(\mathbb{C}); \mathbb{C})$$

by the equation $\Phi(a_0, \cdots, a_j) = \operatorname{Tr}\phi(a_0, \cdots, a_j)$. Since $\phi d = d\phi$ we also have $\Phi d = d\Phi$.

The element we are interested in is

$$\Phi((v-u)^{2k+1} + u^{2k+1} - v^{2k+1}) = \operatorname{Tr}((x+y)^{2k+1} - x^{2k+1} - y^{2k+1})$$

$$= \operatorname{Tr}(w^{2k+1}) = 2i^k(2k+1)!\psi^*(c_{2k+1})$$

Since $d(v-u) = v^2 - u^2 = \frac{1}{2}(v+u)(v-u) + \frac{1}{2}(v-u)(v+u)$ it follows easily that

$$d((v-u)^{2k+1} + u^{2k+1} - v^{2k+1}) = 0.$$

Since $(v-u)^{2k+1} + u^{2k+1} - v^{2k+1}$ has no terms of weight 0 this will imply that

$$(v-u)^{2k+1} + u^{2k+1} - v^{2k+1} = d(\tfrac{1}{w}b((v-u)^{2k+1}))$$

where $w$ is a function, not a constant. Applying the operator $\Phi$ we get the following.

**Proposition 1.3.2.**

$$\psi^*(c_{2k+1}) = \frac{db_{2k}}{2i^k(2k+1)!} \quad \textit{where} \quad b_{2k} = \Phi(\tfrac{1}{w}b((v-u)^{2k+1})).$$

We use the notation $b_{2k} = \operatorname{Tr}(\frac{1}{w}b((x+y)^{2k+1}))$ keeping in mind that $b, w$ are only formally defined. Thus

$$b(x^a y^b x^c \cdots) = (bx^a)y^b x^c \cdots + (-1)^a x^a (by^b) x^c \cdots + \cdots$$

where $bx^a = by^a = 0$ for odd $a$, $bx^{2n} = -x^{2n-1}$ and $by^{2n} = y^{2n-1}$.

This discussion can be summarized by the following theorem.

**Theorem 1.3.3.** *For each $k \geq 0$ and $n \geq 1$ we have a simplicial $2k$-cocycle on the Volodin space $V_n(\mathbb{C})$ given by $D_{2k} = F_{2k} - G_{2k}$ where $F_{2k}, G_{2k}$ are $2k$-cochain on $V_n(\mathbb{C})$ given by*

$$F_{2k}(\gamma) = \frac{1}{2i^k(2k+1)!}\int_{\Delta^{2k}\times I} \operatorname{Tr}\left((h(t,-u)dh(t,u))^{2k+1}\right)$$

$$G_{2k}(\gamma) = \frac{1}{2i^k(2k+1)!}\int_{\Delta^{2k}} \operatorname{Tr}\left(\tfrac{1}{w}b\left((X(t)+Y(t))^{2k+1}\right)\right)$$

*where $h(t,u) = (\gamma(t)\gamma(t)^*)^u$, $X(t) = \gamma(t)^{-1}d\gamma(t) = \gamma_0(t)^{-1}d\gamma_0(t)$ and $Y(t) = X(t)^*$ for all $t \in \Delta^{2k}$ and $u \in I$ where we use barycentric coordinates on $\Delta^{2k}$.*

**Remark 1.3.4.** We note that $F_{2k}, G_{2k}$ are both invariant under stabilization $V_n(\mathbb{C}) \to V_{n+1}(\mathbb{C})$. Furthermore, the cochain $G_{2k}$ is left invariant in the sense that $G_{2k}(g\gamma) = G_{2k}(\gamma)$ for any $g \in GL_n(\mathbb{C})$. Consequently, there is a well-defined $2k$-form $\overline{G}_{2k}$ on $X_n^{diff}(\mathbb{C})$ given by $\overline{G}_{2k}[(g_0, \cdots, g_{2k}), \gamma] = G_{2k}(\gamma)$.

As an example take $k = 0$. Then we saw earlier that $F_0(g) = \log|\det g| = D_0(g)$. The correction term $G_0(g)$ is 0 since $b(X+Y) = 0$.

## 1.4 $D_{2k}$ gives the Borel regulator map

In this section we show that the cocycle $D_{2k}$ in Volodin $K$-theory gives the Borel regulator map $K_{2k+1}\mathbb{C} \to \mathbb{R}$ as claimed in Theorem 1.2.1(b) justifying the following definition.

**Definition 1.4.1.** The cohomology class $\tau_k^V(1,n) = [D_{2k}] \in H^{2k}(V_n(\mathbb{C});\mathbb{R})$ will be called the *higher Franz-Reidemeister torsion class* for $V_n(\mathbb{C})$. We also define

$$\tau_k^V(m,n) \in H^{2k}(V_n(M_m(\mathbb{C}));\mathbb{R})$$

to be the restriction of $\tau_k^V(1,nm)$ under the simplicial transfer map, i.e., $\tau_k^V(m,n) = \mathrm{N}_\bullet^* \tau_k^V(1,nm)$. The stable classes $(\tau_k^V(m,\infty))$ will be written $\tau_k^V(m)$.

Our starting point is Theorem 1.1.4 as explained by Suslin [Sus82]. We will assume that $n \geq 4k+3$ so that $K_{2k+1}(\mathbb{C}) \cong \pi_{2k}V_n(\mathbb{C}) \cong \pi_{2k+1}BGL_n(\mathbb{C})^+$.

Let

$$\psi : |V_n^{diff}(\mathbb{C})| \to H_n$$

be the $GL_n(\mathbb{C})$-equivariant map given on each simplex $\gamma \times \Delta^k$ of $|V_n^{diff}(\mathbb{C})|$ by $\psi(\gamma,t) = \gamma(t)\gamma(t)^*$. Then we get a map of fiber bundles

$$\Psi : X_n^{diff}(\mathbb{C}) = |EGL_n(\mathbb{C})| \times_{GL_n} |V_n^{diff}(\mathbb{C})| \to |EGL_n(\mathbb{C})| \times_{GL_n} H_n$$

over $BGL_n(\mathbb{C})$ by $\Psi[x,\gamma] = [x, \gamma(t)\gamma(t)^*]$ where we use the abbreviation $GL_n$ for $GL_n(\mathbb{C})$ when it occurs as a subscript.

Let $\Sigma^{2k}$ be an integral simplicial $2k$-cycle in $V_n^{diff}(\mathbb{C})$ representing the element

$$[\Sigma^{2k}] \in \pi_{2k}(V_n^{diff}(\mathbb{C})) = K_{2k+1}\mathbb{C}.$$

Since $|V_n^{diff}(\mathbb{C})| = \{I_n\} \times |V_n^{diff}(\mathbb{C})|$ is contained in the $2k$-acyclic space $X_n^{diff}(\mathbb{C})$ the integral $2k$-cycle $\Sigma^{2k}$ bounds an integral $(2k+1)$-chain $W_{2k+1}$ in $X_n^{diff}(\mathbb{C})$ and the image $p(W)$ of $W$ in $BGL_n(\mathbb{C})$ is a $(2k+1)$-cycle representing the image in $H_{2k+1}(BGL_n(\mathbb{C})) = H_{2k+1}(BGL_n(\mathbb{C})^+)$ of the element $[\Sigma^{2k}]$ of $K_{2k+1}\mathbb{C}$ under the Hurewicz map

$$K_{2k+1}\mathbb{C} = \pi_{2k+1}BGL_n(\mathbb{C})^+ \to H_{2k+1}(BGL_n(\mathbb{C})).$$

We now consider the image $\Psi(W)$ of $W$ in $|EGL_n(\mathbb{C})| \times_{GL_n} H_n$. Then $\partial\Psi(W) = \Psi(\Sigma^{2k})$ which lies in one fiber $H_n$. Since $H_n$ is contractible the projection

$$|EGL_n(\mathbb{C})| \times_{GL_n} H_n \to BGL_n(\mathbb{C})$$

is a homotopy equivalence and the $(2k+1)$-cycle $p(W)$ in $BGL_n(\mathbb{C})$ lifts to a $(2k+1)$-cycle $\Psi(W) - C\Psi(\Sigma)$ where $C\Psi(\Sigma)$ is the cone on $\Psi(\Sigma)$ inside the fiber $H_n$. (See Figure 1.1.)

The $(2k+1)$-form $c_{2k+1} = \mathrm{Tr}((h^{-1}dh)^{2k+1})$ on $H_n$ is invariant under the action of $GL_n(\mathbb{C})$. Consequently this differential form gives a real $(2k+1)$-cocycle $Z_{2k+1}$ on the simplicial set of all maps $\Delta^j \to |EGL_n(\mathbb{C})| \times_{GL_n} H_n$ which are simplicial in the first coordinate and smooth in the second.

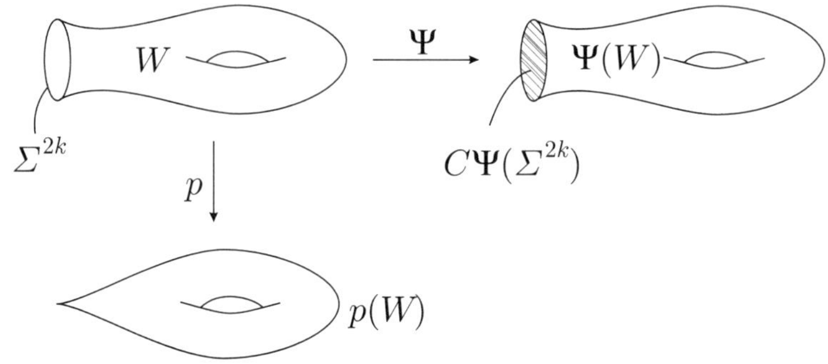


Figure 1.1: $\Sigma^{2k} = \partial W$ in $X_n^{diff}(\mathbb{C})$ and $\Sigma^{2k}$ bounds a cone in $H_n$

**Lemma 1.4.2.** *The value of the Borel regulator map on* $[p(W)] \in H_{2k+1}(GL_n(\mathbb{C}))$ *is given by*

(a) $I_{2k+1}(p(W)) = Z_{2k+1}(\Psi(W) - C\Psi(\Sigma))$.

*Furthermore,*

(b) $Z_{2k+1}(C\Psi(\Sigma)) = F_{2k}(\Sigma)$ *and*

(c) $Z_{2k+1}(\Psi(W)) = G_{2k}(\Sigma)$.

*Proof.* Since the bundle $|EGL_n(\mathbb{C})| \times_{GL_n} H_n$ is fiberwise contractible the value of the cocycle $Z_{2k+1}$ on the $(2k+1)$-cycle $\Psi(W) - C\Psi(\Sigma)$ is equal to its value on $s(p(W))$ where

$$s : BGL_n(\mathbb{C}) \to |EGL_n(\mathbb{C})| \times_{GL_n} H_n$$

is the section of this bundle given by

$$s([g_0, \cdots, g_j], t) = [((g_0, \cdots, g_j), t), \sum t_i (g_i^* g_i)^{-1}].$$

Since $Z_{2k+1}$ is defined to be $c_{2k+1}$ on the second factor we see that $Z_{2k+1}s(p(W)) = I_{2k+1}p(W)$. Consequently,

$$I_{2k+1}p(W) = Z_{2k+1}\Psi(W) - Z_{2k+1}C\Psi(\Sigma).$$

The rest is easy: (b) follows from the definition of $F_{2k}(\Sigma)$ and (c) follows from Proposition 1.3.2 which tells us that the restriction of $\Psi^* Z_{2k+1}$ to the fiber $V_n^{diff}(\mathbb{C})$ of $X_n^{diff}(\mathbb{C})$ is the coboundary of $G_{2k}$. Consequently, $\Psi^* Z_{2k+1}$ is the coboundary of $\overline{G}_{2k}$ (Remark 1.3.4) which implies (c) since $\overline{G}_{2k}(\Sigma) = G_{2k}(\Sigma)$. □

Theorem 1.2.1 follows from Lemma 1.4.2 and Theorem 1.3.3.

## 1.5 The involution in $V_n(\mathbb{C})$

The involution on the $K$-theory of $\mathbb{C}$ given by inverse conjugate transpose is both elementary and well-known. It played a crucial role in the computation of the homotopy type of diffeomorphism spaces ([BL77], [FH78]) as we explain in Chapter VI below. In this section we discuss the involution on $V(\mathbb{C})$.

**Definition 1.5.1.** Suppose that $G$ is a subgroup of the group of units $R^\times$ of a ring $R$. By a *conjugation* on $(R, G)$ we mean an anti-involution on $R$, written $\iota(r) = \overline{r}$ which acts on $G$ by group inverse. Thus

$$\overline{\overline{r}} = r, \quad \overline{rs} = \overline{s}\overline{r} \quad \text{and} \quad \overline{g} = g^{-1}$$

for all $r, s \in R$, $g \in G$. (By an *involution* we mean an automorphism of order 2. An *anti-involution* is an anti-automorphism of order 2.)

When $R = \mathbb{C}$ we always take *complex conjugation* as the conjugation operation (with $G = U(1)$).

Given a conjugation on $(R, G)$ we get a conjugation on $(M_n(R), Mon_n(G))$ by *conjugate transpose*. Here $Mon_n(G) \subseteq M_n(R)^\times$ is the group of $n \times n$ monomial matrices with coefficients in $G$. The operation of inverse conjugate transpose ($g \mapsto (\overline{g}^t)^{-1}$) gives an involution on $GL_n(R)$ for all $n \geq 1$ which is the identity on $Mon_n(G)$. Since this is compatible with stabilization we get an induced automorphism of $GL(R)$ and consequently an involution on $K_i(R)$ for all $i$. We refer to each of these involutions by the symbol $\mathcal{I}ct$.

If $T$ is a $\sigma$-upper triangular matrix then $\mathcal{I}ct(T)$ is a $\sigma^{op}$-upper triangular matrix where $\sigma^{op}$ is the opposite ordering ($\sigma^{op} = \{(x, y) : (y, x) \in \sigma\}$). Consequently, $\mathcal{I}ct$ takes right cosets of $T_n^\sigma(R)$ to right cosets of $T_n^{\sigma^{op}}(R)$. Therefore any conjugation on $(R, G)$ gives an involution $\mathcal{I}ct$ on the Volodin space $V_n(R)$ by

$$\mathcal{I}ct(g_0, \cdots, g_k) = (\mathcal{I}ct(g_0), \cdots, \mathcal{I}ct(g_k)).$$

In the case $(R, G) = (\mathbb{C}, U(1))$ we have an involution on $V_n^{diff}(\mathbb{C})$ given by $\mathcal{I}ct(\gamma)(t) = \mathcal{I}ct(\gamma(t))$. The forgetful map $F : V_n^{diff}(\mathbb{C}) \to V_n(\mathbb{C})$ is compatible with this involution so they are equivalent. (An equivariant inclusion map $V_n(\mathbb{C}) \to V_n^{diff}(\mathbb{C})$ can be given by averaging the linear interpolation map $\lambda$ with $\mathcal{I}ct \circ \lambda \circ \mathcal{I}ct$.)

We note that this involution commutes with transfer over any ring:

$$\mathrm{N}_\bullet \circ \mathcal{I}ct = \mathcal{I}ct \circ \mathrm{N}_\bullet : V_n(M_m(R)) \to V_{nm}(R)$$

and it commutes with the smooth transfer map when $R = \mathbb{C}$:

$$\mathrm{N}_\bullet^{diff} \circ \mathcal{I}ct = \mathcal{I}ct \circ \mathrm{N}_\bullet^{diff} : V_n^{diff}(M_m(\mathbb{C})) \to V_{nm}^{diff}(\mathbb{C}).$$

**Theorem 1.5.2.** *The involution $\mathcal{I}ct$ on $V_n^{diff}(\mathbb{C})$ (and thus on $V_n(\mathbb{C})$) reverses the sign of the higher FR-torsion class, i.e.,*

$$\mathcal{I}ct(\tau_k^V(1, n)) = -\tau_k^V(1, n).$$

**Remark 1.5.3.** This is a well-known fact which follows easily from Borel's description of the continuous cohomology of $GL_n(\mathbb{C})$ as a relative Lie algebra cohomology ([Bor74b]). It was observed by Farrell and Hsiang ([FH78]) who showed that this involution has important applications to diffeomorphisms of manifolds. (See §6.5 below.)

*Proof.* When $\gamma(t)$ is replaced by $(\gamma(t)^*)^{-1}$ then $h = \gamma\gamma^*$ is replaced by $(\gamma^*)^{-1}\gamma^{-1} = h^{-1}$ and $h^{-u}dh^u$ is replaced by $h^u d(h^{-u}) = -(dh^u)h^{-u}$ changing the sign of $\mathrm{Tr}((h^{-u}dh^u)^{2k+1})$ and thus of $F_{2k}(\gamma)$. Also $X(t) = \gamma(t)^{-1}d\gamma(t)$ changes to $\gamma(t)^*d(\gamma(t)^*)^{-1} = -d\gamma(t)^*(\gamma(t)^*)^{-1} = -Y(t)$ and, similarly, $Y(t)$ changes to $-X(t)$. This changes the sign of $\mathrm{Tr}\left(\frac{1}{w}b(X(t)+Y(t))^{2k+1}\right)$ and thus of $G_{2k}(\gamma)$. We conclude that

$$D_{2k}((\gamma^*)^{-1}) = F_{2k}((\gamma^*)^{-1}) - G_{2k}((\gamma^*)^{-1}) = -F_{2k}(\gamma) + G_{2k}(\gamma) = -D_{2k}(\gamma).$$

□

Since $\mathcal{I}ct$ commutes with $\mathrm{N}_\bullet^{diff}$ we have:

**Corollary 1.5.4.** *The involution on $\mathcal{V}_n^{diff}(M_m(\mathbb{C}))$ and thus on $\mathcal{V}_n(M_m(\mathbb{C}))$ reverses the sign of the higher FR-torsion. Thus:*

$$\mathcal{I}ct(\tau_k^V(m,n)) = -\tau_k^V(m,n).$$

## 1.6 Additivity of $\tau_k^V$

Since the cocycle $D_{2k}$ is the integral of a trace it is obviously additive with respect to direct sum of matrices. This means that the higher FR-torsion classes $\tau_k^V(1,n)$, $\tau_k^V(m,n)$ satisfy certain additivity conditions. We will formulate these observations without proof.

The first form of the additivity property is in terms of the *coefficients*. The direct sum map is a ring homomorphism

$$s : M_{m_1}(R) \times \cdots \times M_{m_d}(R) \to M_m(R)$$

(where $m = m_1 + \cdots + m_d$) for any ring $R$. Therefore, for any $n \geq 1$, $s$ induces a simplicial monomorphism

$$s_* : V_n(\prod_{i=1}^d M_{m_i}(R)) \to V_n(M_m(R)).$$

**Proposition 1.6.1 (Additivity of $\tau_k^V(-,n)$).** *In the case $R = \mathbb{C}$ we have:*

$$s^*\tau_k^V(m,n) = \sum_{i=1}^d p_i^*\tau_k^V(m_i,n)$$

*where $p_i : \prod M_{m_i}(\mathbb{C}) \to M_{m_i}(\mathbb{C})$ is the projection to the $i$-th factor.*

This is related to another obvious statement:

**Proposition 1.6.2.** *If $R_1, \cdots, R_d$ are rings and $p_i : \prod R_i \to R_i$ is the projection map then the induced maps $(p_i)_* : V_n(\prod R_i) \to V_n(R_i)$ give a simplicial bijection*

$$\prod (p_i)_* : V_n(\prod R_i) \xrightarrow{\cong} \prod V_n(R_i).$$

*In other words, $V_n$ takes finite products of rings to finite products of simplicial sets.*

*Proof.* The natural isomorphism $GL_n(\prod R_i) \cong \prod GL_n(R_i)$ sends $T_n^\sigma(\prod R_i)$ isomorphically onto $\prod T_n^\sigma(R_i)$ for all partial orderings $\sigma$ of $\{1, 2, \cdots, n\}$. □

The second form of the additivity property increases the index $n$. If $n = n_1 + \cdots + n_d$ and $R$ is any ring let

$$S : \prod V_{n_i}(R) \to V_n(R)$$

be the simplicial map given by direct sum of matrices, i.e.,

$$S_p((g_0^1, \cdots, g_p^1), \cdots, (g_0^d, \cdots, g_p^d)) = (g_0^1 \oplus \cdots \oplus g_0^d, \cdots, g_p^1 \oplus \cdots \oplus g_p^d).$$

Since $g_j^i \in GL_{n_i}(R)$, $g_j^1 \oplus \cdots \oplus g_j^d \in GL_n(R)$.

**Proposition 1.6.3 (Additivity of $\tau_k^V(m, -)$).** *Let $R = M_m(\mathbb{C})$. Then*

$$S^* \tau_k^V(m, n) = \sum_{i=1}^{d} \pi_i^* \tau_k^V(m, n_i)$$

*where $\pi_i : \prod V_{n_i}(R) \to V_{n_i}(R)$ is the projection to the i-th factor.*

## 1.7 Some computations

In this section we give a computation of $F_{2k}$ with respect to a diagonalization of the matrix $h = gg^*$. We also compute the correction term $G_{2k}$ in the special case $k = 1$ for $3 \times 3$ matrices.

The value of the cochain $F_{2k}$ on a $2k$-simplex in $V_n(R)$ is given as an integral over $\Delta^{2k} \times I$ of the trace of a matrix valued $(2k+1)$-form. We will integrate the form and evaluate this trace to obtain a $2k$-form on $\Delta^{2k}$. This $2k$-form will be given at each point $t$ of $\Delta^{2k}$ in terms of the eigenvalues $a_i$ of $h_t$.

Let $t_0$ be a fixed element of $\Delta^{2k}$. Then by conjugation by a fixed unitary matrix we may assume that $A = h_{t_0}$ is a diagonal matrix with positive real diagonal entries $a_i$. For simplicity of notation we translate $t_0$ to the origin and assume that $t_0 = 0$, $A = h_0$. For $t \in \Delta^{2k}$ close to $t_0 = 0$ we have $h_t = A + X(t)$. Suppressing the $t$ from the notation we have $h = A + X$ where $X$ is a hermitian matrix with infinitesimal entries $x_{ij} = x_{ji}$.

**Lemma 1.7.1.** *For $0 \le u \le 1$ let $X_u$ be defined by $h^u = A^u + X_u$. Then*

$$(x_u)_{ij} = \begin{cases} \frac{a_i^u - a_j^u}{a_i - a_j} x_{ij} & \text{if} \quad a_i \neq a_j \\ u a_i^{u-1} x_{ij} & \text{if} \quad a_i = a_j \end{cases}$$

*Proof.* (See [Fla74] or [NH95], Thm 4.8.) It is a general fact that if $f(x)$ is any analytic function so that $f(h)$ commutes with $h$ for all matrices $h$ close to a fixed diagonal matrix $A$ then the directional derivative $D_X f(h)$ at $h = A$ has entries

$$\frac{f(a_i) - f(a_j)}{a_i - a_j} \quad \text{if} \quad a_i \neq a_j$$

$$f'(a_i) x_{ij} \quad \text{if} \quad a_i = a_j$$

Apply this to $f(x) = x^u$. □

Let $d'h^u = dh^u - \partial h^u / \partial u$ be the differential of $h^u$ keeping $u$ fixed. Since $\frac{\partial h^u}{\partial u} = A^u \log A$ commutes with $A$ we have

$$\omega = h_0^{-u/2} dh^u h_0^{-u/2} = \log A du + A^{-u/2} d'h^u A^{-u/2} \qquad (\sim h_0^{-u} dh^u)$$

$$(\omega)_{ij} = (A^{-u/2} d'h^u A^{-u/2})_{ij} = \frac{(a_i/a_j)^{u/2} - (a_j/a_i)^{u/2}}{a_i - a_j}$$

(We understand this notation to represent the limit $(u/a_i)dxij$ in the special case when $a_i = a_j$.)

The term $du$ occurs at $2k+1$ places in the $(2k+1)$-form $\omega^{2k+1}$. Since $\omega$ is conjugate to $h_0^{-u} dh^u$, the $2k$-form we are trying to compute is the integral of $\mathrm{Tr}(\omega^{2k+1})$ as $u$ goes from 0 to 1. Since $\mathrm{Tr}(\omega^{2k+1})$ is invariant under cyclic permutation of the $2k+1$ factors we have

$$\begin{aligned} \mathrm{Tr}(\omega^{2k+1}) &= (2k+1) du \,\mathrm{Tr}(\log A \omega^{2k}) \\ &= -(2k+1) du \,\mathrm{Tr}(\omega \log A \omega^{2k-1}) \\ &= (2k+1) du \,\mathrm{Tr}(\omega^2 \log A \omega^{2k-2}) = \cdots \end{aligned}$$

Taking the average of these terms we get the following where $\sum_\gamma$ stands for the sum over all $2k$-periodic mappings $\gamma : \mathbb{Z} \to \{1, 2, \cdots, n\}$ and $b_{\gamma(i)}, c_{\gamma(i)}$ are the perturbations of $a_{\gamma(i)}, \log a_{\gamma(i)}$ given by $b_{\gamma(i)} = a_{\gamma(i)} \exp((-1)^{i-1}\epsilon)$ and $c_{\gamma(i)} = \log b_{\gamma(i)} = \log a_{\gamma(i)} - (-1)^i \epsilon$ for an infinitesimal $\epsilon > 0$.

$\mathrm{Tr}(\omega^{2k+1}) =$

$$\frac{2k+1}{2k} \sum_\gamma \sum_{j=1}^{2k} (-1)^{j-1} c_{\gamma(j)} du \prod_{i=1}^{2k} \frac{(b_{\gamma(i)}/b_{\gamma(i+1)})^{u/2} - (b_{\gamma(i+1)}/b_{\gamma(i)})^{u/2}}{b_{\gamma(i)} - b_{\gamma(i+1)}} dx_{\gamma(i)\gamma(i+1)}$$

$$\frac{2k+1}{2k} \sum_\gamma \sum_{j=1}^{2k} (-1)^{j-1} c_{\gamma(j)} du \prod_{i=1}^{2k} \frac{(b_{\gamma(i)}/b_{\gamma(i+1)})^{u} - 1}{b_{\gamma(i)} - b_{\gamma(i+1)}} dx_{\gamma(i)\gamma(i+1)}$$

$$\frac{2k+1}{2}\sum_{\gamma}\sum_{j=1}^{2k}(c_{\gamma(1)}-c_{\gamma(2)})du\prod_{i=1}^{2k}\frac{(b_{\gamma(i)}/b_{\gamma(i+1)})^u-1}{b_{\gamma(i)}-b_{\gamma(i+1)}}dx_{\gamma(i)\gamma(i+1)}$$

We extract the portion of this which depends only on $u$ and integrate:

$$\begin{aligned}
&\int_0^1 du\prod_{i=1}^{2k}\left[(b_{\gamma(i)}/b_{\gamma(i+1)})^u-1\right]\\
&=\int_0^1 du\Big\{2+\sum_{1\le i<j\le 2k}(-1)^{j-i}\left[(b_{\gamma(i)}/b_{\gamma(j)})^u+(b_{\gamma(j)}/b_{\gamma(i)})^u\right]\\
&\quad+\sum_{1\le i_1<j_1<i_2<j_2\le 2k}(-1)^{i_1+j_1+i_2+j_2}\left[\left(\frac{b_{\gamma(i_1)}b_{\gamma(i_2)}}{b_{\gamma(j_1)}b_{\gamma(j_2)}}\right)^u+\left(\frac{b_{\gamma(j_1)}b_{\gamma(j_2)}}{b_{\gamma(i_1)}b_{\gamma(i_2)}}\right)^u\right]+\cdots\Big\}\\
&=2\int_0^1 du\Big\{1+\sum_{1\le i<j\le 2k}(-1)^{j-i}\cosh[u(c_{\gamma(i)}-c_{\gamma(j)})]\\
&\quad+\sum_{1\le i_1<j_1<i_2<j_2\le 2k}(-1)^{i_1+j_1+i_2+j_2}\cosh[u(c_{\gamma(i_1)}-c_{\gamma(j_1)}+c_{\gamma(i_2)}-c_{\gamma(j_2)})]+\cdots\Big\}\\
&=2+2\sum_{1\le i<j\le 2k}(-1)^{j-i}\frac{\sinh(c_{\gamma(i)}-c_{\gamma(j)})}{c_{\gamma(i)}-c_{\gamma(j)}}\\
&\quad+2\sum_{1\le i_1<j_1<i_2<j_2\le 2k}(-1)^{i_1+j_1+i_2+j_2}\frac{\sinh(c_{\gamma(i_1)}-c_{\gamma(j_1)}+c_{\gamma(i_2)}-c_{\gamma(j_2)})}{c_{\gamma(i_1)}-c_{\gamma(j_1)}+c_{\gamma(i_2)}-c_{\gamma(j_2)}}+\cdots
\end{aligned}$$

where any term of the form $\frac{\sinh 0}{0}$ must be interpreted as $\cosh 0 = 1$. Thus:

**Proposition 1.7.2.** *The $2k$-form on $\Delta^{2k}$ given by integrating $\mathrm{Tr}(\omega^{2k+1})$ from $u=0$ to $u=1$ is:*

$$\begin{aligned}
&(2k+1)\sum_{\gamma}\frac{c_{\gamma(1)}-c_{\gamma(2)}}{b_{\gamma(1)}-b_{\gamma(2)}}dx_{\gamma(1)\gamma(2)}\left(\prod_{i=1}^{2k}\frac{dx_{\gamma(i)\gamma(i+1)}}{b_{\gamma(i)}-b_{\gamma(i+1)}}\right)\cdot\\
&\qquad\left\{1+\sum_{1\le i<j\le 2k}(-1)^{j-i}\frac{\sinh(c_{\gamma(i)}-c_{\gamma(j)})}{c_{\gamma(i)}-c_{\gamma(j)}}\right.\\
&\left.+\sum_{1\le i_1<j_1<i_2<j_2\le 2k}(-1)^{i_1+j_1+i_2+j_2}\frac{\sinh(c_{\gamma(i_1)}-c_{\gamma(j_1)}+c_{\gamma(i_2)}-c_{\gamma(j_2)})}{c_{\gamma(i_1)}-c_{\gamma(j_1)}+c_{\gamma(i_2)}-c_{\gamma(j_2)}}+\cdots\right\}
\end{aligned}$$

*where $b_i$ are infinitesmal perturbations of the eigenvalues $a_i$ of $A=h_0$ and $c_i=\log b_i$.*

In the special case $k=1$ we have:

**Corollary 1.7.3.** *The integral of* $c_3(\gamma) = \frac{1}{3!2i}\operatorname{Tr}(\omega^3)$ *from* $u = 0$ *to* $u = 1$ *is equal to*

$$\frac{1}{4}\left\{\sum_{i\neq j}\frac{c_i - c_j}{-(b_i-b_j)^2}dx_{ij}dx_{ji}\left(1-\frac{\sinh(c_i = c_j)}{c_i-c_j}\right)\right\}$$

$$= \frac{1}{2}\Im\left\{\sum_{i<j}\frac{-c_i+c_j+\sinh(c_i-c_j)}{(b_i-b_j)^2}dx_{ij}dx_{ji}\right\}$$

$$= \frac{1}{2}\Im\left\{\sum_{i<j}\frac{\sinh\log(a_i/a_j)-\log(a_i/a_j)}{(a_i-a_j)^2}dx_{ij}dx_{ji}\right\}$$

*where* $\frac{\sinh\log(a_i/a_j)-\log(a_i/a_j)}{(a_i-a_j)^2} = 0$ *when* $a_i = a_j$.

The correction term $G_2$ is defined on a two-simplex $\gamma$ to be

$$G_2(\gamma) = \frac{1}{12}\Im\int_{\Delta^2}\operatorname{Tr}\left(\tfrac{1}{w}b(X+Y)^3\right)$$

where $X = \gamma_0^{-1}d\gamma_0 = \gamma^{-1}d\gamma$ and $Y = X^*$. But up to cyclic permutation of the monomials, $(X+Y)^3 = X^3 + 3X^2Y + 3XY^2 + Y^3$. Since $b$ is an anti-derivation with $b(X^{2n}) = -X^{2n-1}$, $b(X^{2n+1}) = b(Y^{2n+1}) = 0$ and $b(Y^{2n}) = Y^{2n-1}$ we get $b(X+Y)^3 = -3XY - 3XY = -6XY$ which we divide by its weight (the number of powers of $X, Y$ in the monomial) $w = 2$ to get

$$G_2(\gamma) = -\frac{1}{4}\Im\int_{\Delta^2}\operatorname{Tr}(XY).$$

For $k = 2, 3$ we have

$$G_4(\gamma) = -\frac{1}{2\cdot 4!}\Re\int_{\Delta^4}\operatorname{Tr}\left(X^3Y + XY^3 + \tfrac{1}{2}(XY)^2\right).$$

$$\begin{aligned} G_6(\gamma) = \frac{1}{2\cdot 6!}\Im\int_{\Delta^6}&\operatorname{Tr}\left(X^5Y + XY^5 + X^3Y^3 + X^3YXY + XYXY^3\right.\\ &\left.+\tfrac{1}{2}XYX^2Y^2 - \tfrac{1}{2}YXY^2X^2 + \tfrac{1}{4}(X^2Y)^2 - \tfrac{1}{4}(XY^2)^2 + \tfrac{1}{3}(XY)^3\right).\end{aligned}$$

We return to the case $k = 1$ and we restrict ourselves to the case $n = 3$. Suppose that $\gamma_0(t)$ is the $3\times 3$ upper triangular matrix

$$\gamma_0(t) = \begin{pmatrix} 1 & x & z \\ 0 & 1 & y \\ 0 & 0 & 1\end{pmatrix}.$$

Then

$$X = \gamma_0(t)^{-1} d\gamma_0(t) = \begin{pmatrix} 1 & -x & xy-z \\ 0 & 1 & -y \\ 0 & 0 & 1 \end{pmatrix} \begin{pmatrix} 0 & dx & dz \\ 0 & 0 & dy \\ 0 & 0 & 0 \end{pmatrix} = \begin{pmatrix} 0 & dx & dz - xdy \\ 0 & 0 & dy \\ 0 & 0 & 0 \end{pmatrix}$$

$Y = X^*$ and

$$\begin{aligned} \mathrm{Tr}(XY) &= dxd\overline{x} + dyd\overline{y} + (dz - xdy)(d\overline{z} - \overline{x}d\overline{y}) \\ &= dxd\overline{x} + dyd\overline{y} + dzd\overline{z} + |x|^2 dyd\overline{y} - xdyd\overline{z} + \overline{x}d\overline{y}dz. \end{aligned}$$

Now suppose that $x = x_1t_1 + x_2t_2, y = y_1t_1 + y_2t_2, z = z_1t_1 + z_2t_2$ where $x_1, x_2, y_1, y_2, z_1, z_2$ are constants and $t_1, t_2$ are barycentric coordinates for $\Delta^2$. Then $dxd\overline{x} = (x_1\overline{x}_2 - \overline{x}_1x_2)dt_1dt_2 = -2i\Im(\overline{x}_1x_2)dt_1dt_2$ , etc., so we have:

**Proposition 1.7.4.** *When $\gamma : \Delta^2 \to T_3(\mathbb{C})$ is a linear simplex (i.e., an element of $V_3(\mathbb{C})[2]$) then*

$$G_2(\gamma) = -\frac{1}{4}\Im \int_{\Delta^2} \mathrm{Tr}(XY)$$

$$= \frac{1}{2}\Im \int_{\Delta^2} \overline{x}_1x_2 + \overline{y}_1y_2 + \overline{z}_1z_2 + \left(|x_1|^2t_1^2 + |x_2|^2t_2^2 + (x_1\overline{x}_2 + \overline{x}_1x_2)t_1t_2\right)\overline{y}_1y_2 \\ + (x_1t_1 + x_2t_2)(y_1\overline{z}_2 - y_2\overline{z}_1)dt_1dt_2$$

$$= \frac{1}{4}\Im\left\{\overline{x}_1x_2 + \overline{y}_1y_2 + \overline{z}_1z_2 + \tfrac{1}{6}\overline{y}_1y_2(|x_1|^2 + |x_2|^2 + \Re(x_1\overline{x}_2)) \\ + \tfrac{1}{3}(x_1 + x_2)(y_1\overline{z}_2 - y_2\overline{z}_1)\right\}$$

For larger $k$ and $n$ we will have a similar but more complicated expression. It will be the integral over a simplex of some rational polynomial in $x_i$'s, $t_i$'s and conjugates of the $x_i$'s. Therefore we can say the following.

**Proposition 1.7.5.** *Let $(g_0, \cdots, g_{2k})$ be a $2k$-simplex in $V_n(\mathbb{C})$. Then the correction term $G_{2k}(g_0, \cdots, g_{2k})$ will be a polynomial with rational coefficients in the real and imaginary parts of the entries of the matrices $(g_0)^{-1}g_i$.*

**Remark 1.7.6.** This is a crucial observation which allows us to ignore the correction term $G_{2k}(\gamma)$ when we compute the coefficient of the polylogarithm term in the value of the higher Reidemeister torsion for circle bundles.

We note that $G_2(\gamma)$ is not the same as the corresponding correction term $\psi$ in [Igu93], Thm 3.2 since the latter is a transcendental function (a dilogarithm) in the entries of the upper triangular matrices $(g_0)^{-1}g_i$.

# Chapter 2

# Spaces of Matrices and Higher Franz-Reidemeister Torsion

## Introduction

In this chapter we extend the formula for the Borel regulator map from the Volodin space of matrices which vary by column operations to spaces of matrices which vary by both row and column operations. Whereas the former is a model for the loop space of the $K$-theory of the coefficient ring, the latter is a model for the Whitehead space (the fiber of the map $BMon(G)^+ \to BGL(R)^+$) where $G$ is a subgroup of the group of units of $R$ and $Mon(G) \subseteq GL(R)$ is the group of monomial matrices with coefficients in $G$. As in Chapter I the formula is a modification of the Kamber-Tondeur-Dupont formula ([KT74],[Dup76]) with correction terms expressed in terms of cyclic homology.

Much of this chapter is devoted to the smooth and continuous versions of the space of matrices. We also use an "isovariant" subcomplex (in §2.3) which consists of those continuous simplices where the isotropy subgroup is equal to the isotropy subgroup of the vertex set.

In the last half of this chapter we apply these formulas to the case of oriented circle bundles. As expected we obtain polylogarithms of roots of unity. We show in §2.8 that the polylogarithms are necessary consequences of the elementary transfer relations between circle bundles which are coverings of each other. Basically a Morse function on any manifold induces a Morse function on any covering of the manifold and the incidence matrices are given by obvious transfer formulas. Then we use the elementary fact that the polylogarithms are, up to a constant, the only smooth functions on $U(1) - \{1\}$ satisfying the transfer formulae. It is the use of the isovariant subcomplex (Sections 2.3 and 2.7) which allows us to extend the definition of the higher Franz-Reidemeister torsion to the cases of holonomy which are not roots of unity.

In §2.9 we show that the space of framed functions on a circle is contractible. This is needed to show that the parametrized Morse theory of a circle bundle is unique up to homotopy and that it is invariant under change of orientation. It was proved in [IK93] that the space of positively framed functions on a circle is contractible. Here we extend this to include negatively framed functions.

Finally, in §2.10 we show that the space of framed functions on a circle is homotopy equivalent to certain combinatorially defined spaces, which are therefore contractible. This is the starting point of the framed graph theorem which we will discuss in Chapter VIII.

The idea of the isovariant subcomplex was inspired by conversations with Richard Palais. I am grateful to Professor Palais for this and the many other ways in which he has helped me through the years.

## 2.1 The space of invertible matrices

### Introduction

The purpose of this section is to discuss a variation of the Volodin definition of the space of invertible matrices over an associative ring $R$. The idea is of course due to Volodin [Vol71]. In fact he considered the more general case of families of chain complexes. We will construct two versions of the space of chain complexes in the next two chapter and show that all three of these spaces are homotopy equivalent. In this chapter we give a formula for the Borel regulator map on the space of invertible matrices.

We will construct this space as a simplicial space of matrices which vary locally by row and column operations including permutations of rows and columns and multiplication of rows and columns by units in the ring. Also these row and column operations will be allowed to change by chain homotopies when there are two or more parameters. This is basically how the incidence matrices in the cellular chain complex of a cell complex vary as the cells slide over each other and the numbering of the cells or their paths to the base point are changed. A family of Morse functions with critical points of indices $i$ and $i+1$ would also produce this kind of family of incidence matrices.

Let $G$ be a subgroup of the group of units of a ring $R$ and let $P$ be a finite set. Then $GL_P(R)$ denotes the group of $P \times P$ invertible matrices over $R$. Equivalently, these are the $R$-automorphisms of the free right $R$-module generated by $P$. We let $Mon_P(G) \subseteq GL_P(R)$ denote the group of $P \times P$ monomial matrices with coefficients in $G$ which we also call *$G$-monomial matrices*.

Suppose that $P$ is a finite *poset*, i.e., $P$ has a partial ordering $\sigma \subseteq P \times P$ (i.e., a transitive and antireflexive relation). Then $T_P(R) = T^\sigma(R) \subseteq GL_P(R)$ will both denote the group of *$P$-upper triangular* or *$\sigma$-upper triangular* $P \times P$ matrices which have the form $I + A$ where $A$ is *strictly $\sigma$-upper triangular* in the sense that $a_{ij} \neq 0$ only when $(i,j) \in \sigma$.

In this chapter we consider *graded posets* $P = P_0 \coprod P_1$ where the elements of $P_i$ are said to have *degree* $i$ and there is no assumed relationship between grading and ordering.

**Definition 2.1.1.** Let $\rho$ be a partial ordering on the graded set

$$B(n) = B_0(n) \coprod B_1(n) = \{y_1, \cdots, y_n\} \coprod \{x_1, \cdots, x_n\}.$$

Let $\sigma, \tau$ be the induced orderings on the subsets $B_0(n)$, $B_1(n)$ respectively. Then for any ring $R$ we define $D_\bullet(R, B(n), \rho)$ to be the simplicial set of acyclic chain complexes based on the graded poset $(B(n), \rho)$. (We also write $\mathcal{D}_\bullet(R, P)$ with $P = (B(n), \rho)$.) The $p$-simplices are triples $(f, E, S)$ where

1. $f = (f(0), \cdots, f(p))$ is a $(p+1)$-tuple of $n \times n$ invertible matrices all belonging to $T^\sigma(R)f(0)T^\tau(R)$ with the property that the $ij$ entry is nonzero only when $y_i < x_j$. We regard each $f(i)$ as the boundary map of an acyclic chain complex $C_*(i) : 0 \to C_1(i) \to C_0(i)$ where $C_d(i)$ is the free right $R$-modules with basis $B_d(n)$.

2. $E = \{E(i,j)\}$ is a family of chain isomorphisms $E(i,j) : C_*(j) \to C_*(i)$ for all $0 \le i < j \le p$ which are $\rho$-upper triangular, or equivalently, $e(i,j) = E(i,j) - I$ is strictly $\rho$-upper triangular (i.e., $e(i,j)(z) = E(i,j)(z) - z$ is an $R$-linear combination of elements of $B(n)$ which are less than $z$ in the partial ordering for all $z \in B(n)$).

3. $S = \{S(i,j,k)\}$ is the (uniquely determined) family of strictly $\rho$-upper triangular chain homotopies $S(i,j,k) : C_0(k) \to C_1(i)$ so that
$$f(i)S(i,j,k) + S(i,j,k)f(k) = E(i,j)E(j,k) - E(i,k)$$
for all $0 \le i < j < k \le p$.

4. The face and degeneracy operators are given in the usual way by deleting and repeating entries of $(f, E, S)$. For degeneracies we need to use the convention that $E(i,i)$ is the identity map (*i.e.*, $e(i,i) = 0$) and $S(i,j,k) = 0$ whenever $i, j, k$ are not distinct.

**Example 2.1.2.** The following is a homotopy commuting diagram of acyclic chain complexes representing an element of $D_2(R, B_2, \rho)$.

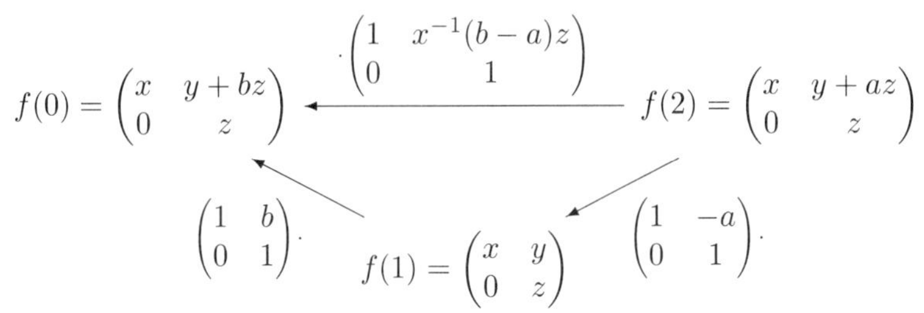


The partial ordering of the generating set $B_2$ is given by $y_1 < x_1 < y_2 < x_2$. The chain maps $E(0,1)$, $E(1,2)$ are the identity in degree 1 and thus the boundary map of their target differs from that of their source by a row operation. The chain map $E(0,2)$ is the identity in degree 0 and in degree 1 it is the elementary matrix
$$E(0,2)_1 = E_{12}(-x^{-1}(b-a)z)$$
so it performs the column operation on the boundary map given by the inverse matrix $E_{12}(x^{-1}(b-a)z)$ as indicated in the diagram. The homotopy is given by $S(0,1,2)(y_2) = x^{-1}(b-a)x_1$.

## Properties of $D_\bullet(R, B(n), \rho)$

Since the upper triangular chain homotopies $S(i,j,k)$ are uniquely determined by $f$ and $E$, it is only their existence which is being postulated.

**Remark 2.1.3.** We can extend the definition of $E(i,j)$ and $S(i,j,k)$ to all $0 \leq i,j,k \leq p$ by setting $E(j,i) = E(i,j)^{-1}, S(i,k,j) = -S(i,j,k)E(k,j)$ and $S(k,i,j) = -E(j,i)S(i,j,k)$ (and using the conventions of 2.1.1(4)).

A uniform notation for $f, E, S$ can be given by $e(i) = f(i)$, $e(i,j) = E(i,j) - id$, $e(i,j,k) = S(i,j,k)$. Then $e(j_0, \cdots, j_m)$ is an endomorphism of $C = R^B$ of degree $m-1$ for each $0 \leq j_0, j_1, \cdots, j_m \leq p$ which is zero if the indices are not distinct and which satisfies the following identity.

$$\sum_{i=0}^{m}(-1)^i e(j_0, \cdots, j_i)e(j_i, \cdots, j_m) = \sum_{i=0}^{m}(-1)^i e(j_0, \cdots, \widehat{j_i}, \cdots, j_m) \tag{2.1}$$

When $m = 0$ this says that $f(i)f(i) = 0$ if we use the convention that $e() = 0$. When $m = 1$ the equations says $f(i)e(i,j) - e(i,j)f(j) = f(j) - f(i)$ or $f(i)E(i,j) = E(i,j)f(j)$. When $m = 2$ it says

$$f(i)S(i,j,k) - e(i,j)e(j,k) + S(i,j,k)f(k) = e(j,k) - e(i,k) + e(i,j)$$

which is equivalent to 2.1.1(3). When $m = 3$ the equation gives the following identity which follows easily from 2.1.1(3)

$$E(i,j)S(j,k,l) - S(i,k,l) + S(i,j,l) - S(i,j,k)E(k,l) = 0 \tag{2.2}$$

We will refer to the collection $\{e(j_0, \cdots, j_m)\}$ as a *system of endomorphisms*. This is a special case of a "system of higher homotopies" which will be defined later.

**Proposition 2.1.4.** *The simplicial set $D_\bullet(R, B(n), \rho)$ is homotopy equivalent to the nerve of the category $\mathcal{D}(R, B(n), \rho)$ given as follows.*

*a) The objects are invertible $n \times n$ matrices $f$ whose $ij$ entries are nonzero only when $y_i < x_j$.*

*b) The morphisms $f \to f'$ are strict $\rho$-upper triangular chain homotopy classes of $\rho$-upper triangular chain isomorphisms between the corresponding acyclic complexes.*

*Proof.* There is the natural simplicial forgetful map $F : D_\bullet \to \mathcal{N}_\bullet\mathcal{D}$ which has a right inverse $G$ given by choosing one representative from every strict upper triangular homotopy class of chain maps. The composition $G \circ F$ is obviously simplicially homotopic to the identity map on $D_\bullet$. □

**Corollary 2.1.5.** *Let $X_\bullet$ be any simplicial set and let $f, g$ be simplicial maps $X_\bullet \to D_\bullet(R, B(n), \rho)$. Then a simplicial homotopy from $f$ to $g$ is given up to (simplicial) homotopy by a family of morphisms $\eta_v : F(f(v)) \to F(g(v))$ in $\mathcal{D}(R, B(n), \rho)$ for all $v \in X_0$ so that the following diagram commutes in*

$\mathcal{D}(R, B(n), \rho)$ *for all* 1*-simplices* $x \in X_1$ *with source* $\partial_0 x = v$ *and target* $\partial_1 x = w$.

$$\begin{array}{ccc} f(v) & \xrightarrow{\eta_v} & g(v) \\ {\scriptstyle E^f(0,1)}\uparrow & & \uparrow{\scriptstyle E^g(0,1)} \\ f(w) & \xrightarrow[\eta_w]{} & g(w) \end{array}$$

**Proposition 2.1.6.** *Take the product* $R \times S$ *of two rings* $R$ *and* $S$*. Then*

*a)* $D_\bullet(R \times S, B(n), \rho) \cong D_\bullet(R, B(n), \rho) \times D_\bullet(S, B(n), \rho)$

*b)* $\mathcal{D}(R \times S, B(n), \rho) \cong \mathcal{D}(R, B(n), \rho) \times \mathcal{D}(S, B(n), \rho)$

### 2.1.1 The space $\mathcal{W}_\bullet(R, G, n)$

The disjoint union of simplicial sets $D_\bullet(R, B)$ where $B \cong (B(n), \rho)$ for some $\rho$ forms the set of objects of a simplicial category $W_\bullet(R, G, n)$ where $G$ is a subgroup of the group of units $R^\times$ of $R$.

First we note that a $P \times Q$ matrix $(f_{pq}), p \in P, q \in Q$ is the same as a right $R$-module homomorphism $f : R^Q \to R^P$ and a $P$-upper triangular matrix corresponds to an endomorphism $f$ of $R^P$ with the property that $f(p) = p + \Sigma_{q<p}\, q f_{qp}$ for all $p \in P$.

One other bit of notation: Suppose that $a : P \to Q$ is an injective/bijective set mapping and $\gamma : P \to G$ is any set map. Then we denote by

$$(\alpha, \gamma)_* : R^P \to R^Q$$

the homomorphism of right $R$-modules given by $(\alpha, \gamma)_*(x) = \alpha(x)\gamma(x)$. We call this a *$G$-monomial monomorphism/isomorphism*. The composition rule is $(\alpha, \gamma)_*(\beta, \gamma')_* = (\alpha \circ \beta, (\gamma \circ \beta)\gamma')_*$. We use the shorthand $\alpha_*$ for $(\alpha, 1)_*$ where 1 sends all of $P$ to $1 \in G$.

**Definition 2.1.7.** Let $\mathcal{W}_\bullet(R, G, n)$ be the simplicial category given as follows.

1. The *objects* of $\mathcal{W}_p(R, G, n)$ are pairs $(C_*, B)$ where $B = B_0 \coprod B_1$ is a graded poset having $n$ elements in each degree and $C_*$ is an object of $D_p(R, B)$.

2. A *morphism* $(C_*, B) \to (C'_*, B')$ is defined to be a degree preserving *closed bijection* $\alpha : B \to B'$ and a function $\gamma : B \to G$ giving a $G$-monomial chain isomorphism

$$(\alpha, \gamma)_* : C_*(i) \to C_*(i)', \quad x \mapsto \alpha(x)\gamma(x)$$

so that $(\alpha, \gamma)_* E(i, j) = E(i, j)'(\alpha, \gamma)_*$ for all $i, j$. Composition is given by $(\alpha, \gamma)_* \circ (\beta, \gamma')_* = (\alpha \circ \beta, (\gamma \circ \beta)\gamma')_*$ where $((\gamma \circ \beta)\gamma')(x) = \gamma(\beta(x))\gamma'(x)$.

3. Face and degeneracy operators are given by deleting and repeating the terms $C_*(i)$ and keeping $B$ the same. (See 2.1.1(4).)

4. Two morphisms $(\alpha,\gamma)_*, (\alpha',\gamma')_* : C \to C'$ will be called *conjugate* if $\alpha = \alpha'$.

**Definition 2.1.8.** We let $\mathcal{W}_\bullet(R,G,n)/_{conj}$ denote the simplicial quotient category given by taking the same simplicial set of objects as $\mathcal{W}_\bullet(R,G,n)$ but with conjugacy classes of morphisms.

To avoid set theoretic difficulties we will assume that the underlying set of the poset $B$ is a subset of some fixed infinite set.

We note that objects and morphisms are given in terms of the poset. Thus, e.g., given any object $(C_*, B)$ in $\mathcal{W}_k(R,G,n)$, any closed bijection $\alpha : B' \to B$ and any map $\gamma : B' \to G$ there exists a *unique* object $(C'_*, B')$ which maps to $(C_*, B)$ by the morphism $(\alpha,\gamma)_*$.

### 2.1.2 The Volodin category $\mathcal{V}_\bullet(R,n)$

The simplicial category $\mathcal{W}_\bullet(R,G,n)$ contains a copy of the Volodin space $V_n(R)$ and a disjoint copy of the classifying space of the group of $n \times n$ $G$-monomial matrices.

**Definition 2.1.9.** Let $\mathcal{V}_\bullet(R,n)$ be the simplicial subcategory of $\mathcal{W}_\bullet(R,1,n)$ with objects all pairs $(C_*, B(n), \rho)$ where each $y_i \in B_0(n)$ is minimal in the partial ordering $\rho$ and with morphisms $(id, 1)$, i.e., the identity mapping on the underlying set of $B(n)$ and $\gamma = 1$ is forced since $G = \{1\}$. Since $\sigma$ is empty in that case, $T^\sigma(R)$ is trivial and the condition on the boundary map $f$ is:

- $f(i) \in f(0)T^\tau(R)$ for all $i$,
- the chain isomorphisms $E_{ij}$ are all equal to the identity on $C_0$ and
- the chain homotopies $S(i,j,k)$ are all zero.

Note that, at the chain level, the morphisms of $\mathcal{V}_\bullet(R,n)$ are those given by $(I_n, I_n)$ where $I_n$ is the $n \times n$ identity matrix. Thus there is a morphism $(C_*, \rho) \to (C'_*, \rho')$ if and only if $C'_* = C_*$ and $\tau' \subseteq \tau$ and this morphism is unique whenever it exists.

**Proposition 2.1.10.** *$|\mathcal{V}_\bullet(R,n)| \simeq |V_n(R)|$ where $V_n(R)$ is the standard Volodin space of $n \times n$ matrices over $R$.*

*Proof.* We consider $V_n(R)$ as a simplicial category with only identity morphisms. Then a simplicial functor $F : \mathcal{V}_\bullet(R,n) \to V_n(R)$ is given by $F(C_*, \rho) = f = (f(0), \cdots, f(p))$ where $f(i)$ is the boundary map of the complex $C_*(i)$. In each degree $p \geq 0$ this map $F_p : \mathcal{V}_p(R,n) \to V_n(R)[p]$ is a homotopy equivalence since the set of all partial orderings $\rho$ compatible with any given $(p+1)$-simplex $f$ in $V_n(R)$ has a unique minimal element $\rho_0$ making $(C_*, B(n), \rho_0)$ the terminal object of $F/f$. Therefore $|F| : |\mathcal{V}_\bullet(R,n)| \to |V_n(R)|$ is a homotopy equivalence by Quillen's Theorem A [Qui73]. □

The Volodin category $\mathcal{V}(R,n)$ is contained in the simplicial full subcategory $\mathcal{W}_\bullet(R,B(n))$ of $\mathcal{W}_\bullet(R,G,n)$ with objects $(C_*,B(n),\rho)$ (fixing the graded set $B(n)$ with variable partial ordering $\rho$). Since every object of $\mathcal{W}_\bullet(R,G,n)$ is isomorphic to an object in $\mathcal{W}_\bullet(R,B(n))$ the simplicial inclusion functor is a homotopy equivalence:

$$\mathcal{W}_\bullet(R,B(n)) \simeq \mathcal{W}_\bullet(R,G,n).$$

Every morphism in $\mathcal{W}_p(R,B(n))$ is given by a pair of $G$-monomial matrices $A,B \in Mon_n(G)$ so that $Af'(i) = f(i)B$ for all $0 \le i \le p$. If we take only the first monomial matrix $A$ we get a forgetful functor from each $W_p(R,B(n))$ to $\mathcal{B}Mon_n(G)$, the groupoid with one object whose automorphism group is the group $Mon_n(G)$ of $n \times n$ $G$-monomial matrices. Since the simplicial operators commute with this forgetful functor we get a map on classifying spaces

$$|W_\bullet(R,B(n))| \to |\mathcal{B}Mon_n(G)|.$$

This map is trivial on $\mathcal{V}_\bullet(R,n)$. Furthermore, it has a left inverse $\mathcal{B}Mon_n(G) \to W_0(R,G,n)$ given by sending the unique object of $\mathcal{B}Mon_n(G)$ to the identity matrix with the minimal partial ordering of $B(n)$ and the monomial matrix $A$ to $(A,A)$.

### 2.1.3 The involution on $\mathcal{W}_\bullet(R,G,n)$

Given a conjugation on $(R,G)$ (Definition 1.5.1) we get an involution on the simplicial category $\mathcal{W}_\bullet(R,G,n)$ by conjugate transpose on objects and inverse conjugate transpose on morphisms. Since the involution on the Volodin space is given by inverse conjugate transpose on objects, the inclusion functor is not compatible with the two involutions. Instead we will see that (stably) they differ by a sign.

We note first that, for any homomorphism of f.g. based free $R$-modules $f: R^P \to R^Q$, the conjugate transpose $f^*$ is a homomorphism of $R$-modules going the other way $f^*: R^Q \to R^P$. Consequently, conjugate transpose gives an anti-involution *ct* on the category of f.g. based free $R$-modules and inverse conjugate transpose gives an involution $\mathcal{I}ct$ on the subcategory of isomorphisms.

**Definition 2.1.11.** Given a conjugation on $(R,G)$ we get a simplicial involution

$$ct: \mathcal{W}_\bullet(R,G,n) \to \mathcal{W}_\bullet(R,G,n)$$

given on objects by $ct(C_*,B) = (C_*^{op}, B^{op})$ where

1. $B^{op} = \{\overline{x} : x \in B\}$ with $\deg \overline{x} = 1 - \deg x$ and $\overline{x} < \overline{y} \Leftrightarrow x > y$, (i.e., $B^{op}$ is $B$ with both degree and partial ordering reversed).

2. $C_*^{op}$ is the free $R$-complex with graded basis set $B^{op}$ and system of endomorphisms $e^{op}$ given by

$$e^{op}(j_0,\cdots,j_p) = e(j_p,\cdots,j_0)^*$$

where $(\ )^*$ denotes conjugate transpose and $e(j_p, \cdots, j_0)$ is defined by Remark 2.1.3.

On morphisms we take inverse conjugate transpose.

Note that $e^{op}$ reverses the order of the vertices. When we put them in the correct order using Remark 2.1.3, the upper triangular chain isomorphisms $E(i,j)$ are inverted. Therefore, the involution $ct$ is conjugate transpose only "vertically." In the "horizontal" direction, i.e., the upper triangular chain isomorphisms and the $G$-monomial morphisms, it is inverse conjugate transpose.

Although the boundary map $f(i)$ is always an isomorphism, the operation of inverting the boundary map is not a functor on $\mathcal{W}_\bullet(R, G, n)$. It is however a functor on the simplicial full subcategory $\mathcal{V}^2_\bullet(R, G, n)$ whose objects $(C_*, B)$ have the property that each element of $B_0$ is less than each element of $B_1$.

**Definition 2.1.12.** The *inversion* functor $\mathcal{I}$ on $\mathcal{V}^2_\bullet(R, G, n)$ is defined on objects $(C_*, B)$ by $\mathcal{I}(C_*, B) = (\mathcal{I}C_*, \mathcal{I}B)$ where

1. $\mathcal{I}B$ is the same poset as $B$ with degrees reversed (so that $\mathcal{I}B_i = B_{1-i}$ for $i = 0, 1$).

2. $\mathcal{I}C_*$ is given by inverting the boundary maps and keeping the same chain isomorphisms (re-indexed). Thus $\mathcal{I}f(i) = f(i)^{-1}$ and $\mathcal{I}E_i(j,k) = E_{1-i}(j,k)$.

3. The morphism are the same (re-indexed).

We have the following obvious observation.

**Proposition 2.1.13.** *The inclusion functor $\mathcal{V}_\bullet(R, n) \to \mathcal{V}^2_\bullet(R, G, n)$ commutes with the involution given by inverse conjugate transpose ($\mathcal{I}ct$ on $\mathcal{V}_\bullet(R, n)$ and $\mathcal{I} \circ ct$ on $\mathcal{V}^2_\bullet(R, G, n)$).*

It would simplify the exposition if we could show that the involution $ct$ is homotopic to the identity on $\mathcal{W}_\bullet(M_m(\mathbb{C}), U(m), n)$. However this is not true. A more convoluted argument is required to prove what we really need which is that $ct$ leaves the higher FR-torsion invariant. Since $\mathcal{I}ct$ reverses the sign of the Borel regulator map we first need to show that $\mathcal{I}$ reverses sign. As a consequence of the obvious additivity of the Borel regulator/higher FR-torsion map with respect to direct sum, this is equivalent to showing that the direct sum of the inversion map and the identity map is homotopic to a constant map.

**Proposition 2.1.14.** *The simplicial functor*

$$J \oplus \mathcal{I} : \mathcal{V}^2_\bullet(R, G, n) \to \mathcal{W}_\bullet(R, G, 2n)$$

*given by the direct sum of the inclusion functor $J$ and the inversion functor $\mathcal{I}$ (followed by inclusion) is homotopic to a constant simplicial functor.*

*Proof.* Let $(g, E, B)$ be one object of $\mathcal{V}_p^2(R, G, n)$. Then $g = (g_0, \cdots, g_p)$ is a $(p+1)$-tuple of invertible matrices and $E = \{E(j,k)\}$ is a system of $B$-upper triangular chain isomorphisms

$$E(j,k) : (R^B, g(k)) \to (R^B, g(j)).$$

Each of these is given by two $n \times n$ matrices $E_i(j,k)$ which are $B_i$-upper triangular having the property that

$$E_0(j,k)g(k) = g(j)E_1(j,k)$$

for all $j, k \in [p]$.

Let

$$F_1 : \mathcal{V}_\bullet^2(R, G, n) \to \mathcal{V}_\bullet^2(R, G, 2n)$$

be the simplicial functor given by

$$F_1(g, E, B) = (g \oplus \mathcal{I}g, E \oplus \mathcal{I}E, B \coprod \mathcal{I}B).$$

The first step is to show that $F_1$ is homotopic to another simplicial functor $F_2$ given by

$$F_2(g, E, B) = (F_2 g, F_2 E, F_2 B)$$

where

- $F_2 g(j) = \begin{pmatrix} g(j) & I \\ 0 & g(j)^{-1} \end{pmatrix}$
- $F_2 E = F_1 E = E \oplus \mathcal{I}E$, i.e., $F_2 E_0 = E_0 \oplus E_1$ and $F_2 E_1 = E_1 \oplus E_0$.
- $F_2 B = B \coprod \mathcal{I}B$ with the partial ordering altered so that each element of $B_0$ is less than each element of $\mathcal{I}B_0$. (The subsets $B_0, \mathcal{I}B_0, B_1, \mathcal{I}B_1$ have the same partial ordering as before and all elements of $F_2 B_0$ are less than all elements of $F_2 B_1$ as required.) We abbreviate these conditions by writing just: $B_0 < \mathcal{I}B_0$.

The homotopy $F_1 \simeq F_2$ is given first by changing the partial ordering of $B \coprod \mathcal{I}B$ to $F_2 B$. This gives a natural transformation $(id, 1)_* : F_1' \to F_1$ where $F_1'(g, E, B) = (F_1 g, F_1 E, F_2 B)$. Next we take the family of $F_2 B$-upper triangular chain isomorphisms $f(j) : F_1 g(j) \to F_2 g(j)$ given by

$$f(j)_0 = \begin{pmatrix} I & g(j) \\ 0 & I \end{pmatrix}$$

and $f(j)_1 = id$. This gives a simplicial homotopy $F_1' \simeq F_2$ by Proposition 2.1.5 which is obviously natural.

Next we take the simplicial functor $F_3 : \mathcal{V}_\bullet^2(R, G, n) \to \mathcal{V}_\bullet^2(R, G, 2n)$ given by

- $F_3 g(j) = \begin{pmatrix} 0 & I \\ -I & g(j)^{-1} \end{pmatrix}$

- $F_3E = F_1E = E \oplus \mathcal{I}E$.
- $F_2B = B \coprod \mathcal{I}B$ with partial ordering $B_1 > \mathcal{I}B_1$.

We indicate this by writing $F_3 = (F_3, F_1, B_1 > \mathcal{I}B_1)$. Again we first change the partial ordering using this time two natural transformations:

$$(F_2, F_1, B_0 < \mathcal{I}B_0) \leftarrow (F_2, F_1, \{B_0 < \mathcal{I}B_0, B_1 > \mathcal{I}B_1\}) \rightarrow (F_2, F_1, B_1 > \mathcal{I}B_1)$$

to change the partial ordering $F_2 B to F_3 B$. This gives a functor $F_2' = (F_2, F_1, F_3)$. Then we take a simplicial homotopy $F_2' \simeq F_3$ induced by the family of chain isomorphisms $f_2(j) : F_2g(j) \rightarrow F_3g(j)$ given by

$$f_2(j)_1 = \begin{pmatrix} I & 0 \\ g(j) & I \end{pmatrix}$$

and $f_2(j)_0 = id$.

Next, we take the simplicial functor $F_4 = (F_4, F_1, B_1 < \mathcal{I}B_1)$ where

$$F_4g(j) = \begin{pmatrix} 0 & I \\ -I & 0 \end{pmatrix}$$

for each $j$. The simplicial homotopy is given by the family of chain isomorphisms

$$f_3(j)_1 = \begin{pmatrix} I & -g(j) \\ 0 & I \end{pmatrix}$$

with $f_3(j)_0 = id$.

Finally, we note that the chain isomorphisms

$$E(j,k) \oplus \mathcal{I}E(j,k) : F_4g(k) \rightarrow F_4g(j)$$

are homotopic to the identity by the chain homotopy

$$S = \begin{pmatrix} 0 & I - E_1(j,k) \\ E_0(j,k) - I & 0 \end{pmatrix}$$

which is upper triangular with respect to the ordering $y_1 < \overline{y}_1 < y_2 < \overline{y}_2 < \cdots < \overline{y}_n, \overline{x}_1 < x_1 < \overline{x}_2 < x_2 \cdots < x_n$. Consequently the composition

$$\mathcal{V}^2_\bullet(R, G, n) \xrightarrow{F_4} \mathcal{V}^2_\bullet(R, G, 2n) \hookrightarrow \mathcal{W}_\bullet(R, G, 2n)$$

is homotopic to a constant simplicial functor. □

### 2.1.4 Functorial properties of $\mathcal{W}_\bullet(R, G, n)$

We note that the simplicial category $\mathcal{W}_\bullet(R, G, n)$ varies naturally with the pair $(R, G)$. If $\phi : R \rightarrow S$ is a ring homomorphism which takes $G \subseteq R^\times$ into $H \subseteq S^\times$ then we have an induced simplicial functor

$$\phi_* : \mathcal{W}_\bullet(R, G, n) \rightarrow \mathcal{W}_\bullet(S, H, n)$$

sending $(C_*, B)$ to $(C_* \otimes S, B)$ and taking morphisms $(\alpha, \gamma)$ to $(\alpha, \phi \circ \gamma)$.

**Proposition 2.1.15.** *Suppose that $\phi, \psi : R \to S$ are two ring homomorphisms taking $G$ into $H$ which are conjugate by a fixed element $h$ of $H$ ($\psi(r) = h\phi(r)h^{-1}$ for all $r \in R$). Then*

$$\phi_*, \psi_* : \mathcal{W}_\bullet(R, G, n) \to \mathcal{W}_\bullet(S, H, n)$$

*are simplicially homotopic.*

*Proof.* Conjugation by $h$ gives a simplicial natural transformation $\phi_* \simeq \psi_*$. □

**Definition 2.1.16 (Transfer).** There is a natural *simplicial "transfer" functor*

$$\mathrm{N} : \mathcal{W}_\bullet(M_m(R), Mon_m(G), n) \to \mathcal{W}_\bullet(R, G, nm)$$

sending $(C_*, B)$ to $(\mathrm{N}\, C_*, mB)$ where

1. $mB = \{(x, i) : x \in B, 1 \leq i \leq m\}$ is the disjoint union of $m$ copies of $B$ with partial ordering $(x, i) < (y, j)$ if $x < y$ (disregarding $i, j$) and grading $\deg(x, i) = \deg(x)$.
2. $\mathrm{N}\, C_* = (\mathrm{N}\, f, \mathrm{N}\, E, \mathrm{N}\, S)$ where $(\mathrm{N}\, f)_{(x,i),(y,j)}$ is the $(i, j)$ entry of $f_{xy} \in M_m(R)$ and similarly for $\mathrm{N}\, E, \mathrm{N}\, Mon$. ($f$ maps to $\mathrm{N}\, f$ under the isomorphism $M_n(M_m(R)) \cong M_{nm}(R)$.)
3. $\mathrm{N}$ is given on morphisms by the monomorphism $Mon_n(Mon_m(G)) \subseteq Mon_{nm}(G)$.

Suppose that $S = End_R(M_R)$ is the endomorphism ring of some right $R$-module $M_R$. Then an object of $\mathcal{W}_\bullet(End_R(M_R), G, n)$ can be viewed as a family of $R$-chain complexes $0 \rightarrowtail M^Q \to M^P$ together with upper triangular chain maps and (unique) chain homotopies. We will sometimes denote this simplicial category by $\mathcal{W}_\bullet(M, G, n)$. Since $R = End_R(R_R)$ this is consistent notation. Since $M_n(R)$ is the endomorphism ring of the $R$-module $R^n$, this means that

$$\mathcal{W}_\bullet(R^m, G, n) = \mathcal{W}_\bullet(M_m(R), G, n).$$

## 2.2 $\mathcal{W}_\bullet^{cont}(\mathbb{C}^m, n)$ and $\mathcal{W}_\bullet^{diff}(\mathbb{C}^m, n)$

### Introduction

Since $M_m(\mathbb{C})$ is a smooth manifold, we can define continuous and smooth versions of the simplicial category $\mathcal{W}_\bullet(\mathbb{C}^m, G, n) = \mathcal{W}_\bullet(M_m(\mathbb{C}), G, n)$ for any subgroup $G$ of the unitary group $U(m)$.

**Definition 2.2.1.** If $G$ is any subgroup of $U(m)$ let $\mathcal{W}_\bullet^{diff}(\mathbb{C}^m, G, n)$ denote the following simplicial category. ($\mathcal{W}_\bullet^{cont}(\mathbb{C}^m, G, n)$ is defined analogously.)

1. The objects of $\mathcal{W}_k^{diff}(\mathbb{C}^m, G, n)$ are pairs $(C_*, B)$ where $B$ is a poset on $2n$ elements $B = B_0 \coprod B_1$ as before and $C_*(t), t \in \Delta^k$ is a smooth $\Delta^k$-family of acyclic chain complexes
$$C_*(t) : 0 \to \mathbb{C}^{mB_1} \xrightarrow{f(t)} \mathbb{C}^{mB_0}$$
whose boundary maps $f(t)$ form a smooth mapping $f : \Delta^k \to GL_{nm}(\mathbb{C})$ together with a smooth $(\Delta^k)^2$-family of $B$-upper triangular chain isomorphisms
$$E(s,t) : C_*(t) \to C_*(s)$$
with the property that there exists a smooth $(\Delta^k)^3$-family of strict $B$-upper triangular chain homotopies
$$S(s,t,u) : E(s,u) \simeq E(s,t)E(t,u)$$
so that $S(s,t,u) = 0$ if $s,t,u$ are not distinct. (In particular, $E(t,t) = id$ and $E(t,s) = E(s,t)^{-1}$ for all $s,t \in \Delta^k$.)

2. A morphism $(C_*, B) \to (C'_*, B')$ is a pair $(\alpha, \gamma)$ where $\alpha : B \to B'$ is a closed bijection and $\gamma : B \to G$ is a mapping so that $(\alpha, \gamma)$ gives a $G$-monomial chain isomorphism
$$(\alpha,\gamma)_* : C_*(t) \to C'_*(t), \qquad x \mapsto \alpha(x)\gamma(x)$$
for all $t \in \Delta^k$ so that the following diagram commutes for all $(s,t) \in \Delta^k \times \Delta^k$.
$$\begin{array}{ccc} C_*(t) & \xrightarrow{(\alpha,\gamma)_*} & C'_*(t) \\ {\scriptstyle E(s,t)}\downarrow & & \downarrow{\scriptstyle E'(s,t)} \\ C_*(s) & \xrightarrow{(\alpha,\gamma)_*} & C'_*(s) \end{array}$$

3. Face and degeneracy operators are defined in the obvious way.

4. We define $\mathcal{W}_k^{diff}(\mathbb{C}^m, G, n)/_{conj}$ as before. ($(\alpha,\gamma)$ is *conjugate* to $(\beta,\gamma')$ if and only if $\alpha = \beta$.)

When $G = U(m)$ we will leave it out of the notation and simply write $\mathcal{W}_\bullet^{diff}(\mathbb{C}^m, n)$ or $\mathcal{W}_\bullet^{cont}(\mathbb{C}^m, n)$.

Let $W_\bullet^{cont}(B(n), \rho)$ be the simplicial subset of the simplicial set of objects of $\mathcal{W}_\bullet^{cont}(\mathbb{C}^m, G, n)$ consisting of all pairs $(C_*, B)$ where $B = B(n)$ with a fixed partial ordering $\rho$. Let $W_\bullet^{diff}(B(n), \rho)$ be defined analogously.

**Lemma 2.2.2.** *The simplicial inclusion map $J$ and forgetful map $F$ are homotopy equivalences:*
$$W_\bullet^{diff}(B(n), \rho) \xrightarrow{J} W_\bullet^{cont}(B(n), \rho) \xrightarrow{F} D_\bullet(\mathbb{C}^m, B(n), \rho).$$

*Proof.* We show that the composition $F \circ J : W_\bullet^{diff}(B(n), \rho) \to D_\bullet(\mathbb{C}^m, B(n), \rho)$ is a homotopy equivalence. A similar argument shows that $F$ is a homotopy equivalence.

(2.2.2.1) First we pass to the nerve of the category $\mathcal{D}(\mathbb{C}^m, B(n), \rho)$ of 2.1.4 and consider the simplicial forgetful map

$$W_\bullet^{diff}(B(n), \rho) \to \mathcal{N}_\bullet \mathcal{D}(\mathbb{C}^m, B(n), \rho).$$

Since these two simplicial sets have the same set of vertices it suffices to show that, for any simplicial map $\Delta^k \to \mathcal{N}_\bullet \mathcal{D}(\mathbb{C}^m, B(n), \rho)$, any lifting of this map to $W_\bullet^{diff}(B(n), \rho)$ over the boundary of $\Delta^k$ extends to a lifting over all of $\Delta^k$. This problem can be expressed as follows.

(2.2.2.2) We have a smooth family of chain complexes $C_*(t), t \in \partial\Delta^k$, with $C_0(t) = C_1(t) = \mathbb{C}^{nm}$ and variable boundary map $f(t) : \mathbb{C}^{nm} \to \mathbb{C}^{nm}$. We also have a smooth family of upper triangular chain isomorphisms

$$E(s, t) : C_*(t) \to C_*(s)$$

defined on the subset $DD$ of $\partial\Delta^k \times \partial\Delta^k$ of all pairs $(s, t)$ where $s, t$ lie in the same proper face of $\Delta^k$. These satisfy the following properties whenever the terms are defined.

1. $E(s, t) = E(t, s)^{-1}$ $(\Rightarrow E(t, t) = id.)$

2. $E(s, t)E(t, r) \simeq E(s, r)$ by a strict $\rho$-upper triangular homotopy (which is necessarily unique).

In the case $k = 1$, the chain isomorphism $C_*(v_0) \to C_*(v_1)$ is defined up to strict upper triangular homotopy for $v_0, v_1 \in \partial\Delta^1$.

(2.2.2.3) If we fix a vertex $v_0$ then a $\rho$-upper triangular chain isomorphism $E : C_*(v_0) \to C_*(t)$ is given by a pair of upper triangular matrices $E_0 \in T^\sigma$ and $E_1 \in T^\tau$ where $\sigma, \tau$ are the partial orderings of $B_0(n) = \{y_1, \cdots, y_n\}$, $B_1(n) = \{x_1, \cdots, x_n\}$. These determine $f(t)$ since $f(t) = E_0 f(v_0) E_1^{-1}$. Given a homotopy $h : C_0(v_0) \to C_1(t)$ we let $S = hE_1^{-1} : C_0(v_0) \to C_1(v_0)$. Then $E = (E_0, E_1)$ is homotopic to $(E_0(I + f(v_0)S), E_1(I + Sf(v_0)))$.

Let $T_\rho (\neq T^\rho)$ be the set of $n \times n$ matrices $S$ whose entries $S_{ij}$ are $m \times m$ complex matrices which are allowed to be nonzero only if $x_i < y_j$. We define a multiplication on $T_\rho$ by $S * S' = S + S' + Sf(v_0)S'$. Then $T_\rho$ becomes a contractible Lie group which acts freely and smoothly on the right on the contractible manifold $T^\sigma \times T^\tau$ by:

$$(E_0, E_1)S = (E_0(I + f(v_0)S), E_1(I + Sf(v_0)).$$

The orbit space $(T^\sigma \times T^\tau)/T_\rho$ is a contractible smooth manifold which maps to the double coset space $T^\sigma f(v_0) T^\tau$ by $g(E_0, E_1) = E_0 f(v_0) E_1^{-1}$ and the elements of the fiber over $f(t)$ correspond to the strict upper triangular chain homotopy classes of upper triangular chain isomorphisms $C_*(v_0) \to C_*(t)$.

(2.2.2.4) The given data gives us a smooth mapping $h : \partial\Delta^k \to (T^\sigma \times T^\tau)/T_\rho$ sending $v_0$ to the base point $(I, I)$. Since the target is contractible this mapping extends to a smooth map $H$ on all of $\Delta^k$. (See Lemma 2.2.3 below.) Composing with the map $g$ to $T^\sigma f(v_0)T^\tau$ we get a smooth family of chain complexes $C_*(t)$ with boundary maps $f(t)$ defined for all $t \in \Delta^k$.

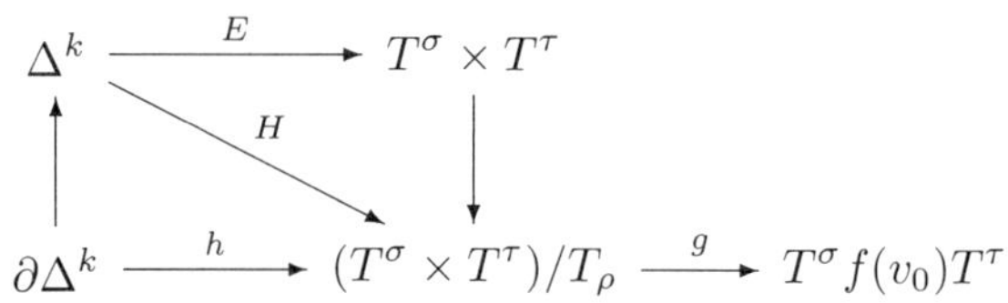


(2.2.2.5) The map $H : \Delta^k \to (T^\sigma \times T^\tau)/T_\rho$ has a given smooth lifting to $T^\sigma \times T^\tau$ over the union $L$ of all proper faces of $\Delta^k$ which meet the vertex $v_0$. Since $T_\rho$ is contractible we can extend this lifting to all of $\Delta^k$ to get $E : \Delta^k \to T^\sigma \times T^\tau$. Then $E(t)$ is a chain isomorphism $C_*(v_0) \to C_*(t)$ for all $t \in \Delta^k$ and $E(t) = E(t, v_0)$ if the latter is defined.

(2.2.2.6) Let $S : \Delta^k \times \Delta^k \to T_\rho$ be any smooth mapping so that

$$f(s)S(s,t) + S(s,t)f(v_0) = E(s,t)E(t) - E(s)$$

(i.e., $S(s,t) = [E_1(s,t)E_1(t) - E_1(s)]f(v_0)^{-1}$) for all $(s,t) \in DD$, (when $s,t$ lie in the same proper face of $\Delta^k$). Then this equation can be used to define $E(s,t)$ for all $(s,t) \in \Delta^k \times \Delta^k$ so that it agrees with the existing terms when $(s,t) \in DD$.

(2.2.2.7) The condition $E(s,t) = E(t,s)^{-1}$ may not be satisfied. However, this can be corrected by repeating $\log_2 n$ times the following recursive replacement which does not alter $E(s,t)$ on $DD$:

$$E(s,t) := \frac{E(s,t) + E(s,t)^{-1}}{2}.$$

□

**Lemma 2.2.3.** *Let $M$ be a smooth contractible manifold and let $f : \partial\Delta^k \to M$ be a continuous mapping which is smooth on each proper face of $\Delta^k$. Then $f$ extends to a smooth mapping on $\Delta^k$.*

*Proof.* Using a partition of unity it suffices to prove this locally in the case $M = \mathbb{R}$. Thus we are given compatible smooth mappings $f_i : H_i \to \mathbb{R}$ for $i = 0, \cdots, k$ where $H_i = \{(x_0, \cdots, x_k) \in [0,\infty)^{k+1} : x_i = 0\}$ and we want to extend $\cup f_i : \cup H_i \to \mathbb{R}$ to $[0,\infty)^{k+1}$. The answer is:

$$f(x) = \sum_{p=1}^{k} (-1)^{p-1} \sum_{0 \le i_1 < \cdots < i_p \le k} f_{i_1}(\pi_{i_1}\pi_{i_2}\cdots\pi_{i_p}(x))$$

where $\pi_i$ is the projection onto $H_i$. □

**Theorem 2.2.4.** *For any subgroup $G$ of $U(m)$ the simplicial inclusion and forgetful functors*

$$\mathcal{W}_\bullet^{diff}(\mathbb{C}^m, G, n) \xrightarrow{\simeq} \mathcal{W}_\bullet^{cont}(\mathbb{C}^m, G, n) \xrightarrow{\simeq} \mathcal{W}_\bullet(\mathbb{C}^m, G, n)$$

*are homotopy equivalences.*

*Proof.* We consider the continuous case and note that the analogous statement holds for $\mathcal{W}_\bullet^{diff}$ by virtue of the fact that Lemma 2.2.2 holds in the smooth case. We use the fact that $\mathcal{W}_\bullet^{cont}(\mathbb{C}^m, G, n)$ has, as a deformation retract, the full subcategory $\mathcal{W}_\bullet^{cont}(B(n))$ with simplicial set of objects

$$W_\bullet^{cont}(B(n)) = \coprod_\rho W_\bullet^{cont}(B(n), \rho)$$

which is homotopy equivalent to the simplicial set

$$D_\bullet(\mathbb{C}^m, B(n)) = \coprod_\rho D_\bullet(\mathbb{C}^m, B(n), \rho)$$

by Lemma 2.2.2.

We want to show that the simplicial forgetful functor $F : \mathcal{W}_\bullet^{cont}(B(n)) \to \mathcal{W}_\bullet = \mathcal{W}_\bullet(\mathbb{C}^m, G, n)$ is a homotopy equivalence. Passing to the nerve, it suffices to show that, for each $q \geq 0$, the simplicial map

$$F_q : \mathcal{N}_q \mathcal{W}_\bullet^{cont}(B(n)) \to \mathcal{N}_q \mathcal{W}_\bullet$$

is a homotopy equivalence.

The simplicial sets $\mathcal{N}_q \mathcal{W}_\bullet^{cont}(B(n))$ and $\mathcal{N}_q \mathcal{W}_\bullet$ have the same set of vertices. These are sequences of morphisms

$$(C^0, P^0) \to (C^1, P^1) \to \cdots \to (C^q, P^q)$$

which are $G$-monomial chain isomorphisms given by closed poset bijections $P^j \to P^{j+1}$ and maps $\gamma_j : P^j \to G$. Thus the posets $P^j$ may be viewed as a decreasing sequence of partial orderings on the same underlying set $P^0$ or, equivalently, as a sequence of refinements of the partial ordering on $P^q$. The maps $\gamma_j : P^j \to G$ for $j = 0, \cdots, q-1$ give a $q$-tuple of maps $P^0 \to G$ (or $P^q \to G$). Furthermore, any choice of $(C^q, P^q)$ and any sequence of closed degree 0 bijections $P^0 \to P^1 \to \cdots \to P^q$ and any $q$-tuple of maps $P^q \to G$ gives a uniquely determined sequence of chain complexes $C^j$ forming a sequence $(C^*, P^*) = (C^0, P^0) \to \cdots \to (C^q, P^q)$ as above.

The components of the simplicial sets $\mathcal{N}_q \mathcal{W}_\bullet^{cont}(B(n))$ and $\mathcal{N}_q \mathcal{W}_\bullet$ containing the sequence $(C^*, P^*)$ are simplicially equivalent to the components of $\mathcal{N}_0 \mathcal{W}_\bullet^{cont}(B(n))$ and $\mathcal{N}_0 \mathcal{W}_\bullet$ containing the last object $(C^q, P^q)$. This is because any admissible $\Delta^k$-family $(C^q(t), E^q(s,t))$ for $(C^q, P^q)$ gives a unique $\Delta^k$-family for $(C^j, P^j)$ for each $j$. Consequently the forgetful functor is a homotopy equivalence on these components by Lemma 2.2.2. □

## Variations of Theorem 2.2.4

(2.2.5.0) We would like to generalize Theorem 2.2.4 using the same proof. To do this we will first review the proof above and make a list of the properties of the simplicial category $\mathcal{W}_\bullet^{cont}(\mathbb{C}^m, G, n)$ that we use.

(2.2.5.1) The first property is that the simplicial set of objects of $\mathcal{W}_\bullet^{cont}$ is homotopy equivalent to that of $\mathcal{W}_\bullet$.

(2.2.5.2) The two simplicial categories have the same category of vertices, i.e., $\mathcal{W}_0^{cont} \cong \mathcal{W}_0$, or equivalently $\mathcal{N}_q\mathcal{W}_\bullet^{cont}$ and $\mathcal{N}_q\mathcal{W}_\bullet$ have the same set of vertices.

(2.2.5.3)(*unique refinement property*) Given any object $(C, B)$ of $\mathcal{W}_k^{cont}$ (or $\mathcal{W}_k$), any closed poset bijection $\alpha : B' \to B$ and any map $\gamma : B' \to G$ there is a unique object $(C', B')$ in $\mathcal{W}_k^{cont}$ (or $\mathcal{W}_k$) with a morphism $(\alpha, \gamma)_* : (C', B') \to (C, B)$.

**Theorem 2.2.5.** *Suppose that $\mathcal{W}_\bullet(X, G), \mathcal{W}_\bullet^{cont}(X, G)$ are simplicial full subcategories of $\mathcal{W}_\bullet(\mathbb{C}^m, G, n), \mathcal{W}_\bullet^{cont}(\mathbb{C}^m, G, n)$ resp. satisfying the following properties.*

- *The forgetful functor $F : \mathcal{W}_\bullet^{cont}(X, G) \to \mathcal{W}_\bullet(X, G)$ induces a homotopy equivalence on the subset of the simplicial set of objects on which the second factor is a fixed graded poset $B$.*
- *$F_0 : \mathcal{W}_0^{cont}(X, G) \to \mathcal{W}_0(X, G)$ is an isomorphism of categories.*
- *$\mathcal{W}_\bullet^{cont}(X, G)$ and $\mathcal{W}_\bullet(X, G)$ have the unique refinement property.*

*Then $F : \mathcal{W}_\bullet^{cont}(X, G) \to \mathcal{W}_\bullet(X, G)$ induces a homotopy equivalence on realizations.*

## Example

**Definition 2.2.6.** Suppose that $m = m_1 + m_2 + \cdots + m_d$. Then we define $\mathcal{W}_\bullet^{diff}(\prod M_{m_i}(\mathbb{C}), \prod U(m_i), n)$ to be the simplicial full subcategory of $\mathcal{W}_\bullet^{diff}(\mathbb{C}^m, \prod U(m_i), n)$ consisting of objects $(C_*, B)$ where the entries of the $P \times Q$ matrices $f(t), E(s,t), S(s,t,u)$ all lie in $\prod M_{m_i}(\mathbb{C}) \subseteq M_m(\mathbb{C})$.

**Corollary 2.2.7.** *The forgetful functor*

$$\mathcal{W}_\bullet^{diff}(\prod M_{m_i}(\mathbb{C}), \prod U(m_i), n) \to \mathcal{W}_\bullet(\prod M_{m_i}(\mathbb{C}), \prod U(m_i), n)$$

*induces a homotopy equivalence on realizations.*

*Proof.* First we need the analogue of Lemma 2.2.2. But this is immediate since the simplicial set of objects $W_\bullet^{diff}(B)$ of $\mathcal{W}_\bullet^{diff}(\prod M_{m_i}(\mathbb{C}), \prod U(m_i), n)$ with a fixed poset $B$ is isomorphic to the product of $W_\bullet^{diff}(B, m_i)$ which are the simplicial sets of objects of $\mathcal{W}_\bullet(M_{m_i}(\mathbb{C}), U(m_i), n)$ with poset $B$ so

$$|W_\bullet^{diff}(B)| \cong \prod |W_\bullet^{diff}(B, m_i)| \simeq \prod |W_\bullet(B, m_i)| \cong |W_\bullet(B)|$$

by Lemma 2.2.2. The other conditions of Theorem 2.4.4 are obvious. □

## A parametrized version of Lemma 2.2.2

For other applications of Theorem 2.2.5 we need to know under what conditions Lemma 2.2.2 holds. If we look at the proof we see that we can replace the coefficient ring $M_m(\mathbb{C})$ by any finite dimensional algebra over the real numbers, for example, $\prod M_{m_i}(\mathbb{C})$ as in 2.2.7. However we need to know Lemma 2.2.2 in the case of the infinite dimensional algebra $C^\infty(I)$ of all smooth complex valued functions on the closed unit interval $I$.

If we try to use matrices with coefficients in $C^\infty(I)$ we have a problem at one point in the proof of Lemma 2.2.2, namely, how to show that the map $T^\sigma \times T^\tau \to (T^\sigma \times T^\tau)/T_\rho$ is a fiber bundle and that the quotient is contractible. To avoid this problem we replace maps $\Delta^k \to C^\infty(I)$ by their adjoints $\Delta^k \times I \to \mathbb{C}$ and we construct a smooth family of acyclic chain complexes parametrized by $\Delta^k \times I$.

**Lemma 2.2.8.** *Choose a fixed partial ordering of the set $B(n)$ and suppose that we have a smooth family of acyclic chain complexes $C_*(t,u)$ parametrized over $\partial\Delta^k \times I$ with boundary maps $f(t,u) \in GLn(\mathbb{C})$ so that $C_*(t,u) \in D_\bullet(\mathbb{C}, B(n), \rho)$. Suppose that $E(s,t,u) : C_*(t,u) \to C_*(s,u)$ is a smooth family of upper triangular chain isomorphisms parametrized by $\partial\Delta^k \times \partial\Delta^k \times I$ satisfying the following conditions (analogous to the ones in 2.2.2).*

1. *$E(s,t,u) = E(t,s,u)^{-1}$ ($\Rightarrow E(t,t,u) = id$).*
2. *$E(s,t,u)E(t,r,u) \simeq E(s,r,u)$ by a strict upper triangular homotopy.*

*Then we can extend $C_*(t,u)$ smoothly to all of $\Delta^k \times I$ and we can extend $E(s,t,u)$ to all of $\Delta^k \times \Delta^k \times I$ so that conditions (1),(2),(3) are satisfied.*

*Proof.* The proof is analogous to the proof of 2.2.2. We will go over just the first step corresponding to 2.2.2.3.

Fix a vertex $v_0$ in $\Delta^k$. Define $T^\sigma, T^\tau$ as before. The underlying set of $T_\rho$ is the same as before but it has a one parameter family of group structures given by

$$S *_u S' = S + S' + Sf(v_0,u)S'.$$

We view this as a 1-parameter family of groups, giving a smooth bundle over $I$. There is a fiberwise action of this group bundle on $T^\sigma \times T^\tau$ given by

$$(E_0, E_1)S = (E_0(I + f(v_0,u)S), E_1(I + Sf(v_0,u)).$$

and the fiberwise orbit space $(T^\sigma \times T^\tau)/T_\rho$ is a smooth manifold which fibers over $I$ with contractible fibers. The proof continues in this way with all maps $h, H, S, E$ being fibered over $I$. □

## 2.3 The isovariant subcomplex of $\mathcal{W}_\bullet^{cont}(\mathbb{C}^m, G, n)$

Although $\mathcal{W}_\bullet(\mathbb{C}^m, G, n)$ and $\mathcal{W}_\bullet^{cont}(\mathbb{C}^m, G, n)$ are homotopy equivalent there is a problem when we compare the quotients with respect to conjugation. The problem arises because the conjugation action on a $k$-parameter family of acyclic chain complexes does not have the same stabilizer subgroup as the restriction of this action to the vertices. The remedy is to consider only those simplices in which this condition holds. This gives us the isovariant subcomplex of $\mathcal{W}_\bullet^{cont}(\mathbb{C}^m, G, n)$ which works very well.

**Definition 2.3.1.** (1) We define the *diagonal automorphisms* of any object of $\mathcal{W}_k(\mathbb{C}^m, G, n)$ or $\mathcal{W}_k^{cont}(\mathbb{C}^m, G, n)$ to be those automorphisms which are conjugate to the identity automorphism. They are given by pairs of diagonal matrices $(D_0, D_1)$ with coefficients in $G$ with the property that

(a) $D_0 f(t) = f(t) D_1$, i.e., $(D_0, D_1)$ gives a chain automorphism $D_*$ of $C_*(t)$ for each $t \in \Delta^k$ and

(b) $D_* E(s,t) = E(s,t) D_*$ for all $s, t \in \Delta^k$.

For $\mathcal{W}_k(\mathbb{C}^m, G, n)$ these conditions are only required to hold over the vertices of $\Delta^k$.

(2) The *isovariant subcomplex* $\mathcal{W}_\bullet^{iso-c}(\mathbb{C}^m, n)$ is defined to be the simplicial full subcategory of $\mathcal{W}_\bullet^{cont}(\mathbb{C}^m, n)$ consisting in degree $k$ of all objects $(C_*, B)$ whose group of diagonal automorphisms is equal to the group of diagonal automorphisms of the restriction of $(C_*, B)$ to the vertices of $\Delta^k$, i.e., any pair $(D_0, D_1)$ satisfying (a),(b) above at all vertices of $\Delta^k$ should also satisfy these conditions over all points of $\Delta^k$ in order for the object to be isovariant.

(3) The *smooth isovariant subcomplexes* $\mathcal{W}_\bullet^{iso-d}(\mathbb{C}^m, n)$ and $\mathcal{W}_\bullet^{iso-d}(\mathbb{C}^m, G, n)$ and the subcomplex $\mathcal{W}_\bullet^{iso-c}(\mathbb{C}^m, G, n)$ are defined analogously, i.e., they are the simplicial full subcategories of $\mathcal{W}_\bullet^{diff}(\mathbb{C}^m, n)$,$\mathcal{W}_\bullet^{diff}(\mathbb{C}^m, G, n)$ and $\mathcal{W}_\bullet^{cont}(\mathbb{C}^m, G, n)$, respectively, whose objects are isovariant.

The isovariant subcomplex satisfies the unique refinement property 2.2.5.3 and has the same category of vertices as $\mathcal{W}_\bullet(\mathbb{C}^m, n)$. To invoke Theorem 2.2.5 we also need an isovariant version of Lemma 2.2.2 or rather of the statement made in 2.2.2.1. For this we take a fixed partial ordering $\rho$ on $B(n)$ and we let $W_\bullet^{iso-c}(B(n), \rho)$ be the simplicial subset of the simplicial set of objects of $\mathcal{W}_\bullet^{iso-c}(\mathbb{C}^m, n)$ consisting of pairs $(C_*, B)$ where $B = (B(n), \rho)$. We note that $W_\bullet^{iso-c}(B(n), \rho)$ is the largest simplicial subset of the simplicial set $W_\bullet^{cont}(B(n), \rho)$ defined in Lemma 2.2.2 on which the forgetful map $W_\bullet^{cont}(B(n), \rho) \to \mathcal{N}_\bullet\mathcal{D}(\mathbb{C}^m, B(n), \rho)$ is isovariant with respect to diagonal automorphisms in the sense that each object in the domain has the same set of diagonal automorphisms as its image.

**Lemma 2.3.2.** *The simplicial forgetful map*

$$F : W^{iso-c}(B(n), \rho) \to \mathcal{N}\mathcal{D}(\mathbb{C}^m, B(n), \rho)$$

*has the property that any simplicial map $\Delta^k \to \mathcal{N}\mathcal{D}(\mathbb{C}^m, B(n), \rho)$ and any lifting of this map to $W^{iso-c}(B(n), \rho)$ over the boundary of $\Delta^k$ extends to a lifting over all of $\Delta^k$. Furthermore the same holds for $W^{iso-d}(B(n), \rho)$, the simplicial set of smooth simplices in $W^{iso-c}(B(n), \rho)$.*

The proof follows the same outline as the proof of Lemma 2.2.2. paying particular attention to the isovariance condition. The details are given at the end of this section. The main advantage of the isovariant version is the following.

**Theorem 2.3.3.** *For any subgroup $G$ of $U(m)$ we have the following homotopy equivalence given by simplicial inclusion and forgetful functors.*

*(a)* $|\mathcal{W}_\bullet^{iso-d}(\mathbb{C}^m, G, n)| \simeq |\mathcal{W}_\bullet^{iso-c}(\mathbb{C}^m, G, n)| \simeq |\mathcal{W}_\bullet(\mathbb{C}^m, G, n)|$

*(b)*

$$\begin{aligned} |\mathcal{W}_\bullet^{iso-d}(\mathbb{C}^m, G, n)/_{conj}| &\simeq |\mathcal{W}_\bullet^{iso-c}(\mathbb{C}^m, G, n)/_{conj}| \\ &\simeq |\mathcal{W}_\bullet(\mathbb{C}^m, G, n)/_{conj}| \end{aligned}$$

*Proof.* (a) follows from Theorem 2.2.5 and Lemma 2.3.2. The proof of (b) is also analogous to the proof of Theorem 2.3 and goes as follows in the case of $\mathcal{W}_\bullet^{iso-c}(\mathbb{C}^m, G, n)$.

It suffices to show that for each fixed $q \geq 0$ the simplicial map on degree $q$ nerves $F_q : \mathcal{N}_q\mathcal{W}_\bullet^{iso-c}(\mathbb{C}^m, G, n)/_{conj} \to \mathcal{N}_q\mathcal{W}_\bullet(\mathbb{C}^m, G, n)/_{conj}$ is a homotopy equivalence. By definition these simplicial sets have the same set of vertices. These are sequences of conjugacy classes of morphisms $(C^0, P^0) \to (C^1, P^1) \to \cdots \to (C^q, P^q)$. The posets $P^0, P^1, \cdots, P^q$ may be viewed as refinements of the partial ordering of $P^q$. The chain complexes $C^i$ are related to $C^q$ by the existence of unspecified chain isomorphisms $C^i \cong C^q$ given by diagonal matrices with coefficients in $G$ in each degree, i.e., the boundary maps are elements of a double coset $f_i \in Df_qD$ for each $i$ where $D$ is the group of diagonal matrices with coefficients in $G$. Furthermore any choice of $(C^q, P^q)$, any sequence of closed graded bijections $P^0 \to P^1 \to \cdots \to P^q$ and any $q$-tuple of elements $f_i$ of the double coset $Df_qD$ gives a unique sequence $(C^*, P^*)$ as above.

If we replace $(C^q, P^q)$ with a $k$-simplex in $\mathcal{N}_q\mathcal{W}^{iso-c}(\mathbb{C}^m, G, n)$ then the additional structure given by the posets $P^i$ and elements $f_i$ of $Df_qD$ are the same in both $\mathcal{N}_q\mathcal{W}_\bullet(\mathbb{C}^m, G, n)$ and $\mathcal{N}_q\mathcal{W}^{iso-c}(\mathbb{C}^m, G, n)$ because the isovariance condition implies that the stabilizer of $f_q$ in $D \times D$ is the same as that of the restriction of $f_q$ to the vertices of $\Delta^k$ and thus the double cosets are equivalent. This implies that the additional structure automatically lifts from $\mathcal{N}_q\mathcal{W}_\bullet(\mathbb{C}^m, G, n)$ to $\mathcal{N}_q\mathcal{W}^{iso-c}(\mathbb{C}^m, G, n)$ and the lifting on $(C^q, P^q)$ is given by the lemma. Therefore $F_q$ is a homotopy equivalence. □

*Proof of 2.3.2.* We follow the proof of 2.2.2.1. Paragraph 2.2.2.2 describes the given data as before but we would add the following remarks.

Let $D$ denote the group of all diagonal automorphisms of the system $\{(C_*(t), E(s,t))\}$ restricted to the vertices of $\Delta^k$. Then the elements of $D$ are pairs of diagonal matrices $(D_0, D_1)$ with diagonal entries in $G \subseteq U(m)$ so that $D_0 f(t) = f(t) D_1$ for all $t \in \partial\Delta^k$ and $D_i E_i(s,t) = E_i(s,t) D_i$ for $i = 0, 1$ and for all $(s,t) \in DD$, i.e., if $s, t$ lie in the same proper face of $\Delta^k$. Furthermore $D_1 S(s,t,u) = S(s,t,u) D_0$ whenever $s, t, u$ lie in the same proper face of $\Delta^k$.

When we come to paragraph 2.2.2.3, the proof needs to be modified. The contractible groups $T^\sigma, T^\tau, T_\rho$ need to be replaced with subgroups $J^\sigma, J^\tau, J_\rho$ defined as follows.

$$J^\sigma = \{E_0 \in T^\sigma \, : \, D_0 E = E D_0 \text{ for all } (D_0, D_1) \in D\}$$

$$J^\tau = \{E_1 \in T^\tau \, : \, D_1 F = F D_1 \text{ for all } (D_0, D_1) \in D\}$$

$$J_\rho = \{S \in T_\rho \, : \, D_1 S = S D_0 \text{ for all } (D_0, D_1) \in D\}$$

Then $D_0 f = f D_1$ for all $f \in J^\sigma f(v_0) J^\tau$. Also $S*S' = S+S'+Sf(v_0)S' \in J_\rho$ for all $S, S' \in J_\rho$ and the star inverse of any element $S$ of $J_\rho$ is $-S(I-Sf(v_0))^{-1} \in J_\rho$ so $J_\rho$ is a subgroup of $T_\rho$. Since the conditions defining $J^\sigma, J^\tau, J_\rho$ are all linear and $J^\sigma, J^\tau, J_\rho$ are nonempty we see that these subgroups are all convex and thus contractible. The Lie group $J_\rho$ acts freely and smoothly on $J^\sigma \times J^\tau$ as in 2.2.2.3 and the orbit space $(J^\sigma \times J^\tau)/J_\rho$ is a contractible manifold which maps to the double coset $J^\sigma f(v_0) J^\tau$ so paragraphs 2.2.2.4-2.2.2.6 work as before.

When we come to 2.2.2.7 we note that $J^\sigma, J^\tau$ are convex groups and thus $E'(s,t) = (E(s,t) + E(t,s)^{-1})/2$ lies in $J^\sigma \times J^\tau$ so the recursion works to complete the proof. □

## 2.4 Higher FR-torsion for $\mathcal{W}_\bullet^{diff}(\mathbb{C}^m, U(m), n)$

For each $k \geq 0$ we construct a $2k$-cochain $F_{2k}$ on $\mathcal{W}_\bullet^{diff}(\mathbb{C}^m, n)$ and a correction term $G_{2k}$ whose difference gives a cocycle representing the Borel regulator map and which extends the $2k$-cochains with the same names defined on $V_n(\mathbb{C})$. Note that the second statement implies the first since we have already shown that $F_{2k} - G_{2k}$ gives the Borel regulator map on $V_n(\mathbb{C})$.

Let $(C_*(t), E(s,t), B)$ be an object of $\mathcal{W}_{2k}^{diff}(\mathbb{C}^m, n)$. Then $C_i = C_i(t) = (\mathbb{C}^m)^{B_i} = \mathbb{C}^{nm}$ is the same for each $t$ and the boundary map $f(t) : C_1 \to C_0$ and its (complex) conjugate transpose $f(t)^* : C_0 \to C_1 \in GL_{nm}(\mathbb{C})$ vary smoothly with $t \in \Delta^{2k}$.

**Definition 2.4.1.** Let $F_{2k}$ be the $2k$-cochain on $\mathcal{W}_\bullet^{diff}(\mathbb{C}^m, n)$ given by

$$F_{2k}(C_*(t), E(s,t), B) = \frac{1}{(2k+1)!2i^k} \int_{\Delta^{2k} \times I} \operatorname{Tr}\left((h^{-1}dh)^{2k+1}\right)$$

where $h$ is the continuous $\Delta^{2k} \times I$ family of endomorphisms of $C_{even}$ given by $h(t,u) = (f(t)f(t)^*)^u$.

Note that $F_{2k}$ is trivial with respect to morphisms in the sense that its value is the same on the source and target of any morphism in $\mathcal{W}_{2k}^{diff}(\mathbb{C}^m, n)$. Thus, to make it into a cocycle, we need to construct another cochain $G_{2k}$ which is also trivial with respect to morphisms so that $F_{2k} - G_{2k}$ is zero on the boundary of any object of $\mathcal{W}_{2k+1}^{diff}(\mathbb{C}^m, n)$.

If $\omega = h^{-1}dh$ then $d\,\mathrm{Tr}(w^{2k+1}) = -\,\mathrm{Tr}(w^{2k+2}) = 0$ so the value of the coboundary of $F_{2k}$ on any object of $\mathcal{W}_{2k+1}^{diff}(\mathbb{C}^m, n)$ is given by

$$\delta F_{2k}(C_*(t), P) = \int_{\Delta^{2k+1}} \psi^*(c_{2k+1})$$

where $\psi : \Delta^{2k+1} \to H$ is given by $\psi(t) = h(t,1)$ and $c_{2k+1}$ is the closed $(2k+1)$-form on $H$, the space of $nm \times nm$ Hermitian matrices, given as before by

$$c_{2k+1} = \frac{1}{(2k+1)!2i^k} \mathrm{Tr}\left((h^{-1}dh)^{2k+1}\right)$$

Note that

$$\psi(t)^{-1}d\psi(t) = (f(t)^*)^{-1}f(t)^{-1}df(t)f(t)^* + f(t)^{-1}df(t)^*$$

which is conjugate to $f(t)^{-1}df(t) + (f(t)^{-1}df(t))^*$. Thus

$$\mathrm{Tr}((h^{-1}dh)^{2k+1}) = \mathrm{Tr}((X+Y)^{2k+1})$$

where $X = f(t)^{-1}df(t)$ and $Y = X^*$.

Let $K_{2k}$ be the $2k$-cochain on the simplicial set of objects of $\mathcal{W}_\bullet^{diff}$ given by

$$K_{2k}(C_*(t), E(s,t), B) = \frac{1}{(2k+1)!2i^k} \int_{\Delta^{2k}} \mathrm{Tr}\left(\tfrac{1}{\ell} b\left((X+Y)^{2k+1}\right)\right) \tag{2.3}$$

where the notation is as in Theorem 1.3.3 of Chapter I. For example when $k = 2$ we have

$$K_4 = -\frac{1}{2(5!)} \int_{\Delta^4} \mathrm{Tr}\left(X^3Y + XY^3 + \tfrac{1}{2}(XY)^2\right)$$

and in general $K_{2k}$ is a linear combination of integrals of traces of products of $X$'s and $Y$'s. Since

$$d\,\mathrm{Tr}(b(X+Y)^{2k+1}) = \mathrm{Tr}(db(X+Y)^{2k+1}) = \mathrm{Tr}((X+Y)^{2k+1} - X^{2k+1} - Y^{2k+1})$$

it follows that

$$\delta F_{2k} - \delta K_{2k} = \frac{1}{(2k+1)!2i^k} \int_{\Delta^{2k+1}} \mathrm{Tr}\left(X^{2k+1} + Y^{2k+1}\right)$$

Note that this is always real since $\mathrm{Tr}(Y^{2k+1}) = (-1)^k\,\mathrm{Tr}((X^{2k+1})^*)$.

We will now construct a $2k$-cochain $L_{2k}$ on $\mathcal{W}_\bullet^{diff}$ so that

$$\delta L_{2k} = \frac{1}{(2k+1)!2i^k} \int_{\Delta^{2k+1}} \mathrm{Tr}\left((f(t)^{-1}df(t))^{2k+1}\right) \tag{2.4}$$

Then $F - K - L - (-1)^k\overline{L}$ will be the desired cocycle and the correction term will be $G_{2k} = K + L + (-1)^k\overline{L} = K + 2\Re(L)$ or $K + 2\Im(L)$ depending on whether $k$ is even or odd.

The $2k$-cochain $L_{2k}$ will be given as a linear combination of integrals of traces of products of $X$ and the matrix 1-forms $U, V$ and the matrix 2-form $W$ given by

$U = d_1 E_0(t,t) = \Sigma \partial_i E_0(t,t) dt_i$

$V = d_1 E_1(t,t) = \Sigma \partial_i E_1(t,t) dt_i$

$W = d_1 d_2 S(t,t,t) f(t) = \Sigma \partial_i \partial'_j S(t,t,t) f(t) dt_i dt_j$

where $d_k$ denotes the differential with respect to the $k$-th $t$-coordinate and $\partial_i, \partial'_i, \partial''_i$ denote differentiation with respect to the $i$-th component $t_i$ of the 1st, 2nd, 3rd $t$-coordinates.

Since $E(t,t) = I$ and $S(s,t,u) = 0$ when $s,t,u$ are not distinct we get

$$U = d_1 E_0 = -d_2 E_0$$

$$V = d_1 E_1 = -d_2 E_1$$

$$W f^{-1} = d_1 d_2 S = -d_3 d_2 S = d_2 d_3 S = -d_1 d_3 S = d_3 d_1 S = -d_2 d_1 S$$

If we differentiate the equation $E_0(s,t) = f(s) E_1(s,t) f(t)^{-1}$ with respect to $s$ at $s = t$ we get

$$U = df f^{-1} + f V f^{-1} = f(X + V) f^{-1}$$

If we differentiate the equation

$$E(s,t)E(t,u) = E(s,u) + f(s)S(s,t,u) + S(s,t,u)f(u)$$

with respect to $s$ and $t$ and evaluate at $s = t = u$ we get:

0. $dX = -X^2$
1. $dU = U^2 - f W f^{-1}$
2. $dV = V^2 - W$
3. $dW = VW - WV$

*Proof of (3):* Since $d^2 V = 0$ we have $dW = dV^2 = (V^2 - W)V - V(V^2 - W) = VW - WV$.

We will construct a combination of $U, V, W, X$ using the formal boundary map in cyclic homology on a 4-dimensional algebra over $\mathbb{C}$ with basis elements $u, v, w, x$ corresponding to $-f^{-1}Uf, V, W, -X$ respectively.

### 2.4.1 Cyclic homology

Let $A = A_0 \oplus A_1$ be the 4-dimensional graded algebra over $\mathbb{C}$ given by

$$A_0 = \left\{ \begin{pmatrix} * & 0 & 0 \\ 0 & * & 0 \\ 0 & 0 & * \end{pmatrix} \right\}, \quad A_1 = \left\{ \begin{pmatrix} 0 & * & 0 \\ 0 & 0 & 0 \\ 0 & 0 & 0 \end{pmatrix} \right\}$$

Thus $A$ has 3 orthogonal idempotents $u, v, x \in A_0$ and one basis element $w$ of degree $|w| = 1$ so that $uw = wv = w$ and all other products with $w$ are zero. ($A$ is a quotient of the upper triangular matrix algebra $T_3(\mathbb{C})$.)

For each $j \geq 0$ let $C_j$ be the quotient of the $(j+1)$-fold tensor power $C_j = (A \otimes A \otimes \cdots \otimes A)/(t-1)$ where $t$ is the graded cyclic operator

$$t(a_0, a_1, \cdots, a_j) = (-1)^{(1+|a_j|)(1+|a_0,a_1,\cdots,a_{j-1}|)}(a_j, a_0, \cdots, a_{j-1})$$

where we use the notation $|a_0, a_1, \cdots, a_i| = i + |a_0| + |a_1| + \cdots + |a_i|$ (as if each comma had weight 1).

Let $b : C_j \to C_{j-1}$ be the graded cyclic boundary operator given by

$$\begin{aligned} b(a_0, a_1, \cdots, a_j) &= (-1)^{|a_0|}(a_0a_1, a_2, \cdots, a_j) + (-1)^{|a_0,a_1|}(a_0, a_1a_2, \cdots, a_j) \\ &+ (-1)^{|a_0,a_1,a_2|}(a_0, a_1, a_2a_3, \cdots, a_j) + \cdots \\ &+ (-1)^{|a_j|+(1+|a_j|)(1+|a_0,a_1,\cdots,a_{j-1}|)}(a_ja_0, a_1, \cdots, a_{j-1}) \end{aligned}$$

Then $b^2 = 0$ and the homology of $(C_*, b)$ is the cyclic homology of $A$. Since $x$ annihilates $u, v, w$ the cyclic complex $(C_*, b)$ breaks up into a direct sum of weighted subcomplexes $C_*^\ell$ if we define the *color weight* $\ell$ of a monomial $(a_0, a_1, \cdots, a_j)$ to be the number of indices $i$ so that $a_i, a_{i+1}$ (indices are modulo $j+1$) have different *colors* where $u, v, w$ have one color, say blue, and $x$ is red. The positive weight subcomplexes are easily seen to be acyclic but we need an explicit chain contraction.

We let $d : C_j \to C_j \oplus C_{j+1}$ with homogeneous components $d' : C_j \to C_{j+1}$ and $d'' : C_j \to C_j$ be given as follows. First:

0. $dx = (x, x)$
1. $du = (u, u) + (u, v) + (v, u) + w$
2. $dv = (v, v) - w$
3. $dw = (v, w) - (w, v)$

Next, on any $(a_0, a_1, \cdots, a_j) \in C_j$ let $d = d' + d''$ be given by

4. $$\begin{aligned} d(a_0, a_1, \cdots, a_j) &= (da_0, a_1, \cdots, a_j) + (-1)^{1+|a_0|}(a_0, da_1, \cdots, a_j) \\ &+ \ (-1)^{1+|a_0,a_1|}(a_0, a_1, da_2, \cdots, a_j) + \cdots \\ &+ \ (-1)^{1+|a_0,a_1,\cdots,a_{j-1}|}(a_0, a_1, \cdots, da_j) \end{aligned}$$

Then $d^2 = 0$ and

**Lemma 2.4.2.** *$(bd + db)(a) = \ell(a)a$ where $\ell(a)$ is the color weight of $a$. In other words, $bd' + d'b$ is multiplication by $\ell$ on $C_*^\ell$ and $bd'' + d''b = 0$.*

*Proof.* This is a consequence of Lemma 4.7 below and the following two formulas which I carefully checked on all 16 possible cases.

a) $b'd'(a) = a$ for $a = u, v, w, x$ where $b' : A \otimes A \to A$ is given by $b'(a, c) = (-1)^{|a|}ac$.

b) $d(ac) = (da)c + (-1)^{|a|}adc + (-1)^{|a|}k(a, c)$ where $k = -1$ or $0$ depending on whether $a, c$ have the same color or not. □

**Lemma 2.4.3.** *Suppose that $\{a_i\}$ is a $K$-basis for a graded $K$-algebra $A$ and that*

(a) $b'd(a_i) = f'(a_i)a_i$ *for some* $f'(a_i) \in K$ *for each* $i$.

(b) $d(a_i a_j) = (da_i)a_j + (-1)^{|a_i|}a_i da_j + (-1)^{|a_i|}k(a_i, a_j)(a_i, a_j)$

*for some* $k(a_i, a_j) \in K$ *for all* $i, j$. *Then*

1. $(b'd + db')(a) = f'(a)a$ *for all* $a = (a_{i_1}, a_{i_2}, \cdots, a_{i_m}) \in A^{\otimes m}$ *where* $f'(a)$ *is defined recursively by*

$$f'(a, c) = f'(a) + f'(c) + k(a, c)$$

($k(a, c) = k$(*last tensor factor of* $a$, *first tensor factor of* $c$).)

2. $(bd + db)(a) = f(a)a$ *where* $f(a) = f'(a) + k(a, a)$ *if* $m \geq 2$.

*Proof.* By induction on $m$. □

Let $\phi$ be the $\mathbb{C}$-linear function on $A$ given on basis elements by

$$\phi(u) = -X - V = -f^{-1}Uf$$

$$\phi(x) = -X, \quad \phi(v) = V, \quad \phi(w) = W$$

Then $\phi$ gives a $\mathbb{C}$-linear map $\Phi$ on $C_j(A)$ by

$$\Phi(a_0, a_1, \cdots, a_j) = \mathrm{Tr}(\phi(a_0)\phi(a_1)\cdots\phi(a_j))$$

and $\Phi d = d\Phi$. Consequently it suffices to find a sequence of elements $z_{(2k+1)} \in C(A)$ so that $dz = 0, \Phi(z) = 0$ and $z_{(2k+1)} - x^{2k+1}$ has no terms of weight zero. (We use $x^{2k+1}$ to denote the tensor power.) Then

$$z_{(2k+1)} - x^{2k+1} = d\left(\tfrac{1}{\ell}b(z_{(2k+1)} - x^{2k+1})\right) = d\left(\tfrac{1}{\ell}b(z_{(2k+1)})\right)$$

$$-\Phi(x^{2k+1}) = \mathrm{Tr}\left((f^{-1}df)^{2k+1}\right) = d\Phi(\tfrac{1}{\ell}b(z_{(2k+1)})) \tag{2.5}$$

since $b(x^{2k+1}) = x^{2k} = 0$. So we can let

$$L_{2k}(C_*(t), E(s,t), B) = \frac{1}{(2k+1)!2i^k}\int_{\Delta^2 k}\Phi\left(\tfrac{1}{\ell}b(z_{(2k+1)})\right) \tag{2.6}$$

In order to construct the element $z_{(2k+1)}$ we need to use a different multiplication on the algebra $A$ producing a new boundary map $b^*$ on $C_*(A)$. So let $*$ be the new commutative multiplication on $A$ given by

1. $x * a = a$ for all $a \in A$.

2. $u * u = u$

3. $v * v = v$

4. $u * w = w$

5. $u * v = v * w = w * w = 0$

Define the new weight $\ell^*$ on a monomial $(a_0, \cdots, a_j)$ to be the number of indices $i$ so that $a_i = w$ and $a_{i-1} \neq w$ (indices are modulo $j+1$ so that $a_{-1} = a_j$).

**Lemma 2.4.4.** *$(b^* d' + d' b^*)(a) = -\ell^*(a) a$ for all $a \in A^{\otimes m}$*

*Proof.* This follows from Lemma 2.4.3 and the following observations.

(a) $(b^*)' d'(w) = 0$ and $(b^*)' d'(a) = a$ for $a = u, v, x$,

(b) $d'(a * c) = (d'a) * c + (-1)^{|a|} a * d'c + (-1)^{|a|} k(a,c)(a,c)$ for $a, c = u, v, x, w$

where $k(x,x) = 0$ and $k(a,c) = -1$ if $(a,c) \neq (x,x)$. □

**Lemma 2.4.5.** *Let $z = \Sigma(-1)^i z_i = z_0 - z_1 + z_2 - \cdots \pm z_k$ where $z_0, z_1, \cdots \in C(A)$ are given recursively by*

*(a) $z_0 = v^{2k+1} + (x - v)^{2k+1}$*

*(b) $z_{i+1} = \frac{b^*}{\ell^*} d'' z_i$*

*where the exponents denote tensor powers. Then $dz = 0, \Phi(z) = 0$ and $z - x^{2k+1}$ has no weight zero component. Consequently, $\Phi(\frac{1}{\ell} b z_{(k)}) = \mathrm{Tr}(X^{2k+1})$.*

*Proof.* The terms $z_i$ are computable by an easy algorithm since $d''v = -w, d''x = 0, v * v = v, (x - v) * (x - v) = x - v, w * w = 0$ and $w * (x - v) = (x - v) * w = w$. Thus

$$d'' z_0 = -(2k+1) w v^{2k} + (2k+1) w (x - v)^{2k}$$
$$z_1 = -(2k+1) w v^{2k-1} - (2k+1) w (x - v)^{2k-1}$$

$z_i$ is a sum of monomials of the form $w^i v^{2k-2i-1}$ and permutations of these letters plus the same with $v$ replaced with $(x - v)$. It follows immediately that the weight zero terms in $z_i$ cancel except for the term $x^{2k+1}$ in $z_0$.

Next we claim by induction that $d' z_i = -d'' z_{i-1}$. This holds for $i = 0$ since $z_{-1} = 0$. Since $z_1, z_2$ have no terms of $*$ -weight zero (each term has at least one $w$ and an odd number of other letters) so

$$d' z_{i+1} = -d'' z_i - \frac{b^*}{\ell^*} d' d'' z_i = -d'' z_i + \frac{b^*}{\ell^*} d'' d' z_i$$
$$= -d'' z_i + \frac{b^*}{\ell^*} d'' d'' z_{i-1} = -d'' z_i$$

since $d'd'' + d''d' = 0$ and $d''d'' = 0$. Therefore

$$dz = \Sigma d' z_i + \Sigma d'' z_i = 0.$$

Finally we come to $\Phi(z)$. Recall that

$$f(w) = W = (d_1 d_2 S) f$$

$$f(v) = V = d_1 E_0$$

$$f(x-v) = f(u) = -f(t)^{-1} U f(t) = -f^{-1} d_1 E_1 f$$

The terms $d_1 d_2 S, f, V = d_1 E_0, U = d_1 E_1$ are all strictly upper triangular. The only term which is not upper triangular is the $f^{-1}$ term which occurs in $\phi(x-v)$. But $x-v$ always has on its left (up to cyclic order) the term $w$ or $x-v$ so the $f^{-1}$ terms are always cancelled by the $f$ on the right side of $\phi(w)$ or $\phi(x-v)$. Therefore $\Phi(z_i) = 0$ for all $i$. □

**Example 2.4.6.** For $k = 1,2,3,4$ we have the following where the terms in the square brackets are the same as the other terms with each $v$ replaced by $x-v$. The salient feature is that $\Phi(z_{(2k+1)}) = 0$ since $\phi(v), \phi(w), f\phi(x-v)f^{-1}, f\phi(w)f^{-1}$ are strictly upper triangular.

1. $z_{(3)} = v^3 - 3wv + [(x-v)^3 - 3w(x-v)] = x^3 - 3x^2v + 3xv^2 - 3wx$

   $bz_{(3)} = 3(x+v)(x-v) = 6vx$

   $\Phi(\frac{1}{\ell} b_{z(3)}) = 3\,\mathrm{Tr}(XV)$

   $d\Phi(\frac{1}{\ell} bz_{(3)}) = 3\,\mathrm{Tr}(-X^2V - XV^2 + XW) = \mathrm{Tr}(X^3)$

2. $z_{(5)} = v^5 - 5wv^3 + 10w^2v + \left[(x-v)^5 - 5w(x-v)^3 + 10w^2(x-v)\right]$

   $bz_{(5)} = 10v(x-v)^3 - 5wx^2 - 5wvx + 15wxv$

   $\Phi(\frac{1}{\ell} bz_{(5)}) = 5\,\mathrm{Tr}(X^3V - \frac{1}{2}VXVX + XV^3 - \frac{1}{2}WX^2 + \frac{1}{2}WVX - \frac{3}{2}WXV)$

3. $z_{(7)} = v^7 - 7wv^5 + 14w^2v^3 + 7wvwv^2 - 35w^3v + [\cdots]$

   $bz_{(7)} = 14v(x-v)^5 - 7w(x-v)^4 - 28w(vx - xv)^2$

   $+7(-w^2x^2 + 5w^2xv - 3w^2vx + 2wxwx - 2wvwx)$

4. $z_{(9)} = v^9 - 9wv^7 + 18w^2v^5 + 9wvwv^4 + 9wv^2wv^3 - 45w^3v^3$

   $-18w^2vwv^2 - 18w^2v^2wv - 3wvwvwv + 126w^4v + [\cdots]$

**Theorem 2.4.7.** *$D_{2k} = F_{2k} - G_{2k}$ is a $2k$-cocycle on $\mathcal{W}_\bullet^{diff}(\mathbb{C}^m, n)$ which restricts to the Borel regulator class on $V_n(\mathbb{C})$ where $G_{2k} = K_{2k} + L_{2k} + (-1)^k \overline{L}$*

$$K_{2k}(C_*(t), E(s,t), B) = \frac{1}{(2k+1)!2i^k} \int_{\Delta^{2k}} \mathrm{Tr}\left(\tfrac{1}{\ell} b\left((X+Y)^{2k+1}\right)\right)$$

$$L_{2k}(C_*(t), E(s,t), B) = \frac{1}{(2k+1)!2i^k} \int_{\Delta^{2k}} \Phi\left(\tfrac{1}{\ell} b(z_{(2k+1)})\right)$$

*In particular, $D_{2k}$ is a linear combination of integrals of traces of products of $X, U, V, W$ and their adjoints.*

We note that $b, w$ have different meanings in the expressions for $K_{2k}$ and $L_{2k}$. In both cases $b$ is the formal boundary operator in cyclic homology. For $K_{2k}$, the algebra is two dimensional and this is explained in Part I. For $L_{2k}$ we use the four dimensional algebra $A$ with its noncommutative multiplication and $w$ is the color weight. (The commutative $*$-multiplication on $A$ is used to define $b^*$ which occurs in the construction of $z_{(2k+1)}$.)

We also note that, since $F_{2k}, G_{2k}, L_{2k}$ are constant with respect to morphisms, they induce cochains on $\mathcal{W}_\bullet^{diff}(\mathbb{C}^m, G, n)/_{conj}$, $\mathcal{W}_\bullet^{iso-d}(\mathbb{C}^m, G, n)$ and $\mathcal{W}_\bullet^{iso-d}(\mathbb{C}^m, G, n)/_{conj}$:

**Proposition 2.4.8.** *$D_{2k}$ is a well-defined cocycle on $\mathcal{W}_\bullet^{diff}(\mathbb{C}^m, G, n)/_{conj}$, $\mathcal{W}_\bullet^{iso-d}(\mathbb{C}^m, G, n)$ and $\mathcal{W}_\bullet^{iso-d}(\mathbb{C}^m, G, n)/_{conj}$.*

By Theorems 2.2.4, and 2.3.3 $[D_{2k}]$ comes from unique $2k$-cohomology classes in $\mathcal{W}_\bullet(\mathbb{C}^m, n)$ and $\mathcal{W}_\bullet(\mathbb{C}^m, n)/_{conj}$.

**Definition 2.4.9.** We define the *higher Franz-Reidemeister torsion classes*

$$\tau_k(m,n) \in H^{2k}(\mathcal{W}_\bullet(\mathbb{C}^m, n); \mathbb{R})$$

$$\tau_k(m,n)/_{conj} \in H^{2k}(\mathcal{W}_\bullet(\mathbb{C}^m, n)/_{conj}; \mathbb{R})$$

to be the cohomology classes corresponding to $[D_{2k}]$ in $H^{2k}(\mathcal{W}_\bullet^{diff}(\mathbb{C}^m, n); \mathbb{R})$ and $H^{2k}(\mathcal{W}_\bullet^{diff}(\mathbb{C}^m, n)_{conj}; \mathbb{R})$ under the isomorphisms given by the forgetful functors (Thm. 2.2.4, and 2.3.3).

### 2.4.2 Properties of $\tau_k(m, n)$

First we restate Theorem 2.4.7 and Proposition 2.4.8 in terms of $\tau_k(m,n)$.

**Proposition 2.4.10.** *(a) $\tau_k(m,n)|\mathcal{V}(M_m(\mathbb{C}), n) = \tau_k^V(m,n)$.*

*(b) $\tau_k(m,n)/_{conj}|\mathcal{W}_\bullet(\mathbb{C}^m, n) = \tau_k(m,n)$.*

**Proposition 2.4.11 (Transfer).** *Let*

$$J : \mathcal{W}_\bullet(M_m(M_d(\mathbb{C})), Mon_m(U(d)), n) \to \mathcal{W}_\bullet(M_{dm}(\mathbb{C}), U(dm), n)$$

*be the simplicial inclusion map and let*

$$\mathrm{N} : \mathcal{W}_\bullet(M_m(M_d(\mathbb{C})), Mon_m(U(d)), n) \to \mathcal{W}_\bullet(M_d(\mathbb{C}), U(d), mn)$$

*be the simplicial transfer functor described in 2.1.16. Then*

$$J^*(\tau_k(dm,n)) = \mathrm{N}^*(\tau_k(d,mn)).$$

*Proof.* By Theorem 2.2.4 we can pass to the smooth version of these maps:

$$J^{diff} : \mathcal{W}_\bullet^{diff}(\mathbb{C}^{dm}, Mon_m(U(d)), n) \to \mathcal{W}_\bullet^{diff}(\mathbb{C}^{dm}, U(dm), n)$$

$$\mathrm{N}^{diff} : \mathcal{W}_\bullet^{diff}(\mathbb{C}^{dm}, Mon_m(U(d)), n) \to \mathcal{W}_\bullet^{diff}(\mathbb{C}^{d}, U(d), mn).$$

Since $J^{diff}$ and $\mathrm{N}^{diff}$ preserve the differential matrix forms $X, U, V, W$ in the formula for $D_{2k}$, they give the same cocycle $(J^{diff})^* D_{2k} = (\mathrm{N}^{diff})^* D_{2k}$. □

**Proposition 2.4.12 (Additivity).** (a) *Let* $m = m_1 + m_2 + \cdots + m_d$ *and let* $p_i : \prod M_{m_i}(\mathbb{C}) \to M_{m_i}(\mathbb{C})$ *be the projection maps. Let* $s : \prod M_{m_i}(\mathbb{C}) \to M_m(\mathbb{C})$ *be the direct sum map* $s(A_1, A_2, \cdots, A_d) = A_1 \oplus A_2 \oplus \cdots \oplus A_d$. *Then*

$$s^*(\tau_k(m,n)) = \sum p_i^*(\tau_k(m_i,n))$$

*in* $H^{2k}(\mathcal{W}_\bullet(\prod M_{m_i}(\mathbb{C}), \prod U(m_i), n); \mathbb{R})$.

(b) *Let* $n = n_1 + \cdots + n_d$ *and let* $S : \prod \mathcal{W}_\bullet(M_m(\mathbb{C}), n_i) \to \mathcal{W}_\bullet(M_m(\mathbb{C}), n)$ *be given by direct sum. Then*

$$S^*\tau_k(m,n) = \sum \pi_i^* \tau_k(m, n_i)$$

*where* $\pi_i : \prod \mathcal{W}(M_m(\mathbb{C}), n_i) \to \mathcal{W}(M_m(\mathbb{C}), n_i)$ *is the projection to the i-th factor.*

*Proof.* We use Theorem 2.2.4 and Corollary 2.2.7 to go to the smooth version where the matrix forms $X, U, V, W$ in the definition of the cocycle for $s^*(\tau_k(m,n))$ and $S^*(\tau_k(m,n))$ are direct sums of component forms $X_i, U_i, V_i, W_i$ occurring in the cocycles for $p_i^*(\tau_k(m_i,n))$ and $\pi_i^*(\tau_k(m,n_i))$, respectively. Since $\mathrm{Tr}(A_1 \oplus A_2 \oplus \cdots \oplus A_d) = \sum \mathrm{Tr}(A_i)$ the proposition holds at the cochain level. □

As we claimed to earlier in the subsection on the involution (2.1.3), the additivity of $\tau_k(m,n)$ implies that it is invariant with respect to the involution $ct$ given by conjugate transpose, at least on the image of the Volodin category $\mathcal{V}_\bullet(\mathbb{C}, n)$.

**Theorem 2.4.13 (Involution).** *The higher FR-torsion classes* $\tau_k(m,n)$ *and* $ct^*\tau_k(m,n)$ *become equal when restricted to the Volodin category* $\mathcal{V}_\bullet(M_m(\mathbb{C}), n)$. *I.e.,* $J^*\tau_k(m,n) = J^* ct^* \tau_k(m,n)$ *where*

$$J^* : H^{2k}(\mathcal{W}_\bullet(\mathbb{C}^m, n); \mathbb{R}) \to H^{2k}(\mathcal{V}_\bullet(M_m(\mathbb{C}), n); \mathbb{R})$$

*is the restriction map in cohomology.*

*Proof.* This is just a combination of previous observations.

First we note that by additivity the simplicial functor

$$J \oplus \mathcal{I} : \mathcal{V}_\bullet^2(M_m(\mathbb{C}), U(m), n) \to \mathcal{W}_\bullet(\mathbb{C}^m, 2n)$$

has the property that

$$(J \oplus \mathcal{I})^* \tau_k(m, 2n) = \tau_k(m,n)|\mathcal{V}_\bullet^2 + \mathcal{I}\left(\tau_k(m,n)|\mathcal{V}_\bullet^2\right).$$

Combining this with Proposition 2.1.14 we get:

$$\mathcal{I}\left(\tau_k(m,n)|\mathcal{V}_\bullet^2\right) = -\tau_k(m,n)|\mathcal{V}_\bullet^2.$$

Since $ct$ commutes with restriction to $\mathcal{V}_\bullet^2(M_m(\mathbb{C}), U(m), n)$ we get the following.

$$\left[ct\mathcal{I}\left(\tau_k(m,n)|\mathcal{V}_\bullet^2\right)\right]|\mathcal{V}_\bullet = -\left[ct\left(\tau_k(m,n)|\mathcal{V}_\bullet^2\right)\right]|\mathcal{V}_\bullet = -ct\left(\tau_k(m,n)\right)|\mathcal{V}_\bullet$$

By Proposition 2.1.13 and Corollary 1.5.4 this gives:

$$\left[ct\mathcal{I}\left(\tau_k(m,n)|\mathcal{V}_\bullet^2\right)\right]|\mathcal{V}_\bullet = \mathcal{I}ct\left(\tau_k(m,n)|\mathcal{V}_\bullet\right) = \mathcal{I}ct(\tau_k^V(m,n)) = -\tau_k(m,n)|\mathcal{V}_\bullet$$

The theorem follows. □

## 2.5 Stabilization of $\mathcal{W}_\bullet(R, G, n)$

We return for a moment to the general case of $\mathcal{W}_\bullet(R, G, n)$. We want to stabilize our matrices by adding an element of the group $G$ in the lower right corner.

$$A \mapsto \begin{pmatrix} A & 0 \\ 0 & g \end{pmatrix}$$

For an acyclic chain complex this corresponds to taking direct sum with the acyclic complex $0 \to R \to R$ where the boundary map is left multiplication by some $g \in G$. This adds two basis elements $x^-, x^+$ to $C_0, C_1$ respectively so that $f(x^+) = x^- g$. Conversely, if we add such new basis elements to $C_0, C_1$ with the condition that they be unrelated in the partial ordering to the other elements of the basis $B$ then they will not be incident to any other elements of $B$ and they will generate a direct summand of the chain complex.

We note that, under stabilization, an isovariant $\Delta^k$ family of chain complexes will remain isovariant. Also, the conjugate transpose of the stabilization is the stabilization of the conjugate transpose (using $g^{-1}$).

**Definition 2.5.1.** A *G-expansion pair* for an object $(C_*, B)$ of $\mathcal{W}_p(R, G, n)$ or $\mathcal{W}_p^{cont}(\mathbb{C}^m, G, n)$ (where $G \subseteq U(m)$) is defined to be a pair of elements $x^- \in B_0, x^+ \in B_1$ so that

- $x^- < x^+$ but $x^-, x^+$ are unrelated to all other elements of the poset $B$.
- $f(x^+) = x^- g$ for some $g \in G$.

**Definition 2.5.2.** Let $(C_*, B)$ be an object of $\mathcal{W}_p(R, G, n)$ (or $\mathcal{W}_p^{cont}$ $(\mathbb{C}^m, G, n))$ and let $(C'_*, B')$ be an object of $\mathcal{W}_p(R, G, n')$ (or $\mathcal{W}_p^{cont}(\mathbb{C}^m, G, n')$) where $n' \geq n$. Then by a *G-expansion* $(C_*, B) \to (C'_*, B')$ we mean a degree 0 poset embedding $\psi : B \to B'$ so that

- The homomorphism $\psi_* : C_*(t) \to C'_*(t)$ induced by $\psi$ (i.e., $\psi_*(\sum x_i r_i) = \sum \psi(x_i) r_i$) is a chain monomorphism for each vertex $t$ in $\Delta^p$ (for each $t \in \Delta^p$ in the continuous case).
- The complement $B' - \psi(B)$ of the image of $\psi$ is a disjoint union of $G$-expansion pairs.

**Definition 2.5.3.** Let $\mathcal{W}_\bullet(R, G)$ (or $\mathcal{W}_\bullet^{cont}(\mathbb{C}^m, G)$, $\mathcal{W}_\bullet^{diff}(\mathbb{C}^m, G)$, $\mathcal{W}_\bullet^{iso-c}$ $(\mathbb{C}^m, G)$, etc.) denote the following simplicial bicategory. In each degree $p \geq 0$, $\mathcal{W}_p(R, G)$ will be the bicategory of squares of the following form.

$$\begin{array}{ccc} (C_*, A) & \xrightarrow{(\alpha,g)_*} & (C'_*, A') \\ \psi \downarrow & & \downarrow \psi' \\ (D_*, B) & \xrightarrow{(\beta,h)_*} & (D'_*, B') \end{array} \qquad (2.7)$$

where

1. The *horizontal arrows* $(\alpha, g)_*$, $(\beta, h)_*$ are morphisms in $\mathcal{W}_p(R, G, n)$ and $\mathcal{W}_p(R, G, n')$ respectively (or in $\mathcal{W}_p^{cont}(\mathbb{C}^m, G, *)$, $\mathcal{W}_p^{diff}(\mathbb{C}^m, G, *)$, etc.).

2. The *vertical arrows* are $G$-expansions.

3. $\beta \circ \psi = \psi' \circ \alpha$

**Definition 2.5.4.** The *stable Volodin category* $\mathcal{V}_\bullet(R)$ is defined to be the simplicial subbicategory of $\mathcal{W}_\bullet(R, 1)$ with objects and horizonal arrows the same as $\coprod_{n \geq 1} \mathcal{V}_\bullet(R, n)$ and vertical arrows given by the inclusion maps $B(n) \hookrightarrow B(m)$ for $n \leq m$. Let $\mathcal{V}_\bullet^2(R, G)$ denote the simplicial full subbicategory of $\mathcal{W}_\bullet(R, G)$ with object set $\coprod_{n \geq 1} \mathcal{V}_\bullet^2(R, G, n)$. We note that the inversion functor $\mathcal{I}$ extends to $\mathcal{V}_\bullet^2(R, G)$.

**Proposition 2.5.5.** *The stable Volodin category $\mathcal{V}_\bullet(R)$ is homotopy equivalent to the direct limit of the unstable categories $\mathcal{V}_\bullet(R, n)$. Consequently, its realization is homotopy equivalent to the stable Volodin space: $|\mathcal{V}_\bullet(R)| \simeq V(R)$.*

**Definition 2.5.6.** Given a conjugation on $(R, G)$ we define the *involution ct* on $\mathcal{W}_\bullet(R, G)$ to be the simplicial bifunctor given by conjugate transpose on objects, inverse conjugate transpose on horizontal arrows and on vertical arrows we take inverse conjugate transpose on the dual of the image, using the fact that conjugate transpose respects direct sum. In other words, the involution on

$$\psi_* : C_*(t) \to C'_*(t) = \psi_*(C_*(t)) \oplus X$$

is

$$\mathcal{I}ct(\psi_*) : C_*(t)^* \to C'_*(t)^* = \psi_*(C_*(t))^* \oplus X^*.$$

We observe that, by definition, the inclusion functors $\mathcal{V}_\bullet^2(R, G) \to \mathcal{W}_\bullet(R, G)$ and $\mathcal{W}_\bullet(R, G, n) \to \mathcal{W}_\bullet(R, G)$ commutes with the involution $ct$ and that the inclusion $\mathcal{V}_\bullet(R) \to \mathcal{V}_\bullet^2(R, G)$ commutes with $\mathcal{I}ct$.

## Functorial properties

A ring map $\phi : R \to S$ sending $G$ into $H$ induces a simplicial functor $\phi_* : \mathcal{W}_\bullet(R, G) \to \mathcal{W}_\bullet(S, H)$ and we have the following.

**Lemma 2.5.7 (Conjugation lemma).** *If $\phi, \psi : R \to S$ are conjugate by a fixed element $h$ of $H$ then $\phi_*, \psi_*$ give homotopic maps on realizations.*

*Proof.* Conjugation by $h \in H$ gives a natural horizontal transformation from $\phi_*$ to $\psi_*$. ☐

**Remark 2.5.8.** In Diagram 2.7 we observe that $(C'_*, A')$ exists and is uniquely determined up to natural isomorphism given $(C_*, A) \to (D_*, B) \to (D'_*, B')$ and similarly $(D_*, B)$ is determined by $(C_*, A) \to (C'_*, A') \to (D'_*, B')$. Therefore we can form a category where the morphisms $(C_*, A) \to (D'_*, B')$ are isomorphism classes of square diagrams as in (2.7). (To compose we insert unique-up-to-isomorphism objects into the upper right and lower left corners of the larger square created by adjoining two smaller squares.) The result is a simplicial category, which we will refer to as the *diagonal category* of $\mathcal{W}_\bullet(R, G)$, whose realization is homotopy equivalent to that of the simplicial bicategory $\mathcal{W}_\bullet(R, G)$.

The above remark holds for any simplicial full subbicategory of $\mathcal{W}_\bullet(R, G)$ or $\mathcal{W}_\bullet^{cont}(\mathbb{C}^m, G)$ which is closed under expansions and collapses in the sense that for any $G$-expansion in $\mathcal{W}_\bullet(R, G)$ or $\mathcal{W}_\bullet^{cont}(\mathbb{C}^m, G)$ if either the source or target lies in the subbicategory then so does the other.

**Theorem 2.5.9.** *The cocycle $D_{2k} = F_{2k} - G_{2k}$ on $\mathcal{W}_\bullet^{diff}(\mathbb{C}^m, G, n)$ extends to $\mathcal{W}_\bullet^{diff}(\mathbb{C}^m, G), \mathcal{W}_\bullet^{iso-d}(\mathbb{C}^m, G)$ and $\mathcal{W}_\bullet^{iso-d}(\mathbb{C}^m, G)/_{conj}$ in the obvious way using the same formula giving higher FR-torsion classes*

$$\tau_k(m) \in H^{2k}(\mathcal{W}_\bullet^{diff}(\mathbb{C}^m, G); \mathbb{R})$$

*which restrict to $\tau_k(m, n) \in H^{2k}(\mathcal{W}_\bullet^{diff}(\mathbb{C}^m, G, n); \mathbb{R})$ for any $n \geq 1$.*

*Proof.* If $g \in U(m)$ then $gg^* = I_m$ and

$$\begin{pmatrix} A & 0 \\ 0 & g \end{pmatrix} \begin{pmatrix} A & 0 \\ 0 & g \end{pmatrix}^* = \begin{pmatrix} AA^* & 0 \\ 0 & I_m \end{pmatrix}$$

Therefore the cochains $F_{2k}, G_{2k}$ are both stationary with respect to horizontal and vertical arrows. □

**Remark 2.5.10.** We note that the pull-back of $\tau_k(m)$ along the linear interpolation functor

$$\lambda : \mathcal{V}_\bullet(M_m(\mathbb{C})) \to \mathcal{W}_\bullet^{diff}(\mathbb{C}^m, U(m))$$

agrees with the cohomology class $\tau_k^V(m)$ of the first chapter.

**Theorem 2.5.11.** *We have the following homotopy equivalences induced by inclusion and forgetful functors.*

*a)* $|\mathcal{W}_\bullet^{diff}(\mathbb{C}^m, G)| \simeq |\mathcal{W}_\bullet^{cont}(\mathbb{C}^m, G)| \simeq |\mathcal{W}_\bullet(\mathbb{C}^m, G)|$,

*b)* $|\mathcal{W}_\bullet^{iso-d}(\mathbb{C}^m, G)| \simeq |\mathcal{W}_\bullet^{iso-c}(\mathbb{C}^m, G)| \simeq |\mathcal{W}_\bullet(\mathbb{C}^m, G)|$ *and*

*c)* $|\mathcal{W}_\bullet^{iso-d}(\mathbb{C}^m, G)/_{conj}| \simeq |\mathcal{W}_\bullet^{iso-c}(\mathbb{C}^m, G)/_{conj}| \simeq |\mathcal{W}_\bullet(\mathbb{C}^m, G)/_{conj}|$

**Remark 2.5.12.** The point is that the cohomology class $\tau_k(m) = [F_{2k} - G_{2k}]$ defined on the smooth version of our simplicial bicategories extend uniquely to the other versions using this proposition (and consequently to $\mathcal{V}_\bullet^2(M_m(\mathbb{C}), U(m))$ and $\mathcal{V}_\bullet(M_m(\mathbb{C}))$ by restriction). Theorem 2.5.11 basically says that we can take any isovariant smoothing of the discrete data and apply the smooth formula.

*Proof.* We will show that the forgetful functor $F : \mathcal{W}_\bullet^x(\mathbb{C}^d, G) \to \mathcal{W}_\bullet(\mathbb{C}^d, G)$ induces a homotopy equivalence on realizations for $x = cont, diff, etc.$. It suffices to show that in any fixed bidegree $(n, m)$ the simplicial map $F_{nm} : \mathcal{N}_{nm}\mathcal{W}_\bullet^x(\mathbb{C}^d, G) \to \mathcal{N}_{nm}\mathcal{W}_\bullet(\mathbb{C}^d, G)$ is a homotopy equivalence.

The simplicial sets $\mathcal{N}_{nm}\mathcal{W}_\bullet^x(\mathbb{C}^d, G)$ and $\mathcal{N}_{nm}\mathcal{W}_\bullet(\mathbb{C}^d, G)$ have the same set of vertices. These are rectangular arrays of morphisms

$$\begin{array}{ccccccc}
(C^{00}, P^{00}) & \longrightarrow & (C^{01}, P^{01}) & \longrightarrow & \cdots & \longrightarrow & (C^{0m}, P^{0m}) \\
\downarrow & & \downarrow & & & & \downarrow \\
(C^{10}, P^{10}) & \longrightarrow & (C^{11}, P^{11}) & \longrightarrow & \cdots & \longrightarrow & (C^{1m}, P^{1m}) \\
\downarrow & & \downarrow & & & & \downarrow \\
\vdots & & \vdots & & \cdots & & \vdots \\
\downarrow & & \downarrow & & & & \downarrow \\
(C^{n0}, P^{n0}) & \longrightarrow & (C^{n1}, P^{n1}) & \longrightarrow & \cdots & \longrightarrow & (C^{nm}, P^{nm})
\end{array}$$

The horizontal morphisms are the $G$-monomial chain isomorphisms given by closed poset bijections $P^{ij} \to P^{i,j+1}$ and maps $P^{ij} \to G$. Thus the posets $P^{0j}$ may be viewed as a sequence of refinements of the partial ordering on $P^{0m}$. We note that the chain complexes $C^{ij}$ are determined up to canonical isomorphism by $C^{0j}$ which in turn are determined up to isomorphism by $C^{0m}$. Thus all $C^{ij}$ are determined by $C^{0m}$ but not canonically. However, for $k$-simplices in $\mathcal{N}_{nm}\mathcal{W}_\bullet^x(\mathbb{C}^d, G)$ and $\mathcal{N}_{nm}\mathcal{W}_\bullet(\mathbb{C}^d, G)$, the additional data contained in the $C^{ij}$ for $(i,j) \neq (0,m)$ is the same in all cases. Consequently it follows that Lemma 2.2.2 that $F_{nm}$ is a homotopy equivalence for all $x$. This proves (a) and (b).

In case (c) we have a different set of additional data. However, just as in the proof of Theorem 2.3.3, this additional data is, by definition, the same for $\mathcal{W}_\bullet(\mathbb{C}^d, G)/_{conj}$, $\mathcal{W}_\bullet^{iso-c}(\mathbb{C}^d, G)/_{conj}$ and $\mathcal{W}_\bullet^{iso-d}(\mathbb{C}^d, G)/_{conj}$. Thus $F_{nm}/_{conj}$ is a homotopy equivalence for $x = iso{-}c$ or $iso{-}d$. □

## Variants of Theorem 2.5.11

As in the case of Theorem 2.2.4 we can extract the elements of proof and obtain other cases of Theorem 5.6.

**Proposition 2.5.13.** *Suppose that $\mathcal{W}_\bullet^{cont}$ and $\mathcal{W}_\bullet$ are simplicial full subbicategories of $\mathcal{W}_\bullet^{cont}(\mathbb{C}^m, G)$ and $\mathcal{W}_\bullet(\mathbb{C}^m, G)$ satisfying the following conditions.*

- *a) The forgetful functor $F : \mathcal{W}_\bullet^{cont}(\mathbb{C}^m, G) \to \mathcal{W}_\bullet(\mathbb{C}^m, G)$ maps $\mathcal{W}_\bullet^{cont}$ to $\mathcal{W}_\bullet$ and induces a homotopy equivalence on the subsets of the simplicial sets of objects of $\mathcal{W}_\bullet^{cont}, \mathcal{W}_\bullet$ with a fixed poset $B$.*
- *b) $F_0 : \mathcal{W}_0^{diff} \to \mathcal{W}_0$ is an isomorphism of bicategories.*
- *c) $\mathcal{W}_\bullet^{cont}, \mathcal{W}_\bullet$ satisfy 2.2.5.3 with respect to horizontal morphisms, i.e., given any object $(C_*, B)$ of either and any closed poset map $a : B' \to B$ and any set map $g : B' \to G$ then the unique object $(C'_*, B')$ in $\mathcal{W}_\bullet^{cont}(\mathbb{C}^m, G)$ or $\mathcal{W}_\bullet(\mathbb{C}^m, G)$ which maps to $(C_*, B)$ by the horizontal morphism $(a, g)_*$ lies in $\mathcal{W}_\bullet^{cont}$ or $\mathcal{W}_\bullet$.*

*d) If the source of any horizontal morphism in $\mathcal{W}_\bullet^{cont}(\mathbb{C}^m, G)$ or $\mathcal{W}_\bullet(\mathbb{C}^m, G)$ lies in $\mathcal{W}_\bullet^{cont}$ or $\mathcal{W}_\bullet$ then so does its target.*

*Then $|F| : |\mathcal{W}_\bullet^{cont}| \to |\mathcal{W}_\bullet|$ is a homotopy equivalence. If, in addition to the above properties, the objects of $\mathcal{W}_\bullet^{cont}$ are isovariant then $|\mathcal{W}_\bullet^{cont}/_{conj}| \simeq |\mathcal{W}_\bullet/_{conj}|$.*

We note that condition (d), which we call *closure with respect to expansion*, and isovariance are the only additional conditions (compared to Thm. 2.2.5).

## 2.5.1 Properties of $\tau_k(m)$

The properties of $\tau_k(m, n)$ carry over to properties of $\tau_k(m)$.

**Corollary 2.5.14 (Additivity lemma).** (a) *Let $p_i : M_m(\mathbb{C})^d \to M_m(\mathbb{C})$ be the projection to the i-th factor and let $s : M_m(\mathbb{C})^d \to M_{dm}(\mathbb{C})$ be the direct sum map. Then*

$$s^*(\tau_k(dm)) = \sum p_i^*(\tau_k(m))$$

*in $H^{2k}(\mathcal{W}_\bullet(M_m(\mathbb{C})^d, U(m)^d); \mathbb{R})$.*

(b) *Let $S : \prod \mathcal{W}_\bullet(M_m(\mathbb{C}), U(m))^d \to \mathcal{W}_\bullet(M_m(\mathbb{C}), U(m))$ be given by direct sum. Then*

$$S^* \tau_k(m) = \sum \pi_i^* \tau_k(m)$$

*where $\pi_i : \prod \mathcal{W}(M_m(\mathbb{C}), U(m))^d \to \mathcal{W}(M_m(\mathbb{C}), U(m))$ is the projection to the i-th factor.*

*Proof.* As in 2.2.7,$\mathcal{W}_\bullet(M_m(\mathbb{C}), U(m))$ and $\mathcal{W}_\bullet(M_m(\mathbb{C})^d, U(m)^d)$ are equivalent to their smooth versions by Proposition 2.5.13. In the smooth version the additivity lemma is obvious since $\mathrm{Tr}(\oplus A_i) = \Sigma\, \mathrm{Tr}(A_i)$. □

We need to extend the transfer functor of 2.1.16 to the simplicial bicategory $\mathcal{W}_\bullet$ in such a way that it preserves expansion.

**Definition 2.5.15 (Transfer with expansions).** We define the simplicial bifunctor

$$\mathrm{N} : \mathcal{W}_\bullet(M_m(R), Mon_m(G)) \to \mathcal{W}_\bullet(R, G)$$

by sending $(C_*, B)$ to $(\mathrm{N}\, C_*, mB)$ where

1. $\mathrm{N}\, C_* = (\mathrm{N} f, \mathrm{N} E, \mathrm{N} S)$ as in Def. 2.1.16.

2. $mB = \{(x, i) : x \in B, 1 \leq i \leq m\}$ is the disjoint union of $m$ copies of $B$ with partial ordering $(x, i) < (y, j)$ whenever one of the following holds.

   a) $x < y$ and $\{x, y\}$ is not an expansion pair or

   b) $\{x, y\}$ forms an expansion pair with $\partial y = xg$, $g \in Mon_m(G)$ and $i = \pi(g)(j)$ where $\pi(g)$ is the underlying permutation of $g$.

3. N is given on horizontal ($G$-monomial) morphisms by the monomorphism $Mon_n(Mon_m(G)) \subseteq Mon_{nm}(G)$ and on vertical ($G$-expansion) morphisms $B \to B'$ by the induced $m$-fold $G$-expansion $mB \to mB'$.

**Corollary 2.5.16 (1st transfer lemma).** *Let*

$$\mathrm{N} : \mathcal{W}_\bullet(M_m(M_d(\mathbb{C})), Mon_m(U(d))) \to \mathcal{W}_\bullet(M_d(\mathbb{C}), U(d))$$

*be the bisimplicial transfer functor defined above and let*

$$J : \mathcal{W}_\bullet(M_m(M_d(\mathbb{C})), Mon_m(U(d))) \to \mathcal{W}_\bullet(M_{dm}(\mathbb{C}), U(dm))$$

*be the inclusion functor: Then*

$$\mathrm{N}^*(\tau_k(d)) = J^*(\tau_k(dm))$$

*in* $H^{2k}(\mathcal{W}_\bullet(M_m(M_d(\mathbb{C})), Mon_m(U(d))); \mathbb{R})$

*Proof.* Same as 2.4.11. □

**Proposition 2.5.17.** *Let* $J : \mathcal{V}_\bullet(M_m(\mathbb{C})) \to \mathcal{W}_\bullet(M_m(\mathbb{C}), U(m))$ *be the simplicial inclusion functor. Then*

$$J^*(ct(\tau_k(m))) = \tau_k^V(m) = J^*(\tau_k(m))$$

*for all* $m, k$.

*Proof.* This follows from Theorem 2.4.13. □

## 2.6 Higher FR-torsion of oriented $S^1$-bundles

### Introduction

Suppose that $B$ is a compact smooth manifold embedded in Euclidean space so that it has a metric. Let

$$U(1) \to E \to B$$

be a smooth principal $U(1)$-bundle. Then $U(1)$ acts freely on $E$ and $B = E/U(1)$. If $Z_n \subseteq U(1)$ is the cyclic group of order $n$ consisting of the $n$-th roots of unity then we get another oriented $S^1$-bundle

$$S^1 \to E/Z_n \to B$$

where $S^1 = U(1)/Z_n$.

Given any $n$-th root of unity $\zeta$ there is a flat complex line bundle over $E/\mathbb{Z}_n$ whose vertical holonomy is given by multiplication by $\zeta$. If $\zeta \neq 1$ then the fibers of $E/Z_n \to B$ are acyclic with respect to these coefficients and therefore there are higher FR-torsion invariants $\tau_k(E/Z_n, \zeta) \in H^{2k}(B; \mathbb{R})$. In the case $\zeta = 1$ the fibers $S^1$ are not acyclic but the fiberwise homology (the flat bundle over $B$

whose fibers are $H_*(S^1;\mathbb{C})$) is trivial so the higher $R$-torsion invariants are still defined as we will later see.

The definition of the higher FR-torsion invariants in general is discussed in later chapters. The special case of oriented $S^1$-bundles is explained in the case of $k = 1$ in [IK93] and we will follow the treatment given there. The definition of $\tau_k(E/Z_n, \zeta)$ depends only on the restriction of $E/Z_n$ to the 2-skeleton of a sufficiently fine triangulation of $B$ and therefore the description of two parameter Morse theory given in [IK93] will be sufficient for our purposes.

The result will be a map $B \to |\mathcal{W}_\bullet(\mathbb{C})|$. To actually compute the higher FR-torsion invariants we would have to lift this map up to $|\mathcal{W}_\bullet^{diff}(\mathbb{C}^d)|$. By Theorem 2.5.11 this lifting is unique up to homotopy.

### 2.6.1 Morse theory

Let $f : E/Z_n \to \mathbb{R}$ be a "fiberwise positively oriented GMF" (PGMF), i.e., a smooth map whose restriction to every fiber has the property that its third derivative is positive at each degenerate critical point. Since the space of all PGMF's on $E/Z_n$ is contractible by [IK93] Thm 1.7 (and Theorem 2.10.15 below) we can choose one without adding any indeterminacy.

Let $b \in B$ be a generic point. Then the restriction $f_b$ of $f$ to the $S^1$-fiber of $E/Z_n$ over $b$ is a Morse function on $S^1$ with alternating maxima and minima $x_1, y_1, x_2, y_2, \cdots, x_m, y_m$ counterclockwise around $S^1$, i.e., following the orientation. Each maximum $x_i$ is allowed to cancel the next minimum $y_i$ since the cancellation deformation would produce a positive cubic singularity. However $x_i$ is not allowed to cancel $y_{i-1}$ since this would give a negative cubic singularity. Therefore $x_i, y_i$ are paired and the components of the set of nondegenerate critical points of $f$ containing $x_i, y_i$ lie over the same open subset $U$ of $B$. As in [IK93] we may assume that $U$ is an open ball (or an open half ball) and that these Morse point components map homeomorphically onto $U$. Over the boundary of $U$ the paired critical points corresponding to $x_i, y_i$ come together at a birth-death point. This implies that the *bifurcation set* of $f$, i.e., the set of all $b \in B$ over which $f_b$ is not Morse, is a union of spheres and half spheres with boundary in $\partial B$. We may assume that these meet transverse.

Since the Morse point components for $f$ are all contractible we may choose a path to a base point in $E/Z_n$ and we obtain an incidence matrix $g(b) \in GL_m(\mathbb{C})$ for each generic point $b \in B$ (where $m$ is a function of $b$). As in [IK93] we make these choices so that the incidence number between paired critical points is $+1$. Then $g(b)$ has the following form where each $a_i = \zeta^{k_i}$ is a power of $\zeta$ with $\Sigma k_i$ congruent to $-1$ modulo $n$ so that $\prod a_i = \zeta^{-1}$. Thus $\det g(b) = 1 - \zeta^{-1}$. g(b) = This can also be written as $g(b) = I_m - DP$ where $D = diag(a_1, \cdots, a_m)$ and $P$ is a cyclic permutation matrix $Pe_i = e_{i-1}$ (or $e_i^t P = e_{i+1}^t$) and $Pe_1 = e_m$.

As $b \in B$ varies in the complement of the bifurcation set the incidence matrix $g(b)$ does not change. However, when $b$ crosses the bifurcation set, $g(b)$ changes discontinuously to a larger or smaller square matrix of a similar shape. As explained geometrically in [IK93] (and justified algebraically in 5.1.7 below), this discontinuous process can be broken up algebraically into three steps consisting of an elementary expansion, one column and one row operation, e.g.,

$$A = \begin{pmatrix} 1 & -1 \\ -u & 1 \end{pmatrix} \to \begin{pmatrix} 1 & -1 & 0 \\ -u & 1 & 0 \\ 0 & 0 & 1 \end{pmatrix} \to \begin{pmatrix} 1 & -1 & 0 \\ -u & 1 & 0 \\ -u & 0 & 1 \end{pmatrix}$$

$$\to \begin{pmatrix} 1 & -1 & 0 \\ 0 & 1 & -1 \\ -u & 0 & 1 \end{pmatrix} = E_{23}(-1)(A \oplus I)E_{31}(-u)$$

where $u = \zeta^{-1}$.

**Lemma 2.6.1.** *To each sufficiently small generic path $\gamma$ in $B$ we can associate a 1-simplex in $\mathcal{W}_\bullet(\mathbb{C})$.*

*Proof.* First we say what we mean by "sufficiently small." Let $U$ be the image in $B$ of a Morse point component of $f$. As $b \in U$ moves to the boundary of $U$ two paired critical points of $f_b$ come together and cancel at a cubic singularity. When $b$ is just outside the boundary of $U$ this cubic singularity persists as an isolated nonsingular inflection point of $f_b$ which we refer to as a "ghost." We choose a closed collar neighborhood of $\partial U$ for each $U$ so that these ghosts exist over each point in the collar outside $\partial U$. We say that the ghosts are created along $\partial U$ and they vanish along the exterior boundary of the collar. We assume that the path $\gamma$ has length much smaller than the minimum width of these collars.

For any $b$ in the image of $\gamma$ let $m$ be the sum of the number of ghosts over $b$ which do not vanish along $\gamma$ plus the number of cubic singularities of $f_b$ plus half the number of critical points of $f_b$. Then $m$ is constant.

For each $t \in I = \Delta^1$ let

$$x_1(t) \leq y_1(t) \leq x_2(t) \leq y_2(t) \leq \cdots \leq x_m(t) \leq y_m(t)$$

in cyclic order be the ghosts, degenerate and nondegenerate singularities of $f_{\gamma(t)}$ where each ghost and degenerate singularity is counted twice as $x_i(t) = y_i(t)$. If we refer to these non-vanishing ghosts as degenerate nonsingular points then the word "degenerate" will encompass both ghosts and cubic singularities.

Let $P$ be the graded poset $P = P_0 \coprod P_1$ where $P_0 = \{y_1, y_2, \cdots, y_m\}$, $P_1 = \{x_1, x_2, \cdots, x_m\}$ with partial ordering given by $a > b$ if either $a, b$ are paired $(a = x_i, b = y_i)$ or $f_{\gamma(t)}(a(t)) > f_{\gamma(t)}(b(t))$ for all $t \in I$, e.g., this necessarily holds when $a = y_i, b = x_{i+1}$.

We need one more assumption on the size of $\gamma$. Let $M$ be the minimum difference between the critical values of unpaired incident critical points of $f$. Let $N$ be the maximum rate of change of the critical values of $f_b$ with respect to $b$. (The critical values of $f_b$ are $C^1$ but not $C^2$ functions of $b$ so $N$ is finite.) Let $\epsilon$ be half the ratio: $\epsilon = \frac{1}{2}M/N$.

If the length of $\gamma$ is less than $\epsilon$ then $a$ will be greater than $b$ in $P$ whenever $a(t)$ is incident to $b(t)$ for any $t \in I$.

Since $\gamma$ is generic, $f_{\gamma(0)}, f_{\gamma(1)}$ are Morse functions with well-defined incidence matrices $g(t)$ which we stabilize to be $m \times m$ matrices by adding degenerate pairs of basis elements $x_i, y_i$ with incidence number 1. As $t$ passes through the bifurcation set the incidence matrix will change by row and column operations as explained in [IK93] and these are upper triangular by the previous paragraph. □

**Lemma 2.6.2.** *To each sufficiently small generic map $\sigma : \Delta^p \to B$ we can associate a $p$-simplex in $\mathcal{W}_\bullet(\mathbb{C})$.*

*Proof.* First we assume that $p = 2$. We take the same size restrictions as in the last lemma. Then over each vertex $b_0 = \sigma(v_0), b_1 = \sigma(v_1), b_2 = \sigma(v_2)$ we have an $m \times m$ incidence matrix $g(b_i)$ which changes by row and column operations as $b$ moves along each edge. In the interior of the 2-simplex we will have isolated points where two birth-death lines cross. At these points there is algebraically an exchange point which gives a chain homotopy between the composition $E(b_2, b_1)E(b_1, b_0)$ and $E(b_2, b_0)$. (See [IK93] for details.) This chain homotopy will be strictly upper triangular if the diameter of $\sigma(\Delta^2)$ is less than $\epsilon = \frac{M}{2N}$.

If $p \geq 3$ then we simply note that the $m \times m$ incidence matrices over the vertices, the upper triangular chain isomorphisms over the edges and the (uniquely determined) strictly upper triangular chain homotopies over the 2-dimensional faces define a $p$-simplex in $\mathcal{W}_\bullet(\mathbb{C})$. □

We need to consider the collection of all $p$-simplices in $\mathcal{W}_\bullet(\mathbb{C})$ associated with $\sigma$.

**Definition 2.6.3.** Let $\sigma : \Delta^p \to B$ be a small smooth simplex. Then by an *admissible simplex* over $\sigma$ we mean the $p$-simplex constructed above with the following allowed alterations.

1. If $x_i = y_i$ is a ghost for all $t \in \Delta^p$ then we may choose to delete the pair $x_i, y_i$ from the poset $P$.
2. We may choose to strengthen the partial ordering on $P$.

The morphisms between these admissible simplices given by closed inclusion maps on posets and the identity element of $G = U(1)$ will be called *admissible morphisms*.

**Lemma 2.6.4.** *There are only a finite number of admissible simplices and admissible morphisms over $\sigma$ and they form a contractible subcategory of the diagonal of $\mathcal{W}_p(\mathbb{C})$ with the property that there is at most one morphism between any two objects.*

**Remark 2.6.5.** The diagonal category is constructed in Remark 2.5.8. The subcategory defined above will be called $\mathcal{C}_p(\sigma)$. We note that this is not a full subcategory of the diagonal of $\mathcal{W}_p(\mathbb{C})$ since we are taking only admissible morphisms.

*Proof.* The category of admissible simplices and admissible morphisms over $\sigma$ has a terminal object, namely the one given by the original construction in 2.6.2. □

**Theorem 2.6.6.** *These admissible simplices fit together to form a map from a space homotopy equivalent to $B$ to the realization $|\mathcal{W}_\bullet(\mathbb{C})|$ of $\mathcal{W}_\bullet(\mathbb{C})$.*

*Proof.* We can either choose a triangulation of $B$ with small smooth simplices or we can take the simplicial set of all small smooth simplices in $B$. In either case we need to take the same size restrictions as in the lemmas above. The proof is the same in both cases so we take the second, more natural construction.

Let $\operatorname{simp} B$ be the category of small smooth simplices in $B$. The objects are pairs $([p], \sigma : \Delta^p \to B)$ and morphisms $([p], \sigma) \to ([q], \tau)$ are simplicial operators $a : [p] \to [q]$ s.t. $\sigma = a^*(\tau)$. We note that $\sigma = a^*(\tau)$ is small for any small $\tau : \Delta^q \to B$ and any simplicial operator $a : [p] \to [q]$. We also have the well-known fact that $|\operatorname{simp} B| \simeq B$.

For each $([p], \sigma)$ in $\operatorname{simp} B$ we consider the contractible finite subcategory $\mathcal{C}_p(\sigma)$ of the diagonal of $\mathcal{W}_p(\mathbb{C})$ given in 2.6.5. If $a : [q] \to [p]$ is a simplicial operator then each admissible simplex $C$ produces an admissible simplex $a^*C$ over $([q], \tau)$ and therefore we have a simplicial category $\mathcal{C}$ which in degree $p$ is the disjoint union of the finite categories $\mathcal{C}_p(\sigma)$. There is a simplicial projection functor from $\mathcal{C}$ to $\operatorname{simp} B$ if we consider $\operatorname{simp} B$ as a simplicial category with only identity morphisms. By Lemma 2.6.4 this is a homotopy equivalence in every degree.

On the other hand there is a natural inclusion functor $\mathcal{C} \to \mathcal{W}_\bullet(\mathbb{C})$. (But the image in not full.) The inclusion functors $\mathcal{C}_p \to \mathcal{W}_p(\mathbb{C})$ do not commute with simplicial operators $a : [p] \to [q]$ since $a^*(C, P) = (a^*C, a^*P)$ in $\mathcal{C}$ but $a^*(C, P) = (a^*C, P)$ in $\mathcal{W}_\bullet(\mathbb{C})$. However, the posets $a^*P, P$ are naturally isomorphic and this isomorphism gives a natural transformation

$$a^+ : (a^*C, P) \to (a^*C, a^*P)$$

in $\mathcal{W}_p(\mathbb{C})$ which is simplicial in the sense that for any simplicial operator $b : [r] \to [p]$ we have

$$(ab)^+ = b^+ \circ b^*(a^+) : (b^*a^*C, P) \to (b^*a^*C, a^*P) \to (b^*a^*C, b^*a^*P).$$

This condition implies that we have a well defined functor from the category of simplices in $\mathcal{C}$ to the category of simplices in the diagonal of $\mathcal{W}_\bullet(\mathbb{C})$ mapping the boundary operator $a^*$ in $\mathcal{C}$ which is a morphism in $\operatorname{simp} \mathcal{C}$ to the composition $a^+ \circ a^*$ in $\operatorname{simp} \mathcal{W}_\bullet(\mathbb{C})$. □

**Definition 2.6.7.** Let $\tau_k(E/Z_n, \zeta) \in H^{2k}(B; \mathbb{R})$ be the pull-back of the universal class $\tau_k = [F_{2k} - G_{2k}] \in H^{2k}(|\mathcal{W}_\bullet(\mathbb{C})|; \mathbb{R})$ by the map given by Theorem 2.6.6.

## 2.7 Higher FR-torsion in the isovariant case

In the previous section we started with an oriented circle bundle $S^1 \to E \to B$ and a flat Hermitian coefficient bundle over $E$. In this section we generalize this using $\mathcal{W}_\bullet^{iso-d}(\mathbb{C}^d)/_{conj}$.

**Theorem 2.7.1.** *Given any oriented $S^1$-bundle and any unitary matrix $X \in U(d)$ with the property that $I_d - X$ is nonsingular, we can associate higher FR-torsion invariants $\tau_k(E, X) \in H^{2k}(B; \mathbb{R})$ with the following properties.*

1. (naturality) *Suppose $f^*E$ is the pull-back of $E$ along a map $f : B' \to B$. Then $f^*\tau_k(E, X) = \tau_k(f^*E, X)$*

2. (orientation reversal) *Suppose that $E^{-1}$ is the same as $E$ with the orientation of the fibers reversed. Then $\tau_k(E^{-1}, X) = \tau_k(E, X^{-1})$.*

3. (compatibility) *If $X = \zeta \in U(1)$ is an $n$-th root of unity then $\tau_k(E/Z_n, X)$ is the same as $\tau_k(E/Z_n, \zeta)$ as defined in the previous section.*

4. (smoothness for $d = 1$) *Suppose that $d = 1$. Then $\tau_k(E, X)$ is a smooth function of $X \in U(1)$ in the sense that its value on any fixed homology class in $H_{2k}(B)$ is smooth in $X$.*

**Remark 2.7.2.** Although our proof uses many specific properties of circle bundles we claim that this result holds in the following more general case.

We assume that we have a smooth bundle $E \to B$ with compact fiber $F$ and $\rho : \pi_1 F \to U(d)$ is a unitary representation satisfying two conditions.

- *(acyclicity)* $F$ is acyclic with respect to $\rho$, i.e., $H_*(F; \mathbb{C}^d) = 0$ if we take $\rho$-twisted coefficients.

- *(trivial holonomy)* There exists a subbundle $E' \to B$ with connected fiber $F'$ so that the action of $\pi_1 E'$ on $\pi_1 F$ leaves the representation $\rho$ invariant. ($\pi_1 E'$ acts on $\pi_1 F$ and thus on $\operatorname{Hom}(\pi_1 F, U(d))$ by the homotopy lifting property of the bundle $E \to B$.)

Then we claim that there is a map $w : B \to |\mathcal{W}_\bullet(\mathbb{C}^d)/_{conj}|$ which is well defined up to homotopy and therefore we can pull back the universal torsion classes in $|\mathcal{W}_\bullet(\mathbb{C}^d)/_{conj}|$ to get higher FR-torsion classes $\tau_k(E, \rho) \in H^{2k}(B; \mathbb{R})$.

In the case $F = S^1$ the trivial holonomy condition is satisfied for $E' = E$ if $E$ is an oriented $S^1$-bundle (or if $X = X^{-1}$). It is also obviously satisfied whenever $E$ is simply connected. Since $\pi_1 S^1 = \mathbb{Z}$, a unitary representation $\pi_1 S^1 \to U(d)$ is given by a single unitary matrix $X$.

*Proof.* We will construct a map from the category of small simplices in $B$ to the simplicial bicategory $\mathcal{W}_\bullet(\mathbb{C}^d)/_{conj}$. In the case $X = \zeta$ this map will lift to $\mathcal{W}_\bullet(\mathbb{C})$. Conditions (1) and (3) will then be obvious. To prove (4) we will do the construction again, this time smoothly with respect to $X$ in the case $d = 1$, using Lemma 2.2.8 and $\mathcal{W}_\bullet^{diff}(\mathbb{C})/_{conj}$. This is a special argument using the commutativity of $M_d(\mathbb{C})$ when $d = 1$ which gets around that fact that we were unable to prove the isovariant version of Lemma 2.2.8.

In order to prove (3) we need to reverse the orientation of the fibers of $E$. Then the positive generalized Morse functions (GMF's) become negative GMF's so we need to use a space of functions which incorporates both.

A *fiberwise framed function* is a fiberwise GMF $f$ on $E$ together with a framing of the negative eigenspace of $D^2 f$ at the critical points so that the framings on two cancelling critical points match at each birth-death point. (See [Igu87] for details.) For $S^1$ this simply means that at each maximum there is a nonzero tangent vector which points towards the minimum which it is allowed to cancel. In the last two sections we will show that the space of framed functions on $S^1$ is contractible. This implies that the space of fiberwise framed functions on $E$ is connected, i.e., a fiberwise framed function on $E$ is unique up to framed homotopy.

Since the definition of a framed function does not use the ambient orientation of the fibers we can reverse the orientation and keep the same function. In fact nothing changes except for the notation: the cohomology class $\tau_k(E, X)$ is now called $\tau_k(E^{-1}, X^{-1})$. This proves (2) (once the terms are defined).

Let $f : E \to \mathbb{R}$ be a framed function. Then we have positive and negative cubic points which give positive and negative ghosts which are isolated inflection points of $f$. The set of critical points of $f$ which are minima form a manifold immersed in $B$ with boundary the set of cubic singularities. We use ghosts to form immersed exterior collars.

We use the same notions of generic small smooth simplices in $B$ and for each such small simplex $\sigma$ we choose a base point in $E'|\sigma \cong \sigma \times S^1$ and a path from this base point to each component of the extended Morse point sets over $\sigma$. If $\sigma = b$ is a vertex then we obtain an $md \times md$ incidence matrix $g(b)$ whose $m$ diagonal entries are $I_d$ and whose off diagonal entries are either 0 or $-X^j$ for some integer $j$. We take the same poset $P$ as before. Then $(C, P)$ is well-defined up to conjugation. Note that the trivial holonomy condition on $E'$ implies in general that we have a well-defined homomorphism $\rho : \pi_1(F_b, x) \to U(d)$ for any $x \in E'$.

If $\sigma$ is a $p$-simplex then we again obtain the contractible subcomplex $\mathcal{C}_p(s)$ of $\mathcal{W}_\bullet(\mathbb{C}^d)$ but the simplicial operators are only well defined up to conjugation. The only difference between Lemma 2.6.2 and this case is that in the former case a uniform choice of base points and paths was made for all simplices. For any single small simplex, however, the construction is identical. When we mod out by conjugation the subcomplexes $\mathcal{C}_p(s)$ fit together to form a functor $\mathcal{C} \to \mathcal{W}_\bullet(\mathbb{C}^d)/_{conj}$ which is not quite simplicial but can be fixed as in 2.6.6.4-2.6.6.6. Therefore we get a map $B \to |\mathcal{W}_\bullet(\mathbb{C}^d)/_{conj}|$ which is well-defined up to homotopy and we can pull-back the universal higher FR-torsion invariants from $|\mathcal{W}_\bullet^{iso-d}(\mathbb{C}^d)/_{conj}| \simeq |\mathcal{W}_\bullet(\mathbb{C}^d)/_{conj}|$ to obtain $\tau_k(E, \rho) \in H^{2k}(B; \mathbb{R})$.

Now we will do everything again, this time smoothly filling in the interiors of the simplices using Lemma 2.2.8. We assume that $d = 1$ and that $X$ is a smooth function $I \to U(1) - \{1\}$. At each vertex $b$ in $B$ the incidence matrix $g(b)$ of $C(b)$ varies smoothly with $X$ for each choice of base points and paths. If we change the base point and paths then we get a new complex $C(b)'$ which is conjugate, i.e. horizontally isomorphic, to $C(b)$ by a horizontal isomorphism $(\alpha, \gamma)_* : C(b)' \to C(b)$ where $\alpha$ is the identity map (since the poset $P$ is intrinsically defined) and $\gamma : P \to U(1)$ is given on each critical point $x$ as the value of $\rho$ on the loop given by the path between the base points and the path from each base point to $x$.

Suppose that $\lambda$ is a small generic smooth 1-simplex in $B$. Then we have a base point and choice of paths associated to $\lambda$ and this gives us two chain complexes over the end points $b_0, b_1$ of $\lambda$ which are conjugates of admissible complexes already constructed over these vertices. There is also a upper triangular chain homotopy between these two chain complexes and by Lemma 2.2.8 we can extend this to a 1-parameter family of chain complexes along $\lambda$ which varies smoothly with $X$.

Next suppose that $\sigma : \Delta^2 \to B$ is a generic smooth small 2-simplex. Then over each edge of $\sigma$ we have a conjugate of the already given 1-parameter family of chain complexes. These agree over the vertices because of the trivial holonomy condition. What unavoidably happens is that the two paths from the chosen base point for $\sigma$ to the one over the vertex $v_0$ going through the two adjacent edges may not be homotopic. However, since $\pi_1 F'$ acts trivially on the representation $\rho : \pi_1 F \to U(1)$, we get the same chain complex and consequently we have a smooth family of chain complexes over $\partial\Delta^2$. Using Lemma 2.2.8 again we can extend this to the interior of $\sigma$ smoothly with respect to $X$.

We need to do one more step to see how isovariance is needed and how we get around it. Let $\tau$ be a generic small smooth 3-simplex in $B$. Then we need to know that the given smooth families of chain complexes over the faces of $\tau$ agree along the edges. We know that they agree on the vertices by the trivial holonomy condition. If we had an isovariant family we would get agreement over the interior by definition. We can do without this condition when $d = 1$ since $M_d(\mathbb{C})$ is abelian and the two versions of the 1-parameter family over the edge differ by conjugation by a diagonal matrix all of whose diagonal entries are the same element of $U(1)$. Since such a matrix is central the two families agree and we get a well defined smooth family over $\partial\Delta^3$ which extends smoothly over $\Delta^3$ by Lemma 2.2.8.

For any fixed $X$ we can use Lemma 2.3.2 to do everything isovariantly and therefore the two constructions agree by Theorem 2.5.11 and all conditions are satisfied. □

## 2.8 Transfer and polylogarithms

In this section we derive a transfer formula for the higher Franz-Reidemeister torsion invariants and obtain a computation for circle bundles as a consequence. The transfer formula holds whenever one bundle is a fiberwise covering space of another. However, since we have so far only defined the higher FR-torsion in the case of families of invertible matrices, and not in the more general acyclic complex case, we really only have the formula for $S^1$-bundles. The transfer in Morse theory is given by the simple fact that any Morse function on a smooth manifold $M$ pulls back to a Morse function on any covering space of $M$.

Since we will be considering several circle bundles related by transfer and torsion classes corresponding to different Hermitian coefficient bundles we need some notation. Suppose that $G$ is a cyclic subgroup of the group of units of a ring $R$ and $g \in G$ is a generator. Assume $g \neq 1$ and $g^n = 1$. Thus $|G| \neq 1$ divides $n$. Given any smooth principal $U(1)$-bundle $E \to B$ we have a twisted $R$-coefficient bundle over $E/Z_n$ with vertical holonomy given by multiplication by $g$. This gives a map

$$w(E/Z_n, g) : B \to |\mathcal{W}_\bullet(R, G)|$$

defined up to homotopy.

Let $\phi : R \to M_d(\mathbb{C})$ be a representation of $R$ on $\mathbb{C}^d$ so that $\phi(G) \subseteq U(d)$. Then $\phi$ induces a map

$$\phi_* : \mathcal{W}_\bullet(R, G) \to \mathcal{W}_\bullet(\mathbb{C}^d)$$

so we can pull back the universal FR-torsion class to get

$$\tau_k(E/Z_n, \phi(g)) = w(E/Z_n, g)^* |\phi_*|^* (\tau_k(d)) \in H^{2k}(B; \mathbb{R}).$$

For the most part it will suffice to consider only the universal case $E = S^{2N+1}$, $B = \mathbb{C}P^N$ for $N \geq 2k$.

With the above notation the conjugation lemma (2.5.7) and additivity lemma (2.5.14) can be expressed as follows.

**Lemma 2.8.1 (Conjugation lemma).** *Let $X, Y$ be conjugate elements of $U(m)$ so that $X^n = Y^n = I_m$ and $X - I, Y - I$ are nonsingular. Then*

$$\tau_k(E/Z_n, X) = \tau_k(E/Z_n, Y).$$

**Lemma 2.8.2 (Additivity lemma).** *Let $D \in U(m)$ be a diagonal matrix with diagonal entries $\xi_1, \xi_2, \cdots, \xi_m$ which are nontrivial $n$-th roots of unity. Then*

$$\tau_k(E/Z_n, D) = \sum_{i=1}^{m} \tau_k(E/Z_n, \xi_i).$$

*Proof.* Let $S = \mathbb{Q}[x]/(1 + x + x^2 + \cdots + x^{n-1})$. Then $x^n = 1$ and $x - 1$ is invertible in $S$. Therefore we have a map

$$w(E/Z_n, x) : B \to |\mathcal{W}_\bullet(S, \langle x \rangle)|.$$

Let $\delta : S \to M_m(\mathbb{C})$ be the unique ring map which sends $x$ to the diagonal matrix $D$ and let $\rho_i : S \to \mathbb{C}$ be the ring map sending $x$ to $\xi_i$. Then $\delta$ sends all of $S$ into the ring of diagonal matrices. Therefore, by the first additivity lemma (2.5.14(a)), $\delta^*(\tau_k(m)) = \sum \rho_i^*(\tau_k(i))$ so

$$\tau_k(E/Z_n, D) = w(E/Z_n, x)^* \delta(\tau_k(m)) = \sum w(E/Z_n, x)^* \rho_i^*(\tau_k(i))$$

$$= \sum \tau_k(E/Z_n, \xi_i).$$

□

**Lemma 2.8.3 (2nd transfer lemma).** *Suppose that $\xi^m$ is a nontrivial $n$-th root of unity. Then*

$$\tau_k(E/Z_n, \xi^m) = \sum_{\zeta^m=1} \tau_k(E/Z_{nm}, \xi\zeta)$$

*Proof.* We take the universal case $E = \mathbb{C}P^N$. We also consider the various possible values of $\xi$ to be representations of the same universal element $x$ in the ring

$$R = \mathbb{Q}[x]/(1 + x^m + x^{2m} + \cdots + x^{(n-1)m})$$

The polynomial $1 + x^m + x^{2m} + \cdots + x^{(n-1)m} = (x^{nm} - 1)/(x^m - 1)$ is the product of all cyclotomic polynomials $\phi_d(x)$ where $d$ ranges over the factors of $nm$ which don't divide $m$. Thus $R$ is the product of all cyclotomic fields $\mathbb{Q}(\zeta_d)$ where $\zeta_d$ is a primitive $d$-th root of unity. We note that $x^{nm} = 1$ in $R$ and $1 - x, 1 - x^m$ are invertible in $R$ since they map to invertible elements of $\mathbb{Q}(\zeta_d)$ for each $d$. Consequently we have a map

$$w(E/Z_{nm}, x) : B \to |\mathcal{W}_\bullet(R, \langle x \rangle)|.$$

If $\xi$ is a root of unity so that $\xi^m$ is a nontrivial $n$-th root of unity and $\zeta = \zeta_m$ then $\rho_i(x) = \xi\zeta^i$ defines a ring epimorphism $\rho_i : R \to \mathbb{Q}(\xi\zeta^i)$ and we have

$$\tau_k(E/Z_{nm}, \xi\zeta^i) = w(E/Z_{nm}, x)^* \rho_i^*(\tau_k).$$

We will consider the nontrivial $n$-th root of unity $\xi^m$ to be a representation of the generator $y$ in the ring $S = \mathbb{Q}[y]/(1+y+\cdots+y^{n-1})$ which is the product of all number fields $\mathbb{Q}(\zeta_s)$ where $s \neq 1$ divides $n$. Note that $y^n = 1$ and $1 - y$ is invertible in $S$ so we have a map $w(E/Z_n, y) : B \to |\mathcal{W}_\bullet(S, \langle y \rangle)|$. Let $\lambda : S \to \mathbb{C}$ be the ring map given by $\lambda(y) = \xi^m$.

As an $S$-module, $R$ is free of rank $m$ with basis $1, x, \cdots, x^{m-1}$. This gives a ring homomorphism $\beta : R \to M_m(S)$ sending $x$ to the $m \times m$ monomial matrix X =

$$X = \begin{pmatrix} 0 & & & y \\ 1 & 0 & & \\ & 1 & 0 & \\ & & \cdots & 0 \end{pmatrix}$$

Let $f : E/Z_{nm} \to \mathbb{R}$ be a fiberwise framed function. Then we get a family of incidence matrices $g(b), b \in B$, with entries powers of $x$ (up to sign) in $R$ which are related by upper triangular chain isomorphisms $E(b, b')$. The composition $f \circ p : E/Z_n \to E/Z_{nm} \to \mathbb{R}$ gives the family of matrices $\beta(g(b))$ with coefficients in $S$ related by chain isomorphisms $\beta(E(b, b'))$ by the usual transfer argument.

Since the fiberwise framed function on the oriented circle bundle $E/Z_n \to B$ is unique up to admissible homotopy this implies that the composition

$$\mathrm{N} \circ \beta^* : \mathcal{W}_\bullet(R, \langle x \rangle) \to \mathcal{W}_\bullet(M_n(S), \langle X \rangle) \to \mathcal{W}_\bullet(S, \langle y \rangle)$$

of $\beta^*$ with the transfer functor N sends $w(E/Z_{nm}, x)$ to $w(E/Z_n, y)$. In other words, the left hand square in the following diagram commutes up to homotopy.

$$\begin{array}{ccccc}
B & \xrightarrow{w(E/Z_n,y)} & |\mathcal{W}_\bullet(S,\langle y\rangle)| & \xrightarrow{\lambda} & |\mathcal{W}_\bullet(\mathbb{C})| \\
\Big\downarrow{\scriptstyle w(E/Z_{nm},x)} & & \Big\uparrow{\scriptstyle \mathrm{N}} & & \Big\uparrow{\scriptstyle \mathrm{N}} \\
|\mathcal{W}_\bullet(R,\langle x\rangle)| & \xrightarrow[b^*]{} & |\mathcal{W}_\bullet(M_n(S),\langle X\rangle)| & \xrightarrow[M_n(\lambda)]{} & |\mathcal{W}_\bullet(\mathbb{C}^n, Mon_n(U(1)))|
\end{array}$$

Since the right hand square obviously commutes, we have by the first transfer lemma (2.5.16) that

$$\tau_k(E/Z_n, \xi^m) = \tau_k(E/Z_{nm}, \lambda(X)).$$

However, $\lambda(X)$ has an eigenvalue $\xi\zeta^i$ for $0 \le i < m$ with corresponding eigenvector $(\xi^m\zeta^{im}, \cdots, \xi\zeta^i)$. Since these are orthogonal, the matrix $\lambda(X)$ is conjugate in $U(m)$ to the diagonal matrix with entries $\xi\zeta^i$. Thus

$$\tau_k(E/Z_{nm}, \lambda(X)) = \sum_{i=0}^{m-1} \tau_k(E/Z_{nm}, \xi\zeta^i)$$

by Lemmas 2.8.1 and 2.8.2. $\square$

**Theorem 2.8.4.** *Let $E \to B$ be a smooth principal $U(1)$-bundle, $n \ge 1, k \ge 0$ and $\xi \in U(1) - \{1\}$. Then*

$$\tau_k(E/Z_n, \xi) = a_k n^k \Re\left(\frac{1}{i^k}\mathcal{L}_{k+1}(\xi)\right) c_1(E)^k$$

*where $\mathcal{L}_{k+1}(\xi)$ is the classical polylogarithm*

$$\mathcal{L}_{k+1}(\xi) = \sum_{m=1}^{\infty} \frac{\xi^m}{m^{k+1}}$$

*and $a_k \in \mathbb{R}$ depends only on $k$.*

**Remark 2.8.5.** We will see later (in 7.10.2) that

$$a_k = -\frac{1}{k!}$$

and that the theorem also holds when $\zeta = 1$ (5.7.13).

*Proof.* When $k = 0$ this is the classical Reidemeister torsion with $a_0 = -1$ so we assume that $k \ge 1$. By naturality it suffices to show this formula in the universal case $B = \mathbb{C}P^\infty$ which is a limit of smooth manifolds. In this case $H_{2k}(B;\mathbb{R}) = \mathbb{R}$ is generated by $c_1(E)^k$ so

$$\tau_k(E/Z_n, \xi) = T_k(n,\xi)c_1(E)^k$$

for some family of real numbers $T_k(n,\xi)$ which are smooth functions of $\xi$ for every $n \geq 1$ and $k \geq 0$ by Theorem 2.7.1.

Since $E/Z_n$ is an oriented circle bundle over $B = \mathbb{C}P^\infty$ it is classified by a map $\mathbb{C}P^\infty \to \mathbb{C}P^\infty$. This map is multiplication by $n$ on $H^2$ and therefore by the ring structure must be multiplication by $n^k$ on $H^{2k}(B)$. Therefore $\tau_k(E/Z_n,\xi) = n^k\tau_k(E,\xi)$ so $T_k(n,\xi) = n^kT_k(1,\xi)$. Now we use the following observation which we prove later.

**Lemma 2.8.6 (Characterization of polylogarithms).** *Suppose that $k \geq 0$ and $L_k$ is a $C^{k+1}$ real valued function on $S^1 - \{1\}$ such that for any integer $n$ and any $x$ in $S^1$ with $x^n \neq 1$ (in particular $n \neq 0$ and $x \neq 1$) we have*

$$L_k(x^n) = n^{k-1}\sum_{z^n=1} L_k(xz) \tag{2.8}$$

*Then $L_k(x)$ is a constant multiple of the function $\Re\left(\frac{1}{i^{k-1}}\mathcal{L}_k(x)\right)$*

Remark: If condition (2.8) holds for roots of unity then it hold for all $x$ in $S^1 - 1$ by continuity. If it holds for $n = -1$ and for positive integers $n$ then it holds for all nonzero integers $n$.

Lemma 2.8.3 (with $n, m$ switched) says that

$$T_k(m,\xi^n) = \sum_{\zeta^n=1} T_k(nm,\xi\zeta)$$

Dividing both sides by $m^k$ we get

$$T_k(1,\xi^n) = \sum T_k(n,\xi\zeta) = \sum n^kT_k(1,\xi\zeta)$$

so $T_{k-1}(1,\xi)$ satisfies (2.8) for all roots of unity $x$ with $\xi^n \neq 1$. It remains to do the case $n = -1$ but this follows from Theorem 2.7.1.2. Changing the sign of $x$ is equivalent to reversing the orientation of the fiber. This is given by an involution of $\mathbb{C}P^\infty$ which is multiplication by $-1$ on $H^2$ and thus multiplication by $(-1)^k$ on $H^{2k}(\mathbb{C}P^\infty)$. This proves (2.8) for $T_{k-1}(1,\xi)$ and therefore $T_k(1,\xi) = a_k\Re\left(\frac{1}{i^k}\mathcal{L}_{k+1}(\xi)\right)$, completing the proof of the theorem. □

*Proof of 2.8.6.* Let $f_k$ be the periodic $C^{k+1}$ function on $\mathbb{R}$ given by $f_k(t) = L_k(e^{2\pi it})$. Then, separating the cases $n = -1$ and $n > 0$ in (2.8) we have:

$(1)_k$ $f_k(-t) = f_k(1-t) = (-1)^{k-1}f_k(t)$ if $t \notin \mathbb{Z}$

$(2)_k$ $f_k(nt) = n^{k-1}\sum_{j=1}^n f_k(t+\frac{j}{n})$ if $nt \notin \mathbb{Z}$ and $n > 0$

The lemma states that, as a consequence of these conditions, $f_k(t)$ is a constant multiple of

$$F_k(t) = \frac{1}{(2\pi)^{k-1}}\Re\left(\frac{1}{i^{k-1}}\mathcal{L}_k(e^{2\pi it})\right) = \Re\left(\frac{1}{(2\pi i)^{k-1}}\sum_{n=1}^{\infty}\frac{e^{2\pi int}}{n^k}\right)$$

Note that the derivative of $F_k$ is $F_{k-1}$ and for $k = 1, 0$ we have:

$$F_1(t) = -\log|1-e^{2\pi it}| = -\frac{1}{2}\log(2-2\cos 2\pi t),$$

$$F_0(t) = F_1'(t) = \frac{\pi\sin 2\pi t}{\cos 2\pi t - 1} = -\pi\cot\pi t$$

Suppose first that $k \geq 1$ and the theorem holds for $k-1$. Then the derivative $f_k'$ of $f_k$ must be a constant multiple of $F_{k-1}$ since it is a $C^k$ function satisfying $(1)_{k-1}$ and $(2)_{k-1}$. This means that $f_k(t) = aF_k(t) + b$ for some constants $a, b$. Since $(1)_k$ and $(2)_k$ are linear conditions held by both $f_k$ and $F_k$ they are also satisfied by the constant function $b = f_k - aF_k$. This implies that $b = 0$ and the theorem holds for $k$.

We are thus reduced to the case $k = 0$. Let $f = f_0$ be a periodic differentiable function on $\mathbb{R} - \mathbb{Z}$ satisfying $(1)_0$ and $(2)_0$. Then we must show that $f(t)$ is a constant multiple of $F_0 = -\pi\cot\pi t$.

Suppose first that $f$ extends to a continuous periodic function on all of $\mathbb{R}$. Then

$$f(t) = \frac{1}{n}\sum_{j=1}^{n} f\left(\frac{t}{n}+\frac{j}{n}\right)$$

is a Riemann sum which converges to the integral of $f$ over $[0,1]$ as $n \to \infty$. Consequently $f$ is a constant which must be zero by $(1)_0$.

In general we claim that $tf(t) = -tf(-t)$ *converges to some constant* $C$ *plus terms of the order of* $t^2$ *as* $t \to 0$. In the case $f(t) = F_0(t) = -\pi\cot\pi t$ this constant is $-1$. Therefore the function $g = f + CF_0$ has properties $(1)_0, (2)_0$ and the additional property that $tg(t) \to 0$ as $t \to 0$. Thus, by the claim, $|tg(t)|$ is bounded by a constant times $t^2$ and $|g(t)|$ is bounded by a constant times $t$. So $g(t)$ converges to zero as $t \to 0$. By the previous argument, $g(t) = 0$ for all $t$. Therefore our original function must be $f = -CF_0$.

It remains to prove the claimed estimate for $tf(t)$. Take any positive odd integer $N = 2m+1$ and any $0 < t < \frac{1}{N}$. Then

$$f\left(t+\tfrac{N-j}{N}\right) = f\left(t-\tfrac{j}{N}\right) = -f\left(\tfrac{j}{N}-t\right)$$

so by $(2)_0$ we have:

$$Ntf(Nt) = tf(t) + \sum_{j=1}^{m} t\left(f\left(\tfrac{j}{N}+t\right) - f\left(\tfrac{j}{N}-t\right)\right) = tf(t) + t^2 g_N(t)$$

where $g_N$ is a continuous function on $(-\frac{1}{N}, \frac{1}{N})$ with $g_N(0) = 2\sum_{1\le j\le m} f'(\frac{j}{N})$. Solving the above equation for $tf(t)$ we get

$$\begin{aligned} tf(t) &= Ntf(Nt) - t^2 g_N(t) \\ &= N^2 tf(N^2 t) - (Nt)^2 g_N(Nt) - t^2 g_N(t) \\ &= N^s tf(N^s t) - \sum_{i=0}^{s-1} (N^i t)^2 g_N(N^i t) \\ &= uf(u) - \sum_{j=1}^{s} (N^{-j} u)^2 g_N(N^{-j} u) \end{aligned}$$

for all $t \in (0, N^{-s})$, $u = N^s t \in (0,1)$. Since the series

$$\sum_{j=1}^{\infty} (N^{-j} u)^2 g_N(N^{-j} u) = u^2 \sum_{j=1}^{\infty} N^{-2j} g_N(N^{-j} u)$$

converges uniformly for $u$ in the closed interval $J = [\frac{1}{N+1}, \frac{N}{N+1}]$, the sequence of functions

$$tf(t) = N^{-s} uf(N^{-s} u)$$

converges to some continuous function $h(u)$ on the interval $J$. Since the end points of $J$ differ by a factor of $N$, we must have $h(\frac{1}{N+1}) = h(\frac{N}{N+1})$. In other words, $tf(t)$ converges to a continuous function of $\log t$ of period $\log N$ as $t \to 0$. Since this is true for all odd $N$, the function $h$ must be equal to a constant $C$. This implies that $tf(t) \to C$ as $t \to 0$ and

$$uf(u) = C + u^2 \sum_{j=1}^{\infty} N^{-2j} g_N(N^{-j} u)$$

as claimed. □

## 2.9 Framed functions on $S^1$

In this section we will show that the space of framed functions on $S^1$ is contractible. This was needed in the proof of 2.7.1.2 which was used to show that higher FR-torsion satisfies the case $n = -1$ in the characterization of polylogarithms. We will first show that the space of framed functions on the interval $I$ is contractible. Then we will modify the argument to do $S^1$. We leave for the next section the proof of intuitively obvious facts such as the fact that framed functions are given up to contractible choice by their configuration of critical points (with framing data) considered either as subsets of $I$ of $S^1$ or simply as finite sequences of symbols $0, +, -$.

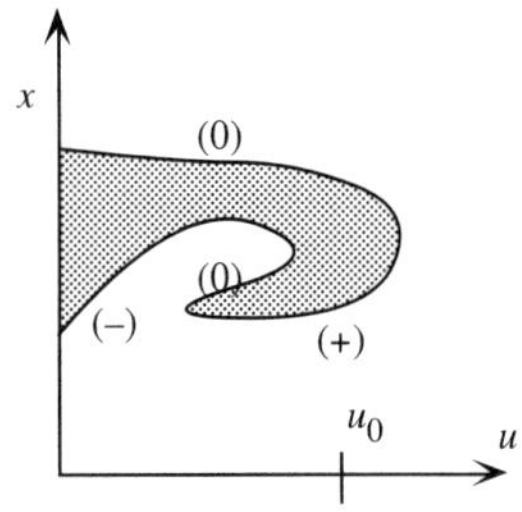


Figure 2.1: The non-ascending set of $f_u$

**Definition 2.9.1.** Let $F(I)$ be the space of all framed functions on $I$ which are nonsingular with positive derivatives on the two end points. If $f_t : I \to \mathbb{R}, t \in D^k$, is a generic smooth family of framed functions then we define the *non-ascending set* of $f$ to be the set:

$$NA(f) = (x, t) \in I \times D^k \, : \, Df_t(x) \leq 0$$

This is the union of the critical set $\Sigma(f)$ and the set of points on which $f$ is decreasing. For simplicity we sometimes refer to a positively or negatively framed maximum as simply $(+)$ or $(-)$. We will refer to a minimum as a $(0)$. The space of all positively framed functions, i.e. those without $(-)$ points, will be denoted $F_+(I)$.

**Example 2.9.2.** Let $f : I \to \mathbb{R}$ be the function $f(x) = 6x3 - 9x^2 + 4x$. This has one maximum at $x = 1/3$ and one minimum at $x = 2/3$. If the maximum is $(+)$ then it is allowed to cancel the minimum directly. If it is (-) then we need to do a more involved deformation to eliminate the critical points in an admissible way. One way is to create a (+)-(0) pair just below the (-) point, cancel the new (0) with the (-) and then cancel the remaining (+)-(0) pair. The deformation sequence is thus:

$$(-)(0) \to (+)(0)(-)(0) \to (+)(0) \to \emptyset \tag{2.9}$$

The non-ascending set in the deformation $f_u$ is given by the shaded region in Figure 2.1.

We note that this deformation has the property that the non-ascending set of $f_u$ remains within $NA(f)$ for all values of the deformation parameter $u$. Note that this property is maintained by moving maxima up and minima down and being careful where we create (+)-(0) pairs. Furthermore if we stop at the point $u_0$ when the (-) point has been exchanged for a (+) point then we can perform the deformation without changing the value of $f$ on the end points $0, 1$. (It could happen that $f(0) > f(1)$ in which case the final step could not be taken without changing these values.)

**Theorem 2.9.3.** *The space of framed functions $F(I)$ is contractible. Moreover, given any generic family of framed functions $f_t$ in $F(I)$ for $t \in S^k$ the critical points can be eliminated by a framed deformation $f_{tu}$ with the property that for each value of $u$ the subset $NA(f_u)$ of $S^k \times I$ is contained in $NA(f)$. Furthermore the (-) points can be eliminated by a deformation which does not change the values of $f_t$ at the end points $0, 1$.*

*Proof.* If there are no (-) points then we have a family of positive GMF which can be easily cancelled as in [IK93]. (See the next section for details.) Therefore it suffices to eliminate the (-) points by a framed deformation. If we make sure that this deformation never removes all the critical points for parameters $t$ over which the original function $f_t$ has critical points then we can keep the values of $f_t$ on 0 and 1 fixed for all $t$. The procedure is as follows.

a) Push up the (-) sheets except in a very thin collar around the boundary fold points. (Figure 2.2)

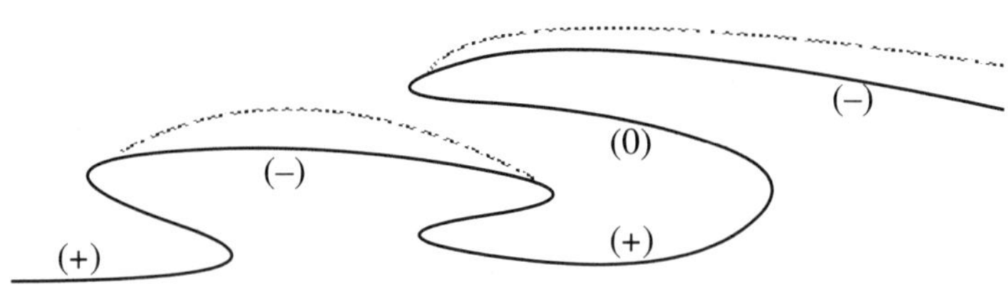


Figure 2.2: Push up the (-) sheets

b) Create a cancelling (+)-(0) pair in the region just underneath the (-) sheets inside the original non-ascending set $NA(f)$. (Figure 2.3)

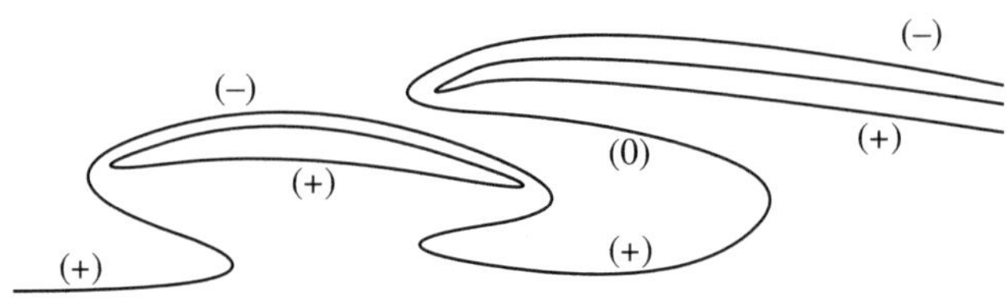


Figure 2.3: Create cancelling pairs under the (-) sheets

c) Cancel each (-) sheet with the new (0) sheet which has been created under it. This produces a thin strip of (-) points. (Figure 2.4)

d) We now have a thin strip of (-) points which lies along the original set of (0)-(-) fold points which forms a $k-1$ manifold $W^{k-1}$ which is immersed in $S^k$ and normally oriented. More precisely we have the shape indicated in Figure 2.5.

e) At this point we proceed by induction on $m$ the maximum number of sheets of $W^{k-1}$ which lies over the same point in $S^k$. By transversality $m \leq k$. Let $V \subseteq S^k$ be the set of points which are covered the maximum number of

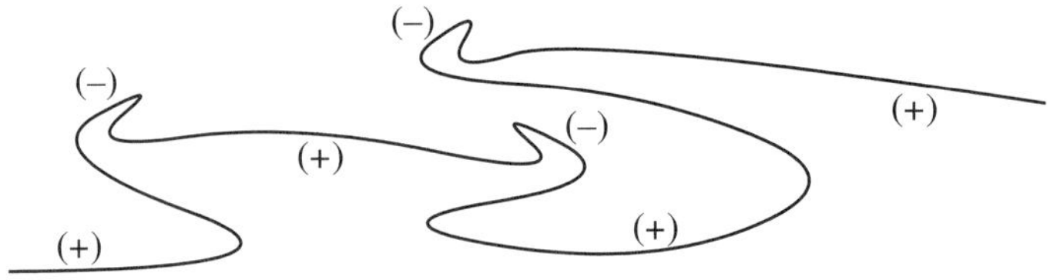


Figure 2.4: Cancel the (-) sheets with the new (0) sheets

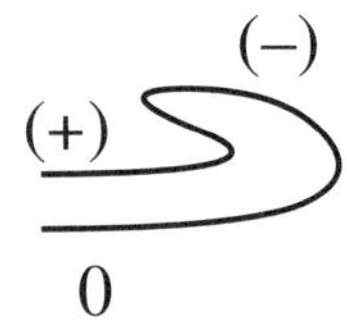


Figure 2.5: We get this shape $\times W^{k-1}$

times by $W$. Then by transversality $V$ will be a submanifold of $S^k$ of dimension $k-m$. Let $M^{k-m} \subseteq W$ be the set of points in $\pi^{-1}(V) \cap W$ which are minimal in the $I$-direction.

Underneath the points in $M$ there must be a sheet of (+) points. This is because the functions $f_t$ all have positive derivative at $x = 0$ so there must be a maximum or negative cubic below every minimum. By construction of $M$ there cannot be a negative cubic or a (-) point below it so there must be an uninterrupted (+) sheet below the points of $M$. Thus we obtain the shape given in (Figure 2.6) $\times M^{k-m} \times D^{m-1}$ where $D^{m-1}$ represents the normal disk of the (maybe nontrivial) normal bundle of $M$ in $W$.

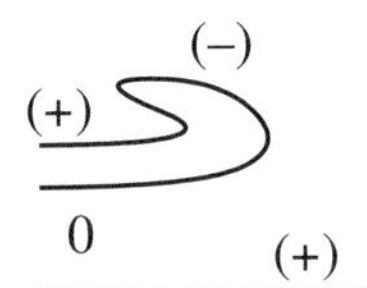


Figure 2.6: We get this shape $\times M^{k-m} \times D^{m-1}$

f) Now perform the deformation shown in Figure 2.7 in the center of each normal disk $D^{m-1}$. This reduces the number $m$ by 1 so by induction we can eliminate the entire (-) set.

g) Figure 2.8 shows drawings of the crucial step (f) in the case $m = 2$. By spinning these drawing around the central horizontal axis we get the general case. The regions are labelled by the names of the sheets in order.

The heavy lines in Figure 2.8 represent the (0)-(-) cancellations which will form the set $W$ in the next induction cycle. The fact that it has been removed from the central horizontal line at the end means that $m$ has been reduced by one. $\square$

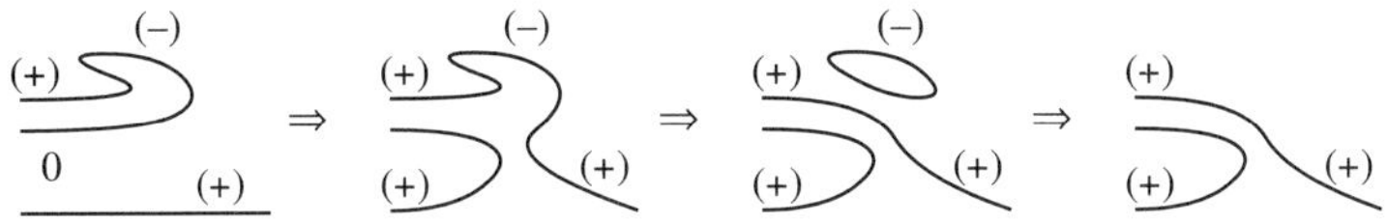


Figure 2.7: Eliminate what is left of the (-) set

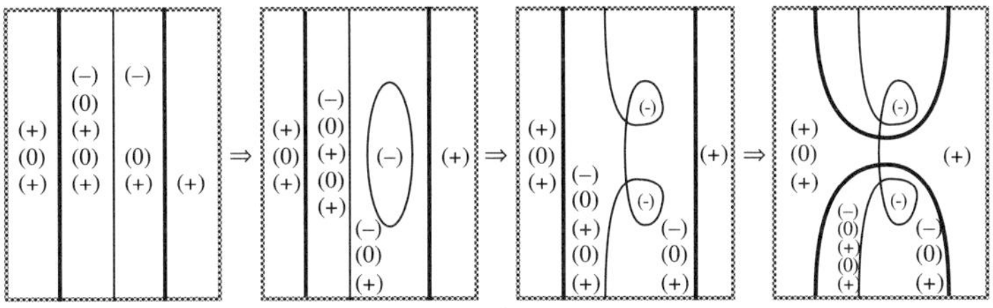

Figure 2.8: The crucial step (f) in the case $m = 2$

**Corollary 2.9.4.** *The space of framed functions on the circle $F(S^1)$ is contractible.*

*Proof.* We use the fact proved in [IK93] (and in the next section) that the space of positively framed functions $F_+(S^1)$ is contractible and the technical lemma also proved below that $F(S^1)$ has the homotopy type of a CW-complex. Then it suffices to show the (-) points can be eliminated from any family of framed functions. So we take a generic smooth family of framed functions on the circle $f_t \in F(S^1)$ parametrized by the disk $D^k$ so that $f_t \in F_+(S^1)$, i.e., has no (-) points or negative cubic points when $t \in \partial D^k$. We need to eliminate the (-) points without changing $f_t$ for $t$ near $\partial D^k$.

Let $J$ be the space of all $t \in D^k$ so that $f_t$ has either a (-) point or a negative cubic point. Then $J$ is a compact subset of the interior of $D^k$. For each $t$ in $J$ choose a point $x$ in $S^1$ so that $f_t$ is increasing at $x$. Then in some small open ball neighborhood $B(t)$ of $t$ in the interior of $D^k, x \times B(t)$ will be disjoint from $NA(f)$, i.e., $Df_s(x) > 0$ for all $s \in B(t)$. Choose a finite subcovering $B(t_i)$ of $J$. We can assume the points $x(t_i)$ are distinct.

We take a smooth generic triangulation of $D^k$ so that the diameter of each simplex is less than the Lebesgue number of the covering of $D^k$ given by the $B(t_i)$ and the complement of $J$. Then we proceed by induction on dimension to eliminate all the (-) points from these smooth simplices. When we come to each simplex we assume by induction that we have eliminated the (-) points over the boundary. Since the simplex is small it lies in one of the open sets $B(t_i)$. We "cut" the circle along $x(t_i) \times B(t_i)$ where the derivative of $f_t$ is positive and we have a family of framed functions on an interval. Theorem 9.3 tells us that we can eliminate the (-) points and still keep the same value of $f$ along $x(t_i) \times B(t_i)$. In other words we still have functions on the circle. We also keep the non-ascending set within the original set $NA(f)$. This means that each of the sets $x(t_j) \times B(t_j)$ remain outside of $NA(f)$ so we don't mess up the set up for the next step in the induction. □

## 2.10 Combinatorial framings

This section follows the approach to framed functions given in [IK93]. We will construct two combinatorial versions of the space of the space of framed functions both on $S^1$ and on $I$. There are two purposes. First, we want to fill in technical details in the proofs in the last section where we assumed that $F(S^1)$ had the homotopy type of a CW-complex and that $F_+(S^1), F_+(I)$ were contractible. We also implicitly assumed that our drawings showing deformations of critical points could always be realized as a deformation of the framed functions. This follows from the fact that the natural mapping sending a framed function to its configuration of critical points is a homotopy equivalence (Prop 2.10.14). The second and perhaps more important reason for these constructions is that they are needed in a later chapter to define framed structures on arbitrary graphs.

The idea is based on the fact that, up to a contractible choice, a framed function $f$ on $I$ or $S^1$ is determined by its configuration of critical points $\Sigma(f)$ together with labels given as follows.

**(0)** for Morse points of index 0, i.e., minima.

**(+)** for Morse points of index 1 with positive framing.

(−) for Morse points of index 1 with negative framing.

**(+0)** for cubic singularities with positive third derivative.

(0−) for cubic singularities with negative third derivative.

As indicated by the notation, a (+)-(0) critical point pair will converge to a (+0) cubic singularity which in turn will disappear eliminating the pair. However, to make a Hausdorff space we must make the cubic singularities disappear slowly by assigning a mass to it which starts at 2 and goes to 0. There are two versions of the space of such configurations depending on whether we take the configuration $\Sigma(f)$ as a subset of the domain of $f$ or simply as an ordered or cyclically ordered set. This makes no difference in the homotopy type and both models are useful. In all cases an admissible configuration will be a sequence of symbols alternating between a sign (+/−) and zero (0) grouped into disjoint singletons and pairs.

**Definition 2.10.1.** Let $CF(I) = CF(\mathbb{R})$ denote the space of all *linearly ordered configurations* $(a_*, \mu)$ consisting of

a) a finite sequence

$$(a_1, a_2, \cdots, a_{2n})$$

with $n \geq 0$ where each $a_{even} = 0$ and $a_{odd} = +$ or $-$ together with

b) a partitioning of the sequence into disjoint singletons and consecutive pairs $(+, 0)$ or $(0, -)$ which we call *doublets* (the pairs $(0, +), (-, 0)$ not being allowed) and

c) a mass function $\mu = (\mu(a_1), \mu(a_2), \cdots, \mu(a_{2n}))$ so that the mass of each singleton is 1 and $0 < \mu(a_i) = \mu(a_{i+1}) < 1$ for each doublet $(a_i, a_{i+1})$.

Note that the partition of the configuration is uniquely determined by the mass function $\mu$. The topology on $CF(I)$ is given by letting the mass vary continuously and deleting points when the mass goes to zero. In particular this means that a doublet becomes unpaired when its mass goes to 2.

**Proposition 2.10.2.** *$CF(I)$ has the structure of a CW-complex and the subspace $CF_+(I)$ of configurations without $(-)$ points is a contractible subcomplex.*

*Proof.* The space $CF(I)$ has an obvious cell structure with $k$-cells given by fixed configurations with $k$ doublets. The subcomplex $CF_+(I)$ has an action of the multiplicative monoid $I$ given by $t(a_*, \mu) = (a_*, t\mu)$ and is therefore contractible. □

**Definition 2.10.3.** Let $CF(S^1)$ denote the space of all *cyclically ordered configurations* which we define to be equivalence classes of triples $(a_*, \mu, \theta)$ where some cyclic permutation of $(a_*, \mu)$ lies in $CF(I)$ with at least one singleton (point with $\mu = 1$) and $\theta \in \mathbb{R}/2\pi\mathbb{Z}$ modulo the equivalence relation

$$((a_1, a_2, \cdots, a_{2n}), \mu, \theta) \sim ((a_2, a_3, \cdots, a_{2n}, a_1), \mu', \theta')$$

where $\mu'$ is cyclically permuted from $\mu$ so that the points $a_i$ keeps the same mass ($\mu'_i = \mu'(a_{i+1}) = \mu(a_{i+1}) = \mu_{i+1}$) and

$$\theta' = \theta + \frac{2\pi\mu(a_1)}{\sum_{i=1}^{2n} \mu(a_i)}$$

**Proposition 2.10.4.** *$CF(S^1)$ admits the structure of a CW-complex so that the subspace $CF_+(S^1)$ of configurations without $(-)$ is a subcomplex.*

**Remark 2.10.5.** Note that $I$ does not act on $CF(S^1)$ since cyclically ordered configurations must, by definition, have at least one (and thus at least two) points of mass 1. This corresponds to the fact that any continuous function on $S^1$ has at least one max and at least one min.

*Proof.* Each element of $CF(S^1)$ has a unique representative $(a_*, \mu, \theta_0)$ where

$$0 \leq \theta_0 < \frac{2\pi\mu(a_{2n})}{\sum_{i=1}^{2n} \mu(a_i)}$$

This gives $CF(S^1)$ an obvious cell structure with $k$-cells given by configurations with either $\theta_0 = 0$ and $k$ doublets or $\theta_0 \neq 0$ and $k - 1$ doublets. □

When we take our ordered configurations and embed them in $\mathbb{R}$ or $S^1 = \mathbb{R}/2\pi\mathbb{Z}$ we will get spaces of embedded configurations which are easily seen to be homotopy equivalent to both the configuration space and the corresponding framed function space. We will also see that the embedded version of $CF_+(S^1)$ is contractible.

**Definition 2.10.6.** The space $EF(I)$ consists of triples $(a_*, \mu, \psi)$ where $(a_*, \mu)$ is a linearly ordered configuration and $\psi : \{1, 2, \cdots, 2n\} \to (0, 1)$ is a mapping which is *ordered* in the sense that $\psi(i) \leq \psi(j)$ if $i \leq j$ and which also has the property that the inverse image of each point in its image is either a singleton or admissible doublet ($(+, 0)$ or $(0, -)$) of points of equal mass. (We allow doublets of total mass 2.) The topology is the obvious one letting $\psi, \mu$ vary continuously. This also means we allow doublets of vanishing mass to converge in $I$.

We also define $EF(\mathbb{R})$ analogously by replacing $(0, 1)$ with $\mathbb{R}$ in the definition above. If we choose a homeomorphism $(0, 1) \cong \mathbb{R}$ we get a homeomorphism $EF(I) \cong EF(\mathbb{R})$.

**Proposition 2.10.7.** *The forgetful mapping $\alpha : EF(I) \to CF(I)$ given by $\alpha(a_*, \mu, \psi) = (a_*, \mu)$ and the induced map $EF_+(I) \to CF_+(I)$ are homotopy equivalences.*

*Proof.* Identifying $EF(I)$ with $EF(\mathbb{R})$, the reverse mapping

$$\beta : CF(I) \to EF(\mathbb{R})$$

is given by $\beta(a_*, \mu) = (a_*, \mu', \psi)$ where the mass function $\mu'$ is given by $\mu'(a_i) = min(1, 2\mu(a_i))$ and $\psi(i) \in \mathbb{R}$ are uniquely determined by the following properties.

1. $\psi(i + 1) - \psi(i) = max(0, 2\mu - 2)$ if $(a_i, a_{i+1})$ is a doublet with mass $\mu = 2\mu(a_i)$.
2. $\psi(i + 1) - \psi(i) = \mu(a_i) + \mu(a_{i+1})$ if $(a_i, a_{i+1})$ is not a doublet.
3. $\sum_{i=1}^{2n} \psi(i)\mu(a_i) = 0$

Since $\alpha\beta(a_*, \mu) = (a_*, \mu')$ a homotopy $h_t : \alpha \circ \beta \simeq id$ is given by $h_t(a_*, \mu) = (a_*, \mu_t)$ where

$$\mu_t(a_i) = min(1, (1 + t)\mu(a_i))$$

(so $\mu_t = min(2, (1 + t)\mu)$ on doublets).

Since $\beta\alpha(a_*, \mu, \psi) = (a_*, \mu', \psi')$ with $\psi'$ satisfying (1),(2),(3) above, a homotopy $\beta \circ \alpha \simeq id$ is given by $(a_*, \mu_t, \psi_t)$ where $\mu_t$ is as above and

$$\psi_t(i) = (1 - t)\psi(i) + \phi_t(i)$$

where $\phi_t(i) \in \mathbb{R}$ are given uniquely by

1. $\phi_t(i + 1) - \phi_t(i) = max(2, (1 + t)\mu) - 2$ if $(a_i, a_{i+1})$ is a doublet of mass $\mu$.

2. $\phi_t(i+1) - \phi_t(i) = t\mu_t(a_i) + t\mu_t(a_{i+1})$ if $(a_i, a_{i+1})$ is not a doublet.

3. $\sum_{i=1}^{2n} \phi_t(i)\mu(a_i) = 0$

Since these mappings and deformations do not create new points in the configurations they also induce a homotopy equivalence $EF_+(I) \simeq CF_+(I)$. □

In the very last step we assumed that 0 was the center of mass of $\phi_t$ with respect to the mass $\mu$ instead of $\mu_t$ in order for the result to be invariant under cyclic permutation and thus applicable to the proof that $EF(S^1) \simeq CF(S^1)$ below.

**Definition 2.10.8.** We define $EF(S^1)$ to be the space of embedded configurations $(a_*, \mu, \psi)$ in $S^1$ with at least two singletons. Although $(a_*, \mu)$ is only well defined up to cyclic permutation, the embedding $\psi$ picks out a unique representative so that

$$0 \leq \psi(1) \leq \psi(2) \leq \cdots \leq \psi(2n) < 2\pi$$

This representative has the property that $(a_{2n}, a_1)$ does not form a doublet since doublets are required to have the same image under $\psi$.

**Proposition 2.10.9.** *Let $\alpha_0 : EF(S^1) \to CF(S^1)$ be the forgetful mapping given by $\alpha_0(a_*, \mu, \psi) = (a_*, \mu, \theta)$ where $\theta$ is the center of mass:*

$$\theta = \frac{\sum \psi(i)\mu(a_i)}{\sum \mu(a_i)}$$

*Then $\alpha_0$ induces homotopy equivalences $EF(S^1) \simeq CF(S^1)$ and $EF_+(S^1) \simeq CF_+(S^1)$ where the + denotes the subspaces with no $(-)$ points.*

*Proof.* First we note that $\alpha_0$ is well defined since if $(a_*, \mu)$ is cyclically permuted then $\psi(1)$ increases by $2\pi$ and $\theta$ increases by $2\pi\mu(a_1)/\sum \mu(a_i)$.

The reverse mapping $\beta_0 : CF(S^1) \to EF(S^1)$ will be constructed using the mapping $\beta : EF(I) \to CF(\mathbb{R})$ in the proof of Proposition 2.10.7. If $\beta(a_*, \mu) = (a_*, \mu', \psi)$ then let $\beta_0(a_*, \mu, \theta) = (a_*, \mu', \psi_0)$ where $\psi_0$ is given by $\psi_0(i) = \theta + c\psi(i)$ where

$$c = \frac{2\pi}{\mu(a_1) + \mu(a_{2n}) + \psi(2n) - \psi(1)}$$

is chosen so that the spacing between $\psi_0(1) + 2\pi$ and $\psi_0(2n)$ is given in the same manner as the other spacings. The mapping $\beta_0$ is well defined since $\theta$ is the center of mass with respect to $\mu$ instead of $\mu'$.

Since $\alpha_0 \circ \beta_0$ is given by $\alpha_0\beta_0(a_*, \mu, \theta) = (a_*, \mu', \theta')$ a homotopy $\alpha_0\beta_0 \simeq id$ is given by $(a_*, \mu_t, \theta_t)$ where $\mu_t$ is as before $(\mu_t(a_i) = min(1, (1+t)\mu(a_i))$ and

$$\theta_t = \frac{\sum \psi(i)\mu_t(a_i)}{\sum \mu_t(a_i)}$$

The composition $\beta_0 \circ \alpha_0$ is given by $\beta_0\alpha_0(a_*, \mu, \psi) = \beta_0(a_*, \mu, \theta) = (a_*, \mu', \psi')$ where $\psi'$ spreads out $(a_*, \mu')$ evenly over the circle so a homotopy $\beta_0 \circ \alpha_0 \simeq id$

can be given by taking the homotopy $(1-t)\psi(i)+\phi_t(i)$ in the proof of Proposition 2.10.7 and spreading it out evenly over the circle. Thus the homotopy is $h_t(a_*,\mu,\psi)=(a_*,\mu_t,\psi_t)$ where $\mu_t$ is as before and

$$\psi_t(i)=\theta+c_t(1-t)\psi(i)+c_t\phi_t(i)$$

where $c_0=0$ and

$$c_t=\frac{2\pi t}{t\mu_t(a_1)+t\mu_t(a_{2n})+\phi_t(2n)-\phi_t(1)}$$

Since these mappings and deformations do not create new points in the configurations they also induce a homotopy equivalence $EF_+(S^1)\simeq CF_+(S^1)$. ☐

**Proposition 2.10.10.** *$CF_+(S^1)$ and $EF_+(S^1)$ are contractible.*

*Proof.* Since $CF_+(S^1)\simeq EF_+(S^1)$ is a CW-complex it suffices to show that any mapping $C:\partial D^n\to EF_+(S^1)$ extends continuously to $D^n$ for all $n\geq 0$.

The configuration $C(t), t\in D^n$, will be constructed as a union of a finite number of disjoint configurations

$$C(t)=A(t)\amalg B_1(t)\amalg\cdots B_n(t)$$

each of which varies continuously with $t\in D^n$ but only some of which will have points of mass 1 for each $t$. The first, $A(t)$, is given as follows.

1. For $t\in\partial D^n$ let $A(t)=C(t)$.
2. As $u$ goes from 0 to 1 we let $A((1-\frac{u}{3})t)$ be obtained from $A(t)=C(t)$ by moving each (+) points towards the next (0) point $u$ part of the way. Then $A(\frac{2}{3}t)$ will be a union of doublets (and thus not admissible).
3. As $w$ goes from 1 to 0 we let $A(\frac{w+1}{3}t)$ be obtained from $A(\frac{2}{3}t)$ by multiplying the mass of each doublet by $w$.
4. Let $A(rt)$ will be the empty configuration for all nonnegative $r\leq\frac{1}{3}$ and $t\in\partial D^n$.

Next we construct the configurations $B_i(t)$. For each $t_0$ in the $n$-disk $D^n(0,\frac{2}{3})$ of radius $\frac{2}{3}$ about 0 in $D^n$ the configuration $A(t_0)$ is a finite set of (+0) doublets and therefore there is a point $x_0$ in $S^1$ disjoint from $A(t_0)$. By continuity, $x_0$ will be disjoint from $A(t)$ for all $t$ close to $t_0$. Therefore we can choose a finite covering of $D^n(0,\frac{2}{3})$ by small $n$-disks $V_i$ and find points $x_i\in S^1$ all distinct so that $x_i$ is disjoint from $A(t)$ for all $t$ in $V_i$. Choose $\epsilon>0$ so that the closed intervals $[x_i,x_i+\epsilon]$ are disjoint from each other and from $A(t)$ for all $t\in V_i$. Then we can take $B_i(t)$ to be the configuration with exactly two points (+) at $x_i$ and (0) at $x_i+\epsilon$ which converge and cancel over a small neighborhood of $\partial V_i$. Then $A(t)\amalg B_1(t)\amalg\cdots\amalg B_n(t)$ will be the desired null homotopy of $C$. ☐

Finally we show that these combinatorial framings are equivalent to the corresponding smooth framings.

**Proposition 2.10.11.** *a) There is a mapping $\gamma : F(I) \to EF(I)$ which sends $F_+(I)$ to $EF_+(I)$ and has the property that the critical set $\Sigma(f)$ of any framed function $f$ is equal to the image in $I$ of the set of all points of mass 1 in $\gamma(f)$.*

*b) Furthermore, any such mapping induces homotopy equivalences $F(I) \simeq EF(I)$ and $F_+(I) \simeq EF_+(I)$.*

*c) Consequently, $F_+(I)$ is contractible and $F(I) \simeq EF(I) \simeq CF(I)$.*

*Proof.* A mapping $\gamma : F(I) \to EF(I)$ fitting the above description can be constructed by letting $\gamma(f)$ be the configuration in $I$ consisting of the critical points of $f$ and the set of "ghosts" of $f$ which we take to be the points $z$ in the interior of $I$ so that $f''(x) = 0$ and $f'(x)f'''(x) > 0$. The signs of all nondegenerate critical points of $f$ are given by the framing of $f$ and the ghosts are $(+0)$ or $(0-)$ depending on whether $f'(x) > 0$ or $< 0$ respectively. The mass of each ghost is given in terms of a positive real number $\delta(f)$ which we define to be the minimum of $f'(0), f'(1)$ and the minimum distance from the 3-jet $(f'(x), f''(x), f'''(x))$ to the origin in $\mathbb{R}^3$. The formula is:

$$\mu(z) = max(1 - \frac{|f'(z)|}{\delta(f)}, 0)$$

If $\mu(z) \neq 0$ then $|f'(z)| < \delta(f)$ so $z \neq 0, 1$ and $f'''(z) \neq 0$ so by the implicit function theorem the position of $z$ in $(0, 1)$ varies continuously with $f$. This proves (a).

Now suppose that $\gamma : F(I) \to EF(I)$ is any mapping given by (a). A reverse mapping is given by sending each embedded configuration $C \subset I$ with singletons $x_1, \cdots, x_{2n}$ and doublets $z_1, \cdots, z_m$ of masses $2w_1, \cdots, 2w_m$ to the function $f$ given by $f(0) = 0$ and

$$f'(t) = \prod_{i=1}^{2n}(t - x_i) \prod_{j=1}^{m}(w_j(t - z_j)^2 + 1 - w_j)$$

This clearly varies continuously with $C$ and therefore defines a continuous mapping $\lambda : EF(I) \to F(I)$ by $\lambda(C) = f$. The composition $\lambda \circ \gamma : F(I) \to F(I)$ is homotopic to the identity as is the induced map $F_+(I) \to F_+(I)$ since $\lambda(\gamma(f))$ has the same configuration of critical points as $f$. The homotopy is given by

$$uf + (1 - u)\lambda(\gamma(f))$$

Since $EF_+(I)$ is contractible we conclude that $F_+(I)$ is also contractible and $F_+(I) \simeq EF_+(I)$.

To show that $F(I) \simeq EF(I)$ we need to show that the composition $\gamma \circ \lambda : EF(I) \to EF(I)$ is also homotopic to the identity map. This follow from the following lemma applied to the composition $\alpha \circ \gamma \circ \lambda \circ \beta : CF(I) \to CF(I)$ where $\alpha : EF(I) \to CF(I)$ and its homotopy inverse $\beta$ are as given in Proposition 2.10.7. ☐

**Lemma 2.10.12.** *Let $\phi : CF(I) \to CF(I)$ be any mapping with the property that*

$$\phi(a_*, \mu) = (b_*, \mu')$$

*where the subsequence of $a_*$ of points of mass $\mu \geq \frac{1}{2}$ is equal to the subsequence of $b_*$ of points of mass $\mu' = 1$. Then $\phi$ is homotopic to the identity.*

*Proof.* Since $\alpha \circ \beta \simeq id$ is a mapping having these properties it suffices to construct a homotopy of each such $\phi$ to a fixed mapping. Such a homotopy $\phi_t(a_*, \mu) = (b_*, \mu_t)$ is given by

$$\mu_t(b_i) = \begin{cases} (1-t)\mu'(b_i) & \text{if} \quad \mu'(b_i) < 1 \\ 1 - 2t(1-\mu(a_j)) & \text{if} \quad \mu'(b_i) = 1 \text{ and } b_i \text{ corresponds to } a_j \end{cases}$$

□

We need the following analogous lemma to prove that $F(S^1) \simeq EF(S^1)$.

**Lemma 2.10.13.** *Any mapping $\phi : EF(S^1) \to EF(S^1)$ which fixes all points of mass 1 is homotopic to the identity.*

*Proof.* Let $\eta : EF(S^1) \to EF(S^1)$ be the map given by $\eta(a_*, \mu, \psi) = (a_*, \mu', \psi)$ where $\mu'(a_i) = min(1, 2\mu(a_i))$. Then $\eta$ is homotopic to the identity since $\mu_t(a_i) = min(1, (1+t)\mu(a_i))$ deforms $\mu$ to $\mu'$.

Now we claim that $\eta \simeq \phi \circ \eta$. If $\phi\eta(a_*, \mu, \psi) = (b_*, \mu'', \psi')$ then $(b_*, \mu'')$ agrees with $(a_*, \mu')$ on all points of mass 1. If $b_i$ corresponds to $a_j$ with $\mu(a_j) > \frac{1}{2}$ then in a neighborhood of $(a_*, \mu, \psi)$ we will have a point close to $a_j$ with $\mu > \frac{1}{2}$ so there will be a point close to $b_i$ with mass $\mu'' = 1$. Consequently, $b_i$ is not a limit of points of mass $\mu'' < 1$ so we can define a deformation $(b_*, \mu''_t, \psi')$ by

$$\mu''_t(b_i) = \begin{cases} (1-t)\mu''(b_i) & \text{if} \quad \mu''(b_i) < 1 \\ 1 - 2t(1-\mu(a_j)) & \text{if} \quad \mu''(b_i) = 1 \text{ and } b_i \text{ corresponds to } a_j \end{cases}$$

This gives a homotopy from $\phi \circ \eta$ to a mapping independent of $\phi$ so in particular $\phi \circ \eta \simeq id \circ \eta = \eta$. □

**Proposition 2.10.14.** *a) There is a mapping $\gamma : F(S^1) \to EF(S^1)$ which sends $F_+(S^1)$ to $EF_+(S^1)$ and has the property that the critical set $\Sigma(f)$ of any framed function $f$ is equal to the image in $S^1$ of the set of all points of mass 1 in $\gamma(f)$.*

*b) Furthermore, any such mapping induces homotopy equivalences $F(S^1) \simeq EF(S^1)$ and $F_+(S^1) \simeq EF_+(S^1)$.*

*c) Consequently, $F_+(S^1)$ is contractible and $F(S^1) \simeq EF(S^1) \simeq CF(S^1)$.*

*Proof.* The proof is analogous to the proof of Proposition 2.10.11 using Lemma 2.10.13 in place of Lemma 2.10.12 and $\lambda(C) = f : \mathbb{R}/2\pi\mathbb{Z} \to \mathbb{R}$ for configurations $C \subset \mathbb{R}/2\pi\mathbb{Z}$ given by the following formulas.

$$f(x) = \int_0^x e^{cg(t)} g(t)\, dt$$

$$g(t) = \prod_{i=1}^{2n} \sin \frac{1}{2}(t - x_i) \prod_{j=1}^{m} (w_j \sin^2 \frac{1}{2}(t - z_j) + 1 - w_j)$$

where $c = c(g)$ is uniquely determined by the condition that $f(2\pi) = 0$. (See [IK93] for more details.) $\square$

To summarize, we have the following.

**Theorem 2.10.15.** *The spaces* $CF(S^1)$, $EF(S^1)$, $F(S^1)$, $CF_+(S^1)$, $EF_+(S^1)$, $F_+(S^1)$ *are all contractible as are the corresponding spaces for* $I$.

*Proof.* $CF_+(S^1), EF_+(S^1)$ are in Proposition 2.10.10. $F_+(S^1)$ is in Prop. 2.10.14. Corollary 2.9.4 shows that $F(S^1)$ is weakly contractible but the proof assumes it has the homotopy type of a CW-complex. But Propositions 2.10.4 and 2.10.14 above show that $F(S^1)$ has the homotopy type of a CW-complex. Thus $F(S^1) \simeq CF(S^1) \simeq EF(S^1)$ are all contractible.

The analogous argument holds for $I$ by Propositions 2.10.2, 2.10.11 and Theorem 2.9.3. $\square$

# Chapter 3

# A Model for the Whitehead Space

## Introduction

This chapter constructs a model for the Whitehead space

$$\mathcal{W}h^h_\bullet(R,G) = \text{fiber}(f_* : Q(BG_+) \to K(R)).$$

where $f_*$ is induced by a homomorphism $f$ from a group $G$ into the group of units of a ring $R$. This model is a simplicial category whose objects are multiple mapping cylinders of acyclic chain complexes.

This construction is the natural generalization of the space of matrices discussed in Chapters I and II. It is, in spirit, the same as the space of chain complexes proposed by Volodin [Vol71]. However, in Chapter V we prove that this space has the correct homotopy type and we show that, with some care, it can be used to define the higher FR-torsion invariants.

Section 3.1 contains a description of the objects of our simplicial category. This simplicial set $\mathcal{M}_\bullet(R,P)$ resemble the nerve of the category of based free chain complexes with graded basis set $P$ and chain isomorphisms which are upper triangular with respect to this basis.

Section 3.2 describes the different classes of morphism between the objects of $\mathcal{M}_\bullet(R) = \coprod \mathcal{M}(R,P)$ both in detail and concisely in one equation (3.1). The *G-monomial morphisms* (Definition 3.2.7) define the unstable version of $\mathcal{W}h_\bullet(R,G)$ in which the number of basis elements is fixed. We define *G-monomial functors* $\xi$ on simplicial sets $B$ so that they correspond to families of chain complexes parametrized by $B$. In section 3.3 we show that, as expected, homotopic morphisms give simplicially homotopic mapping cones.

Section 3.4 contains the key lemma (3.4.9) which shows that if a $G$-monomial functor $\xi$ has *constant free homology* then we can kill the fiberwise homology in a canonical way up to simplicial homotopy. In Theorem 3.4.11 this is extended to the case of *upper triangular homology*.

Section 3.5 shows that both of these constructions are independent of all choices (Theorems 3.5.10 and 3.5.11). However, in the upper triangular case we need to assume that $R$ is a field. We also prove invariance under stabilization by elementary expansions (the Stabilization Lemma 3.5.8).

In Section 3.6 we stabilize our Whitehead space by adding elementary expansions. We define a *G-expansion functor* on a simplicial set $B$ so that it corresponds to a family of based free chain complexes over $B$ which vary by both upper triangular basis change and *elementary G-expansions* (adding a pair of basis elements which are incident by an element of $G$). The main theorem (3.5.12) is that the key lemmas for $G$-monomial functors also hold for $G$-expansion functors.

## 3.1 Multiple mapping cylinders

In this section we construct a simplicial set whose objects will eventually become the set of objects of our simplicial category model for the Whitehead space. A $k$-simplex in our simplicial set will consist of a $(k+1)$-tuple of based free chain complexes $C(0), \cdots, C(k)$ together with chain maps $f_{ji}: C(i) \to C(j)$ for $0 \leq i < j \leq k$, chain homotopies $f_{kj} \circ f_{ji} \simeq f_{ki}$ and higher chain homotopies indexed by decreasing sequences $k \geq i_0 > i_1 > \cdots > i_n \geq 0$. This is equivalent to an iterated or multiple mapping cylinder as we now describe.

**Topology.** First let's discuss topological multiple mapping cylinders. The mapping cylinder of a continuous map $F_1 : X_0 \to X_1$ is a space $Z(F_1) = (X_0 \times I) \cup X_1$ with a projection $p : Z(F_1) \to I$ so that $p^{-1}(0) = X_0, p^{-1}(1) = X_1$. Given a homotopy commutative diagram

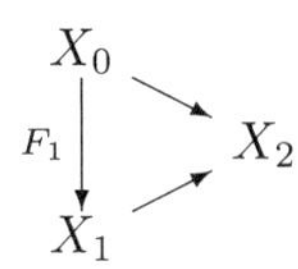


we obtain a map $F_2 : Z(F_1) \to X_2$ whose mapping cylinder $Z(F_2) = (X_0 \times \Delta^2) \cup (X_1 \times \Delta^1) \cup X_2$ has a projection to $\Delta^2$ where $X_0, X_1, X_2$ lie over the three vertices. Continuing in this way we obtain a $k$-fold mapping cylinder $Z(F_k) = \cup_{0 \leq i \leq k} X_i \times \Delta^{k-i}$ which is the mapping cylinder of a map from the $(k-1)$-fold mapping cylinder $Z(F_{k-1})$ to $X_k$.

If the $X_i$ are cell complexes then $Z(F_k)$ has a cell decomposition with cells (of dimension $n + dim(e)$) given by pairs $(v, e)$ where $v = (v_0, \cdots, v_n)$ is a decreasing sequence $k \geq v_0 > v_1 > \cdots > v_n \geq 0$ representing an $n$-face of $\Delta^k$ and $e$ is a cell in $X_{v_n}$.

**Algebra.** In the case of chain complexes, $Z_1 = Z(F_1)$ is the mapping cylinder of a chain map $F_1 : C(0) \to C(1)$. In general $Z_k$ will be the mapping cylinder of a chain map $F_k : Z_{k-1} \to C(k)$ which will be constructed by induction.

As a graded abelian group $Z_k$ will be the direct sum of an $n$-fold suspended copy $\Sigma^n C(i)$ of $C(i)$ for every decreasing sequence of integers $k \geq v_0 > v_1 > \cdots > v_n = i \geq 0$. We will call this $\Sigma^v$. Note that $i = v_n$ is determined by $v = (v_0, \cdots, v_n)$. We also note that $\oplus_{u \subseteq v} \Sigma^u(C)$ is a subcomplex of $Z_k$ which is also an $n$-fold mapping cylinder.

**Morse Theory.** The idea comes from parametrized Morse theory. If $p : E \to B$ is a smooth bundle and $f : E \to \mathbb{R}$ is an oriented fiberwise generalized

Morse function then $f$ induces a Morse function $f_b$ on almost all fibers $p^{-1}(b) = M_b$ of $p$ and we obtain a cellular free chain complex $C(b)$ with basis elements corresponding to critical points of $f_b$ and thus partially ordered by critical value. If $b'$ is close to $b$ than $C(b')$ will be chain isomorphic to $C(b)$ (up to stabilization by elementary expansions which we will deal with later.)

If $b, b', b''$ are three close points in $B$ then we will have a homotopy commuting diagram of chain isomorphisms

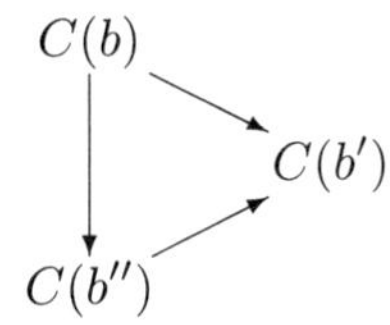


This gives a 2-fold mapping cylinder. Similarly we get a $k$-fold mapping cylinder over every small $k$-simplex in $B$.

Since $C(b), C(b'), C(b'')$, etc. will all be free based chain complexes with the same partially ordered basis $P$ (ignoring expansions) the boundary map with coefficients in a ring $R$ in the multiple mapping cylinder will be a family of maps $e(v)$ between shifted copies of the free $R$-module $R^P$ generated by $P$.

**Higher chain homotopies.** Now we will develop this idea abstractly.

Suppose that $P$ is a *finite graded poset*, i.e., a finite poset whose underlying set is graded: $P = P_0 \amalg P_1 \amalg \cdots \amalg P_m$. If $R$ is a ring then we will take the set of all free $R$-complexes $0 \to C_m \to C_{m-1} \to \cdots \to C_0 \to 0$ where $C_i = R^{P_i}$ is a free right $R$-module with basis $P_i$ so that the boundary of each element $x \in P_i$ is a linear combination of elements $y < x$. (Thus, if $P_{i+1}$ contains no elements less than $x$ then $\partial x = 0$.) This is the same as saying that the boundary map is an endomorphism of $R^P = \oplus R^{P_i}$ which is homogeneous of degree $-1$ with respect to the grading and whose matrix is strictly upper triangular with respect to the partial ordering. These chain complexes form the vertices of a simplicial set $\mathcal{M}_\bullet(R, P)$ defined as follows.

**Definition 3.1.1.** An element of $\mathcal{M}_k(R, P)$ is defined to be a system of endomorphisms called *higher homotopies* $e(v_*) = e(v_0, \cdots, v_n)$ of $R^P$ indexed by the nonincreasing sequences $k \geq v_0 \geq v_1 \geq \cdots \geq v_n \geq 0$ satisfying the following.

1. $e(v_0, \cdots, v_n) = 0$ if $v_*$ is *degenerate* (i.e., $v_i$ are not distinct).

2. $e(v_0, \cdots, v_n)$ is homogeneous of degree $n - 1$.

3. $e(v_*)$ is strictly upper triangular with respect to the partial ordering of $P$, i.e., for each $x \in P$, $e(v_*)(x)$ is an $R$-linear combination of elements $y < x$ of $P$.

4. $e(\emptyset) = 0$. (This is the case $|v_*| := n = -1$.)

5.
$$\sum_{i=0}^{n}(-1)^i e(v_0,\cdots,v_i)e(v_i,\cdots,v_n)=\sum_{i=0}^{n}(-1)^i e(v_0,\cdots,\widehat{v_i},\cdots,v_n).$$

(In the notation of 3.1.13 below this is written $e' * e = \delta e$.)

The sets $\mathcal{M}_k(R,P)$ form a simplicial set $\mathcal{M}_\bullet(R,P)$ with the obvious simplicial operators $a^* : \mathcal{M}_k(R,P) \to \mathcal{M}_j(R,P)$ given by

$$(a^*e)(v_0,v_1,\cdots,v_n) := e(a(v_0),a(v_1),\cdots,a(v_n))$$

for all non-decreasing maps

$$a : [j] = \{0,\cdots,j\} \to [k] = \{0,\cdots,k\}.$$

An important special case is the *constant* system of higher homotopies $d \in \mathcal{M}_k(R,P)$ which have the property that $d(v_*) = 0$ for all admissible sequences $v_*$ with $|v_*| \geq 1$. We will sometimes consider the more general case where $R^P$ is replaced by a direct sum of $R$-modules indexed by $P$.

*What is the meaning of Definition 3.1.1?*

For $n = 0$ conditions (4),(5) say $e(v)^2 = 0$ so we have *chain complexes* $C(v) = (R^P, e(v))$ with the same underlying graded $R$-module $R^P$.

If $n = 1$ then $e(v_0)e(v_0,v_1) - e(v_0,v_1)e(v_1) = e(v_1) - e(v_0)$ or $d(1+e(v_0,v_1)) = (1+e(v_0,v_1))d$ so

$$E_{01} = 1 + e(v_0,v_1) : C(v_1) \to C(v_0)$$

is a chain map which is necessarily a *chain isomorphism* since $e(v_0,v_1)$ is a nilpotent. In the constant case, $E_{01} = 1$ so $C(v_1) = C(v_0) = (R^P,d)$ for all $v_0$, $v_1$.

If $n = 2$, using the shorthand notation $f_{123\cdots} = e(v_1,v_2,v_3,\cdots)$, we have $f_0 f_{012} - f_{01} f_{12} + f_{012} f_2 = f_{12} - f_{02} + f_{01}$ or

$$\partial f_{012} + f_{012}\partial = E_{01}E_{12} - E_{02}$$

so $f_{012} = e(v_0,v_1,v_2)$ is a *chain homotopy* $E_{01}E_{12} \simeq E_{02}$.

The next "higher chain homotopy" (for $n = 3$) satisfies:

$$\partial f_{0123} - f_{0123}\partial = E_{01}f_{123} - f_{023} + f_{013} - f_{012}E_{23}.$$

We note that if the partial ordering of $P$ *respects degree* in the sense that $x < y$ whenever $\deg x < \deg y$ then the homotopies $f_{012} = e(v_0,v_1,v_2)$ and higher homotopies must all be zero since they raise degree.

**Example 3.1.2.** We will construct a system of higher homotopies out of a single chain complex and a $(k+1)$-tuple of upper triangular matrices. Suppose that $(C,\partial)$ is a based free $R$-complex and suppose that the basis $P_n$ of $C_n$ has a partial ordering. Let $T(P_n,R)$ be the group of $R$-homomorphisms $f : R^{P_n} \to$

$R^{P_n}$ which are "upper triangular" in the sense that $f(x)$ equals $x$ plus an $R$-linear combination of elements $y < x$ for all $x \in P_n$. (The matrix of $f$ is upper triangular with 1's on the diagonal.)

For $k \geq 0$ fixed suppose we are given

$$f_n(0), f_n(1), \cdots, f_n(k) \in T(P_n, R)$$

for each $n$. Then we have an element $e \in \mathcal{M}_k(R, P)$, where $P$ is the finite graded poset $P = \coprod P_n$ with ordering respecting degree, given as follows.

1. $e(i) = f(i) \circ \partial \circ f(i)^{-1}$ for $0 \leq i \leq k$
2. $E(j, i) = f(j) \circ f(i)^{-1}$ for $0 \leq i < j \leq k$
3. $e(v_*) = 0$ if $|v_*| \geq 2$.

Here $f(i) = \oplus f_n(i) \in T(P, R)$. This is a system of higher homotopies because:

1. $e(i)^2 = 0$ for all $0 \leq i \leq k$
2. $e(j)E(j, i) = E(j, i)e(i)$ for all $0 \leq i < j \leq k$
3. $E(\ell, j)E(j, i) = E(\ell, i)$ for all $0 \leq i < j < \ell \leq k$

and because $e(i)$ and $e(j, i) = E(j, i) - 1$ behave correctly with respect to the partial ordering. This concludes the example.

Returning to the general case, we will show that the endomorphisms $e(v_*)$ which make up a system of higher homotopies are exactly the data necessary to form a multiple mapping cylinder $Z_k$ given recursively as a mapping cylinder of a chain map $Z_{k-1} \to C(k)$ with $Z_{-1} = 0$.

**Definition 3.1.3.** If $e = \{e(v_*)\} \in \mathcal{M}_k(R, P)$ let $Z_k(e)$ be the based free chain complex over $R$ given as follows.

1. As a graded $R$-module $Z_k(e)$ is the direct sum

   $$Z_k(e) = \bigoplus_{v_* \neq \emptyset} \Sigma^{|v_*|} R^P$$

   of the iterated suspensions $\Sigma^n R^P$ of $R^P$ for all nondegenerate $v_* = (v_0, v_1, \cdots, v_n)$ where $0 \leq n \leq k$. (We exclude the empty set.) We let

   $$z(v_*) : R^P \to Z_k(e)$$

   denote the monomorphism $R^P \to Z_k(e)$ of degree $|v_*|$ onto the component $\Sigma^{|v_*|} R^P$ of $Z_k(e)$. (And let $z(\emptyset) = 0$.)

2. The boundary map $\partial$ on $Z_k(e)$ is given by the formula

   $$\begin{aligned}\partial z(v_0, \cdots, v_n) &= \textstyle\sum_{i=0}^n (-1)^i z(v_0, \cdots, v_i) e(v_i, \cdots, v_n) \\ &\quad - \textstyle\sum_{i=0}^n (-1)^i z(v_0, \cdots, \widehat{v}_i, \cdots, v_n)\end{aligned}$$

   (or $\partial z = z' * e - \delta z$ in the notation of 3.1.14 below.)

For example, when $n = 1$ and $(v_0, v_1) = (1, 0)$ we have

$$\partial z(1,0) = z(1)e(1,0) - z(1,0)e(0) + z(1) - z(0)$$

or

$$\partial z(1,0) + z(1,0)\partial = z(1)E(1,0) - z(0),$$

i.e., $z(1,0)$ is a homotopy

$$z(1,0) : z(0) \simeq z(1)E(1,0).$$

**Remark 3.1.4.** If we write the generators of $Z_k(e)$ as

$$c(v_0, \cdots, v_j) = z(v_0, \cdots, v_j)c$$

where $c \in P$ then the boundary map on $Z_k(e)$ is given by

$$\partial c_\beta(v_0, \cdots, v_j) = \partial z(v_*)c_\beta = (z' * e)(v_*)c_\beta - \delta z(v_*)c_\beta$$

$$= \sum_{i=0}^{j}(-1)^i z(v_0, \cdots, v_i)e(v_i, \cdots, v_j)c_\beta - \sum_{i=0}^{j}(-1)^i z(v_0, \cdots, \widehat{v_i}, \cdots, v_j)c_\beta$$

$$= \sum_{i=0}^{j}(-1)^i c_\alpha(v_0, \cdots, v_i)e_{\alpha\beta}(v_i, \cdots, v_j) - \sum_{i=0}^{j}(-1)^i c_\beta(v_0, \cdots, \widehat{v_i}, \cdots, v_j)$$

where $e_{\alpha\beta}(v_*) \in R$ is the coefficient of $c_\alpha$ in $e(v_*)c_\beta$. This formula can also be written:

$$\partial c_\beta(\sigma) = \sum_{i=0}^{j}(-1)^i c_\alpha(b_i(\sigma))e_{\alpha\beta}(f_{j-i}(\sigma)) - \sum_{i=0}^{j}(-1)^{j-i} c_\beta(\partial_i\sigma).$$

If $e \in \mathcal{M}_k(R, P)$ the poset $P$ determines a lattice of subcomplexes $Z_k^A(e)$ of $Z_k(e)$ which forms a crucial part of the structure of the multiple mapping cylinder.

**Definition 3.1.5.** If $A \subseteq P$ is *closed* in the sense that it contains any element of $P$ less than an element of $A$ then we define $Z_k^A(e)$ to be the subcomplex of $Z_k(e)$ generated by all $z(v_*)(a)$ where $a \in A$ and $v_* = (v_0, \cdots, v_n)$.

If we write $\Lambda P$ for the lattice of closed subsets of $P$ then we can write the above collection of complexes as

$$\{Z_k^A(e) \subseteq Z_k(e) \; : \; A \in \Lambda P\}.$$

We make the following trivial observations.

**Proposition 3.1.6.** *For all $A, B \in \Lambda P$ we have:*

1. $Z_k^{A\cap B} = Z_k^A(e) \cap Z_k^B(e)$

2. $Z_k^{A\cup B} = Z_k^A(e) + Z_k^B(e)$

3. $Z_k^{\emptyset}(e) = 0$

4. $Z_k^P(e) = Z_k(e)$.

*In other words, the assignment* $A \mapsto Z_k^A(e)$ *is a morphism of lattices (from* $\Lambda P$ *to the lattice of subcomplexes of* $Z_k(e)$*).*

**Theorem 3.1.7.** $Z_k(e)$ *is the mapping cylinder of the chain map*

$$F_k : Z_{k-1}(\partial_k e) \to C(k)$$

*given on generators* $z(v_*)$ *with* $k > v_0 > v_1 > \cdots > v_n \geq 0$ *by*

$$F_k \circ z(v_0, \cdots, v_n) = E(k, v_0, \cdots, v_n)$$

*where* $E = e$ *for* $n > 0$ *and* $E = 1 + e$ *when* $n = 0$.

Note that when $n = 0$, the map $F_k \circ z(v_0) = E(k, v_0) : C(v_0) \to C(k)$ is the chain isomorphism we talked about earlier. Since a mapping cylinder ($Z_k(e)$) is always homotopy equivalent to the target complex and $E(k, v_0) \simeq E(k, v_1)E(v_1, v_0)$ we have:

**Corollary 3.1.8.** *The inclusion maps*

$$z(v) : C(v) \to Z_k(e)$$

*are chain homotopy equivalences for all* $0 \leq v \leq k$ *and are related by* $z(v_0) \simeq z(v_1)E(v_1, v_0)$.

If $a : [j] \to [k]$ is a non-decreasing map then there is a chain map

$$a_* : Z_j(a^*(e)) \to Z_k(e)$$

given by $a_* \circ z(v_0, \cdots, v_n) = z(a(v_0), \cdots, a(v_n))$. By Corollary 3.1.8 this is a homotopy equivalence. Consequently, $Z$ gives a functor from the category of simplices of $\mathcal{M}_\bullet(R, P)$ to the category of based free $R$-complexes and chain homotopy equivalences. [Recall that the category of simplices simp $K$ of a simplicial set $K$ has objects $([n], x)$ where $n \geq 0$ and $x \in K_n$ and morphisms

$$a_* : ([n], x) \to ([m], y)$$

are given by non-decreasing maps $a : [n] \to [m]$ so that $x = a^*(y)$.]

Thus if $B$ is a simplicial set and $\xi : B \to \mathcal{M}_\bullet(R, P)$ is a simplicial map then $Z(\xi)$ is a functor from the category of simplices of $B$ to the category of quasi-isomorphisms of based free $R$-complexes. Composing with a homology functor we get the "homology bundle" associated to $\xi$. This is related to the "total complex" of $\xi$ by a Serre spectral sequence.

**Definition 3.1.9.** For $n \geq 0$ the $n$-th *homology bundle*

$$H_n(\xi) : \operatorname{simp} B \to iso\text{Mod-}R$$

is defined to be the composition of $Z(\xi)$ with the $n$-th homology functor, i.e., $H_n(\xi)(\sigma) = H_n(Z(\xi(\sigma)))$. Similarly, for any right $R$-module $M$ we have the cohomology bundle $H^n(\xi, M)$ defined by $H^n(\xi, M)(\sigma) = H^n(Z(\xi(\sigma)), M)$.

We note that a functor $\operatorname{simp} B \to iso\text{Mod-}R$ is equivalent to a locally trivial coefficient bundle of $R$-modules over the geometric realization $|B|$ of $B$. A contravariant functor $\operatorname{simp} B \to iso\text{Mod-}R$ will also work since $f \mapsto f^{-1}$ gives an isomorphism $iso\text{Mod-}R^{op} \cong iso\text{Mod-}R$. Thus the cohomology bundle $H^n(\xi, M)$ also gives a coefficient bundle (of abelian groups) over $B$.

**Definition 3.1.10.** Let $F$ be any functor from the category of simplices of $B$ to the category of $R$-complexes. Then the *total complex* $E(F)$ of $F$ is defined to be the quotient of

$$\bigoplus_{k \geq 0} \bigoplus_{x \in B_k} F(x)$$

given by identifying the elements $s \in F(a^*(x))$ with their images $a_*(s) \in F(x)$ for all $a : [j] \to [k]$.

Since every element of every simplicial set is uniquely expressible as a degeneracy of a nondegenerate element, the total complex of our functor $Z(\xi)$ is given by:

$$E(Z(\xi)) = \bigoplus \Sigma^n R^P$$

where the sum runs over all nondegenerate simplices $([n], x)$ of $B$. Imitating the notation for $Z_k(e)$, we denote the inclusion maps for the summands of $E(Z(\xi))$ as

$$z([n], x) : R^P \to E(Z(\xi)).$$

Then the boundary homomorphism is given by

$$\partial \circ z([n], x) = \sum_{p+q=n} (-1)^q z([q], b_q^*(x)) e([p], f_p^*(x))$$

$$- \sum_{i=0}^{n} (-1)^{n-i} z([n-1], \partial_i^*(x))$$

where $f_p : [p] \to [n]$, $b_q : [q] \to [n]$ are the front $p$-face ($f(i) = i$) and back $q$-face ($b(j) = p + j$) respectively and both $z, e$ are zero on degenerate simplices.

**Theorem 3.1.11.** *There are natural first quadrant spectral sequences with $E^2$-terms given by*

$$E^2_{p,q} = H_p(B, H_q(\xi)), \qquad E_2^{p,q} = H^p(B, H^q(\xi, M))$$

*converging to $H_{p+q}(E(Z(\xi)))$, $H^{p+q}(E(Z(\xi)), M)$ respectively.*

**Remark 3.1.12.** We will refer to this as the *Serre spectral sequence*. The same name will be applied to generalizations that we will give later.

*Proof.* A basis element of $E(Z(\xi))$ has *bidegree* $(p,q)$ if it has the form $z([p],x)(y)$ where $y \in P$ had degree $q$ and $x \in B_p^{nd}$ is nondegenerate $(nd)$. This gives a bifiltration of $E(Z(\xi))$ and gives a spectral sequence in the usual way, just as in the proof of the usual Serre spectral sequence.

To determine the $E^1$-term in the homology spectral sequence we look at the terms with simplicial degree $n$ in the formula for the boundary map in $E(Z(\xi))$:

$$\partial z([n],x) \equiv (-1)^n z([n],x)e(v_0(x)) \quad (mod\ E(Z(\xi))|B^{n-1})$$

where $v_0(x)$ is the front vertex of $x$. In other words, if $\partial x$ is the union of all of the proper faces of $x$ and $\overline{x}$ is the union of $x$ and $\partial x$ then $z([n],x)$ gives a chain isomorphism

$$E(Z(\xi))|\overline{x}/E(Z(\xi))|\partial x \cong \Sigma^n C(v_0(x)) \cong \Sigma^n Z_n(\xi(x))$$

Consequently, the $E^1$ term is:

$$E^1_{n,m} \cong \bigoplus_{x\in B_n^{nd}} \Sigma^n H_m(Z_n(\xi(x))) = C_n(B, H_m(\xi))$$

If we look at the terms of simplicial degree $n$ and $n-1$ in $\partial z([n],x)$ we see that the differential $d^1 : E^1_{n,m} \to E^1_{n-1,m}$ is the alternating sum of restriction maps to the faces of each $n$-simplex, so

$$E^2_{n,m} \cong H_n(B, H_m(\xi))$$

as claimed. The same bifiltration gives the cohomology spectral sequence. □

To prove Theorem 3.1.7 and other formulas it will be helpful to have some shorthand notation.

**Definition 3.1.13.** Let $f*g, \delta f, f'$ be defined by the following formulas whenever they make sense.

1. $(f*g)(v_0,\cdots,v_n) = \sum_{i=0}^n f(v_0,\cdots,v_i)g(v_i,\cdots,v_n)$
2. $\delta f(v_0,\cdots,v_n) = \sum_{i=0}^n (-1)^i f(v_0,\cdots,\widehat{v}_i,\cdots,v_n)$
3. $f'(v_0,\cdots,v_n) = (-1)^n f(v_0,\cdots,v_n)$

We note that the juxtaposition $\delta f$ does not denote composition of functions as it does in the other cases such as $\partial z = \partial \circ z$.

These symbols satisfy the following easy identities.

**Proposition 3.1.14.** *1. $*$ is associative: $(f*g)*h = f*(g*h)$*

*2. $-'$ distributes over $*$: $(f*g)' = f'*g'$*

3. $-'' = id$, *i.e.*, $(f')' = f$.

4. $\delta^2 = 0$, *i.e.*, $\delta\delta f = 0$.

5. $\delta$ *anticommutes with* $-'$: $(\delta f)' = -\delta(f')$

6. $\delta$ *is a graded derivation:* $\delta(f * g) = \delta f * g + f' * \delta g$

We can use these identities to prove that $Z_k(e)$ is indeed a chain complex:

$$\begin{aligned} \partial\partial z &= \partial(z' * e - \delta z) = (\partial z)' * e - \delta(\partial z) \\ &= (z' * e)' * e - (\delta z)' * e - \delta(z' * e - \delta z) \\ &= z * e' * e - z * \delta e = z * 0 = 0 \end{aligned}$$

**Lemma 3.1.15.** *A chain map* $F : Z_k(e) \to D$ *of* $Z_k(e)$ *into any* $R$*-complex* $D$ *is given by* $F \circ z(v_*) = f(v_*) : R^P \to D$ *where* $f(v_*)$ *is a homogeneous map of degree* $|v_*|$ *for each nondegenerate* $v_*$ *so that*

$$\partial f = f' * e - \delta f.$$

*Proof.* We need to show that $F\partial \circ z = \partial F \circ z$.

$$\begin{aligned} F\partial \circ z &= F(z' * e - \delta z) = f' * e - \delta(F \circ z) \\ &= f' * e - \delta f = \partial f = \partial F \circ z \end{aligned}$$

□

*Proof of Theorem 3.1.7.* By the above lemma our homomorphism $F_k : Z_{k-1} \to C(k)$ is a chain map if the mappings $f(v_*) : R^P \to C(k)$ given by

$$f(v_*) = F_k \circ z(v_*) = E(k, v_0, \cdots, v_n)$$

satisfy the condition $f' * e = \partial f + \delta f$.

For $n = 0$ this is easy to verify: $(f' * e)(v_0) = f(v_0)e(v_0) = E(k, v_0)e(v_0) = e(k)E(k, v_0) = \partial E(k, v_0) = \partial f(v_0) = \partial f(v_0) + \delta f(v_0)$ since $\delta f(v_0) = 0$.

For $n \geq 1$ we have:

$$\begin{aligned} (f' * e)(v_*) &= f(v_0)e(v_0, \cdots, v_n) - f(v_0, v_1)e(v_1, \cdots, v_n) + \cdots \\ &= E(k, v_0)e(v_*) - e(k, v_0, v_1)e(v_1, \cdots, v_n) + \cdots \\ &= -(e' * e)(k, v_*) + e(k)e(k, v_*) + e(v_*) \end{aligned}$$

$$\begin{aligned} \delta f(v_*) &= \textstyle\sum_{i=0}^n (-1)^i f(v_0, \cdots, \widehat{v}_i, \cdots, v_n) \\ &= \textstyle\sum_{i=0}^n (-1)^i E(k, v_0, \cdots, \widehat{v}_i, \cdots, v_n) \\ &= \textstyle\sum_{i=0}^n (-1)^i e(k, v_0, \cdots, \widehat{v}_i, \cdots, v_n) \\ &= e(v_*) - (\delta e)(k, v_*) \end{aligned}$$

$$\begin{aligned} \partial f(v_*) &= \partial e(k, v_0, \cdots, v_n) \\ &= e(k)e(k, v_*) \end{aligned}$$

so $f' * e = \partial f + \delta f$ and $F_k$ is indeed a chain map.

The mapping cylinder of $F_k$ is given by $Z(F_k) = Z_{k-1}(e) \oplus \Sigma Z_{k-1}(e) \oplus C(k)$ with generators denoted by $(a, \sigma z(v_*)b, c) = a + z(k, v_*)b + z(k)c$ and boundary (for $n \geq 1$) given as it should be by:

$$\begin{aligned}\partial \sigma z(v_*) &= z(k)F_k z(v_*) - z(v_*) - \sigma \partial z(v_*) \\ &= z(k)e(k, v_0, \cdots, v_n) - (\sigma z' * e)(v_*) - z(v_*) + \sigma \delta z(v_*) \\ &= (z' * e)(k, v_0, \cdots, v_n) - \delta z(k, v_0, \cdots, v_n)\end{aligned}$$

The case $n = 0$ is not very different. □

## 3.2 Morphisms of multiple mapping cylinders

In this section we define morphisms of multiple mapping cylinders. We first give the intuitive definition and then we give the formal definition (3.2.5) which is much shorter and easier to work with. We begin with $k = 0$.

Suppose that $e_1 \in \mathcal{M}_0(R, P_1)$ and $e_2 \in \mathcal{M}_0(R, P_2)$. Then each $e_i$ represents the boundary map of a free $R$-complex with basis poset $P_i$. These chain complexes are called $Z_0(e_i)$ and each of these has a lattice of subcomplexes $Z_0^A(e_i)$ for all $A$ in the lattice $\Lambda P_i$ of closed subsets of $P_i$.

A *morphism* $f : e_2 \to e_1$ is simply a chain map

$$F : Z_0(e_2) \to Z_0(e_1).$$

Just as $e_1, e_2$ refer to the endomorphisms of $R^{P_i}$ giving the boundary maps of the two chain complexes, $f$ will formally refer to the $R$-module homomorphism $f : R^{P_2} \to R^{P_1}$ corresponding to $F$. In other words, $f$ is a degree 0 mapping satisfying the condition:

$$e_1 \circ f = f \circ e_2.$$

Let $\lambda : \Lambda P_2 \to \Lambda P_1$ be a mapping which commutes with finite unions (Definition 3.2.2(2) below). We say that the morphism $f : e_2 \to e_1$ is *subordinate* to $\lambda$ if the corresponding chain map $F$ sends $Z_k^A(e_2)$ to $Z_k^{\lambda A}(e_1)$ for all $A \in \Lambda P_2$. In terms of the homomorphism $f$ this can be written: $f(R^A) \subseteq R^{\lambda A}$.

This defines a category $\mathcal{M}_0(R)$ having objects all pairs $(e, P)$ where $e \in \mathcal{M}_0(R, P)$ and morphisms

$$(f, \lambda) : (e_2, P_2) \to (e_1, P_1)$$

where $f : e_2 \to e_1$ is a morphism subordinate to $\lambda : \Lambda P_2 \to \Lambda P_1$.

Before we go to $k = 1$ we discuss homotopies in the $k = 0$ case. Suppose that $f, g$ are two morphisms $e_2 \to e_1$ subordinate to the same mapping $\lambda : \Lambda P_2 \to \Lambda P_1$. Let $F, G : Z_0(e_2) \to Z_0(e_1)$ be the corresponding chain maps. Then a *homotopy* $h : f \simeq g$ is defined to be a degree 1 homomorphism $h : R^{P_2} \to R^{P_1}$ giving a chain homotopy $H : F \simeq G$, i.e., $h$ satisfies the condition:

$$e_1 \circ h + h \circ e_2 = f - g.$$

We say that $h$ is *subordinate* to $\lambda : \Lambda P_2 \to \Lambda P_1$ if $h(R^A) \subseteq R^{\lambda A}$ for all $A \in \Lambda P_2$.

Now we go to the case $k = 1$. Suppose that $e_i \in \mathcal{M}_1(R, P_i)$ for $i = 1, 2$. Then each $e_i$ represents two vertices $e_i(0), e_i(1)$ connected by an upper triangular chain map

$$E_i(1,0) : Z(e_i)(0) = (R^{P_i}, e_i(0)) \to (R^{P_i}, e_i(1)) = Z(e_i)(1)$$

given by $E_i(1,0) = 1 + e_i(1,0)$. A morphism $f : e_2 \to e_1$ consists of two chain maps $F(j) : Z(e_2)(j) \to Z(e_1)(j)$ for $j = 0, 1$ together with a chain homotopy

$$F(1,0) : E_1(1,0)F(0) \simeq F(1)E_2(1,0)$$

which we represent by the following diagram:

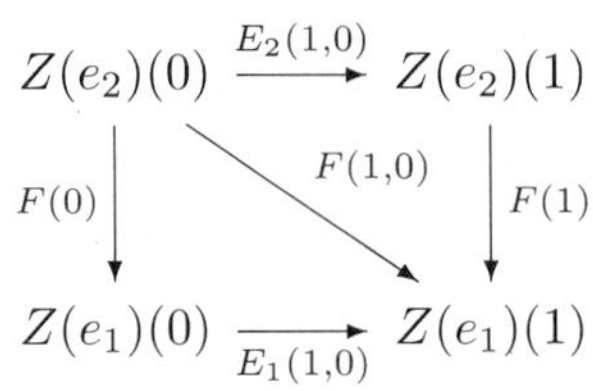


The mappings $F(0), F(1), F(1,0)$ together give a chain map on mapping cylinders:

$$F : Z_1(e_2) \to Z_1(e_1).$$

We say that $f$ is *subordinate* to a mapping $\lambda : \Lambda P_2 \to \Lambda P_1$ if this chain map $F$ sends $Z_1^A(e_2)$ into $Z_1^{\lambda A}(e_1)$ for all $A \in \Lambda P_2$.

Formally we define $f$ to be the triple of $R$-module homomorphisms

$$f(0), f(1), f(1,0) : R^{P_2} \to R^{P_1}$$

of degrees $0, 0, 1$ respectively representing the three mappings $F(0), F(1)$, $F(1,0)$. The conditions required of $F(0), F(1), F(1,0)$ (i.e., that $F(0), F(1)$ be chain maps and $F(1,0)$ be a chain homotopy as described above) can be summarized by the following single equation:

$$\delta f = f' * e_2 - e_1 * f. \tag{3.1}$$

In the case of $f(1,0)$ this gives:

$$f(0) - f(1) = [f(1)e_2(1,0) - f(1,0)e_2(0)] - [e_1(1)f(1,0) + e_1(1,0)f(0)]$$

which is the same as

$$f(1,0)\partial + \partial f(1,0) = f(1)E_2(1,0) - E_1(1,0)f(0)$$

since $E = 1 + e$.

In the case $k = 2$ a morphism $f : e_2 \to e_1$ consists of seven homomorphisms $f(i), f(i,j), f(2,1,0) : R^{P_2} \to R^{P_1}$ represented by the following homotopy commuting diagram.

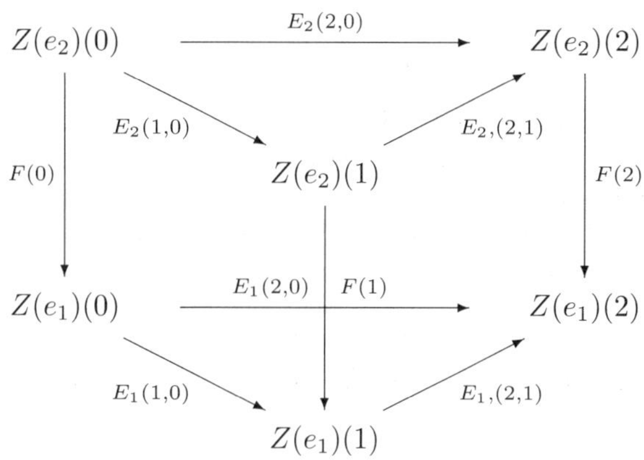


The five sides of this triangular prism have associated homotopies $e_2(2,1,0)$, $F(1,0)$, $F(2,0)$, $F(2,1)$ and $e_1(2,1,0)$. Equation (3.1) for $f(2,1,0)$ gives the following higher homotopy between these five homotopies.

$$\begin{aligned}\delta f(2,1,0) &= f(1,0) - f(2,0) + f(2,1)\\ &= f(2)e_2(2,1,0) - f(2,1)e_2(1,0) + f(2,1,0)e_2(0)\\ &\qquad - e_1(2)f(2,1,0) - e_1(2,1)f(1,0) - e_1(2,1,0)f(0)\end{aligned}$$

$$\begin{aligned}&\partial f(2,1,0) - f(2,1,0)\partial\\ &\quad = f(2)e_2(2,1,0) - f(2,1)E_2(1,0) - E_1(2,1)f(1,0) - e_1(2,1,0) + f(2,0)\end{aligned} \tag{3.2}$$

Equation (3.1) will be used as the formal definition of a morphism $e_2 \to e_1$. (See Definition 3.2.5 below.) However, the elementary description of these morphisms given above in the cases $k = 0, 1, 2$ give us the following important observation.

**Proposition 3.2.1.** (0) *Suppose that $e_i \in \mathcal{M}_0(R, P_i)$ for $i = 1, 2$ have projective homology. Then a morphism $f : e_2 \to e_1$ is given uniquely up to homotopy by a degree* 0 *map in homology $H_*(Z_0(e_2)) \to H_*(Z_0(e_1))$ (i.e., by a sequence of homomorphisms $H_n(Z_0(e_2)) \to H_n(Z_0(e_1))$).*

(1) *Suppose that $e_i \in \mathcal{M}_1(R, P_i)$ for $i = 1, 2$ have projective homology. Then two morphisms $f(0) : e_2(0) \to e_1(0)$ and $f(1) : e_2(1) \to e_1(1)$ extend (not uniquely) to a morphism $f : e_2 \to e_1$ if and only if the following diagram*

*commutes.*

$$\begin{array}{ccc} H_*(Z(e_2)(0)) & \xrightarrow{E_2(1,0)_*} & H_*(Z(e_2)(1)) \\ \downarrow{\scriptstyle f(0)_*} & & \downarrow{\scriptstyle f(1)_*} \\ H_*(Z(e_1)(0)) & \xrightarrow{E_1(1,0)_*} & H_*(Z(e_1)(1)) \end{array}$$

(2) *Suppose that* $e_i \in \mathcal{M}_2(R, P_i)$ *for* $i = 1, 2$ *have the property that the complex* $\mathrm{Hom}_R(Z_2(e_2), Z_2(e_1))$ *has trivial homology in degree* 1. *Then any system of six homomorphisms* $f(i), f(k,j) : R^{P_2} \to R^{P_1}$ *for* $j = 0, 1, 2$ *and* $(k, j) = (1,0), (2,0), (2,1)$ *satisfying (3.1) extends to a morphism* $f : e_2 \to e_1$, *(i.e.,* $f(2,1,0)$ *can be chosen to satisfy (3.2)).*

(3) *The conclusions in* (0), (1) *and* (2) *also hold with the projective homology assumption (resp.* $H_1(\mathrm{Hom}_R(Z_2(e_2), Z_2(e_1))) = 0$ *assumption) replaced by the assumption that* $e_2$ *has no homology below a fixed degree* $n$ *and* $e_1$ *has no homology above degree* $n$.

*Proof.* When a free chain complex has projective homology it is homotopy equivalent to its graded homology considered as a complex with zero boundary map. Consequently, a chain map between such complexes is uniquely determined by its induced map in homology which is arbitrary. Statements (0) and (1) follow from this observation.

To prove (2) we note that the right hand side of (3.1) is a degree 1 map $h : Z(e_2)(0) \to Z(e_1)(2)$ so that $\partial h + h\partial = 0$, i.e., a 1-cycle in the complex $\mathrm{Hom}_R(Z(e_2)(0), Z(e_1)(2)) \simeq \mathrm{Hom}_R(Z_2(e_2), Z_2(e_1))$. By assumption any such 1-cycle is the boundary of a 2-chain $f(2,1,0)$.

Finally, for (3) we note that if $C$ is a free $R$-complex with no homology below degree $n$ then $C$ contains as a deformation retract a projective subcomplex $E$ with the property that $E_i = 0$ for $i < n$. Consequently, if $D$ is another chain complex with no homology above degree $n$, then

$$H_0(\mathrm{Hom}_R(C, D)) \cong \mathrm{Hom}_R(H_n(C), H_n(D))$$

and $H_k(\mathrm{Hom}_R(C, D)) = 0$ for $k \geq 1$. In other words, a chain map $C \to D$ is uniquely determined up to chain homotopy by the induced map on $H_n$ which is arbitrary and all homotopies are unique up to higher homotopy. □

In Proposition 3.2.1 we ignored the subordination condition. However, in the future we will need to pay attention to this condition even when it is vacuous. We begin the formal discussion with our "backward" definition of a poset morphism.

**Definition 3.2.2.** If $P, Q$ are finite posets then we define a *poset morphism* $\alpha : P \to Q$ to be a rule which assigns to each $x \in P$ a closed subset (also called an "order ideal") $\alpha(x) \subseteq Q$ so that $\alpha(x) \subseteq \alpha(y)$ whenever $x \leq y$ in $P$. *Composition* of morphisms is given by $\alpha \circ \beta(x) = \bigcup_{y \in \beta(x)} \alpha(y)$. The *identity morphism* $\iota_P : P \to P$ is given by $\iota(x) = \overline{x}$ where $\overline{x}$ is the closure of $x$ in $P$.

Two equivalent formulations of this definition are as follows.

1. $\alpha : P \to Q$ is the same as an order preserving mapping from $P$ to the lattice $\Lambda Q$ of closed subsets of $Q$.

2. A morphism $\alpha : P \to Q$ is the same as a mapping $\lambda : \Lambda P \to \Lambda Q$ which preserves finite unions (in particular $\lambda(\emptyset) = \emptyset$). The correspondence is given by $\alpha(x) = \lambda(\overline{x})$ and

$$\lambda(C) = \alpha(C) = \bigcup_{x \in C} \alpha(x).$$

Note that an order preserving mapping $f : P \to Q$ induces a reverse morphism $\alpha : Q \to P$ by $\alpha(x) = f^{-1}(\overline{x})$. This is why out definition is "backwards." One can show, however, that the notion of poset isomorphism remains the same.

Two extreme examples of poset morphisms $P \to Q$ are the *minimal* and *maximal* morphisms $\epsilon, \omega$ given by $\epsilon(x) = \emptyset$ and $\omega(x) = Q$ for all $x \in P$. Some other examples are given below.

**Example 3.2.3.** ([IK95]) Suppose that $Q$ is a closed subset of a poset $P$. Then we get the following triangle of poset morphisms.

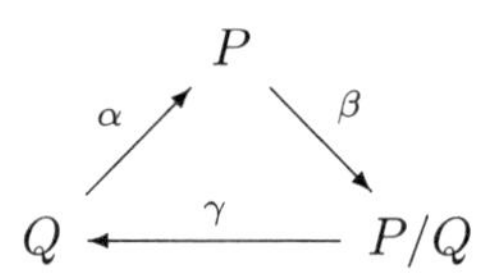


1. The inclusion morphism $\alpha : Q \to P$ given by $\alpha(x) = \overline{x}$.

2. The quotient morphism $\beta : P \to P/Q := P - Q$ given by $\beta(x) = \overline{x} - Q$.

3. A morphism $\gamma : P/Q \to Q$ given by $\alpha(x) = Q \cap \overline{x}$.

The partial ordering of $P$ is determined by the orderings on the subposet $Q$ and quotient poset $P/Q$ and the poset morphism $\gamma : P/Q \to Q$: If $x, y \in P$ both lie in $Q$ or both in $P - Q$ then they are ordered according to the partial ordering of $P$ or $P/Q$ respectively. If $x \in Q$ and $y \notin Q$ then $x < y$ if and only if $x \in \gamma(y)$. This inspires the following definition.

**Definition 3.2.4.** The *mapping cone* $C(\alpha)$ of a poset morphism $\alpha : P_0 \to P_1$ is defined to be the disjoint union $C(\alpha) = P_0 \coprod P_1$ with partial ordering given by $x > y$ if

1. $x, y \in P_1$ with $x > y$,

2. $x, y \in P_0$ with $x > y$ or

3. $x \in P_0, y \in P_1$ with $y \in \alpha(x)$.

In other words, the closed subsets of $C(\alpha)$ have the form $A \cup B$ where $A \in \Lambda P_0, B \in \Lambda P_1$ with $\alpha(A) \subseteq B$. If $P_0, P_1$ are finite graded posets then we increase the degrees of the elements of $P_0$ by 1. Thus the mapping cone $C(\alpha)$

of a morphism of graded posets $\alpha : P_0 \to P_1$ is given as a graded set by $C(\alpha) = \Sigma P_0 \coprod P_1$ where $\Sigma$ is the operation which increases the degrees of all elements by 1 with partial ordering given above. In the (ungraded) example above we note that $P = C(\gamma)$.

To make $C$ into a functor we need to say what it does to morphisms of arrow. First, we define a *morphism of arrows* $\gamma : \alpha \to \beta$ to be a square of poset morphisms:

$$\begin{array}{ccc} P_0 & \xrightarrow{\gamma_0} & Q_0 \\ {\scriptstyle\alpha}\downarrow & & \downarrow{\scriptstyle\beta} \\ P_1 & \xrightarrow{\gamma_1} & Q_1 \end{array}$$

having the property that $\beta \circ \gamma_0 \subseteq \gamma_1 \circ \alpha$, i.e., $\beta\gamma_0(x) \subseteq \gamma_1\alpha(x)$ for every $x \in P_0$. This includes the case when the diagram commutes. Then an induced poset morphism

$$C(\gamma) : C(\alpha) \to C(\beta)$$

is given by $C(\gamma)(y) = \gamma_1(y) \subseteq Q_1$ if $y \in P_1$ and $C(\gamma)(x) = \gamma_0(x) \cup \gamma_1\alpha(x)$ if $x \in P_0$.

**Definition 3.2.5.** Suppose that $P_1, P_2$ are two finite graded posets and, for the same $k$, $e_i \in \mathcal{M}_k(R, P_i)$. A *(general) morphism* $f : e_2 \to e_1$ is defined to be a family of $R$-homomorphisms $f(v_*) : R^{P_2} \to R^{P_1}$ of degree $|v_*|$ for all nondegenerate $v_*$ satisfying Equation (3.1) above. We say that $f$ is *subordinate* to a poset morphism $\alpha : P_2 \to P_1$ if $f(v_*)(x)$ is a linear combination of the elements of $\alpha(x)$ for all $x \in P_2$ and all nondegenerate $v_*$. We sometimes indicate this by writing the morphism as a pair:

$$(f, \alpha) : (e_2, P_2) \to (e_1, P_1)$$

We also write $\mathrm{Hom}_\alpha(e_2, e_1)$ for the group of all morphisms $e_2 \to e_1$ subordinate to $\alpha$. *Composition* of morphisms is defined by $(f, \alpha) \circ (g, \beta) = (f * g, \alpha \circ \beta)$. We will denote by $\mathcal{M}_\bullet(R)$ the simplicial category of systems of higher homotopies and general morphisms.

**Definition 3.2.6.** We will call $f : e_2 \to e_1$ a *simple morphism* if $f(v_*) = 0$ for $|v_*| \neq 0$, i.e., the only nontrivial cases are $f(v)$ where $v$ is a vertex of $\Delta^k$. In that case the formula $\delta f = f' * e_2 - e_1 * f$ is equivalent to the following three conditions (which can be summarized as $f * E_2 = E_1 * f$).

1. $f(v)e_2(v) = e_1(v)f(v)$, i.e.,

$$f(v) : C_2(v) \to C_1(v)$$

   is a chain map for all vertices $v$ in $\Delta^k$.

2. $f(v_0)E_2(v_0, v_1) = E_1(v_0, v_1)f(v_1)$, i.e., we have a commuting diagram of chain maps:

$$\begin{array}{ccc} C_2(v_1) & \xrightarrow{E_2(v_0,v_1)} & C_2(v_0) \\ {\scriptstyle f(v_1)}\downarrow & & \downarrow{\scriptstyle f(v_0)} \\ C_1(v_1) & \xrightarrow{E_1(v_0,v_1)} & C_1(v_0) \end{array}$$

for all $k \geq v_1 > v_0 \geq 0$. (For general morphisms this diagram is only required to commute up to the homotopy $f(v_0, v_1)$.)

3. $f(v_0)e_2(v_0, \cdots, v_n) = e_1(v_0, \cdots, v_n)f(v_n)$ for all admissible $v_*$.

Some examples of simple morphisms, all of which will be subordinate to *graded closed bijections* $\alpha : P_2 \to P_1$, i.e., morphisms given by order preserving, degree preserving bijections going the other way, are given as follows.

The simple morphism $(f, \alpha)$ will be called a *permutation morphism* if $\alpha : P_2 \to P_1$ is a graded closed bijection and $f(v)$ is the permutation $f(v)z(v)(x) = z(v)\alpha(x)$ for all $x \in P_2$. The *identity morphism* $(I, \iota_P)$ on $(e, P)$ is an example of a permutation morphism.

**Definition 3.2.7.** A *G-monomial morphism* (where $G$ is a subgroup of the group of units of $R$) is defined to be a simple morphism $f$ over a graded closed bijection $\alpha : P_2 \to P_1$ so that

$$f(v)z(v)x = z(v)\alpha(x)\gamma(x)$$

for all $x \in P_2$ where $\gamma : P_2 \to G$ is a set mapping independent of $v$. We note that $f$ is uniquely determined by $\alpha$ and $\gamma$ so we sometimes write $(f, \alpha) = (\alpha, \gamma)_*$. Permutation morphisms are special cases of monomial morphisms.

**Definition 3.2.8.** A *homotopy* $h$ between two morphisms $f, g : e_2 \to e_1$ which are subordinate to the same poset morphism $\alpha : P_2 \to P_1$ is defined to be a family of $R$-homomorphism $h(v_*) : R^{P_2} \to R^{P_1}$ which are homogeneous of degree $|v_*| + 1$, which are subordinate to $\alpha$ and which satisfy:

$$h' * e_2 + e_1' * h - \delta h = f - g$$

We let $[e_2, e_1]_\alpha$ be the quotient of $\mathrm{Hom}_\alpha(e_2, e_1)$ modulo homotopy subordinate to $\alpha$.

It is an easy exercise to show that:

1. Given a homotopy $h : f \simeq g$ and a morphism $F : e_1 \to e_0$, $F' * h$ is a homotopy $F * f \simeq F * g$.

2. Similarly, $h * G$ is a homotopy $f * G \simeq g * G$ if $G : e_3 \to e_2$.

3. Finally, for any family of $R$-homomorphisms $h(v_*)$ as above,

$$f = h' * e_2 + e_1' * h - \delta h$$

gives a morphism $f : e_2 \to e_1$ which is, evidently, null homotopic.

**Definition 3.2.9.** If $P$ is a finite graded poset let $\Sigma P$ be the same poset with the degrees of all elements increased by 1. Thus $\Sigma P = \{\sigma x : x \in P\}$ where $\sigma x < \sigma y$ in $\Sigma P$ if and only if $x < y$ in $P$. The *suspension functor*

$$\Sigma : \mathcal{M}_\bullet(R) \to \mathcal{M}_\bullet(R)$$

is the simplicial homotopy functor given by

1. $\Sigma(e) = -\sigma e' \sigma^{-1}$ on objects,
2. $\Sigma(f) = \sigma f' \sigma^{-1}$ on morphisms and
3. $\Sigma(h) = -\sigma h' \sigma^{-1}$ on homotopies.

Thus $\Sigma(e)(v_*)(\sigma x) = -\sigma e'(v_*)(x)$ and similarly for $\Sigma(f), \Sigma(h)$. We will usually suppress $\sigma$ from the notation and simply write $\Sigma(e) = -e'$, $\Sigma(f) = f'$, $\Sigma(h) = -h'$. It is easy to see that $\Sigma$ is a simplicial functor. When we say that it is a "homotopy functor" we simply mean that it preserves homotopies. Thus a homotopy $h : f \simeq g$ gives a homotopy $\Sigma(h) : \Sigma(f) \simeq \Sigma(g)$. We note that $\Sigma^{-1}(e, P)$ is defined whenever $P$ has no elements of degree 0.

**Example 3.2.10.** Suppose that $Q$ is a closed subset of a finite graded poset $P$ and $e \in \mathcal{M}_k(R, P)$. Then we get the *extension sequence*:

$$(e|Q, Q) \xrightarrow{(f,\alpha)} (e, P) \xrightarrow{(g,\beta)} (\overline{e}, P/Q) \xrightarrow{(h,\gamma)} \Sigma(e|Q, Q)$$

The poset morphisms $\alpha, \beta, \gamma$ are as in Example 3.2.3. $\Sigma(e|Q, Q) = (\Sigma(e), \Sigma Q)$ is defined above and the other terms are defined below.

(1) For each $v_*$ the endomorphism $e(v_*)$ of $R^P$ leaves $R^Q$ invariant so we get induced endomorphisms $(e|Q)(v_*)$ of $R^Q$ and $\overline{e}(v_*)$ of $R^{P/Q} \cong R^P/R^Q$.

(2) $f : e|Q \to e$ is the *inclusion morphism*, a simple morphism given by the inclusion map: $R^Q \to R^P$ when $|v_*| = 0$. Then $\delta f = 0, f' = f$ and the equation $f * (e|Q) = e * f$ is equivalent to the commutativity for all $v_*$ of the left hand square in the following diagram.

$$\begin{array}{ccccc} R^Q & \longrightarrow & R^P & \longrightarrow & R^{P/Q} \\ \downarrow{\scriptstyle (e|Q)(v_*)} & & \downarrow{\scriptstyle e(v_*)} & & \downarrow{\scriptstyle \overline{e}(v_*)} \\ R^Q & \longrightarrow & R^P & \longrightarrow & R^{P/Q} \end{array}$$

(3) $g : e \to \overline{e}$ is the *quotient morphism* given by

$$g(v_*) = \begin{cases} \text{quotient map} : R^P \to R^{P/Q} = R^P/R^Q & \text{if } |v_*| = 0 \\ 0 & \text{if } |v_*| \neq 0 \end{cases}$$

Then $\delta g = 0, g' = g$ and the equation $g * e = \overline{e} * g$ is equivalent to the commutativity of the right hand square in the above diagram.

(4) The third morphism $(h, \gamma) : (\bar{e}, P/Q) \to (\Sigma(e|Q), \Sigma Q)$ is given as follows. For each nondegenerate $v_*$ the endomorphism $e(v_*)$ of $R^P$, when written in matrix notation is

$$e(v_*) = \begin{pmatrix} (e|Q)(v_*) & h(v_*) \\ 0 & \bar{e}(v_*) \end{pmatrix}$$

Since $\delta e = e' * e$ we get

$$\begin{aligned} \begin{pmatrix} \delta(e|Q) & \delta h \\ 0 & \delta \bar{e} \end{pmatrix} &= \begin{pmatrix} e'|Q & h' \\ 0 & \bar{e}' \end{pmatrix} * \begin{pmatrix} e|Q & h \\ 0 & \bar{e} \end{pmatrix} \\ &= \begin{pmatrix} e'|Q * e|Q & (e'|Q) * h + h' * \bar{e} \\ 0 & \bar{e}' * \bar{e} \end{pmatrix} \end{aligned}$$

so we see that $\delta h = h' * \bar{e} + (e'|Q) * h = h' * \bar{e} - \Sigma(e|Q) * h$ making this a morphism $h : \bar{e} \to \Sigma(e|Q)$.

**Proposition 3.2.11.** (a) *A morphism $f : e_2 \to e_1$ determines a chain map*

$$F = Z_k(f) : Z_k(e_2) \to Z_k(e_1)$$

*by $F \circ z_2(v_*) = (z_1 * f)(v_*)$.*

(b) *$Z_k$ is a functor, i.e., $Z_k(f) \circ Z_k(g) = Z_k(f * g)$ and $Z_k(I)$ is the identity mapping.*

(c) *A homotopy $h : f \simeq g$ between two morphisms $f, g : e_2 \to e_1$ gives a chain homotopy*

$$H = Z_k(h) : Z_k(f) \simeq Z_k(g)$$

*by $H \circ z_2(v_*) = (z_1' * h)(v_*)$.*

*Proof.* For (a) we want to show that $F\partial \circ z_2 = \partial F \circ z_2$.

$$\begin{aligned} F\partial \circ z_2 &= F \circ (z_2' * e_2 - \delta z_2) \\ &= (F \circ z_2)' * e_2 - \delta(F \circ z_2) \\ &= (z_1 * f)' * e_2 - \delta(z_1 * f) \\ &= \underbrace{z_1' * f' * e_2 - z_1' * \delta f}\, - \delta z_1 * f \\ &= \qquad z_1' * e_1 * f - z_1' * e_1 * f + (\partial z_1) * f \\ &= \partial(z_1 * f) \\ &= \partial F \circ z_2 \end{aligned}$$

Note that the condition in Definition 3.2.5 is used only at the fifth step to show that the underbraced terms in line 4 give $z_1' * e_1 * f$ in line 5. Since $z_1(v_*)$ are monomorphisms with linearly independent images, the identity $F\partial = \partial F$ is equivalent to the condition in Definition 3.2.5. (b) is trivial.

A similar computation proves (c):

$$\begin{aligned} (\partial H + H\partial) \circ z_2 &= \partial(z_1' * h) + H \circ (z_2' * e_2 - \delta z_2) \\ &= (\partial z_1)' * h + (H \circ z_2)' * e_2 - \delta(H \circ z_2) \\ &= (z_1' * e_1)' * h - (\delta z_1)' * h + (z_1' * h)' * e_2 - \delta(z_1' * h) \\ &= z_1 * e_1' * h + z_1 * h' * e_2 - z_1 * \delta h \\ &= z_1 * (f - g) \\ &= (F - G) * z_2 \end{aligned}$$

□

## 3.3 Mapping cones

In analogy with Theorem 3.1.7 the mapping cone of the chain map $F = Z_k(f)$ in Proposition 3.2.11 is the chain complex associated to a system of higher homotopies $C(f)$ which we now define.

**Definition 3.3.1.** Let $f(v_*) : R^{P_2} \to R^{P_1}$ be a morphism $f : e_2 \to e_1$ subordinate to a poset morphism $\alpha : P_2 \to P_1$. Then the *mapping cone* $C(f)$ of $f$ is defined to be the system of higher homotopies

$$e_3(v_*) = \begin{pmatrix} e_1 & f' \\ 0 & -e_2' \end{pmatrix} (v_*) = \begin{pmatrix} e_1(v_*) & (-1)^{|v_*|} f(v_*) \\ 0 & (-1)^{1+|v_*|} e_2(v_*) \end{pmatrix}$$

on the graded poset $C(\alpha)$ which is the disjoint union of $P_1$ and a suspended copy $\Sigma P_2 = \{\sigma x : x \in P_2\}$ of $P_2$ with the elements of $P_1, \Sigma P_2$ ordered as before and $x < \sigma y$ for all $y \in P_2, x \in \alpha(y) \subseteq P_1$. Thus if elements of $R^{C(\alpha)} = R^{P_1} \oplus \Sigma R^{P_2}$ are denoted $(a, \sigma b)$ where $a \in R^{P_1}$ and $b \in R^{P_2}$ then

$$e_3(v_*)(a, \sigma b) = (e_1(v_*)a + f'(v_*)b, -\sigma e_2'(v_*)b).$$

In the special case when the target is zero we get the suspension of the domain: $\Sigma(e) = C(0 : e \to 0)$ is given by $\Sigma(e)(v_*)\sigma x = -\sigma e'(v_*)x$ or $\Sigma(e) = -\sigma e' \sigma^{-1}$.

**Proposition 3.3.2.** (a) *$C(f)$ is a system of higher homotopies for $C(\alpha)$ over $R$*

(b) *The closed subset $P_1$ in $C(\alpha)$ gives the extension sequence:*

$$(e_1, P_1) \rightarrowtail (C(f), C(\alpha)) \twoheadrightarrow \Sigma(e_2, P_2) \xrightarrow{\Sigma(f,\alpha)} \Sigma(e_1, P_1).$$

*Conversely, any extension sequence*

$$(e|Q, Q) \xrightarrow{(f,\alpha)} (e, P) \xrightarrow{(g,\beta)} (\bar{e}, P/Q) \xrightarrow{(h,\gamma)} \Sigma(e|Q, Q)$$

*in which $P/Q$ contains no elements of degree 0 (i.e., is desuspendable) is permutation isomorphic to a sequence as above. (In particular, $(e, P) \cong (C(\Sigma^{-1}(h)), C(\Sigma^{-1}(\gamma)))$ by a permutation isomorphism.)*

(c) *We have a natural chain isomorphism*

$$Z_k(C(f)) \cong C(Z_k(f))$$

*between the k-fold mapping cylinder $Z_k(C(f))$ of $C(f)$ and the mapping cone $C(F)$ of $F = Z_k(f)$.*

(d) *Given a morphism of arrows, $h : f \to g$, i.e., a commuting diagram in the category $\mathcal{M}_k(R)$:*

$$\begin{array}{ccc} e_2 & \xrightarrow{h_2} & d_2 \\ {\scriptstyle f}\downarrow & & \downarrow{\scriptstyle g} \\ e_1 & \xrightarrow{h_1} & d_1 \end{array}$$

*(so that $h_1 * f = g * h_2$), there is an induced morphism of mapping cones:*

$$h_1 \oplus \Sigma(h_2) = \begin{pmatrix} h_1 & 0 \\ 0 & h_2' \end{pmatrix} : C(f) \to C(g)$$

*making $C$ into a simplicial functor $Ar(\mathcal{M}_\bullet(R)) \to \mathcal{M}_\bullet(R)$.*

*Proof.* This is very straightforward. To check (a) we need to verify that $e_3' * e_3 = \delta e_3$. But

$$e_3' * e_3 = \begin{pmatrix} e_1 & f' \\ 0 & -e_2' \end{pmatrix}' * \begin{pmatrix} e_1 & f' \\ 0 & -e_2' \end{pmatrix} = \begin{pmatrix} e_1' * e_1 & -f * e_2' + e_1' * f' \\ 0 & e_2 * e_2' \end{pmatrix}$$

which is equal to $\delta e_3$ since $\delta f = f' * e_2 - e_1 * f$. This proves (a). To prove (b) just compare the definition of the mapping cone with the definition of the extension sequence. (d) is also straightforward.

Finally, we want to show that $Z_k(e_3)$ is the mapping cone of $F = Z_k(f) : Z_k(e_2) \to Z_k(e_1)$. The generators of $Z_k(e_3)$ are the images of

$$z_3(v_*) = \begin{pmatrix} z_1 & 0 \\ 0 & -z_2' \end{pmatrix} (v_*) : R^{P_1} \oplus \Sigma R^{P_2} \to Z_k(e_3)$$

So the boundary map $\partial_3$ on $Z_k(e_3)$ is given by

$$\begin{aligned} \partial_3 z_3 &= z_3' * e_3 - \delta z_3 \\ &= \begin{pmatrix} z_1 & 0 \\ 0 & -z_2' \end{pmatrix}' * \begin{pmatrix} e_1 & f' \\ 0 & -e_2' \end{pmatrix} - \begin{pmatrix} \delta z_1 & 0 \\ 0 & (\delta z_2)' \end{pmatrix} \\ &= \begin{pmatrix} z_1' * e_1 - \delta z_1 & z_1' * f' \\ 0 & (z_2' * e_2 - \delta z_2)' \end{pmatrix} \\ &= \begin{pmatrix} \partial_1 z_1 & -(F \circ z_2)' \\ 0 & (\partial_2 z_2)' \end{pmatrix} \\ &= \begin{pmatrix} \partial_1 & F \\ 0 & -\partial_2 \end{pmatrix} \begin{pmatrix} z_1 & 0 \\ 0 & -z_2' \end{pmatrix} \end{aligned}$$

In other words, the boundary map is given by

$$\partial_3 = \begin{pmatrix} \partial_1 & F \\ 0 & -\partial_2 \end{pmatrix}$$

which is the same as the boundary map for the mapping cone of $F$. □

**Corollary 3.3.3.** *The extension sequence in the proposition above gives a long exact sequence in homology:*

$$\cdots \to H_n(Z_k(e_2) \to H_n(Z_k(e_1)) \to H_n(Z_k(C(f)))$$

$$\to H_{n-1}(Z_k(e_2)) \to \cdots \to H_0(Z_k(C(f))) \to 0$$

There is also a long exact homotopy sequence:

**Theorem 3.3.4.** *Suppose that $P_i$ is a finite graded poset and $e_i \in \mathcal{M}_k(R, P_i)$ for $i = 0, 1, 2$. Let $(f, \alpha) : (e_2, P_2) \to (e_1, P_1)$ be any morphism. Then the sequence of morphisms*

$$e_2 \xrightarrow{f} e_1 \xrightarrow{j} C(f) \xrightarrow{q} \Sigma(e_2) \xrightarrow{\Sigma f} \Sigma(e_1) \to \cdots$$

*induces a long exact homotopy sequence:*

$$[e_0, e_2]_{\sigma^{-1}\bar{\beta}} \xrightarrow{f_*} [e_0, e_1]_{\beta \cap P_1} \xrightarrow{j_*} [e_0, C(f)]_{\beta} \xrightarrow{q_*}$$

$$[e_0, \Sigma(e_2)]_{\bar{\beta}} \xrightarrow{\Sigma f_*} [e_0, \Sigma(e_1)]_{\sigma(\beta \cap P_1)} \to \cdots$$

*for any poset morphism $\beta : P_0 \to C(\alpha)$ where $\bar{\beta} : P_0 \to \Sigma P_2$ is the composition of $\beta$ with the quotient morphism $C(\alpha) \to \Sigma P_2$ and $\beta \cap P_1 : P_0 \to P_1$ is defined by $(\beta \cap P_1)(x) = \beta(x) \cap P_1$.*

*Proof.* We will first prove exactness at $[e_0, e_1]$. Since $j * f = 0$ the composition $j_* \circ f_* = 0$. Thus it suffices to show that $\ker j_* \subseteq \operatorname{im} f_*$. So suppose that $g \in \operatorname{Hom}_{\beta \cap P_1}(e_0, e_1)$ and $h$ is a null homotopy of $j * g$ subordinate to $\beta$. Then

$$h' * e_0 + C(f)' * h - \delta h = j * g$$

In matrix notation this is:

$$\begin{pmatrix} h_1 \\ h_2 \end{pmatrix}' * e_0 + \begin{pmatrix} e_1 & f'\sigma^{-1} \\ 0 & -\sigma e_2' \sigma^{-1} \end{pmatrix}' * \begin{pmatrix} h_1 \\ h_2 \end{pmatrix} - \delta \begin{pmatrix} h_1 \\ h_2 \end{pmatrix} = \begin{pmatrix} I \\ 0 \end{pmatrix} * g = \begin{pmatrix} g \\ 0 \end{pmatrix}$$

This gives two equations:

1. $h_1' * e_0 + e_1' * h_1 + f * \sigma^{-1} h_2 - \delta h_1 = g$
2. $h_2' * e_0 - \sigma e_2 * \sigma^{-1} h_2 - \delta h_2 = 0$

The second equation, after left multiplying by $\sigma^{-1}$, says that $\sigma^{-1} h_2$ is a morphism $e_0 \to e_2$. This morphism is obviously subordinate to $\sigma^{-1}\bar{\beta}$, so $[\sigma^{-1} h_2] \in [e_0, e_1]_{\sigma^{-1}\bar{\beta}}$. The first equation says that $h_1$ gives a homotopy $g \simeq f * \sigma^{-1} h_2$ subordinate to $\beta \cap P_1$. Thus $[g] = f_*[\sigma^{-1} h_2]$ as needed.

Now we show exactness at $[e_0, C(f)]$. Since $q_* \circ j_* = 0$ it suffices to show that $\ker q_* \subseteq \operatorname{im} j_*$. So let $g \in \operatorname{Hom}_{\beta}(e_0, C(f))$ and let $h$ be a null homotopy of $q * g$ subordinate to $\bar{\beta}$. In matrix notation, $q = (0, I), g = (g_1, g_2)^t$ and $q * g = g_2$. Thus we have

$$h' * e_0 + (-\sigma e_2' \sigma^{-1}) * h - \delta h = g_2$$

Then:

$$\begin{pmatrix} 0 \\ h \end{pmatrix}' * e_0 + \begin{pmatrix} e_1 & f'\sigma^{-1} \\ 0 & -\sigma e_2' \sigma^{-1} \end{pmatrix} * \begin{pmatrix} 0 \\ h \end{pmatrix} - \delta \begin{pmatrix} 0 \\ h \end{pmatrix} = \begin{pmatrix} f' * \sigma^{-1} h \\ g_2 \end{pmatrix}$$

so $(0, h)^t$ gives a homotopy

$$\begin{pmatrix} 0 \\ h \end{pmatrix} : \begin{pmatrix} g_1 \\ g_2 \end{pmatrix} \simeq \begin{pmatrix} g_1 - f' * \sigma^{-1} h \\ 0 \end{pmatrix} = \begin{pmatrix} I \\ 0 \end{pmatrix} * (g_1 - f' * \sigma^{-1} h)$$

so $[g] = j_*[g_1 - f' * \sigma^{-1}h]$ proving exactness at the second point.

The remaining case, $[e_0, \Sigma e_2]$, is similar and left to the reader. □

Finally, we need to show that when two morphisms are homotopic their mapping cones are chain isomorphic and simplicially homotopic. This will follow from the following lemma.

**Lemma 3.3.5.** *Let $e_0, e_1$ be elements of $\mathcal{M}_k(R,P)$ and suppose that $f : e_0 \to e_1$ is a morphism which is subordinate to $\iota_P$ so that $X = f - I$ is strictly upper triangular, i.e., $X$ is subordinate to the deleted closure morphism $\beta(x) = \overline{x} - \{x\}$. Then:*

(a) *There is an inverse morphism $g : e_1 \to e_0$ also subordinate to $\iota_P$ so that $g - I$ is strictly upper triangular.*

(b) *$f$ induces a simplicial homotopy*

$$f_* : \Delta^k \times I \to \mathcal{M}_\bullet(R,P)$$

*from $e_0$ to $e_1$. Furthermore, $g, f_*$ are natural with respect to simplicial operators*

$$a^* : \mathcal{M}_k(R) \to \mathcal{M}_j(R)$$

*and with respect to simple morphisms of pairs.*

**Remark 3.3.6.** (1) A morphism $f$ as above will be called *upper triangular*. By definition this will mean that $f - I$ is strictly upper triangular.

(2) By a *simple morphism of pairs*

$$s : (f : e_0 \to e_1) \to (g : d_0 \to d_1)$$

we mean a pair of simple morphisms $s_i : e_i \to d_i$ over the same poset morphism $\alpha : P \to Q$ so that $s_1 * f = g * s_0$.

*Proof.* (a) is easy. Since $f = I + X$ where $X$ is nilpotent we have

$$f^{-1} = I - X + X * X - X * X * X + \cdots$$

To prove (b) we need to construct a natural system of higher homotopies over every simplex in $\Delta^k \times I$ which agree with $e_i$ on $\Delta^k \times i$.

The $n$-simplices in $\Delta^k \times I$ are given by order preserving mappings

$$a : [n] \to [k] \times [1].$$

To match the earlier notation we write: $a(i) = (v_{n-i}, b_{n-i})$. Then $v_*, b_*$ are nonincreasing sequences $k \geq v_0 \geq v_1 \geq \cdots \geq v_n \geq 0$ and $1 \geq b_0 \geq b_1 \geq \cdots \geq b_n \geq 0$. The system of homotopies $e_2$ is given on this simplex by

$$e_2(a(n), \cdots, a(0)) = b_0 e_1(v_*) + (1 - b_0) e_0(v_*)$$

$$+ \sum_{i=0}^{n} (b_0 - b_i) E_1(v_0, \cdots, v_i) X(v_i, \cdots, v_n)$$

Since the summation term is zero when $b_*$ is constant this is clearly a simplicial homotopy from $e_0$ to $e_1$. To see that it is a system of higher homotopies we apply the formulas in Proposition 3.1.14 after writing $e_2$ in index-free notation:

$$e_2 = b * e_1 + (I - b) * e_0 + b * e_1 * X - e_1 * b * X - \delta b * X$$

where $b = b_0 I$ and $X = f - I$.

A simple morphism of pairs $(s_1, s_0) : (e_1, e_0) \to (d_1, d_0)$ induces a simple morphism $s_2 : e_2 \to d_2$ by

$$s_2(v, b) = bs_1(v) + (1 - b)s_0(v)$$

i.e., $s_2 = s_i$ on $\Delta^k \times i$ for $i = 0, 1$, and clearly these constructions are all natural with respect to face and degeneracy operations. □

**Proposition 3.3.7.** *Let $f, g : e_2 \to e_1$ be morphisms subordinate to the same poset morphism $\alpha : P_2 \to P_1$ and let $h : f \simeq g$ be a homotopy subordinate to $\alpha$.*

(a) *We have an upper triangular isomorphism*

$$C(h) : C(f) \to C(g)$$

*subordinate to the identity morphism on the poset $C(\alpha)$.*

(b) *We have a simplicial homotopy:*

$$C(h)_* : \Delta^k \times I \to \mathcal{M}_\bullet(R)$$

*from $C(f)$ to $C(g)$.*

*Furthermore, $C(h)$ and $C(h)_*$ are natural with respect to simplicial operators $a^* : \mathcal{M}_k(R) \to \mathcal{M}_j(R)$ and with respect to pairs of simple morphisms.*

The proof below explains what we mean by "naturality with respect to a pair of simple morphisms."

*Proof.* Let

$$C(h) = \begin{pmatrix} I & h' \\ 0 & I \end{pmatrix}$$

Then $C(h)$ is a morphism $C(h) : C(f) \to C(g)$ subordinate to the identity on $C(\alpha)$. To verify this we simply expand the required equation $\delta C(h) = C(h)' * C(f) - C(g) * C(h)$ in matrix form:

$$\delta \begin{pmatrix} I & h' \\ 0 & I \end{pmatrix} = \begin{pmatrix} I & h' \\ 0 & I \end{pmatrix}' * \begin{pmatrix} e_1 & f' \\ 0 & -e_2' \end{pmatrix} - \begin{pmatrix} e_1 & g' \\ 0 & -e_2' \end{pmatrix} * \begin{pmatrix} I & h' \\ 0 & I \end{pmatrix}$$

In the $(1, 2)$-entry this is the equation:

$$-(\delta h)' = f' - h * e_2' - e_1 * h' - g'$$

which holds since $h$ is a homotopy $h : f \simeq g$. Since $C(h)$ is clearly upper triangular, our proposition follows from Lemma 3.3.5.

For the naturality condition suppose that we also have $f_0, g_0 : d_2 \to d_1$ and $h_0 : f_0 \simeq g_0$ subordinate to $\beta : Q_2 \to Q_1$ and a pair of simple morphisms $(s_i, \gamma_i) : (e_i, P_i) \to (d_i, Q_i)$ so that $s_1 * f = f_0 * s_2$, $s_1 * g = g_0 * s_2$, $s_1 * h = h_0 * s_2$ and $\beta\gamma_1 = \gamma_2\alpha$. Then we have induced simple morphisms $C(s)_f : C(f) \to C(f_0)$ and $C(s)_g : C(g) \to C(g_0)$ both of which are given by $s_1 \oplus \Sigma(s_2)$. Since $s_i' = s_i$ we have

$$\begin{pmatrix} s_1 & 0 \\ 0 & s_2 \end{pmatrix} * \begin{pmatrix} I & h' \\ 0 & I \end{pmatrix} = \begin{pmatrix} I & h_0' \\ 0 & I \end{pmatrix} * \begin{pmatrix} s_1 & 0 \\ 0 & s_2 \end{pmatrix}$$

or

$$C(s)_g * C(h) = C(h_0) * C(s)_f$$

so $C(s) = (C(s)_g, C(s)_f)$ is a simple morphism of pairs $C(h) \to C(h_0)$ and therefore by Lemma 3.3.5 induces a simple morphism $C(h)_* \to C(h_0)_*$ given to be $C(s)_f$ on $\Delta^k \times 0$ and $C(s)_g$ on $\Delta^k \times 1$. □

## 3.4 Morphisms of bundles and the twisted homology bundle

In this section we will look at the problem of constructing morphisms $f : \xi \to \eta$ between two "families of chain complexes." The objects we consider here arise from families of Morse functions, as opposed to the generalized Morse functions we consider elsewhere. If $E \to B$ is a smooth bundle with compact manifold fibers and $f : E \to \mathbb{R}$ is a smooth function which is Morse on every fiber then the number of critical points of $f$ restricted to the fibers is a locally constant function on $B$. For any generic triangulation of $B$ with small simplices, the critical set will be a finite graded poset over each simplex and the partial ordering over a simplex will be refined when restricted to its faces. Thus the reverse mapping (the inclusion of the face into the simplex) gives a closed bijection F of posets. This gives a functor from the category of small generic simplices in $B$ to the category of finite posets. The more difficult task of constructing families of higher homotopies over this poset functor will be discussed later. Here we will give definitions inspired by, but independent of, Morse theory and work purely algebraically.

The set-up is as follows. We take the base $B$ to be a fixed simplicial set. Let $P$ be a functor

$$P : \operatorname{simp} B \to gcb\mathcal{P}$$

from the category of simplices of $B$ to the subcategory $gcb\mathcal{P}$ of the category of finite graded posets consisting of the graded closed bijections. Thus $P$ assigns a finite graded poset $P(\sigma)$ to each $\sigma \in B_k$ and a graded morphism $\pi_{\sigma\tau} : P(\tau) \to P(\sigma)$ induced by an order preserving, degree preserving bijection $P(\sigma) \to P(\tau)$ for all faces (or degeneracies) $\tau$ of $\sigma$. By a morphism $\alpha : P \to Q$ of such poset functors we mean a family of poset morphisms $\alpha_\sigma : P(\sigma) \to Q(\sigma)$ having the property that for all faces (or degeneracies) $\tau \subseteq \sigma$ ($\tau \to \sigma$) we get an induced (vertical) morphism of (horizontal) arrows:

$$\begin{array}{ccc} P(\tau) & \xrightarrow{\alpha_\tau} & Q(\tau) \\ {\scriptstyle \pi^P_{\sigma\tau}} \downarrow & & \downarrow {\scriptstyle \pi^Q_{\sigma\tau}} \\ P(\sigma) & \xrightarrow{\alpha_\sigma} & Q(\sigma) \end{array}$$

i.e, $\alpha_\sigma \pi^P_{\sigma\tau}(A) \subseteq \pi^Q_{\sigma\tau}\alpha_\tau(A)$ for all $A \in \Lambda P(\tau)$. This definition has the property that the mapping cone $C(\alpha)$ given by

$$C(\alpha)(\sigma) = C(\alpha_\sigma : P(\sigma) \to Q(\sigma))$$

is a functor which fits into a sequence of morphisms

$$P \xrightarrow{\alpha} Q \to C(\alpha) \to \Sigma P \xrightarrow{\Sigma\alpha} \Sigma Q \to \cdots$$

We also note that:

**Proposition 3.4.1.** *Given any two poset functors $P, Q$ on $B$ and any two extensions $Q \rightarrowtail E_0 \twoheadrightarrow P, Q \rightarrowtail E_1 \twoheadrightarrow P$ the functors $E_0, E_1$ are simplicially homotopic in the sense that there is a poset functor $F$ on $B \times I'$, where $I'$ is a subdivision of $I$, which is equal to $E_i$ on $B \times i$ for $i = 0, 1$. In particular, the simplicial homotopy class of the mapping cone $C(\alpha)$ depends only on the domain and range of $\alpha$.*

**Remark 3.4.2.** We needed to subdivide the interval $I$ in order to make simplicial homotopy into an equivalence relation. In the proof below we show that $E_1, E_2$ are both simplicially homotopic to a third extension and this defines a poset functor on $B \times I'$ where $I'$ is $I$ subdivided into two subintervals at a third vertex.

*Proof.* Since every extension is the desuspension of a mapping cone it suffices to show that $C(\alpha)$ is independent of $\alpha$ up to simplicial homotopy. We do this by comparing $C(\alpha)$ with $C(\omega)$ where $\omega$ is the maximal morphism $\omega : P \to Q$ given by $\omega(x) = Q$ for all $x \in P$. This case, however, follows from the fact that the natural set mapping $C(\alpha)_\sigma \to C(\omega)_\sigma$ is an order preserving bijection for all $\sigma \in B$ giving a natural transformation $C(\omega) \to C(\alpha)$. This gives a simplicial homotopy $C(\omega) \simeq C(\alpha)$ by taking the trivial homotopy

$$C(\alpha) \circ \pi_B : \operatorname{simp}(B \times I) \to \operatorname{simp} B \to gcb\mathcal{P}$$

and replacing it over $B \times 0$ with $C(\omega) \times 0$. □

Given a poset functor $P$ as above, let $\xi$ be a functor which assigns to each $\sigma \in B_k$ an element

$$\xi(\sigma) \in \mathcal{M}_k(R, P(\sigma))$$

and to each face or degeneracy $\tau \to \sigma$ a $G$-monomial morphism

$$g_{\sigma\tau} : \xi(\tau) \to \xi(\sigma)|\tau$$

subordinate to $\pi_{\sigma\tau}$ so that $(g_{\sigma\tau}|\rho) \circ g_{\tau\rho} = g_{\sigma\rho}$ for all $\rho \subseteq \tau \subseteq \sigma$ ($\rho \to \tau \to \sigma$ in general). We say that $\xi$ is a *$G$-monomial functor*

$$\xi : \operatorname{simp} B \to \operatorname{simp} \mathcal{M}_\bullet(R)$$

*subordinate* to the functor $P$.

If $\eta : \operatorname{simp} B \to \operatorname{simp} \mathcal{M}_\bullet(R)$ is a $G$-monomial functor subordinate to the functor $Q$ then a *morphism* $f : \xi \to \eta$ subordinate to $\alpha : P \to Q$ is defined to be a family of general morphisms $f_\sigma : \xi(\sigma) \to \eta(\sigma)$ subordinate to $\alpha_\sigma : P(\sigma) \to Q(\sigma)$ which are compatible with face and degeneracy operators in the sense that the following diagram commutes for all $\tau \to \sigma$.

$$\begin{array}{ccc} \xi(\tau) & \xrightarrow{f_\tau} & \eta(\tau) \\ {\scriptstyle g^\xi_{\sigma\tau}}\downarrow & & \downarrow{\scriptstyle g^\eta_{\sigma\tau}} \\ \xi(\sigma)|\tau & \xrightarrow{f_\sigma|\tau} & \eta(\sigma)|\tau \end{array}$$

In particular $f_\sigma$ is determined on each faces of $\sigma$ by

$$f_\sigma|\tau = g^\eta_{\sigma\tau} \circ f_\tau \circ (g^\xi_{\sigma\tau})^{-1}$$

We also get an induced map $C(f_\tau) \to C(f_\sigma)$.

**Definition 3.4.3.** If $f : \xi \to \eta$ is a morphism, we define the *fiberwise mapping cone* $C(f)$ to be the $G$-monomial functor subordinate to $C(\alpha)$ given by $C(f)(\sigma) = C(f_\sigma)$.

It follows from Proposition 3.3.7 that, up to simplicial homotopy, $C(f)$ depends only on the fiber homotopy class of $f$ where by a *fiber homotopy* $h : f \simeq g$ we mean a family of homotopies $h_\sigma : f_\sigma \simeq g_\sigma$ compatible with face and degeneracy operations. We assume that all three maps, $f, g, h$ are subordinate to the same morphism $\alpha : P \to Q$.

**Theorem 3.4.4.** *If $h : f \simeq g$ then $C(h)_* : C(f) \simeq C(g)$ is a $G$-monomial simplicial homotopy. In this case, there is a $G$-monomial functor*

$$\mathit{simp}(B \times I) \to \operatorname{simp} \mathcal{M}_\bullet(R)$$

*which is equal to $C(f)$ at one end and $C(g)$ at the other and which is subordinate to $C(\alpha) \circ \pi_B$ where $\pi_B : B \times I \to B$ is the projection map.*

**Remark 3.4.5.** In general a *$G$-monomial simplicial homotopy* is given by a $G$-monomial functor on $B \times I'$ where $I'$ is a subdivision of $I$. As in Remark 3.4.2, this makes $G$-monomial simplicial homotopy into an equivalence relation.

For each $G$-monomial functor $\xi : \operatorname{simp} B \to \operatorname{simp} \mathcal{M}_\bullet(R)$ the $G$-monomial morphisms $g_{\sigma\tau}$ induce chain isomorphisms

$$Z(g_{\sigma\tau}) : Z(\xi(\tau)) \cong Z(\xi(\sigma))|\tau.$$

Consequently, we get a well-defined *homology bundle* $H_n(\xi)$ over $B$ for each $n \geq 0$ given as before by $H_n(\xi)(\sigma) = H_n(Z(\xi(\sigma)))$.

Note that the homology bundle $H_n(\xi)$ is not the same as the homology $H_n(E_\xi)$ of the total complex $E_\xi = E(Z(\xi))$ of the functor $Z(\xi)$. They are related by the Serre spectral sequence (Theorem 3.1.11) whose statement and proof are the same as before.

The question we are interested in is: When does a morphism of homology bundles $H_n(\xi) \to H_n(\eta)$ lift to a morphism $f : \xi \to \eta$? In particular we are interested in the case when $H_*(\xi) \cong H_*(\eta)$ in which case the fiberwise mapping cone $C(f)$ will be acyclic. The local case is given by the following lemma which will be proved later. (Special cases were also addressed in Proposition 3.2.1.)

**Lemma 3.4.6.** *Let $P_1, P_2$ be finite graded posets. Let $e_1 \in \mathcal{M}_k(R, P_1)$, $e_2 \in \mathcal{M}_k(R, P_2)$ be systems of higher homotopies. Suppose also that*

$$H^n(Z(e_2), H_{n+k-1}(Z(e_1))) = 0$$

*for all $n \geq 0$. Then any morphism $f : e_2 \to e_1$ which is defined over $\partial\Delta^k$ can be extended to $\Delta^k$. In other words, if $f(v_*)$ are defined for $|v_*| < k$ then it can be defined for all $v_*$.*

*Furthermore, if we assume that $H^n(Z(e_2), H_{n+k}(Z(e_1))) = 0$ for all $n \geq 0$ and if $f, g$ are two morphisms $e_2 \to e_1$ and $h : f \simeq g$ is a homotopy defined over $\partial\Delta^k$ then $h$ can be extended to $\Delta^k$.*

This solves our problem in an important special case.

**Proposition 3.4.7.** *Suppose that $\xi, \eta : \operatorname{simp} B \to \operatorname{simp} \mathcal{M}_\bullet(R)$ are $G$-monomial functors and $n \geq 0$ is a fixed integer so that*

1. *$H_i(\xi) = 0$ for $i < n$*
2. *$H_j(\eta) = 0$ for $j > n$.*

*Let $\phi : H_n(\xi) \to H_n(\eta)$ be a morphism of coefficient bundles over $B$ (i.e., $\phi$ is a family of compatible homomorphisms $\phi_\sigma : H_n(\xi)(\sigma) \to H_n(\eta)(\sigma)$.) Then there exists a morphism $f : \xi \to \eta$ which induces $\phi$ on $H_n$. Furthermore, $f$ is unique up to fiber homotopy.*

*The relative version also holds. Namely, any morphism $\xi|A \to \eta|A$ inducing $\phi$ over a simplicial subset $A$ of $B$ extends to $B$ and any two such extensions are fiber homotopic relative to $A$. Consequently, taking $B \times I$ relative to its two ends, the fiberwise cone $C(f)$ is unique up to $G$-monomial simplicial homotopy.*

**Remark 3.4.8.** (a) Suppose that $B$ is connected with base point $v_0$ and $A$ is empty. Then the fundamental group $\pi_1 B = \pi_1(|B|, v_0)$ acts on the homology of

$\xi$ and $\eta$ at $v_0$ and a morphism $\phi$ of coefficient bundles over $B$ is equivalent to a homomorphism of (right) $R[\pi_1 B]$-modules. Consequently the proposition says that any homomorphism of $R[\pi_1 B]$-modules

$$\phi_{v_0} : H_n(\xi)(v_0) \to H_n(\eta)(v_0)$$

is induced by a morphism $f : \xi \to \eta$ which is unique up to fiber homotopy.

(b) Since the poset morphism is not mentioned we are using the *maximal morphism* $\omega : P \to Q$ given by $\omega(\sigma)(x) = Q(\sigma)$ for all $\sigma \in B$ and all $x \in P(\sigma)$. Then all morphisms and fiber homotopies will be subordinate to $\omega$.

*Proof.* For each vertex $v \in B_0 - A_0$ we are given a homomorphism

$$\phi_v : H_n(\xi)(v) \to H_n(\eta)(v).$$

By Proposition 3.2.1 there is a chain map $f_v : Z_0(\xi(v)) \to Z_0(\eta(v))$ unique up to homotopy inducing the map $\phi_v$ on $H_n$.

For each edge $\sigma \in B_1 - A_1$ with endpoints $v, w$ we have a commuting diagram whose horizontal arrows are isomorphisms

$$\begin{array}{ccccc} H_n(\xi)(v) & \xrightarrow{\cong} & H_n(\xi)(\sigma) & \xleftarrow{\cong} & H_n(\xi)(w) \\ \downarrow{\scriptstyle \phi_v} & & \downarrow{\scriptstyle \phi_\sigma} & & \downarrow{\scriptstyle \phi_w} \\ H_n(\eta)(v) & \xrightarrow{\cong} & H_n(\eta)(\sigma) & \xleftarrow{\cong} & H_n(\eta)(w) \end{array}$$

Consequently, the following diagram commutes up to a chain homotopy say $h$ which is unique up to higher homotopies.

$$\begin{array}{ccc} Z_0(\xi(v)) & \xrightarrow{E^\xi(w,v)} & Z_0(\xi(w)) \\ \downarrow{\scriptstyle f_v} & & \downarrow{\scriptstyle f_w} \\ Z_0(\eta(v)) & \xrightarrow{E^\eta(w,v)} & Z_0(\eta(w)) \end{array}$$

By Proposition 3.2.1 this implies that $f_v, f_w$ extend to a morphism $f_s : \xi(\sigma) \to \eta(\sigma)$.

By Lemma 3.4.6 there is no obstruction to constructing the higher homotopies $f_\tau$ for $\deg \tau \geq 2$. Also the uniqueness up to homotopy need only be checked over the vertices. □

We use Proposition 3.4.7 in the following situation. Suppose that $B$ is connected with base point $v_0$ and that $\pi_1 B$ acts trivially on $H_n(\xi)(v_0)$ for all $n \geq 0$. Suppose also that $H_n(\xi)(v_0)$ is free with a chosen basis $Q_n$ for each $n \geq 0$. Then each homology bundle $H_n(\xi)$ is an untwisted bundle: $H_n(\xi)(\sigma) \cong R^{Q_n}$ for all $\sigma \in B$. And these can be combined to form a constant $G$-monomial functor

$$H(\xi) : \operatorname{simp} B \to \operatorname{simp} \mathcal{M}_\bullet(R)$$

subordinate to the constant poset functor $Q$ defined as $Q = \coprod_{n\geq 0} Q_n$ with partial ordering inverse to degree (i.e., $x < y$ when $\deg x > \deg y$).

**Theorem 3.4.9 (Killing constant free homology).** *Suppose that $\xi$ is a $G$-monomial functor whose homology bundle is untwisted and free with basis poset $Q$ as described above. (In other words, there is a fixed isomorphism*

$$\phi_n : R^{Q_n} \cong H_n(\xi)(\sigma)$$

*compatible with face and degeneracy operators for all $\sigma \in B$ and all $n \geq 0$.) Then there exist:*

1. *a $G$-monomial functor $\widetilde{H}(\xi, Q) : \operatorname{simp} B \to \operatorname{simp} \mathcal{M}_\bullet(R)$ subordinate to $Q$ and*

2. *a morphism $f : \widetilde{H}(\xi, Q) \to \xi$ inducing an isomorphism of homology bundles:*
$$H_n(\widetilde{H}(\xi, Q)) \cong H_n(\xi)$$
*which agrees with $\phi_n$ at each $\sigma \in B$.*

*Furthermore, $\widetilde{H}(\xi, Q)$ and the fiberwise mapping cone $C(f)$ of $f$ are unique up to upper triangular isomorphism. Also, given $\widetilde{H}(\xi, Q)$, $f$ is unique up to fiber homotopy. These existence and uniqueness statements also hold relative to any simplicial subset $A$ of $B$. Consequently, taking $B \times I$ relative to its ends, $\widetilde{H}(\xi, Q)$ and $C(f)$ are determined up to $G$-monomial simplicial homotopy by the $G$-monomial simplicial homotopy class of $\xi$. Finally, $\widetilde{H}(\xi, Q)$, $f$ and $C(f)$ are natural with respect to ring homomorphisms.*

**Remark 3.4.10.** (a) We will denote $C(f)$ by $C(\xi, Q)$ and call it the *fiberwise cone* of $\xi$ with respect to the basis $Q$. The functor $\widetilde{H}(\xi, Q)$ will be called the *twisted homology bundle* of $\xi$.

(b) This theorem is easily generalized to the projective case. If $\xi$ is a $G$-monomial functor whose homology bundle is constant and projective then we can simply add a fixed free complex $\eta$ whose homology is projective and complementary to the homology of $\xi$. Then $\xi \oplus \eta$ will have constant free homology and the theorem will apply. (See 3.5.12 below.)

(c) This theorem could also be generalized to the case where $\pi_1(B) = \pi_1(|B|, v_0)$ acts on $H_n(\xi)(v_0) \cong R^{Q_n}$ by $G$-monomial matrices. However, this is not a useful generalization since it is not clear how this depends on the choice of basis (Lemma 3.5.3).

(d) Another more useful generalization is the extension to the case when $\pi_1(B)$ acts on $H_n(\xi)(v_0)$ by upper triangular matrices. This is done in Theorem 3.4.11 below.

(e) It follows from a comparison of Serre spectral sequences that $f : \widetilde{H}(\xi, Q) \to \xi$ induces a quasi-isomorphism of total complexes:

$$E(Z(\widetilde{H}(\xi, Q))) \simeq E(Z(\xi)).$$

*Proof.* We construct $\widetilde{H}(\xi, Q)$ and $C(f)$ by downward induction on degree. We begin by letting $m \geq 0$ be maximal so that $H_m(\xi)(v_0) \neq 0$. Let

$$\widetilde{H}_{(m)}(\xi) : \operatorname{simp} B \to \operatorname{simp} \mathcal{M}_\bullet(R)$$

be the unique constant functor subordinate to $Q_{(m)} = Q_m$ and let

$$f_{(m)} : \widetilde{H}_{(m)}(\xi) \to \xi$$

be a morphism which induces the isomorphism

$$(f_{(m)})_*(\sigma) = \phi_m(\sigma) : H_m(\widetilde{H}_{(m)}(\xi))(\sigma) = R^{Q_m} \cong H_m(\xi)(\sigma)$$

for each $\sigma \in B$. Then $f_{(m)}$ exists, agrees with the given morphism on $A$ $(f_{(m)}|A = (f|A)|\widetilde{H}_{(m)}(\xi))$ and is unique up to fiber homotopy rel $A$ by Proposition 3.4.7.

Now take any $m \geq 0$ and suppose that

1. $Q_{(m)} = \coprod_{n \geq m} Q_n$ with partial ordering inverse to degree,

2. $\widetilde{H}_{(m)}(\xi) : \operatorname{simp} B \to \operatorname{simp} \mathcal{M}_\bullet(R)$ is a $G$-monomial functor subordinate to $Q_{(m)}$ and

3. $f_{(m)} : \widetilde{H}_{(m)}(\xi) \to \xi$ is a morphisms inducing the given isomorphism $\phi_n$ in degree $n$ homology for all $n \geq m$.

4. $\widetilde{H}_{(m)}(\xi)$ and $f_{(m)}$ agree with what is given over $A$. $(\widetilde{H}_{(m)}(\xi) = \widetilde{H}(\xi)|Q_{(m)}$ and $f_{(m)} = f|\widetilde{H}_{(m)}(\xi)$.)

Suppose also that $\widetilde{H}_{(m)}(\xi)$ and $C(f_{(m)})$ are unique up to upper triangular isomorphism and $f_{(m)}$ is unique up to fiber homotopy. Then we want to extend $\widetilde{H}_{(m-1)}(\xi)$ and $f_{(m-1)}$, which are already given over $A$, to all of $B$ . If $Q_{m-1}$ is empty then $Q_{(m-1)} = Q_{(m)}$ and we can take $\widetilde{H}_{(m-1)}(\xi) = \widetilde{H}_{(m)}(\xi)$, $f_{(m-1)} = f_{(m)}$. Therefore we may assume that $Q_{m-1} \neq \emptyset$.

By condition (3) the homology bundle of $C(f_{(m)})$ is zero in degrees $\geq m$. Therefore, by Proposition 3.4.7 there is a morphism

$$\eta : H_{m-1}(\xi) \to C(f(m))$$

unique up to fiber homotopy inducing the isomorphism $\phi_{m-1}$ on $H_{m-1}$. Compose this with the quotient morphism

$$\gamma : C(f_{(m)}) \to \Sigma \widetilde{H}_{(m)}(\xi)$$

and we get a morphism

$$\gamma \circ \eta : H_{m-1}(\xi) \to \Sigma \widetilde{H}_{(m)}(\xi)$$

which is subordinate to the maximal poset morphism $\omega : Q_{m-1} \to Q_{(m)}$ which has mapping cone $C(\omega) = \Sigma Q_{(m-1)}$. We let

$$\widetilde{H}_{(m-1)}(\xi) = \Sigma^{-1} C(\gamma \circ \eta).$$

This is a $G$-monomial functor over the constant poset functor $Q_{(m-1)}$. Since the cone of $\eta$ fits into an extension sequence

$$\xi \rightarrowtail C(\eta) \twoheadrightarrow C(\gamma \circ \eta)$$

we see that $C(\eta)$ is the fiberwise mapping cone of a morphism

$$f_{(m-1)} : \widetilde{H}_{(m-1)}(\xi) = \Sigma^{-1} C(\gamma \circ \eta) \to \xi$$

which is unique up to fiber homotopy. ($f_{(m-1)}$ is uniquely determined by $f_{(m)}$ and $\gamma$ which are only unique up to fiber homotopy.) This completes the induction. We take $\widetilde{H}(\xi, Q) = \widetilde{H}_{(0)}(\xi), f = f_{(0)}$. □

The theorem and proof above can be generalized to the situation where the fiber homology is filtered and the associated graded homology is constant and free. More precisely, we assume that $H_n(\xi)(v_0)$ is free for each $n \geq 0$ with a partially ordered basis $P_n$ and that $\pi_1(B) = \pi_1(|B|, v_0)$ acts on $H_n(\xi)(v_0)$ by upper triangular transformations, i.e., the action of each element of $\pi_1(B)$ sends each basis element in $P_n$ is to itself plus a linear combination of lower elements. Let $P = \coprod P_n$ with each element of degree $n$ being less than each element of lower degree. Under these conditions we say that $\xi$ has an *upper triangular homology bundle* with *basis poset* $P$ at $v_0$. ($P$ is a *homology basis poset* for $\xi$. $P$ is always finite.)

**Theorem 3.4.11 (Killing upper triangular homology).** *Suppose that $\xi$ is a $G$-monomial functor on $B$ so that $H_n(\xi)(v_0)$ is free with basis $P_n$ and $\pi_1 B$ acts on $H_n(\xi)(v_0)$ by upper triangular transformations. Then there exists*

1. *a $G$-monomial functor*

$$\widetilde{H}(\xi, P) : \operatorname{simp} B \to \operatorname{simp} \mathcal{M}_\bullet(R)$$

*subordinate to $P$ and*

2. *a morphism $f : \widetilde{H}(\xi, P) \to \xi$ inducing an isomorphism of homology bundles:*

$$H_n(\widetilde{H}(\xi, P)) \cong H_n(\xi)$$

*which is the identity map of the common basis $P_n$ of $H_n(\widetilde{H}(\xi, P))(v_0)$ and $H_n(\xi)(v_0)$.*

*Furthermore, $\widetilde{H}(\xi, P)$ and $C(f)$ are unique up to upper triangular isomorphism and they are determined up to $G$-monomial simplicial homotopy by the $G$-monomial simplicial homotopy class of $\xi$.*

*Proof.* As in the proof of Theorem 3.4.9, it suffices to construct the fiberwise mapping cone $C(f)$ since this has $H_n(\widetilde{H}(\xi, P))$ as a quotient with $f$ as the attaching map. The construction is by induction of the size of $P$.

Let $m \geq 0$ be maximal so that $P_m$ is nonempty and let $S$ be the set of minimal elements of $P_m$. ($S$ is also the set of minimal elements of the entire poset $P$.) Then the elements of $S \subset H_m(\xi)(v_0)$ are fixed under the action of $\pi_1 B$ so there is an $R[\pi_1 B]$-homomorphism

$$\phi : R^S \to H_m(\xi)(v_0)$$

which is the identity on $S$. (Here $R^S$ has the trivial $\pi_1 B$ action.) Let $\eta$ be the unique constant functor ($\eta(\sigma) = R^S$) subordinate to the constant poset functor $S$. Then by Proposition 3.4.7 there is a morphism $f_S : \eta \to \xi$ realizing the map $\phi$ and $f_S$ is unique up to fiber homotopy.

Take the fiberwise mapping cone $C(f_S)$. Then $C(f)$ is obtained by induction from $C(f_S)$ using the filtered basis $P - S$ which is smaller than $P$. □

We end this section by stating, without proof, the dual versions of Proposition 3.4.7 and Theorem 3.4.11, which follow from the same technical lemma (3.4.6). And we give a proof of this lemma.

**Proposition 3.4.12.** *Suppose that* $\xi, \eta : \operatorname{simp} B \to \operatorname{simp} \mathcal{M}_\bullet(R)$ *are G-monomial functors and* $n \geq 0$ *is a fixed integer so that*

1. $H_i(\xi) = 0$ *for* $i < n$
2. $H_j(\eta) = 0$ *for* $j \neq n$.

*Let* $\phi : H_n(\xi) \to H_n(\eta)$ *be a morphism of coefficient bundles over B (i.e.,* $\phi$ *is a family of compatible homomorphisms* $\phi_\sigma : H_n(\xi(\sigma)) \to H_n(\eta(\sigma))$*.) Then there exists a morphism* $f : \xi \to \eta$ *which induces* $\phi$ *on* $H_n$*. Furthermore, f is unique up to fiber homotopy and natural with respect to ring homomorphisms. The relative version also holds.*

**Theorem 3.4.13.** *Suppose that* $\xi$ *is a G-monomial functor with upper triangular homology bundle and homology basis poset P at* $v_0$ *as in Theorem 3.4.11 above. Then there exist:*

1. *a G-monomial functor* $\widetilde{H}'(\xi, P) : \operatorname{simp} B \to \operatorname{simp} \mathcal{M}_\bullet(R)$ *subordinate to P and*
2. *a morphism* $f : \xi \to \widetilde{H}'(\xi, P)$ *inducing an isomorphism of homology bundles:*
$$H_n(\xi) \cong H_n(\widetilde{H}'(\xi, P))$$
*which is the identity on* $P_n$ *at* $v_0$.

*Furthermore,* $\widetilde{H}'(\xi, P)$ *and the fiberwise mapping cone* $C(f)$ *of f are unique up to upper triangular isomorphism. Also, given* $\widetilde{H}'(\xi, P)$*, f is unique up to fiber homotopy and natural with respect to ring homomorphisms. Finally* $\widetilde{H}'(\xi, P)$ *and* $C(f)$ *are uniquely determined up to G-monomial simplicial homotopy by the G-monomial simplicial homotopy class of* $\xi$ *(and P.)*

**Remark 3.4.14.** The composition

$$\widetilde{H}(\xi, P)(v_0) \to \xi(v_0) \to \widetilde{H}'(\xi, P)(v_0)$$

is by construction the identity mapping on the basis poset $P$. It is also an isomorphisms of $R[\pi_1 B]$-modules. Consequently, it is an upper triangular isomorphism subordinate to the identity poset morphism $P \to P$. By Lemma 3.3.5, it is invertible. Therefore, composing with the inverse of this isomorphism, we may assume that $\widetilde{H}'(\xi, P)$ of the above theorem is the same as the twisted homology bundle $\widetilde{H}(\xi, P)$ given by Theorem 3.4.11. We also note that, by this argument, the existence clauses of Theorems 3.4.11 and 3.4.13 imply the uniqueness of $\widetilde{H}(\xi, P)$ up to upper triangular isomorphism.

*Proof of Lemma 3.4.6.* Recall first that a morphism $f : e_2 \to e_1$ is by definition a family of $R$-homomorphisms $f(v_*) : R^{P_2} \to R^{P_1}$ of degree $|v_*|$ for all nondegenerate $v_*$ so that

$$e_1 * f - f' * e_2 + \delta f = 0$$

We are given $f(v_*)$ for all $|v_*| < k$. We want to choose $f(v_*)$ for the *top cell* $v_* = (k, k-1, \cdots, 1, 0)$ so that

$$(e_1 * f - f' * e_2 + \delta f)(v_*) = 0.$$

However, many of the terms in this expression are already defined. The undefined terms are $e_1(v_0)f(v_*) - (-1)^k f(v_*)e_2(v_k)$. If we call the sum of the other terms $\Delta(v_*)$ then we are trying to solve the following equation for the undefined term $f(v_*)$.

$$e_1(v_0)f(v_*) - (-1)^k f(v_*)e_2(v_k) = -\Delta(v_*)$$

Here $v_0 = k$ and $v_k = 0$ but we will keep the generic notation for consistency.

*Claim*:

$$e_1(v_0)\Delta(v_*) + (-1)^k \Delta(v_*)e_2(v_k) = 0$$

Suppose for a moment that this is true. Then $\Delta(v_*)$ gives a chain map

$$\Delta(v_*) : \Sigma^{k-1} C_2(v_k) \to C_1(v_0)$$

The mapping $f(v_*)$ that we are looking for is thus any null homotopy of $\Delta(v_*)$. But all chain maps $\Delta(v_*)$ are null homotopic if the given condition

$$H^n(C_2(v_k), H_{n+k-1}(C_1(v_0))) = 0$$

holds for all $n$. By Corollary 3.1.8, this is equivalent to the assumption of the lemma.

The homotopy $h : f \simeq g$ can be constructed with a similar argument in one higher degree. So we just have to prove the claim.

*Proof of Claim:* To simplify the notation we will add the two undefined terms $e_1(k)f(v_*) - (-1)^k f(v_*)e_2(0)$ back into $\Delta(v_*)$. We view $f(v_*)$ as a variable quantity when $v_*$ is equal to the top cell $(k, \cdots, 0)$. In the expression in our claim we will then be adding four variable terms which cancel since $e_i(v_j)^2 = 0$:

$$\begin{gathered} e_1(v_0)[e_1(v_0)f(v_*) - (-1)^k f(v_*)e_2(v_k)] \\ +(-1)^k[e_1(v_0)f(v_*) - (-1)^k f(v_*)e_2(v_k)]e_2(v_k) = 0 \end{gathered}$$

Therefore the statement of the claim is unchanged. Let $D(v_*)$ be given by

$$D(v_*) = z_1(v_0)\Delta(v_*)$$

Then $D(v_*)$ has two variable terms and, since $z_1(v_0)$ is a monomorphism, it suffices to show that the quantity

$$\partial D(v_*) + (-1)^k D(v_*)e_2(v_0) = e_1(v_0)[e_1(v_0)\Delta(v_*) + (-1)^k \Delta(v_*)e_2(v_k)]$$

is zero. However, according to the proof of Proposition 3.2.11 we have

$$D(v_*) = [F\partial \circ z_2 - \partial F \circ z_2](v_*)$$

where $F : Z_k(e_2) \to Z_k(e_1)$ is given by $F \circ z_2 = z_1 * f$. This is variable on the top cell. Now we compute using the fact that $\partial^2 = 0$:

$$\begin{aligned} \partial D(v_*) &= (\partial F \partial z_2)(v_*) = (\partial F - F\partial)(\partial z_2)(v_*) \\ &= (\partial F - F\partial)(z_2' * e_2 - \delta z_2)(v_*) \\ &= (\partial F - F\partial)\textstyle\sum_{i=0}^k (-1)^i z_2(v_0, \cdots, v_i)e_2(v_i, \cdots, v_k) \\ &-(\partial F - F\partial)\textstyle\sum_{i=0}^k (-1)^i z_2(v_0, \cdots, \widehat{v_i}, \cdots, v_k) \end{aligned}$$

At this point we use the fact that $\partial F = F\partial$ over $\partial\Delta^k$, i.e., $\partial F z_2(w_*) = F\partial z_2(w_*)$ if $|w_*| < n$. So, everything is zero except for the $i = k$ term in the first sum and we get:

$$\partial D(v_*) = (-1)^k(\partial F - F\partial)z_2(v_*)e_2(v_k) = (-1)^k D(v_*)e_2(v_k)$$

as claimed. □

We only used the assumption of the lemma once in the above proof when we used the fact that any chain map $\Delta(v_*) : \Sigma^{k-1}C_2(v_k) \to C_1(v_0)$ is null homotopic if $H^n(C_2(v_k), H_{n+k-1}C_1(v_0)) = 0$. However, the relative version of this statement is also true. Namely, if we are given that

$$H^n(C_2(v_k)/C_0(v_k), H_{n+k-1}C_1(v_0)) = 0$$

for all $n \geq 0$ then any chain contraction of a chain map $\Delta(v_*) : \Sigma^{k-1}C_2(v_k) \to C_1(v_0)$ on the subcomplex $\Sigma^{k-1}C_0(v_k)$ can be extended to $\Sigma^{k-1}C_2(v_k)$. More precisely the condition is that the $k-1$-st homology of the homomorphism (Hom) complex $\mathrm{Hom}(C_2(v_k)/C_0(v_k), C_1(v_0))$ should be zero.

In case this condition is not satisfied we can precisely measure the obstruction by noting that $-\Delta(v_*)$ is a $(k-1)$-cocycle in this Hom complex and that the desired morphism $f : e_2 \to e_1$ exists if and only if this cocycle is a coboundary. Furthermore, different choices of $f$ differ by a $k$-cocycle in this Hom complex. Consequently, we obtain the following refinement of Lemma 3.4.6.

**Proposition 3.4.15.** *Let $P_1, P_2$ be finite graded posets and let $P_0$ be a closed subset of $P_2$. Let $e_i \in \mathcal{M}_k(R, P_i)$ be systems of higher homotopies so that $e_0 = e_2|P_0$. Suppose that we are given a morphism $f : e_2 \to e_1$ defined over $\partial\Delta^k$ and an extension of $f|e_0$ to $\Delta^k$. (In different words, $f$ is defined on $\partial\Delta^k \times P_2 \cup \Delta^k \times P_0$.) Then there exists a well defined cohomology class*

$$o(f) \in H^{k-1}(\mathrm{Hom}(Z(e_2)/Z(e_0), Z(e_1)))$$

*which is equal to zero if and only if $f$ extends to $\Delta^k \times P_2$. Also, any two such extensions, say $f, g$, will differ by a $k$-cocycle whose cohomology class*

$$\theta(f, g) \in H^k(\mathrm{Hom}(Z(e_2)/Z(e_0), Z(e_1)))$$

*is arbitrary (i.e., takes all values for any fixed $f$ or $g$).*

*Furthermore, if $f$ and $g$ are two morphisms $e_2 \to e_1$ defined over $\partial\Delta^k \times P_2 \cup \Delta^k \times P_0$ which agree only on $(\Delta^k)^{k-2} \times P_2 \cup \Delta^k \times P_0$ where $(\Delta^k)^{k-2} \subseteq \partial\Delta^k$ is the $(k-2)$-skeleton of $\Delta^k$, then the obstruction cocycles of $f, g$ differ by an alternating sum of the difference cocycles of $f, g$ restricted to the codimension 1 faces of $\Delta^k$. In fact:*

$$o(f) - o(g) = -\sum_{i=0}^{k} (-1)^i \theta(f_i, g_i)$$

*where $f_i, g_i$ are the restrictions of $f, g$ to the $i$-th face of $\Delta^k$.*

**Remark 3.4.16.** The *obstruction cocycle* is $-\Delta(v_*)$ which, by Corollary 3.1.8, makes the obstruction class equal to:

$$o(f) = -[z_1(v_0)][\Delta(v_*)][z_2(v_k)]^{-1}.$$

The *difference cocycle* is $f(v_*) - g(v_*)$ making

$$\theta(f, g) = [z_1(v_0)][f(v_*) - g(v_*)][z_2(v_k)]^{-1}.$$

*Proof.* The last statement follows from the formula:

$$\Delta(v_*) = E_1(v_0, v_1) f(v_1, \cdots, v_k) + \sum_{i=1}^{k-1} (-1)^i f(v_0, \cdots, \hat{v}_i, \cdots, v_k)$$
$$+ (-1)^k f(v_0, \cdots, v_{k-1}) E_2(v_{k-1}, v_k) + \textit{lower terms}.$$

□

Proposition 3.4.15 gives us the usual obstruction theory for constructing morphisms of bundles over a base $B$ with obstructions in the cohomology of $B$ with coefficients in the cohomology of the appropriate Hom complex. We will not formulate this in general. We only need the following special case.

**Corollary 3.4.17.** *Let $P_0, P_1, P_2$, $e_i \in \mathcal{M}_k(R, P_i)$ be as in Proposition 3.4.15. Let $\Lambda \subseteq \Delta^k$ be the union of all but one of the codimension 1 faces of $\Delta^k$, making $\Lambda$ the cone on $\partial\Delta^{k-1}$. Then any morphism $f : e_2 \to e_1$ defined over $\Lambda \times P_2 \cup \Delta^k \times P_0$ extends to $\Delta^k \times P_2$ and the extension is unique up to homotopy.*

## 3.5 Uniqueness of the fiberwise cone

The twisted homology bundle $\widetilde{H}(\xi, P)$ and the fiberwise cone $C(\xi, P) = C(f)$ depend on the choice of partially ordered basis $P_n$ for each $H_n(\xi)(v_0)$. We need to know how they change with a change of this basis poset. We will start by explaining our notation for basis change.

If $Q = (q_1, q_2, \cdots, q_n)$ is an ordered basis for a free $R$-module $M$ and $A = (a_{ij})$ is an $n \times n$ matrix with coefficients in $R$ we let $QA$ denote the new basis given by the indicated matrix product where we view $Q$ as a row matrix.

$$QA = (q_1, \cdots, q_n) \begin{pmatrix} a_{11} & \cdots & a_{1n} \\ \vdots & \ddots & \vdots \\ a_{n1} & \cdots & a_{nn} \end{pmatrix} = \left( \sum_{i=1}^{n} q_i a_{i1}, \cdots, \sum_{i=1}^{n} q_i a_{in} \right)$$

For example, if $A$ is a permutation matrix then $QA$ is a permutation of the basis $Q$. If $\phi_Q : R^n \to M$ and $\phi_A : R^n \to R^n$ are the isomorphisms of right $R$-modules corresponding to $Q$ and $A$ then the basis $QA$ corresponds to the composition $\phi_Q \circ \phi_A$ since $\phi_Q(Ax) = QAx = \phi_{QA}(x)$. We will also use the notation $QA = \phi_A^*(Q)$ and we note that $(\phi_A \phi_B)^* = (\phi_{AB})^* = \phi_B^* \phi_A^*$.

**Lemma 3.5.1 (Monomial basis change lemma).** *Let $\xi, P$ be as above and suppose that $P_n A_n$ is another basis for $H_n(\xi)(v_0)$ which differs from $P_n$ by a $G$-monomial matrix $A_n$ whose underlying permutation matrix gives a poset automorphism of $P_n$. Then the resulting $G$-monomial functors $\widetilde{H}(\xi, PA)$ and $C(\xi, PA)$ are related to $\widetilde{H}(\xi, P)$ and $C(\xi, P)$ by $G$-monomial simplicial homotopy.*

*Proof.* Take the constant homotopy

$$\operatorname{simp}(B \times I) \to \operatorname{simp} B \xrightarrow{\widetilde{H}(\xi, P)} \operatorname{simp} \mathcal{M}_\bullet(R)$$

and change it over one end $(B \times 1)$ by the given $G$-monomial change of coordinates. This defines a $G$-monomial functor on $B \times I$ and therefore gives a simplicial homotopy $\widetilde{H}(\xi, P) \simeq \widetilde{H}(\xi, PA)$. A similar construction gives the simplicial homotopy $C(\xi, P) \simeq C(\xi, PA)$. □

**Lemma 3.5.2 (Poset refinement lemma).** *Let $P'$ be a homology basis poset for $\xi$ with the same underlying set as $P$ but with a refinement of the partial ordering (so that the identity map $\alpha : P' \to P$ is a closed bijection). Then there are $G$-monomial simplicial homotopies $\widetilde{H}(\xi, P') \simeq \widetilde{H}(\xi, P)$ and $C(\xi, P') \simeq C(\xi, P)$.*

*Proof.* Choose $\widetilde{H}(\xi, P)$ and $C(\xi, P)$. Refining the partial ordering on $P$ gives choices for $\widetilde{H}(\xi, P')$ and $C(\xi, P')$ and the identity morphism on $R^P$ gives a $G$-monomial morphism

$$i : \widetilde{H}(\xi, P') \to \widetilde{H}(\xi, P)$$

subordinate to $\alpha$ inducing $G$-monomial simplicial homotopies $\widetilde{H}(\xi, P') \simeq \widetilde{H}(\xi, P)$ and $C(f \circ i) = C(\xi, P') \simeq C(f) = C(\xi, P)$. □

**Lemma 3.5.3 (Elementary basis change lemma).** *Let $\xi, P$ be as above and suppose that $P_nT_n$ is another basis for $H_n(\xi)(v_0)$ which differs from $P_n$ by a matrix $T_n$ which is upper triangular (with 1's on the diagonal) with respect to the given partial ordering of $P_n$ for each $n \geq 0$. Then the resulting G-monomial functors $\widetilde{H}(\xi, PT)$ and $C(\xi, PT)$ are related to $\widetilde{H}(\xi, P)$ and $C(\xi, P)$ by G-monomial simplicial homotopy.*

**Remark 3.5.4.** This is the lemma which requires the action of $\pi_1(B)$ on $H_*(\xi)(v_0)$ to be upper triangular and not $G$-monomial.

*Proof.* Let $\widetilde{H}'(\xi)$ be obtained from $\widetilde{H}(\xi, P)$ by right conjugating all higher homotopies by $T = \oplus_{n\geq 0} T_n$. Then we claim that $\widetilde{H}'(\xi) \simeq \widetilde{H}(\xi, PT)$ and that $\widetilde{H}'(\xi) \simeq \widetilde{H}(\xi, P)$.

To see that $\widetilde{H}'(\xi) \simeq \widetilde{H}(\xi, PT)$ we note that, by construction, the automorphism $\phi_T$ of $R^P = \oplus R^{P_n}$ corresponding to $T$ gives a morphism

$$\phi_T : \widetilde{H}'(\xi) \to \widetilde{H}(\xi, P)$$

and the composition of this with $f : \widetilde{H}(\xi, P) \to \xi$ gives a morphism

$$f' = f \circ \phi_T : \widetilde{H}'(\xi) \to \xi$$

inducing $\phi_P \circ \phi_T = \phi_{PT}$ in homology. Therefore $\widetilde{H}'(\xi) \simeq \widetilde{H}(\xi, PT)$ and $C(f') \simeq C(\xi, PT)$ by Theorem 3.4.11.

To see that $\widetilde{H}'(\xi) \simeq \widetilde{H}(\xi, P)$ note that the inverse $\phi_T^{-1}$ of the automorphism $\phi_T$ of $R^P$ corresponding to $T$ gives an upper triangular morphism

$$\phi_T^{-1} : \widetilde{H}(\xi, P) \to \widetilde{H}'(\xi)$$

subordinate to the identity on $P$. By Lemma 3.3.5 this induces a simplicial homotopy $\widetilde{H}(\xi, P) \simeq \widetilde{H}'(\xi)$ which is a fortiori $G$-monomial. Also $f' \circ \phi_T^{-1} = f \circ \phi_T \circ \phi_T^{-1} = f$ so $\phi_T^{-1}$ gives an upper triangular morphism $C(f') \to C(f)$ inducing a simplicial homotopy $C(f') \simeq C(f)$ again by Lemma 3.3.5. Composing with the previous homotopies we get a $G$-monomial simplicial homotopy

$$C(\xi, PT) \simeq C(f') \simeq C(f) \simeq C(\xi, P)$$

□

**Definition 3.5.5.** Suppose that $\xi, \eta$ are $G$-monomial functors on $B$ subordinate to poset functors $P, Q$. Then the *(external) direct sum* $\xi \oplus \eta$ is the $G$-monomial functor subordinate to $P \coprod Q$ given by

$$(\xi \oplus \eta)(\sigma) = \xi(\sigma) \oplus \eta(\sigma)$$

Conversely, suppose that $\xi$ is a $G$-monomial functor on $B$ subordinate to a poset functor $P$ and suppose that $P$ is the disjoint union of two closed subfunctors $P = P_1 \coprod P_2$. Then, by definition, all higher homotopies and $G$-monomial morphisms send $R^{P_1}$ to $R^{P_1}$ and $R^{P_2}$ to $R^{P_2}$ so $\xi$ splits as an *(internal) direct sum* $\xi = \xi_1 \oplus \xi_2$ where $\xi_i = \xi | P_i$ is subordinate to $P_i$ for each $i$.

**Lemma 3.5.6.** *Suppose that $\xi, \eta$ have upper triangular homology bundles with basis posets $P, Q$ respectively. Then $\xi \oplus \eta$ has upper triangular homology with homology basis poset $P \coprod Q$ and we have isomorphisms:*

$$\widetilde{H}(\xi \oplus \eta, P \coprod Q) \cong \widetilde{H}(\xi, P) \oplus \widetilde{H}(\eta, Q)$$

$$C(\xi \oplus \eta, P \coprod Q) \cong C(\xi, P) \oplus C(\eta, Q)$$

*and both isomorphisms are natural up to fiber homotopy.*

In the special case (of Definition 3.5.5) when $Q$ contains only two elements $x^+, x^-$ of degree $n, n-1$ respectively with $x^- < x^+$ and $\eta(\sigma)$ is the constant system $(R^Q, d)$ where $dx^+ = x^-$ for all $\sigma \in B$, we say that $\xi \oplus \eta$ is obtained from $\xi$ by *elementary expansion* in degree $n$ and $\xi$ is obtained from $\xi \oplus \eta$ by *elementary collapse.* Conversely, $\xi$ is an elementary expansion of $\xi|Q$ if and only if the complement of $Q$ in $P$ is a free expansion pair as defined below.

**Definition 3.5.7.** A *free expansion pair* for a $G$-monomial functor $(\xi, P)$ is a subfunctor $S$ of $P$ so that

1. $S(\sigma) = \{x^+, x^-\}$ where $x^- < x^+$ for all $\sigma \in B$.
2. $S$ and $P - S$ are closed, i.e., $x^+, x^-$ are unrelated in the partial ordering to all other elements of $P$.
3. $\xi(v)(x^+) = x^-$ for every vertex $v \in B_0$. (In particular, $\deg x^+ = 1 + \deg x^-$.)

We also call $S(\sigma) = \{x^+, x^-\}$ a free expansion pair for $(\xi(\sigma), P(\sigma))$ for each $\sigma \in B$.

We define *stable equivalence* to be the smallest equivalence relation which includes $G$-monomial simplicial homotopy and elementary expansions in all positive degrees. We note that stably equivalent functors have isomorphic homology bundles.

There is another method of stabilization which alters the homology bundle, namely, the addition of a free generator in only one degree. More precisely, let $S$ be a poset functor on $B$ which is constant (i.e., the same finite graded poset at all $\sigma \in B$) unordered ($x < y$ for no pairs $x, y \in S$) and all in one degree ($S = S_n$). Let $\eta$ be the *constant functor* $\eta(\sigma) = R^S$ for all $\sigma \in B$. Then $\eta$ is equal to its own homology bundle and its fiberwise cone $C(\eta, S)$ is stably equivalent to zero since it is the fiberwise mapping cone of the identity morphism on $\eta$. By Lemma 3.5.6 this implies the following.

**Lemma 3.5.8 (Stabilization Lemma).** *Let $\xi$ be a $G$-monomial functor over $B$ with upper triangular homology bundle and homology basis poset $P$ and let $\eta$ be a constant functor as described above. Then the fiberwise cone $C(\xi \oplus \eta, P \coprod S)$ is stably equivalent to $C(\xi, P)$. In other words, each basis poset $P_n$ can be extended without changing the stable equivalence class of $C(\xi, P)$.*

**Definition 3.5.9.** If $G$ is a subgroup of the group of units of a ring $R$, the *Whitehead group* $Wh_1(R, G)$ is defined to be the quotient of $K_1R$ by the image of $\pm G$. Since $K_1R$ is the abelianized quotient of $GL_\infty(R)$ we have:

$$Wh_1(R, G) = \frac{K_1R}{\pm G} \cong \lim_{\rightarrow} \frac{GL_n(R)}{E_n(R)S_n(G)}$$

Here $S_n(G) \subseteq GL_n(R)$ denotes the subgroup of $G$-monomial matrices and $E_n(R) \subseteq GL_n(R)$ is the subgroup generated by elementary matrices.

We note that since

$$\begin{pmatrix} 0 & g \\ -g^{-1} & 0 \end{pmatrix} = E_{12}(g)E_{21}(-g^{-1})E_{12}(g) \in E_2(R)$$

we have $E_n(R)S_n(G) = E_n(R)S_n(\pm G) = E_n(R)M_1(\pm G)$ for all $n \geq 2$.

We recall some well know facts about Whitehead torsion.

1. Every finitely generated based free acyclic R-complex $E$ has a well defined Whitehead torsion $\tau(E) \in Wh_1(R, G)$.

2. Two such chain complexes have the same torsion if and only if they are stably equivalent.

3. If $E = (R^{Q_n}, 0)$ is a based free chain complex with zero boundary map and $f : E \to E$ is a chain isomorphism then the torsion of the mapping cone of $f$ is equal to the torsion of $f$:

$$\tau(C(f)) = \tau(f) := \sum_{n \geq 0} (-1)^n [f_n]$$

where $[f_n]$ is the element of $Wh_1(R, G)$ represented by the matrix of $f_n$.

**Theorem 3.5.10 (Basis Change Theorem).** *Suppose that $\xi$ is a $G$-monomial functor over $B$ with constant free homology and homology basis poset $Q$. Suppose also that*

$$g_n : R^{Q_n} \xrightarrow{\approx} R^{Q_n}$$

*are isomorphisms having the property that*

$$\sum_{n \geq 0} (-1)^n [g_n] = 0 \ \in Wh_1(R, G)$$

*Then the fiberwise cones $C(\xi, Q)$, $C(\xi, g^*(Q))$ are stably equivalent.*

*Proof.* Combining Lemmas 3.5.1, 3.5.3 and the Stabilization Lemma (3.5.8) we see that the theorem holds in the case when

$$[g_n] = 0 \in Wh_1(R, G)$$

for all $n \geq 0$.

To prove the general case we stabilize using the Stabilization Lemma, replacing $Q$ by $Q \coprod S$ where $S$ is constant (the same over all $\sigma \in B$). Using Lemma 3.5.3 the isomorphisms $g_n$ can be shifted to $S$, i.e., we may assume that $g_n$ is the direct sum of the identity on $R^{Q_n}$ and an isomorphisms $h_n$ on $R^{S_n}$ where $\sum(-1)^n[h_n] = 0$ in $Wh_1(R, G)$. Then

$$C(\xi, g^*(Q)) \simeq C(\xi, Q) \oplus C(\eta, h^*(S))$$

so it suffices to show that $C(\eta, h^*(S))$ is stably equivalent to zero.

However, the $G$-monomial functor $\eta$ and its basis $S$ are constant and therefore we may ignore the parameterization by $B$. In that case we are looking at a single based free chain complex $\eta$ whose boundary map is zero. The chain isomorphism $h : R^S \to \eta$ has Whitehead torsion

$$\tau(h) = \sum_{n \geq 0} (-1)^n [h_n] = 0 \ \in Wh_1(R, G)$$

and therefore its mapping cone $C(h) = C(\eta, h^*(S))$ has zero torsion and is thus stably equivalent to zero. □

To extend Theorem 3.5.10 to the upper triangular case we need to assume that the ground ring $R$ is a field.

**Theorem 3.5.11 (Upper triangular basis change theorem).** *Suppose that $R = K$ is a field. Let $\xi$ be a $G$-monomial functor on $B$ with upper triangular homology bundle and let $P, Q$ be two choices for homology basis posets at a vertex $v_0$ of $B$ and let*

$$g_n : K^{P_n} \cong H_n(\xi)(v_0) \cong K^{Q_n}$$

*be the change of basis map. Suppose also that*

$$\sum_{n \geq 0} (-1)^n [g_n] = 0 \ \in Wh_1(K, G) = K^\times / \pm G.$$

*Then the fiberwise cones $C(\xi, P), C(\xi, Q)$ are stably equivalent.*

*Proof.* We first consider the case then $P$ and $Q$ are both "canonical." Once we show that canonical basis posets give stably equivalent fiberwise cones we can compare an arbitrary basis poset to a canonical one.

We are assuming that the $K[\pi_1 B]$-modules $M_n = H_n(\xi)(v_0)$ have composition series with successive quotient being semi-simple. A canonical such series is given by the *socle* series:

$$0 = S^{-1}(M_n) \subset S(M_n) \subset S^2(M_n) \subset S^3(M_n) \cdots \subset S^\ell(M_n) = M_n$$

where each $S^{i+1}(M_n)$ is the inverse image in $M_n$ of the $\pi_1 B$-fixed submodule of $M_n / S^i(M_n)$. We will say that $P$ is a *canonical* basis poset if each $P_n$ is a disjoint union $P_n = \coprod P_n^i$ so that $P_n^1 \coprod \cdots \coprod P_n^i$ is a basis for $S^i(M_n)$ for all $n, i$ and an element of $P_n^i$ is less than an element of $P_m^j$ if and only if $(-n, i) < (-m, j)$ in lexicographic order.

If $P$ and $Q$ are canonical basis posets then the basis change maps $g_n : K^{P_n} \to K^{Q_n}$ have an upper triangular block form with diagonal blocks $g_n^1, \cdots, g_n^\ell$ representing the basis change maps

$$g_n^i : K^{P_n^i} \to K^{Q_n^i}$$

for the layers $S^i(M_n)/S^{i-1}(M_n)$ of $M_n$. If each $[g_n^i] = 0$ in $Wh_1(K,G) = K^\times/\pm G$ then the basis change lemmas (3.5.3, 3.5.1) imply that $C(\xi, P) \simeq C(\xi, Q)$. If $[g_n^i] \neq 0$ for some $n, i$ then we can stabilize using the Stabilization Lemma replacing $P, Q$ by $P \coprod S, Q \coprod S$ where $S$ is constant with filtration analogous to $P$ and $Q$. ($S_n = \coprod S_n^i$ ordered lexicographically by the pair $(-n, i)$.)

Using the poset refinement lemma (3.5.2) we may refine the partial ordering of $P \coprod S$ (and $Q \coprod S$) so that the layers $P_n^i \coprod S_n^i$ are ordered lexicographically by the pair $(-n, i)$. Then, as in the proof of Theorem 3.5.10, we can use the basis change lemmas to replace $P \coprod S$ with $Q \coprod h^*S$ where $\sum(-1)^n[h_n] = 0$ in $Wh_1(K,G)$. Since $C(\eta, h^*S)$ is constant with trivial fiberwise Whitehead torsion it is null homotopic and we have:

$$C(\xi,P) \simeq C(\xi \oplus \eta, P \textstyle\coprod S) \simeq C(\xi \oplus \eta, Q \coprod h^*S) \simeq C(\xi, Q) \oplus C(\eta, h^*S) \simeq C(\xi, Q).$$

Now suppose that $P$ is an arbitrary homology basis poset for $\xi$ at $v_0$. Then by a sequence of moves which do not change the stable equivalence class of $C(\xi, P)$ (upper triangular basis changes and poset refinements) we will transform $P$ into a canonical homology poset basis. By the above discussion this will prove the theorem.

Let $n$ be maximal so that $P_n$ is not canonical. Let $S$ be the set of minimal elements of $P_n$. Then $S$ is a linearly independent subset of $S(M_n)$. If $S$ does not span $S(M_n)$ then we choose a fixed element $x$ of $S(M_n)$ which is not in the span of $S$. We also use the poset refinement lemma (3.5.2) to assume that the elements of $P - S$ are ordered linearly. When $x$ is written as a linear combination of elements of $P_n$ we get:

$$x = ay + \textit{lower terms}$$

where $a \neq 0$. By an upper triangular change of basis we may assume that $x = ay$. Then $y$ is an element of $S(M_n)$ and by the poset refinement lemma we can change the partial ordering of $P_n$ to make $y$ minimal. Proceeding in this way we may assume that $S$ spans $S(M_n)$.

Now strengthen the partial ordering of $P_n$ so that every element of $P_n - S$ is greater than every element of $S$. Then let $S^2$ be the set of minimal elements of $P_n - S$. By the same process as before we can increase $S^2$ until it spans the second layer $S^2(M_n)/S(M_n)$. Proceeding in this way, $P$ will eventually become canonical reducing us to the case we have already settled. □

Since the fiberwise cone $C(\xi, Q)$ has based free acyclic fibers it has a well-defined torsion over every vertex of $B$ which is constant over components. In other words, $\tau(C(\xi, Q))$ is a mapping:

$$\tau(C(\xi, Q)) : \pi_0|B| \to Wh_1(R, G)$$

With this observation, the above theorems can be rephrased as:

1. $(C(\xi,Q), C(\xi,g^*Q)$ are stably equivalent if and only if they have the same Whitehead torsion over every component of $B$.

2. $\tau(C(\xi,g^*Q)) = \tau(C(\xi,Q)) + \tau(g)$

Assume that $B$ is connected and suppose that $\xi$ is a $G$-monomial functor on $B$ with constant *projective homology bundle* $H(\xi)$ (projective as an $R$-module). Then the fiberwise homology can be trivialized in the following way.

Let $\eta$ be any constant $G$-monomial functor over $B$ with complementary projective homology (so that $\xi \oplus \eta$ has constant free homology). One construction of $\eta$ is given as follows. Choose a vertex $v \in B_0$ and let $(R^{P_n}, d) = \xi(v)$. Then $D_n := d(R^{P_{n+1}}) \subseteq R^{P_n}$ is a direct summand since

$$R^{P_n}/D_n \cong D_{n-1} \oplus H_n(\xi(v))$$

is projective by induction on $n$. Thus there is a projection homomorphism $p_n : R^{P_n} \to R^{P_n}$ with kernel $D_n$ and we can let

$$\eta = (R^{P_n} \oplus R^{P_{n-1}}, p_{n-1})$$

Then $H_n(\eta) = \ker(p_{n-1}) \oplus \operatorname{coker}(p_n) = D_{n-1} \oplus D_n$ and $H_n(\xi \oplus \eta) \cong R^{P_n}$.

If we choose any basis $Q$ for the homology of $\xi \oplus \eta$ then we can construct the twisted homology bundle $\widetilde{H}(\xi \oplus \eta, Q)$ and the fiberwise mapping cone $C(\xi \oplus \eta, Q)$.

**Corollary 3.5.12.** *Up to stable equivalence, the fiberwise mapping cone $C(\xi \oplus \eta, Q)$ depends only on its Whitehead torsion $\tau(C(\xi \oplus \eta, Q)) \in Wh_1(R,G)$ (and on $\xi$.)*

*Proof.* Let $\eta', Q'$ be other choices for $\eta, Q$ and suppose that

$$\tau(C(\xi \oplus \eta', Q')) = \tau(C(\xi \oplus \eta, Q)).$$

Then we will show that $C(\xi \oplus \eta', Q')$ and $C(\xi \oplus \eta, Q)$ are stably equivalent.

Choose another constant functor $\xi'$ with the same homology as $\xi$ and choose bases $S, T$ for $\xi' \oplus \eta$, $\xi' \oplus \eta'$ so that $C(\xi' \oplus \eta, S)$ and $C(\xi' \oplus \eta', T)$ have zero torsion and are therefore stably equivalent to zero. Then

$$C(\xi \oplus \eta, Q) \sim C(\xi \oplus \eta, Q) \oplus C(\xi' \oplus \eta', T) \cong C(\xi \oplus \eta \oplus \xi' \oplus \eta', Q \coprod T)$$

which, by the Basis Change Theorem (3.5.10), is stably equivalent to

$$C(\xi \oplus \eta', Q') \sim C(\xi \oplus \eta', Q') \oplus C(\xi' \oplus \eta, S) \cong C(\xi \oplus \eta' \oplus \xi' \oplus \eta, Q' \coprod S)$$

since they have the same torsion. □

## 3.6 A model for the stabilized Whitehead space

In this section we will stabilize the simplicial subcategory of $G$-monomial morphisms in $\mathcal{M}_\bullet(R)$ by adding elementary expansions. The result will be a simplicial category $\mathcal{W}h_\bullet(R,G)$ whose components are the stable equivalence classes of objects of $\mathcal{M}_\bullet(R)$. If $\mathcal{W}h_\bullet^h(R,G)$ denotes the simplicial full subcategory of $\mathcal{W}h_\bullet(R,G)$ consisting of acyclic chain complexes and higher homotopies then the components of $\mathcal{W}h_\bullet^h(R,G)$ will be in $1-1$ correspondence with the group $Wh_1(R,G)$. Thus:

$$\pi_0(|\mathcal{W}h_\bullet^h(R,G)|) \cong Wh_1(R,G)$$

The higher homotopy groups of $\mathcal{W}h_\bullet^h(R,G)$ will be given by Theorem 5.6.1 below which by Corollary 5.6.2 says that there is a homotopy fiber sequence:

$$\Omega BGL(R)^+ \to |\mathcal{W}h_\bullet^h(R,G)| \to Q_0(BG_+)$$

The results of the previous two sections will extend easily to this stabilized Whitehead space, allowing us to construct a functor

$$\operatorname{simp} B \to \operatorname{simp} \mathcal{W}h_\bullet^h(R,G)$$

out of a functor $\operatorname{simp} B \to \operatorname{simp} \mathcal{W}h_\bullet(R,G)$ whose associated homology bundle is free and constant (or upper triangular in the case when $R$ is a field).

**Definition 3.6.1.** For any subgroup $G$ of the group of units of an associative ring $R$ let $\mathcal{W}h_\bullet(R,G)$ be the simplicial subcategory of $\mathcal{M}_\bullet(R)$ consisting of all morphisms $(f,\alpha) : (e,P) \to (e_1,P_1)$ which factor as a composition

$$(e,P) \to (e,P) \oplus (d,S) = (e \oplus d, P \textstyle\coprod S) \to (e_1,P_1)$$

of an expansion morphism $(e,P) \to (e,P) \oplus (d,S)$ (i.e., $S$ is a disjoint union of free expansion pairs in $(e \oplus d, P \coprod S)$) and a $G$-monomial morphism $f$ : $(e \oplus d, P \coprod S) \to (e_1,P_1)$. Composition with a morphism $g : (e_1,P_1) \oplus (d_1,S_1) \to (e_2,P_2)$ is given by

$$g \circ (f \oplus id_{(d_1,S_1)}) : (e,P) \oplus (d,S) \oplus (d_1,S_1) \to (e_1,P_1) \oplus (d_1,S_1) \to (e_2,P_2)$$

We will call these morphisms *$G$-expansion morphisms.*

First we want to rephrase this definition so that it is intrinsic (with no reference to $S$). The question is: How can we recognize the fact that the complement of $\alpha(P)$ in $P_1$ is the image of a union of free expansion pairs under a $G$-monomial morphism? The answer is easy: If $(h,\beta) : (e \oplus D, P \coprod S) \to (e_1,P_1)$ is a $G$-monomial morphism then $\beta$, being a graded closed bijection, sends each free expansion pair $\{x^+,x^-\} \subseteq S$ to a pair of elements $\{y^+,y^-\} \subseteq P_1$ and $h$, being $G$-monomial, sends $x^+,x^-$ to $h(x^+) = y^+g, h(x^-) = y^-g'$. Consequently, each boundary map $e_1(v)$ of $(e_1,P_1)$ sends $y^+$ to $y^-(g'g^{-1})$. This leads to the following definition.

**Definition 3.6.2.** A *free $G$-expansion pair* for an object $(e, P)$ of $\mathcal{M}_k(R)$ is a subset $S$ of $P$ so that

1. $S = \{x^+, x^-\}$ where $x^- < x^+$.
2. $S$ and $P - S$ are closed, i.e., $x^+, x^-$ are unrelated in the partial ordering to all other elements of $P$.
3. There is a $g \in G$ so that $e(v)(x^+) = x^- g$ for every $0 \leq v \leq k$. (In particular, $\deg x^+ = 1 + \deg x^-$.)

Definition 3.6.1 can now be rephrased as follows. A morphism $(f, \alpha) : (e, P) \to (e_1, P_1)$ of $\mathcal{M}_k(R)$ lies in $\mathcal{W}h_k(R, G)$ if $\alpha : P \to \alpha(P)$ is a graded closed bijection, the complement of $\alpha(P)$ in $P_1$ is a union of free $G$-expansion pairs and $f : (e, P) \to (e_1|\alpha(P), \alpha(P))$ is $G$-monomial.

Since a $G$-monomial morphism sends free $G$-expansion pairs to free $G$-expansion pairs, this collection of morphisms is closed under composition. However, we note that the image of the forgetful functor

$$\mathcal{W}h_\bullet(R, G) \to \mathcal{P}$$

is not a subcategory of $\mathcal{P}$ since a graded closed bijection $\alpha : P \to P_1$ could send an expansion pair in $P$ to two closed points in $P_1$. We will fix this problem later.

Now we want to extend the results and constructions of the last two sections to "functors"

$$\xi : \operatorname{simp} B \to \operatorname{simp} \mathcal{W}h_\bullet(R, G)$$

As before $B$ is a simplicial set and we say that $\xi$ is a *$G$-expansion functor* if it sends elements of $B_k$ to objects of $\mathcal{W}h_k(R, G)$ and morphisms $a_* : ([j], \tau) \to ([k], \sigma)$ in $\operatorname{simp} B$ (where $\tau = a^*\sigma$) to morphisms

$$g^\xi_{\sigma\tau} : \xi(\tau) \to \xi(\sigma)|\tau = a^*\xi(\sigma)$$

in $\mathcal{W}h_j(R, G)$ which we will call the *structure maps* of the functor $\xi$. We note that $G$-monomial functors are a special case of $G$-expansion functors.

As before, a *morphism* $(f, \alpha) : \xi \to \eta$ of $G$-expansion functors means a collection of general morphisms

$$f_\sigma : \xi(\sigma) \to \eta(\sigma)$$

subordinate to $\alpha_\sigma : P(\sigma) \to Q(\sigma)$ for all $\sigma \in B$ which are compatible with simplicial operators in the sense that $g^\eta_{\sigma\tau} \circ f_\tau = f_\sigma|\tau \circ g^\xi_{\sigma\tau}$ for all $\tau = a^*\sigma$ and a *fiber homotopy* $h : f \simeq g$ is a collection of homotopies $h_\sigma : f_\sigma \simeq g_\sigma$ for all $\sigma \in B$ which are analogously compatible with simplicial operators $a^*$. We recall that, by a "simplicial homotopy" of such functors, we mean a functor on $\operatorname{simp}(B \times I')$ for some subdivision $I'$ of $I = \Delta^1$.

**Lemma 3.6.3.** *Suppose that $\xi$ and $\eta$ are $G$-expansion functors on $B$ and $\Sigma^{-1}(\xi)$ is a $\pm G$-expansion functor. Let $(f, \alpha) : \Sigma^{-1}(\xi) \to \eta$ be a morphism of $\pm G$-expansion functors so that $\alpha(S) = \emptyset$ for all $\pm G$-expansion pairs in $\Sigma^{-1}(\xi)$ and the image of $\alpha$ contains no $G$-expansion pairs in $Q$. Then the fiberwise mapping cone $C(f) = C(f, \alpha)$ is a $G$-expansion functor on $B$ and a fiber homotopy $h : f \simeq g$ between two such morphisms gives a simplicial homotopy*

$$C(h)_* : \mathrm{simp}(B \times I) \to \mathrm{simp}\,\mathcal{W}h_\bullet(R, G)$$

*from $C(f)$ to $C(g)$.*

*Proof.* It follows from the naturality of $C(f)$ (Proposition 3.3.2(d)) and $C(h)_*$ (Proposition 3.3.7) that the structure maps of these functors are $G$-expansion morphisms provided that $\alpha$ satisfies the stated conditions. □

**Lemma 3.6.4.** *Proposition 3.4.7 holds for $G$-expansion functors.*

*Proof.* Proposition 3.4.7 is proved one simplex at a time using Lemma 3.4.6. This proof works without any problem for $G$-expansion functors. When we go the next dimensional simplices we may have additional direct summands in the domain which come from $G$-expansions. However these are acyclic and therefore, by Lemma 3.4.6, we are free to extend our morphisms arbitrarily on these summands. For example, we can send them to zero. □

**Theorem 3.6.5.** *Suppose $B$ is a connected simplicial set with base point $v_0$ and $\xi$ is a functor on $\mathrm{simp}\,B$ as described above. Suppose also that $\xi$ has upper triangular homology bundle with basis poset $P$ at $v_0$. Then there exist:*

1. *a $G$-monomial functor $\widetilde{H}(\xi, P) : \mathrm{simp}\,B \to \mathrm{simp}\,\mathcal{W}h_\bullet(R, G)$ subordinate to $P$,*
2. *a morphism $f : \widetilde{H}(\xi, P) \to \xi$ inducing an isomorphism of homology bundles:*
$$H_n(\widetilde{H}(\xi, P)) \cong H_n(\xi)$$
*which is the identity on $P_n$ at $v_0$ and*
3. *a morphism $g : \xi \to \widetilde{H}(\xi, P)$ so that $g * f$ is the identity morphism on $\widetilde{H}(\xi, P)$. (However, the composite poset morphism is the maximal morphism $\omega : P \to P$.)*

*Furthermore, $\widetilde{H}(\xi, P)$ and the fiberwise mapping cones $C(f), C(g)$ of $f, g$ are unique up to simplicial homotopy and, given $\widetilde{H}(\xi, P)$, $f, g$ are unique up to fiber homotopy.*

**Remark 3.6.6.** We note that $\widetilde{H}(\xi, P)$ is still a $G$-monomial functor so its suspension is also $G$-monomial. This makes $C(\xi, P) = C(f)$ a $G$-expansion functors on $B$ by Lemma 3.6.3 if we choose the poset morphism $\alpha$ for $f$ to be the maximal morphism which misses all the expansion pairs in $\xi$. The mapping cone $C(g)$ is not a $G$-expansion functor, however any odd suspension of it is. (Take the poset morphism for $g$ that sends all expansion pairs to $\emptyset$ and is maximal everywhere else.)

*Proof.* Since the structure maps for the $G$-expansion functor $\xi$ goes from faces into simplices, (1) and (2) follow from Lemma 3.4.6 in the same was as Theorem 3.4.11.

For each $n \geq 0$ take the decomposition of $P_n$ as

$$P_n = \coprod_{i \geq 1} P_n^i$$

where $P_n^1 = T(P_n)$ is the set of maximal elements of $P_n$, $P_n^2 = T(P_n - P_n^1)$, $P_n^3 = T(P_n - P_n^1 - P_n^2)$, etc., $T(P)$ being the set of maximal elements of a poset $P$. Then

$$r^i(P_n) = \coprod_{j > i} P_n^j$$

is a closed subset of $P_n$ generating an $R[\pi_1 B]$-submodule of $M_n = H_n(\xi)(v_0)$ which we will call $r^i(M_n)$. The corresponding bundle will be written $H_n^i(\xi)$. Note that the quotient bundle $H_n^i(\xi)/H_n^{i+1}(\xi)$ is constant being free on the unordered set $P_n^i$.

In order to prove (3) we construct $C(g)$ by killing the fiber homology of $\xi$ from the bottom up in lexicographic order according to the pair $(n, i)$. If at the $(n, i)$-th stage we have constructed $C(g)_{(n,i)}$ with no fiber homology below degree $n + 1$ and with $n + 1$-st homology

$$H_{n+1}(C(g)_{(n,i)})(v_0) \cong r^i(M_n)$$

then we can construct a epimorphism from $\Sigma^{-1}(C(g)_{(n,i)})$ to the constant functor $H_n^i(\xi)/H_n^{i+1}(\xi)$ by Lemma 3.6.4. The result is the construction of an acyclic $\pm G$-expansion functor $C(g)$ which by Proposition 3.3.2(b) must be the fiberwise mapping cone of some general morphism

$$g : \xi \to \widetilde{H}'(\xi, P).$$

As we observed in Remark 3.4.14, the composition

$$g * f : \widetilde{H}(\xi, P) \to \xi \to \widetilde{H}'(\xi, P)$$

is necessarily an upper triangular isomorphism subordinate to the identity on $P$. (More precisely, the composition is $(g*f, \omega)$ and $(g*f, \iota_P)$ is an upper triangular isomorphism.) Therefore, composing $g$ with the inverse of this isomorphism, we may replace $\widetilde{H}'(\xi, P)$ by $\widetilde{H}(\xi, P)$ and assume that $g*f$ is the identity morphism. □

The fiber of the $G$-expansion functor $C(\xi, P)$ over each vertex of $B$ is a finitely generated based free acyclic complex over $R$ and consequently has a well defined torsion which is constant on each component of $B$. We call this the *fiberwise Whitehead torsion* of $C(\xi, P)$. The basis change lemmas (3.5.1 and 3.5.3), the poset refinement lemma (3.5.2) and the Stabilization Lemma (3.5.8) hold as before and the proof of the basis change theorems (3.5.10 and 3.5.11) carries over to give the following.

**Theorem 3.6.7.** *Suppose that* $\xi : \operatorname{simp} B \to \operatorname{simp} \mathcal{W}h_\bullet(R, G)$ *is a* $G$*-expansion functor and either*

1. $\xi$ *has constant free homology or*

2. $R$ *is a field and* $\xi$ *has upper triangular homology*

*with* $P$ *being a homology basis poset in either case. Then the* $G$*-expansion functor* $C(\xi, P) : \operatorname{simp} B \to \operatorname{simp} \mathcal{W}h^h_\bullet(R, G)$ *given by Theorem 3.6.5 is uniquely determined up to a* $G$*-expansion simplicial homotopy*

$$\operatorname{simp}(B \times I') \to \operatorname{simp} \mathcal{W}h^h_\bullet(R, G)$$

*by the value of its fiberwise Whitehead torsion over each component of* $B$ *and is otherwise independent of the choice of* $P$.

Just as Corollary 3.5.12 follows from Theorem 3.5.10, we get the following analogous statement about $G$-expansion functors $\xi$ with projective homology.

**Corollary 3.6.8.** *If* $\xi$ *is a* $G$*-expansion functor on* $B$ *with constant projective homology bundle and* $\eta$ *is a fixed free* $R$*-complex with complementary projective homology, then the fiberwise mapping cone* $C(\xi \oplus \eta, Q)$ *of* $\xi \oplus \eta$ *with respect to a basis* $Q$ *for its homology is uniquely determined up to* $G$*-expansion simplicial homotopy by the value of its fiberwise Whitehead torsion over each component of* $B$ *and is otherwise independent of both* $\eta$ *and* $Q$.

If the fiberwise Whitehead torsion of $C(\xi \oplus \eta, Q)$ is nonzero then we can add a constant acyclic functor to $\eta$ with the complementary torsion and the new $C(\xi \oplus \eta, Q)$ will have zero Whitehead torsion. This gives a description of the canonical procedure that constructs the cone $C(\xi)$.

**Definition 3.6.9.** Let $\xi$ be a $G$-expansion functor on $B$ with constant projective fiberwise homology. Then the *canonical cone* $C(\xi)$ of $\xi$ is defined to be any functor obtained in the following way.

1. Add a constant $R$-complex $\eta_0$ with complementary projective homology so that $\xi \oplus \eta_0$ has constant free fiber homology.

2. Choose any basis $Q$ for the homology and construct the fiberwise cone $C(\xi \oplus \eta_0, Q)$.

3. Add a locally constant $R$-complex $\eta_1$ which is acyclic with complementary Whitehead torsion over each component of $B$, i.e., $\eta_1$ is fixed on each component of $B$.

The result

$$C(\xi) = C(\xi \oplus \eta_0, Q) \oplus \eta_1$$

is, up to $G$-expansion simplicial homotopy, independent of all choices.

**Remark 3.6.10.** We note that when $R$ is a field and $\xi$ is a $G$-expansion functor with upper triangular homology, steps (1) and (2) above are unnecessary and the *canonical cone* $C(\xi)$ can be defined by

$$C(\xi) = C(\xi, P) \oplus \eta$$

where $\eta$ is given by (3).

We now apply this to the problem of constructing higher Whitehead torsion invariants for smooth bundles. We need to assume the following lemma which follows from the framed function theorem. (See 4.6.4 below.)

**Lemma 3.6.11.** *Suppose that $M \to E \to B$ is a smooth bundle, i.e., $M, E, B$ are smooth manifolds and $p : E \to B$ is a smooth fibration with compact fiber $M$. Then there exists a continuous triangulation of $B$ and a $G$-expansion functor*

$$\xi : \text{simp}\, B \to \mathcal{W}h_\bullet(\mathbb{Z}[\pi_1 E], \pi_1 E)$$

*which is well-defined up to $G$-expansion simplicial homotopy so that the homology of the total complex of $\xi$ is isomorphic to the homology of $E$ with coefficients in $\mathbb{Z}[\pi_1 E]$:*

$$H_*(E(Z(\xi))) \cong H_*(E, \mathbb{Z}[\pi_1 E])$$

The theorem and lemma above immediately give:

**Corollary 3.6.12.** *Let $M \to E \to B$ be as above and suppose that $\rho : \pi_1 E \to F^\times$ is a one-dimensional representation of $\pi_1 E$ over a field $F$ so that $\pi_1 B$ acts on the homology groups $H_n(M, F)$ by upper triangular transformations with respect to some basis. Then there is a continuous mapping*

$$B \to |\mathcal{W}h^h_\bullet(F, G)|$$

*well defined up to homotopy where $G = \rho(\pi_1 E) \subseteq F^\times$.*

**Remark 3.6.13.** Since all modules over the $n \times n$ matrix ring $R = M_n(F)$ are projective, we can generalize this to the case when $\rho$ is an $n$-dimensional representation of $\pi_1 E$ over $F$ to get a continuous mapping

$$B \to |\mathcal{W}h^h_\bullet(M_n(F), G)|$$

well defined up to homotopy where $G = \rho(\pi_1 E) \subseteq GL_n(F)$ assuming that $\pi_1 B$ acts trivially on $H_*(E, F^n)$.

**Example 3.6.14.** (a) Let $E \to B$ be any oriented smooth sphere bundle. Taking $\rho$ to be trivial we get a well defined mapping $B \to |\mathcal{W}h^h_\bullet(R, 1)|$ for any ring $R$. The universal example is the map

$$BDiff^+(S^n) \to |\mathcal{W}h^h_\bullet(\mathbb{Z}, 1)|$$

(b) Let $T_{g,p}$ be the Torelli group for surfaces of genus $g$ with $p$ punctures. Then we get a well-defined homotopy class of maps:

$$BT_{g,p} \to |\mathcal{W}h^h_\bullet(\mathbb{Z}, 1)|$$

We return to the problem of suspension of $G$-expansion functors. Since suspension reverses the signs of all boundary maps making the incidence number of cancelling critical points negative, the suspension of a $G$ expansion functor is a $\pm G$-expansion functor. We can avoid this by using a different sign convention when suspending $G$-expansions.

**Definition 3.6.15.** We define the *alternating suspension* to be the functor

$$\Sigma' : \mathcal{W}h_\bullet(R, G) \to \mathcal{W}h_\bullet(R, G)$$

which raises all degrees by 1 and keeps boundary maps, morphisms and all higher homotopies the same. Thus in the notation of Definition 3.2.9: $\Sigma'(e) = \sigma e \sigma^{-1}$, $\Sigma(f) = \sigma f \sigma^{-1}$, $\Sigma(h) = \sigma h' \sigma^{-1}$

We call $\Sigma'$ the alternating suspension because it is the composition of $\Sigma$ with the involution which reverses the signs of all basis elements in odd degrees. This is the same as conjugating all operations on $R^P$ with the map $\epsilon : R^P \to R^P$ given by $\epsilon(x) = (-1)^{\deg x} x$ for all $x \in P$. The relationship between $\Sigma$ and $\Sigma'$ is then given by $\epsilon e \epsilon = -e'$, $\epsilon f \epsilon = f'$, and $\epsilon h \epsilon = -h'$. We also note that $(\Sigma')^2 = \Sigma^2$, i.e., the two suspensions have the same square.

If we apply conjugation by $\epsilon$ to the mapping cone $C(f, \alpha) = (C(f), C(\alpha))$ of a morphism

$$(f, \alpha) : (e_2, P_2) \to (e_1, P_1)$$

we get the *alternating mapping cone* $C'(f, \alpha) = (C'(f), C(\alpha))$ where

$$C'(f) = \begin{pmatrix} id & 0 \\ 0 & \epsilon \end{pmatrix} \begin{pmatrix} e_1 & f' \\ 0 & -e' \end{pmatrix} \begin{pmatrix} id & 0 \\ 0 & \epsilon \end{pmatrix} = \begin{pmatrix} e_1 & \epsilon f \\ 0 & e_2 \end{pmatrix} \tag{3.3}$$

If $f : E \to D$ is an ordinary chain map then $C'(f) = D \oplus \Sigma(E)$ with boundary map

$$\partial^{C'(f)} = \begin{pmatrix} \partial^D & \epsilon f \sigma^{-1} \\ 0 & \sigma \partial^E \sigma^{-1} \end{pmatrix}$$

i.e., $\partial(y + \sigma x) = (\partial y + (-1)^{\deg x} f(x) + \sigma(\partial x))$.

Figure 3.1: Dragon eating the pictures for Chapter III

# Chapter 4

# Morse Theory and Filtered Chain Complexes

## Introduction

In this chapter we use filtered chain complexes to pass naturally from families of Morse functions to systems of higher homotopies. The idea, due to John Klein, is that the only canonical way to associate a chain complex to a Morse function $f : M \to I$ is to take the total singular complex of $M$ together with the filtration given by the Morse function $f$. The filtration is basically given by taking the total singular complexes of $f^{-1}[0, c_i]$ where $c_i$ are the critical values of $M$.

In §4.1 we prove the basic Lemma 4.1.7 that every filtered chain complex determines a system of higher homotopies uniquely up to contractible choice, i.e., a certain fiber category is contractible. This allows us to show, in §4.2, that the cellular chain complex obtained from a Morse function which is well-known to be unique up to homotopy is in fact unique up to contractible choice.

Given a family of Morse functions, i.e., a smooth function $f : E \to I$ on the total space of a smooth bundle $p : E \to B$ which is Morse on each fiber, we get filtrations for the total singular complexes of the restriction $E|\sigma$ of $E$ to generic small smooth simplices in $B$. By the basic lemma this gives a $G$-monomial functor on the category of small simplices in $B$. This is explained in §4.3 and extended to generalized Morse functions (GMF's) in §4.4.

§4.5 deals with the technical point of what happens at the "birth-death points" of a family of GMF's. We deal with this both topologically and algebraically. Finally in §4.6 we explain the framed function theorem which tells us that there is a canonical choice for a fiberwise GMF on a smooth bundle.

Putting these together we conclude that to every smooth bundle $E \to B$ we can associate a canonically defined functor from the category of small simplices of $B$ into the Whitehead space.

## 4.1 Comparison with filtered chain complexes

We will show that the multiple mapping cylinder construction gives a homotopy equivalence between the category of simplices of $\mathcal{M}_\bullet(R, P)$ and the category of filtered chain complexes over $R$ with fixed basis $P$. Filtered chain complexes are naturally associated to "small" $\Delta^k$-families Morse functions and this theorem will allow us to associate an element of $\mathcal{M}_k(R,P)$ uniquely up to "contractible choice." We begin with the definitions and basic properties of filtered chain complexes (taken from [IK95]).

**Definition 4.1.1.** An *$R\Lambda P$-module* is an $R$-module $M$ together with a submodule $M^A$ for each $A \in \Lambda P$ so that

1. $M^P = M$
2. $M^{A\cap B} = M^A \cap M^B$ for all $A, B \in \Lambda P$.

A morphism of $R\Lambda P$-modules, or *filtered morphism*, is an $R$-homomorphism $f: M \to N$ so that $f(M^A) \subseteq N^A$ for all $A \in \Lambda P$.

Note that condition (2) implies that $M^A \subseteq M^B$ whenever $A \subseteq B$. Also note that $M^\emptyset$ may not be zero.

Given any chain complex $E = (E_n, \partial)$ of $R\Lambda P$-modules and filtered morphisms, we note that $E^A = (E_n^A, \partial|E^A)$ must be a subcomplex of $E$ for every $A \in \Lambda P$. If $e \in \mathcal{M}_k(R,P)$ then the multiple mapping cylinder $Z_k(e)$ is an example of such a chain complex where the filtration is given by letting $E^A$ be the subcomplex of $E = Z_k(e)$ generated by all $z(v_*)(x)$ where $x \in A$.

Another example is given as follows. Let $M$ be any $R\Lambda P$-module and let $N \subseteq M$ be any $R$-submodule. Then there are induced filtrations of $N$ and the quotient $M/N$ given by $N^A = N \cap M^A$ and letting $(M/N)^A$ be the image of $M^A$ in $M/N$ for all $A \in \Lambda P$. Then the short exact sequence $0 \to N \to M \to M/N \to 0$ contains the subsequence

$$0 \to N^A \to M^A \to (M/N)^A \to 0$$

which is exact for all $A \in \Lambda P$. Thus $N \to M \to M/N$ is a chain complex in the category of $R\Lambda P$-modules.

An $R\Lambda P$-complex $(C_*, \partial)$ is called *strongly acyclic* (and the sequence of filtered morphism is called *strongly exact*) if the subcomplex $C^A$ is acyclic for all $A \in \Lambda P$. Thus, in the example above,

$$0 \to N \to M \to M/N \to 0$$

is a strongly exact sequence of filtered modules.

An $R\Lambda P$-module $M$ is called *strongly projective* if the functor $\mathrm{Hom}_{R\Lambda P}(M, -)$ sends strongly exact sequences of $R\Lambda P$-modules to exact sequences of abelian groups. Note that any strongly acyclic complex of strongly projective $R\Lambda P$-modules necessarily splits.

An example of a strongly projective $R\Lambda P$-module is a "strongly free" $R\Lambda P$-module $M$ where, by *strongly free*, we mean free on a basis set $S$ so that $M^A$ is free on a subset $S^A$ of $S$ for each $A \in \Lambda P$. By Definition 4.1.1 we must have $S^{A\cap B} = S^A \cap S^B$ and $S^P = S$. We note that an $R\Lambda P$-module is strongly projective if and only if it is a filtered direct summand of a strongly free module.

We can now give the definition of a filtered chain complex (without the expansion pairs which we will discuss later).

**Definition 4.1.2.** ([IK95]) A *$\Lambda P$-filtered $R$-complex* $(E, \lambda)$ is a chain complex $E$ in the category of $R\Lambda P$-modules together with cohomology classes

$$\lambda_A(x) \in H^d(E^B, E^A; R)$$

for all pairs $A \in \Lambda P, x \in P_d$ so that $B = A \coprod \{x\} \in \Lambda P$ satisfying the following conditions.

1. $E_i$ is *strongly projective* for all $i$.
2. $E^{\emptyset} \simeq 0$
3. $E^A + E^B \simeq E^{A \cup B}$
4. $\lambda_A(x)$ induces an isomorphism $H_{\deg x}(E^B, E^A) \cong R$.
5. $H_i(E^B, E^A) = 0$ for $i \neq \deg(x)$.
6. The cohomology classes $\lambda_A(x)$ are *compatible* in the sense that $\lambda_A(x)$ restricts to $\lambda_{A'}(x)$ whenever $A' \subseteq A$ and $\lambda_{A'}(x)$ is defined.

Let $\mathcal{F}(R, P)$ denote the category of $\Lambda P$-filtered $R$-complexes where the morphisms are defined to be those filtered chain maps $f : D \to E$ for which the induced map in cohomology

$$f^* : H^*(E^B, E^A) \to H^*(D^B, D^A)$$

preserves $\lambda_A(x)$ whenever $B = A \coprod \{x\}$, i.e., $f^* \lambda_A^E(x) = \lambda_A^E(x)$.

**Definition 4.1.3.** If $e \in \mathcal{M}_k(R, P)$ then the multiple mapping cylinder $Z_k(e)$ has the structure of a $\Lambda P$-filtered $R$-complex given as follows.

1. For each $A \in \Lambda P$ let $Z(e)^A$ be the subcomplex of $Z(e)$ generated by $z(v_*)A$ for all $v_* = (v_0, \cdots, v_n)$. (We identify elements of $P$ with their corresponding generators in $R^P$.)
2. The cohomology classes

   $$\lambda_A(x) \in H^{\deg(x)}(Z(e)^{A \cup \{x\}}/Z(e)^A; R)$$

   are given by $\lambda_A(x)(z(v_0)x) = 1$.

Given any simplicial operator $a : [j] \to [k]$ we have a $\Lambda P$-filtered chain map

$$a_* : Z_j(a^*(e)) \to Z_k(e)$$

given by $a_* \circ z(v_0, \cdots, v_n) = z(a(v_0), \cdots, a(v_n))$. This defines a functor

$$Z : \text{simp}\, \mathcal{M}_\bullet(R, P) \to \mathcal{F}(R, P)$$

from the category of simplices in the simplicial set $\mathcal{M}_\bullet(R, P)$ into the category of $\Lambda P$-filtered chain complexes.

In order to show that $Z$ induces a homotopy equivalence on realizations we need the reverse construction which finds in each $\Lambda P$ filtered chain complex a "minimal subcomplex" $L$ which is isomorphic to $Z_0(e)$ for some $e \in \mathcal{M}_0(R, P)$.

**Lemma 4.1.4 (Decomposition theorem [IK95]).** *Let $E$ be a $\Lambda P$-filtered $R$-complex. Then $E$ is a direct sum of two $R\Lambda P$-complexes*

$$E = D \oplus L$$

*where $L$ is* minimal *in the sense that $L^{A\cup\{x\}}/L^A$ is free of rank 1 in degree $\deg x$ and zero in all other degrees for all admissible $(A, x)$. Furthermore $L$ can be taken to be any $\Lambda P$-filtered $R$-subcomplex of $E$ (with compatible cohomology classes) which is minimal in the above sense.*

**Remark 4.1.5.** Given a direct sum decomposition $E = D \oplus L$ which is filtered in the sense that $E^A = D^A \oplus L^A$ it is easy to see that the following are equivalent.

1. $D$ is strongly acyclic.
2. $j : L^B/L^A \to E^B/E^A$ is a quasi-isomorphism for all $A \subseteq B$ closed in $P$.
3. $L$ is a $\Lambda P$-filtered $R$-subcomplex of $E$.

*Proof.* (From [IK95]) By induction on the number of elements of $P$. If $P$ is empty then $E = E^\emptyset$ is acyclic and thus strongly acyclic and $L = 0$ is the unique minimal $\Lambda P$-filtered subcomplex of $E$.

If $P = \{x\}$ is a singleton then $E/E^\emptyset$ has free homology and thus splits as a direct sum

$$E/E^\emptyset \cong C \oplus H_*(E/E^\emptyset)$$

where $C$ is acyclic. If $D$ is the inverse image of $C$ in $E$ then $D^\emptyset = E^\emptyset$ is acyclic as is $D/D^\emptyset \cong C$. Thus $D$ is strongly acyclic. Since the quotient $E/D \cong H_*(E/E^\emptyset)$ is free the sequence $D \to E \to E/D$ splits giving the desired splitting $E = D \oplus L$.

Conversely, if $L \subseteq E$ is a minimal $\Lambda P$-filtered $R$-subcomplex then the projection $L \to E/D$ must be an isomorphism since it induces an isomorphism in cohomology by assumption. Therefore $E = D \oplus L$.

Now suppose that $P$ has at least two elements. Let $x \in P$ be maximal. Then $Q = P - \{x\}$ is a closed subset of $P$ and $E^Q$ is a $\Lambda Q$-filtered $R$-complex. Also $E/E^Q$ is a $\Lambda\{x\}$-filtered $R$-complex so by induction they both split:

$$E^Q = D_1 \oplus L_1, \quad E/E^Q = D_2 \oplus L_2$$

Since $D_2$ is both strongly exact and strongly projective, there is a lifting $D_3 \subseteq E$ of $D_2$ to $E$. (Choose a filtered chain contraction $h$ of $D_2$ and a lifting $\tilde{h} : D_2 \to E$ of $h$ to $E$ and let $D_3$ be the image of the chain map $\partial\tilde{h} + \tilde{h}\partial : D_2 \to E$.) Since $L_2$ is free of rank 1 it also has a lifting $L_3 \subseteq E$. Then $D = D_1 \oplus D_3, L = L_1 \oplus L_3$ satisfy the lemma.

In the case when $L$ is already given we may assume by induction that $L_1 = L^Q = L \cap E^Q$. Since $L$ maps onto $L_2 \subseteq E/E^A$ we may choose the lifting $L_3$ to lie in $L$. Then $L_1 \oplus L_3$ in the above construction will be equal to the given $L$. □

**Theorem 4.1.6.** *The functor*

$$Z : \operatorname{simp} \mathcal{M}_\bullet(R, P) \to \mathcal{F}(R, P)$$

*given above induces a homotopy equivalence on realizations.*

*Proof.* By Quillen's Theorem A it suffices to show that the fiber category over any $E \in \mathcal{F}(R, P)$ is contractible. This follows from the following lemma. □

**Lemma 4.1.7.** *Let $E$ be a $\Lambda P$-filtered $R$-complex and let $Z_\bullet/E$ denote the simplicial set of all pairs $(e, \phi)$ where $e \in \mathcal{M}_k(R, P)$ and $\phi : Z_k(e) \to E$ is a $\Lambda P$-filtered chain map preserving the cohomology classes $\lambda_A(x)$. Then $Z_\bullet/E$ is a simplicial cone on each of its vertices.*

*Proof.* Choose a fixed decomposition $E = D \oplus L$ where $D$ is strongly acyclic and $L$ is minimal. Then $L$ is isomorphic to the filtered chain complex $Z_0(e_0)$ for some $e_0 \in \mathcal{M}_0(R, P)$ so we can choose a fixed $\Lambda P$-morphism $\phi_0 : Z_0(e_0) \to E$ whose image is $L$. In particular $Z_\bullet/E$ is nonempty.

Conversely let $e_0 \in \mathcal{M}_0(R, P)$ be any vertex and let $\phi_0 : Z_0(e_0) \to E$ be any $\Lambda P$-filtered chain map preserving the cohomology classes $\lambda_A(x)$. Then the image $L = \phi_0(Z_0(e_0))$ is necessarily a minimal $\Lambda P$-filtered $R$-subcomplex of $E$ so $E = D \oplus L$ where $D$ is strongly acyclic.

In either case let $\psi : E \to Z_0(e_0)$ be the projection map $E \to L$ composed with the isomorphism $\phi_0^{-1} : L \cong Z_0(e_0)$. Then $\phi_0 \circ \psi : E \to E$ is a projection map with image $L$. Since $D$ is strongly acyclic and strongly projective it has a $\Lambda P$-null homotopy. Thus $\phi_0 \circ \psi$ is homotopic to the identity map on $E$ by a $\Lambda P$-filtered chain homotopy which is zero on $L$. Choose a fixed such chain homotopy $H$.

Now for any $(e, \phi)$ in $Z_\bullet/E$ consider the following diagram.

$$\begin{array}{ccc} Z_k(e(v_*)) & \xrightarrow{\psi \circ \phi} & Z_0(e_0) \\ & {\scriptstyle\phi}\searrow \quad \swarrow{\scriptstyle\phi_0} & \\ & E & \end{array}$$

This diagram commutes up to a $\Lambda P$-filtered chain homotopy given by $H \circ \phi$. By Theorem 3.1.7 the mapping cylinder of $\psi \circ \phi$ is the $(k+1)$-fold multiple mapping cylinder of a $(k+1)$-simplex $c(e) \in \mathcal{M}_{k+1}(R, P)$. The chain homotopy $H \circ \phi$ together with $\phi$ and $\phi_0$ gives a natural $\Lambda P$-morphism $Z_{k+1}(c(e)) \to E$ and we have constructed a simplicial chain contraction of the entire fiber complex $Z_\bullet/E$ to the single vertex $(e_0, \phi_0)$ which we chose arbitrarily. □

We will later need to assume that our filtered chain complexes satisfy the additional condition $E^\emptyset = 0$. However, it is easy to see that this does not affect the homotopy type of the space of filtered chain complexes.

**Proposition 4.1.8.** *Let $\mathcal{F}_\emptyset(R, P)$ be the full subcategory of $\mathcal{F}(R, P)$ of $\Lambda P$-filtered $R$-complexes $(E, \lambda)$ so that $E^\emptyset = 0$. Then the inclusion functor*

$$\mathcal{F}_\emptyset(R, P) \to \mathcal{F}(R, P)$$

*is a homotopy equivalence on realizations.*

**Remark 4.1.9.** Note that the image of $Z : \operatorname{simp} \mathcal{M}_\bullet(R,P) \to \mathcal{F}(R,P)$ lies in $\mathcal{F}_\emptyset(R,P)$. Consequently, $Z$ gives a homotopy equivalence

$$|\operatorname{simp} \mathcal{M}_\bullet(R,P)| \simeq |\mathcal{F}_\emptyset(R,P)|.$$

*Proof.* We have a retraction $r : \mathcal{F}(R,P) \to \mathcal{F}_\emptyset(R,P)$ given by $r(E,\lambda) = (E/E^\emptyset, \lambda)$ and the quotient map $E \to E/E^\emptyset$ gives a homotopy $id \simeq |r|$. □

We will later need to compare different filtrations of the same chain complex. To do this we take the alternating mapping cone of the identity map with the associated filtration. The (obvious) definitions are given, in more general terms, as follows.

**Definition 4.1.10.** Suppose that $E_i \in \mathcal{F}(R,P_i)$ for $i = 1,2$, $\alpha : P_1 \to P_2$ is a morphism of posets and $f : E_1 \to E_2$ is a morphism of $R$-complexes so that $f(E^A) \subseteq E_2^{\alpha(A)}$ for all $A \in \Lambda P_1$. Then the *alternating mapping cone* $C'(f,\alpha) = (C'(f), C(\alpha))$ is the filtered chain complex with total complex the alternating cone $C'(f)$ of $f$ (3.3) and poset the cone $C(\alpha)$ of $\alpha$ with filtration $C'(f)^{A \cup B} = C'(E_1^A \to E_2^B)$. The orientation classes $\lambda_A(x)$ are induced by the inclusion $(E_2, P_2) \to C'(f,\alpha)$ and projection $C'(f,\alpha) \to \Sigma'(E_1, P_1)$.

## 4.2 Morse theory and filtered chain complexes

In this section we use filtered chain complexes as an intermediate step to associate a cellular chain complex to a Morse function in a canonical way up to higher homotopies. We will describe this process first in the case of a finite relative cell complex $(Y, X)$ and then in the case of a manifold relative to part of its boundary $(M, \partial_0 M)$ using two different methods. We also look at the cellular mapping cone associated to a map between two pairs of spaces or between two cellular chain complexes for the same manifold pair given by two different Morse functions. We call these the *comparison case.*

Given a relative cell complex

$$Y = X \cup e_1 \cup \cdots \cup e_n$$

let $P$ be the set of cells partially ordered by their order of attachment. Thus for every closed subset $A$ of the poset $P$ we have a subcomplex $Y^A$ ($X$ union the cells of $A$). Also, $e_i$ is minimal in $P - A$ if and only if the cell $e_i$ is attached on $Y^A$.

Suppose that $\mathcal{F}$ is a coefficient sheaf of right $R$-modules on $Y$ with stalk $\mathcal{F}_*$ over a fixed base point $* \in Y$. Assuming that $Y$ is connected this means that $\mathcal{F}_x \cong R_R$ for every $x \in Y$.

Now choose:

1. A relative homeomorphism $\psi_i : (D^j, \partial D^j) \to (Y^{A \cup e_j}, Y^A)$ where $Y^A$ is any subcomplex on which $e_i$ is attached. ($\psi_i$ is called the *characteristic map* of $e_i$.)

2. A path $\gamma_i$ from the center $b_i = \psi_i(0)$ of each cell $e_i$ to the base point $* \in Y$ (giving an isomorphism $\gamma_i^* : R \cong \mathcal{F}_{b_i}$.)

If $(Y, X)$ is a relative CW-complex then these choices determine the *cellular chain complex* of $(Y, X)$ with coefficients in $\mathcal{F}$. This is a based free $R$-complex $C(Y, X; \mathcal{F})$ with one free generator $[e_i]$ for each $e_i \in P$ whose homology is equal to $H_*(Y, X; \mathcal{F})$.

In the comparison case we have a map of pairs

$$f : (Y', X') \to (Y, X)$$

and a poset morphism $\alpha : P' \to P$ having the property that for each $x \in P'$, the cell in $Y'$ corresponding to $x$ maps into the subcomplex of $Y$ corresponding to $\alpha(x) \in \Lambda P$. If $f$ is a cellular map of CW-pairs and if $\alpha(x)$ does not contain elements of degree greater than $\deg x$ then we get an induced map of cellular chain complexes

$$f_* : C(Y', X'; f^*\mathcal{F}) \to C(Y, X; \mathcal{F})$$

and we can define the cellular mapping cone $C'(f, \alpha)$ to be the alternating mapping cone of $f_*$ with filtration given by $C(\alpha)$. If $f$ is a homotopy equivalence then $C'(f_*)$ is acyclic and its torsion is, by definition, the torsion of $f$.

Now suppose that the attachment order does not respect the dimension, i.e., cells are attached on other cells of higher dimension or that the mapping $f$ is not cellular. Then the cellular chain complexes $C(Y, X; \mathcal{F})$ and $C'(f)$ are not well-defined. However we can form the following canonically defined filtered chain complexes and choose any minimal subcomplex as our cellular chain complex.

Let $E = S_*(Y, X; \mathcal{F})$ be the total relative singular complex of the pair $(Y, X)$ with coefficients in $\mathcal{F}$. For every $A \in \Lambda P$ let

$$E^A = S_*(Y^A, X; \mathcal{F}).$$

Then the characteristic map $\psi_i$ of each $e_i \in P$ gives a generator $[e_i]$ of $H_j(Y^{A \cup e_i}, Y^A; R)$ (where $j = \dim e_i$) which maps to $H_j(Y^{A \cup e_i}, Y^A; \mathcal{F})$ by $\gamma_i^*$. We let $\lambda_A(e_i) \in H^j(E^{A \cup e_i}/E^A; R)$ be the cohomology class which is dual to $\gamma_i^*[e_i]$.

**Theorem 4.2.1.** *$(E, \{\lambda_A(x)\})$ is a $\Lambda P$-filtered $R$-complex. Furthermore, if $(Y, X)$ is a CW-complex then the cellular chain complex $C(Y, X; \mathcal{F})$ is isomorphic as a based free $R$-complex to every minimal subcomplex of $E$.*

*Proof.* It is easy to verify that $(E, \{\lambda_A(x)\})$ is a $\Lambda P$-filtered $R$-complex. We will show only the second part.

Since the minimal subcomplex of $E$ and the cellular chain complex $C(Y, X; \mathcal{F})$ are free on the same set of generators it suffices to show that they every pair of cells in $Y$ with consecutive dimensions has the same incidence (in $R$) in both chain complexes. Thus it suffices to consider the special case when $P$ has only two elements $e_1 < e_2$ of dimension $j, j+1$.

In that case, the boundary map $d : R = C_{j+1} \to C_j = R$ of the cellular chain complex of $(Y, X)$ is given by the connecting homomorphism:

$$\partial : H_{j+1}(E, E^{e_1}) \to H_j(E^{e_1}, E^{\emptyset})$$

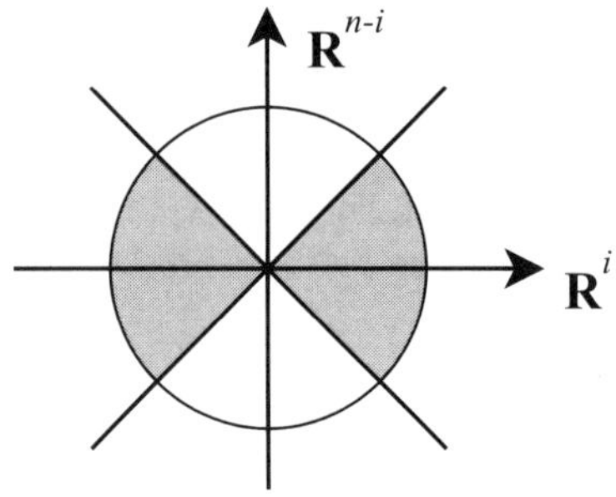


Figure 4.1: $\phi_x(U_x \cap f^{-1}[0,c])$

using the generators $\gamma_2^*[e_2]$ and $\gamma_1^*[e_1]$. Thus if $\partial\gamma_2^*[e_2] = \gamma_1^*[e_1]r$ with $r \in R$ then $d$ is left multiplication by $r$. Since the cohomology classes $\lambda^E(e_i)$ are dual to $\gamma_i[e_i]$ they must be related by $\lambda^E(e_2) = r\delta(\lambda^E(e_1))$. Since the cohomology classes of any minimal subcomplex $L$ are the restrictions of the cohomology classes on $E$ we have: $\lambda^L(e_2) = r\delta(\lambda^L(e_1))$. Consequently the two generators of $L$ are also incident by $r$. □

In the comparison case, $(f, \alpha)$ induces a map of total relative singular chain complexes

$$f_* : D = S_*(Y', X'; \mathcal{F}') \to E = S_*(Y, X; \mathcal{F})$$

sending $D^A$ into $E^{\alpha(A)}$ for all $A \in \Lambda P'$. Consequently we get an filtered chain complex $(C'(f_*), C(\alpha))$ by (4.1.10).

**Corollary 4.2.2.** *If $f$ is cellular and $\alpha$ preserves degree then every minimal subcomplex of $(C'(f_*), C(\alpha))$ is isomorphic to the alternating mapping cone of the map of cellular chain complexes induced by $f$.*

Now we will describe two different filtered chain complexes associated to a Morse function using the "standard" and "minimal" partial orderings of the singular set. We will need both of these later.

Suppose that $M$ is a compact smooth manifold with boundary $\partial M = \partial_0 M \coprod \partial_1 M$. Then by a *Morse function* on $M$ we mean a smooth function $f : M \to I$ with nondegenerate critical points in the interior of $M$ so that $f^{-1}(i) = \partial_i M$ for $i = 0, 1$.

Suppose that $x$ is a nondegenerate critical point of $f$ of index $i$ and critical value $f(x) = c$. Then there is a system of smooth coordinates $u_1, \cdots, u_n$ for $M$ defined in a neighborhood $U_x$ of $x$ so that

$$f|U_x = c - u_1^2 - \cdots - u_i^2 + u_{i+1}^2 + \cdots + u_n^2$$

The coordinate functions $u_1, \cdots, u_n$ define a smooth mapping $\phi_x : U_x \to \mathbb{R}^n$ which we may assume to be a diffeomorphism onto an open ball $B_\epsilon(0)$ centered at the origin. We will call $U_x$ a *standard ball neighborhood* of $x$ in $M$. The shaded region in Fig. 4.1 shows the image $\phi_x(U_x \cap f^{-1}[0,c])$ of all point $y \in U_x$ with $f(y) \leq f(x) = c$ which is given by

$$u_1^2 + \cdots + u_i^2 \geq u_{i+1} + \cdots + u_n^2$$

From this description the following lemma is clear.

**Lemma 4.2.3.** *1.* $H_i(f^{-1}[0,c], f^{-1}[0,c] - \{x\}) \cong \mathbb{Z}$.

*2.* $H_j(f^{-1}[0,c], f^{-1}[0,c] - \{x\}) = 0$ *if* $j \neq i$.

*3.* $f^{-1}[0,c]$ *is a deformation retract of* $f^{-1}[0,c] \cup U_x$ *for any standard ball neighborhood* $U_x$ *of* $x$ *in* $M$.

We need to choose a generator

$$[x] \in H_{\text{ind}(f,x)}(f^{-1}[0,c], f^{-1}[0,c] - \{x\}) \cong \mathbb{Z}$$

for each $x$ in the singular set $\Sigma(f)$. The Morse function $f$ together with a choice of generators $[x]$ for each $x \in \Sigma(f)$ will be called an *oriented Morse function.*

The *standard filtered chain complex* of $M$ associated to an oriented Morse function $f$ is given as follows. The graded poset is the singular set $\Sigma(f)$ of $f$ with degree given the index ($\deg x = \text{ind}(f,x)$) with the *standard partial ordering* given by $x < y$ when $f(x) < f(y)$. We let $\Sigma_s(f)$ denote $\Sigma(f)$ with the standard partial ordering. For every closed subset $A$ of $\Sigma_s(f)$ let $c = c(A)$ be given by

$$c(A) = \min(1, \min\{f(x) : x \in \Sigma_s(f) - A\})$$

Thus $c(\Sigma_s(f)) = 1$ and $c(A) = f(x)$ for any minimal element $x$ of $\Sigma_s(f) - A$ when $A \neq \Sigma_s(f)$. Let

$$M_s^A = \{y \in M : f(y) \leq c(A) \text{ and } y \notin \Sigma_s(f) - A\}$$

**Lemma 4.2.4.** *1.* $M_s^{\Sigma_s(f)} = M$

*2.* $M_s^{A \cap B} = M_s^A \cap M_s^B$ *for all* $A, B \in \Lambda\Sigma_s(f)$.

*3.* $H_{\text{ind}(f,x)}(M_s^{A\cup\{x\}}, M_s^A) \cong \mathbb{Z}$ *and* $H_j(M_s^{A\cup\{x\}}, M_s^A) = 0$ *for* $j \neq \text{ind}(f,x)$ *whenever* $A \in \Lambda\Sigma_s(f)$ *and* $x$ *is minimal in* $\Sigma_s(f) - A$.

*Proof.* If $c(A \cup \{x\}) = c(A)$ then $M_s^{A\cup\{x\}} = M_s^A \cup \{x\}$ and this lemma follows from the previous lemma. If $d = c(A \cup \{x\}) > c = c(A)$ then there are no critical points in $f^{-1}(c,d]$ and consequently $f^{-1}[0,c] = M_s^A \cup \{x\}$ is a deformation retract of $M_s^{A\cup\{x\}} = f^{-1}[0,d] - (\Sigma_s(f) - A - \{x\})$ so

$$H_j(M_s^{A\cup\{x\}}, M_s^A) \cong H_j(f^{-1}[0,c], f^{-1}[0,c] - \{x\}) \cong \mathbb{Z}$$

if $j = \text{ind}(f,x)$ and is zero otherwise. □

From this lemma we immediately get the following.

**Proposition 4.2.5.** *Given an oriented Morse function $f$ a $\Lambda\Sigma_s(f)$-filtered $\mathbb{Z}$-complex can be given as follows.*

1. $E^A = S_*(M_s^A, \partial_0 M)$, *the total relative singular chain complex of the pair* $(M_s^A, \partial_0 M)$, *for all* $A \in \Lambda\Sigma_s(f)$.

2. $\lambda_A(x) \in H^i(M_s^{A\cup\{x\}}, M_s^A; \mathbb{Z})$ *is the cohomology class dual to the orientation class* $[x] \in H_i(M_s^{A\cup\{x\}}, M_s^A)$ *whenever* $A, A\coprod\{x\} \subseteq \Sigma_s(f)$ *are closed and* $i = \operatorname{ind}(f, x)$.

**Remark 4.2.6.** The *cellular chain complex* of $(M, \partial_0 M)$ corresponding to $f$ is any minimal subcomplex of $E$. By Lemma 4.1.7 this is uniquely determined up to an upper triangular isomorphism which in turn is unique up to all higher homotopies. If the orientation of $f$ is changed then this chain complex will change by a $G$-monomial isomorphism where $G = \{\pm 1\}$. As in Theorem 4.2.1, this cellular chain complex is equal to the usual cellular chain complex associated to the Morse function when both are defined.

**Remark 4.2.7.** If $f, g : M \to I$ are two oriented Morse functions with $f^{-1}(0) = \partial_0 M = g^{-1}(0)$ then we get two filtered chain complexes $E_f, E_g$ with the same total complex $S_*(M, \partial_0 M)$. The identity map on this complex gives a filtered map $E_f \to E_g$ subordinate to the poset morphism $\alpha : \Sigma_s(f) \to \Sigma_s(g)$ which sends $A \in \Lambda\Sigma_s(f)$ to the smallest $B \in \Sigma_s(g)$ having the property that $M_s^A(f) \subseteq M_s^B(g)$.

Now we generalize this construction in two ways. First we use partial orderings of $\Sigma(f)$ other than the standard ordering and second we use twisted coefficients. To allow nonstandard partial orderings of $\Sigma(f)$ we need a Riemannian metric on $M$. We choose this metric so that the coordinate chart $\phi_x : U_x \to \mathbb{R}^n$ described above is an isometry on some standard ball neighborhood $U_x$ of every $x \in \Sigma(f)$. Then the gradient vector field $\nabla f$ with respect to such a metric will be equal to the standard vector field on each $U_x$:

$$\nabla f | U_x = (D\phi_x)^{-1}(-2u_1, \cdots, -2u_i, 2u_{i+1}, \cdots, 2u_n)$$

A new minimal partial ordering of $\Sigma(f)$ will be generated by the incidence relation with respect to the gradient flow.

For any $x, y \in M$ we say that $x$ is *incident to* $y$ (or $y$ is *incident under* $x$) if $f(x) \geq f(y)$ and there exists a trajectory of the smooth flow generated by $\nabla f$ which contains both $x$ and $y$ in its closure. *Multiple incidence* is defined to be the transitive relation on $M$ generated by incidence with respect to $\nabla f$. For each $x \in M$ we let $C(x)$ be the set of all points in $M$ which are multiply incident to $x$. If $U_x$ is a standard ball neighborhood of $x$ in $M$ then $C(x) \cap U_x = \phi_x^{-1}(0 \times \mathbb{R}^n)$ is an open $(n-i)$-ball.

**Proposition 4.2.8.** *Multiple incidence is a closed condition, i.e., the set of all pairs $(x, y)$ where $x$ is multiply incident to $y$ is a closed subset of $M \times M$.*

*Proof.* We will show that the complement $W$ of this set is open. So suppose $(x, y) \in W$, i.e., $x$ is not multiply incident to $y$. In particular $x \neq y$.

Case 1. Suppose $a = f(x) < b = f(y)$. In this case

$$(x, y) \in f^{-1}[0, c) \times f^{-1}(c, 1] \subseteq W$$

where $c = (a + b)/2$.

Case 2. Suppose $a \geq b$ and the closed interval $[b, a]$ contains at most one critical value of $f$, say $c$. (Let $c = (a + b)/2$ if $f$ has no critical points in this interval.) Then there is an $\epsilon > 0$ so that $c$ is the only (possible) critical value of $f$ in the open interval $(b - \epsilon, a + \epsilon)$. Every point in $f^{-1}[c, a + \epsilon)$ will be incident to a unique point in $L = f^{-1}(c)$ and every point in $f^{-1}(b - \epsilon, c]$ will be incident under a unique point in $L$. In particular, there will be unique points $x_0, y_0$ so that $x$ is incident to $x_0$ and $y_0$ is incident to $y$. Since $x$ is not multiply incident to $y$, $x_0 \neq y_0$ and we can choose disjoint open neighborhoods $U, V$ for $x_0, y_0$ in $L$. Then

$$(x, y) \in (U^+ \cup U^-) \times (V^+ \cup V^-) \subseteq W$$

where $U^+$ (resp. $U^-$) is the set of all points in $f^{-1}(b-\epsilon, a+\epsilon)$ which are incident to (resp. under) $U$ and similarly for $V^+, V^-$.

Case 3. Suppose that $a > b$ and the closed interval $[b, a]$ contains at least two critical values of $f$. Choose $\epsilon > 0$ as in Case 2 and let $c$ be a regular value of $f$ which lies between two of the critical values of $f$ in $[b, a]$. Let $L = f^{-1}(c)$, let $A$ be the set of all points in $L$ incident under $x$ and let $B$ be the set of all points in $L$ which are incident to $y$. Then, by induction on the number of critical values of $f$, $A, B$ are disjoint closed subsets of $L$. Let $U, V$ be disjoint open neighborhoods of $A, B$ in $L$. Then $x \times (L - U)$ and $(L - V) \times y$ are compact subsets of $W$. Again by induction on the number of critical values of $f$ we can conclude that both of these compact sets lie in the interior of $W$. This implies that there are open neighborhoods $U_0, V_0$ of $x, y$ in $M$ so that $U_0 \times (L - U)$ and $(L - V) \times V_0$ are contained in $W$. Since $(L - U) \cup (L - V) = L$ every point in $L$ is either not incident to any point in $V_0$ or not incident under any point in $U_0$. Thus $(x, y) \in U_0 \times V_0 \subseteq W$. □

Now take the partial ordering of $\Sigma(f)$ given by multiple incidence. We call this the *minimal* partial ordering and write $\Sigma^{min}(f)$ for $\Sigma(f)$ with this ordering. For each closed subset $A$ of $\Sigma^{min}(f)$ let $M^A(f)$ (or $M^A$ for short) be the (open) set of all elements of $M$ which are not multiply incident to any point in $\Sigma^{min}(f) - A$. Thus

$$M - M^A(f) = \bigcup_{x \in \Sigma(f) - A} C(x).$$

So, $M^{\Sigma(f)} = M$ and $M^{A \cup B} = M^A \cup M^B$.

**Lemma 4.2.9.** *Let $A \subseteq \Sigma^{min}(f)$ be closed in the minimal ordering and let $x$ be a minimal element of $\Sigma^{min}(f) - A$. Then for any coefficient sheaf $\mathcal{F}$ over $M$ we* *have* $H_i(M^{A\cup\{x\}}, M^A; \mathcal{F}) \cong \mathcal{F}_x$ *for* $i = \text{ind}(f,x)$ *and* $H_j(M^{A\cup\{x\}}, M^A; \mathcal{F}) = 0$ *for* $j \neq i$. *Furthermore, the orientation of* $f$ *gives a generator*

$$[x] \in H_i(M^{A\cup\{x\}}, M^A; R) \cong R$$

*for trivial ring coefficients* $R$.

*Proof.* First note that $M^{A\cup\{x\}} - M^A$ is the union of $\{x\}$ and the set of all regular points of $f$ which are incident to $x$. Thus the flow generated by $-\nabla f$ will deform $M^{A\cup\{x\}}$ into $M^A \cup U_x$ where $U_x$ is any standard ball neighborhood of $x$ inside $M^{A\cup\{x\}}$. Thus $(M^{A\cup\{x\}}, M^A)$ is homotopy equivalent to $(M^A \cup U_x, M^A)$ which, by excision, has the same homology as $(U_x, U_x - \phi_x^{-1}(\mathbb{R}^{n-i})) \simeq (\mathbb{R}^i, \mathbb{R}^i - \{0\})$. Thus

$$H_*(M^{A\cup\{x\}}, M^A; \mathcal{F}) \cong H_*(\mathbb{R}^i, \mathbb{R}^i - \{0\}; \mathcal{F}_x).$$

The integral orientation class $[x]$ gives a geometric orientation of $\mathbb{R}^i$ which specifies a generator of $H_i(M^{A\cup\{x\}}, M^A; \mathcal{F})$ for each generator of $\mathcal{F}_x$. For trivial $R$ coefficients we take the generator corresponding to $1 \in R$. □

Now suppose, as before, that $f$ is an oriented Morse function on $M$, we have a Riemannian metric on $M$ which is standard in a neighborhood of each critical point of $M$ and $\mathcal{F}$ is a coefficient sheaf of right $R$-modules over $M$ with $\mathcal{F}_* = R$ over the base point $* \in M$. We choose a homotopy class of paths $\gamma_x$ from each $x \in \Sigma(f)$ to the base point $*$. This fixes an isomorphism $\gamma_x^* : R \cong \mathcal{F}_x$ for each $x \in \Sigma(f)$ (given by lifting the path $\gamma_x$ to $\mathcal{F}$ in all possible ways). By Lemma 4.2.9 this determines a generator

$$\gamma_x^*[x] \in H_i(M^{A\cup\{x\}}, M^A; \mathcal{F}) \cong R$$

**Proposition 4.2.10.** *Let* $E = S_*(M, \partial_0 M; \mathcal{F})$ *be the total relative singular complex of* $(M, \partial_0 M)$ *with coefficients in* $\mathcal{F}$. *Then* $E$ *has the structure of a* $\Lambda\Sigma^{min}(f)$*-filtered* $R$*-complex given by:*

1. $E^A = S_*(M^A, \partial_0 M; \mathcal{F})$ *for all* $A \in \Lambda\Sigma^{min}(f)$.
2. $\lambda_A(x) \in H^i(E^{A\cup\{x\}}/E^A; R)$ *is the cohomology class dual to the generator* $\gamma_x^*[x] \in H_i(E^{A\cup\{x\}}/E^A) = H_i(M^{A\cup\{x\}}, M^A; \mathcal{F}) \cong R$ *given above.*

**Remark 4.2.11.** As in 4.2.6 any minimal subcomplex of $E$ gives a vertex in the simplicial set $\mathcal{M}_\bullet(R, \Sigma^{min}(f))$ representing the cellular chain complex of $(M, \partial_0 M)$. If the paths to the base point are changed this chain complex will change by a $G$-monomial isomorphism where $G$ is the image of $\pi_1(M)$ in $R$ given by the holonomy $\rho : \pi_1(M) \to Aut(\mathcal{F}_*) = R^\times$ of $\mathcal{F}$.

Suppose that $f, g : M \to I$ are two oriented Morse functions on the same manifold $M$ with $f^{-1}(0) = \partial_0 M = g^{-1}(0)$. Let $E_f, E_g$ be the total singular complex $S_*(M, \partial_0 M; \mathcal{F})$ with filtrations given by $f$ and $g$. For each $A \in \Lambda\Sigma^{min}(f)$ let $\alpha(A) \in \Lambda\Sigma^{min}(g)$ be the smallest element so that $M^A(f) \subseteq M^{\alpha(A)}(g)$. In other words, so that the unstable cone $C(y)$ with respect to the gradient of $g$ for each $y \in \Lambda\Sigma^{min}(g) - \alpha(A)$ lies in the union of the unstable cones $C(x)$ with respect to the gradient of $f$ for all $x \in \Lambda\Sigma^{min}(f) - A$.

**Proposition 4.2.12.** *The alternating mapping cone of the identity mapping* $(id, \alpha) : E_f \to E_g$ *is an acyclic filtered chain complex. Any minimal subcomplex has zero Whitehead torsion.*

*Proof.* This is well-know in a different language. However, the proof uses the results of the next two sections. Basically, any two oriented Morse functions are homotopic by a family of oriented generalized Morse functions $f_t$, $t \in I$. By Theorem 4.4.5, the alternating mapping cone construction on this family gives a $G$-expansion homotopy from the cone of $D \to E$ to the cone of the identity on $E$ which obviously has zero Whitehead torsion. □

## 4.3 The $G$-monomial functor of a family of Morse functions

Suppose that $p : E \to B$ is a smooth fiber bundle with fiber $M$ where $E, B, M$ are compact connected smooth manifolds. Let $E_0, E_1 \subseteq E$ be disjoint subbundles with fibers $\partial_0 M, \partial_1 M$ so that $M = \partial_0 M \coprod \partial_1 M$. Suppose that $f : E \to I$ is a smooth function with $f^{-1}(i) = E_i$ for $i = 0, 1$ and suppose that the restriction $f_t = f|M_t$ of $f$ to the fiber $M_t = p^{-1}(t)$ is a Morse function for all $t \in B$. Then the singular set

$$\Sigma(f) = \bigcup_{t \in B} \Sigma(f_t)$$

is a covering space of $B$ and its normal bundle in $E$ is the restriction of the *vertical tangent bundle* (the subbundle of the tangent bundle of $E$ consisting of vectors tangent to the fibers).

We will assume in this section that $f$ is an *oriented* fiberwise Morse function. This means that $f_t$ is oriented for each $t \in B$ and the orientation varies continuously with $t$. In other words, the negative eigenspaces of the critical points of $f$ form an oriented vector bundle over $\Sigma(f)$.

**Lemma 4.3.1 (Parameterized Morse lemma).** *For each point $x$ in $\Sigma(f)$ there is a connected neighborhood $W(x)$ of $p(x)$ in $B$ and coordinate functions $u_1, \cdots, u_n$ defined in a neighborhood $U_{W(x)}$ of the component of $x$ in $\Sigma(f)|W(x)$ in $p^{-1}(W(x))$ so that $\phi_{W(x)} = (p, u_1, \cdots, u_n) : U_{W(x)} \to W(x) \times \mathbb{R}^n$ is a smooth embedding and*

$$f|U_{W(x)} = c \circ p - u_1^2 - \cdots - u_i^2 + u_{i+1}^2 + \cdots + u_n^2$$

*where $c : W(x) \to \mathbb{R}$ is a smooth function. Furthermore, for all $x, y$ in $\Sigma(f)$ with the same index $i$ there is a smooth function*

$$g : W(x) \cap W(y) \to SO(i) \times O(n - i)$$

*so that $\phi_W(y) = \hat{g} \circ \phi_{W(x)}$ on $U_{W(x)} \cap U_{W(y)}$ where $\hat{g}(a, b) = (a, g(a)(b))$.*

**Remark 4.3.2.** For any connected subset $W$ of $W(x)$ and any sufficiently small $\epsilon > 0$ we will call $U_W = \phi_{W(x)}^{-1}(W \times B_\epsilon(0))$ a *uniform standard ball neighborhood* of the component of $x$ in $\Sigma(f)|W$.

*Proof.* See, e.g., the appendix to [Igu88]. □

Choose a Riemannian metric on $E$ so that the fibers $M_t$ are totally geodesic and so that the vertical gradient of $f$ is standard near $\Sigma(f)$. (By the *vertical gradient* we mean the component $\nabla_{\mathsf{v}} f$ of the gradient of $f$ in direction of the vertical tangent bundle, or equivalently, the gradient of the restricted functions $f_t$. By *standard near* $\Sigma(f)$ we mean that $\phi_W : U_W \to W \times B_\epsilon(0)$ is an isometry on each fiber so that

$$\nabla_{\mathsf{v}} f = (D\phi_W)^{-1}(-2u_1, \cdots, -2u_i, 2u_{i+1}, \cdots, 2u_n)$$

on $U_W$.) Then multiple incidence is defined fiberwise and is a closed relation as in 4.2.8. In other words, the set

$$\bigcup_{y \in E} C(y) \times y$$

of all pairs $(x, y)$ in $E$ so that $p(x) = p(y)$ and $x$ is multiply incident to $y$ is a closed subset of $E \times E$.

Let $K$ be the set of all pairs $(x, y)$ of distinct points in $\Sigma(f)$ so that $p(x) = p(y)$. Then $K$ is a finite covering space of $B$ and is therefore compact. Let $K_+$ be the subset of $K$ of all pairs $(x, y)$ where $x$ is multiply incident to $y$. Then $f(x) > f(y)$ for all $(x, y) \in K_+$. Since $K_+$ is compact, there is an $\delta > 0$ so that $f(x) > f(y) + \delta$ for all $(x, y) \in K_+$. In other words, when two critical points of $f_t : M_t \to I$ have critical value within $\delta$ they are unrelated in the partial ordering given by multiple incidence.

The critical values of $f_t$ vary smoothly with $t \in B$. Consequently, the difference between two critical values of $f_t$ will change by less than $\delta$ on all "small" subsets of $B$. More precisely:

**Definition 4.3.3.** A subset $U$ of $B$ will be called *small* if it satisfies the following properties.

1. $U$ is evenly covered by $\Sigma(f)$, i.e., there is a diffeomorphism

   $$\psi : \Sigma(f_{t_0}) \times U \to \Sigma(f)|U$$

   for some $t_0 \in U$ so that $p(\psi(x, t)) = t$.

2. $U$ is path connected. (So that $\psi$ is well defined.)

3. For any $x \neq y$ in $\Sigma(f_{t_0})$, the function

   $$g(t) = f_t(\psi(x, t)) - f_t(\psi(y, t))$$

   varies by less than $\delta$ on $U$ (i.e., $\max(g) - \min(g) < \delta$).

Now suppose that $\sigma : \Delta^k \to B$ is a smooth mapping with small image. Let

$$M_\sigma = \{(x, t) \in E \times \Delta^k | p(x) = \sigma(t)\}$$

be the pull-back of $E$ to $\Delta^k$ and let $f_\sigma : M_\sigma \to I$ be the composition $M_\sigma \to E \to_f I$. Then the singular set $\Sigma(f_\sigma) \subseteq M_\sigma$ of $f_\sigma$ is the pull-back of $\Sigma(f)$ along $\sigma$. For each $t \in \Delta^k$ let $M_t = M_{\sigma(t)} \times t$ be the fiber of $M_\sigma$ over $t$ and let $f_t = f_\sigma | M_t$.

Let $\Sigma_\sigma = \Sigma(f_b) \subseteq M_b$ be the fiber of $\Sigma(f_\sigma)$ over the barycenter $b \in \Delta^k$ and let

$$\psi : \Sigma_\sigma \times \Delta^k \to \Sigma(f_\sigma)$$

be as in the definition above. Then $\Sigma_\sigma$ has two partial ordering. One is the *standard* partial ordering given by $x > y$ if $f(\psi(x,t)) > f(\psi(y,t))$ for all $t \in \Delta^k$. The other is the *minimal* partial ordering which is defined to be the transitive closure of the relation: $x > y$ if $\psi(x,t)$ is multiply incident to $\psi(y,t)$ in $M_t$ for some $t \in \Delta^k$. In other words, $x > y$ if there is a sequence of elements $x = z_0, z_1, \cdots, z_n = y$ in $\Sigma_\sigma$ and corresponding elements $t_1, \cdots, t_n$ in $\Delta^k$ so that $\psi(z_{i-1}, t_i)$ is multiply incident to $\psi(z_i, t_i)$ for $i = 1, \cdots, n$. Since $\sigma(\Delta^k)$ is small, $f(\psi(x,t)) > f(\psi(z_1,t)) > \cdots > f(\psi(y,t))$ for all $t \in \Delta^k$. Consequently the standard ordering is a refinement of the minimal ordering.

If $\tau$ is a face (or degeneracy) of $\sigma$ then the mapping $\Sigma_\tau \to \Sigma_\sigma$ induced by the inclusion $\Sigma(f_\tau) \subseteq \Sigma(f_\sigma)$ is an order preserving bijection in the minimal ordering and a closed bijection in the standard ordering. The following diagram indicates the direction of the natural poset morphisms.

$$\begin{array}{ccc} \Sigma_\tau & \longrightarrow & \Sigma_\sigma \\ \downarrow & & \downarrow \\ \Sigma_\tau^{min} & \longleftarrow & \Sigma_\sigma^{min} \end{array}$$

Therefore $\Sigma$ with the standard ordering defines a poset functor on the category of small simplices of $B$.

To each small simplex $\sigma$ of $B$ we will be associating a cellular chain complex for $M_\sigma$. This means that each such $\sigma$ behaves like a vertex. In other words we will be working with the first barycentric subdivision of $B$. More precisely we will work with

$$SdB = \mathcal{N}_\bullet \operatorname{simp}^{op} B$$

which is the simplicial nerve of the opposite of the category of smooth small simplices in $B$. To avoid some clumsy terminology we will choose a fixed smooth triangulation of $B$ with small simplices. The vertices of $SdB$ are the simplices of $B$ and the $k$-simplices of $SdB$ are $(k+1)$-tuples of simplices $\sigma_0 \supseteq \sigma_1 \supseteq \cdots \supseteq \sigma_k$. (We call $SdB$ the *first barycentric subdivision* of $B$ since a homeomorphism $|SdB| \cong |B|$ is given by sending each vertex $\sigma$ of $SdB$ to the barycenter $b_\sigma$ of $\sigma$ and extending linearly.) The poset functor $\Sigma^{sd}$ on $\operatorname{simp} SdB$ is defined on this by

$$\Sigma^{sd}(\sigma_0, \sigma_1, \cdots, \sigma_k) = \Sigma_{\sigma_0}$$

with the standard partial ordering. Over each simplex $\sigma$ however we will use the minimal ordering. We can always recover the correct structure using the forgetful functors:

$$\mathcal{F}(R, \Sigma_\sigma^{min}) \to \mathcal{F}(R, \Sigma_\sigma)$$

$$\mathcal{M}_\bullet(R, \Sigma_\sigma^{min}) \to \mathcal{M}_\bullet(R, \Sigma_\sigma)$$

For every closed subset $A$ of $\Sigma_\sigma^{min}$ let

$$M_\sigma^A = M_\sigma - \bigcup_{t \in \Delta^k} \left( \bigcup_{x \in \Sigma(f_t) - A} C(x) \times t \right)$$

be the set of all elements of $M_\sigma$ which are not multiply incident to any point in $\Sigma(f_\sigma) - \psi(A \times \Delta^k)$. Over every $t \in \Delta^k$ this agrees with the definition in the previous section. In particular note that:

1. $M_\sigma^{\Sigma_\sigma} = M_\sigma$
2. $M_\sigma^{A \cap B} = M_\sigma^A \cap M_\sigma^B$ for all $A, B \in \Lambda\Sigma_\sigma^{min}$.

We also have the following statement analogous to Lemma 4.2.9 with the same proof.

**Lemma 4.3.4.** *Let $A \subseteq \Sigma_\sigma^{min}$ be closed and let $x$ be a minimal element of $\Sigma_\sigma^{min} - A$. Then for any coefficient sheaf $\mathcal{F}$ over $E$ we have $H_i(M_\sigma^{A \cup \{x\}}, M_\sigma^A; \mathcal{F}) \cong \mathcal{F}_x$ if $i = \deg x$ and $H_j(M_\sigma^{A \cup \{x\}}, M_\sigma^A; \mathcal{F}) = 0$ for $j \neq i$. Furthermore, the orientation of $f$ gives a generator*

$$[x] \in H_i(M_\sigma^{A \cup \{x\}}, M_\sigma^A; R) \cong R$$

*for trivial ring coefficients $R$.*

*Proof.* The flow generated by $-\nabla_v f$ will deform $M_\sigma^{A \cup \{x\}}$ into $M_\sigma^A \cup U_x$ where $U_x$ is any uniform standard ball neighborhood of $\psi(x \times \Delta^k)$ inside $M_\sigma^{A \cup \{x\}}$. Thus $(M_\sigma^{A \cup \{x\}}, M_\sigma^A)$ is homotopy equivalent to $(M_\sigma^A \cup U_x, M_\sigma^A)$ which, by excision, has the same homology as $(U_x, U_x - \phi_x^{-1}(\Delta^k \times \mathbb{R}^{n-i})) \simeq (\mathbb{R}^i, \mathbb{R}^i - \{0\})$. ☐

This time we choose a base point $* \in E$ and a homotopy class of paths $\gamma_{\sigma x}$ from $x \in \Sigma(f_{\sigma(b)}) \subseteq M_{\sigma(b)}$ to $*$ in $E$ where $b \in \Delta^k$ is the barycenter. Assuming that $\mathcal{F}$ is a coefficient sheaf of right $R$-modules so that $\mathcal{F}_* = R$ this gives an isomorphism $\gamma_{\sigma x}^* : R \cong \mathcal{F}_x$ for each $x \in \Sigma_\sigma$ and thus a generator

$$\gamma_{\sigma x}^*[x] \in H_i(M_\sigma^{A \cup \{x\}}, M_\sigma^A; \mathcal{F}) \cong R$$

Since $\Delta^k$ is contractible, we get a continuous family of paths from (the image in $E$ of) the points of $\Sigma(f_\sigma)$ to the base point of $E$. However, since these paths cannot in general be chosen continuously up to homotopy for all points in $\Sigma(f)$ the choice of paths for $\sigma$ will not in general agree with the choice of paths on the faces $\tau$ of $\sigma$. They will differ by an element of $\pi_1(E)$ and the isomorphisms $R \cong \mathcal{F}_x$ will differ by an element of the image $G$ of $\pi_1(E)$ in $Aut(\mathcal{F}_*) = R^\times$.

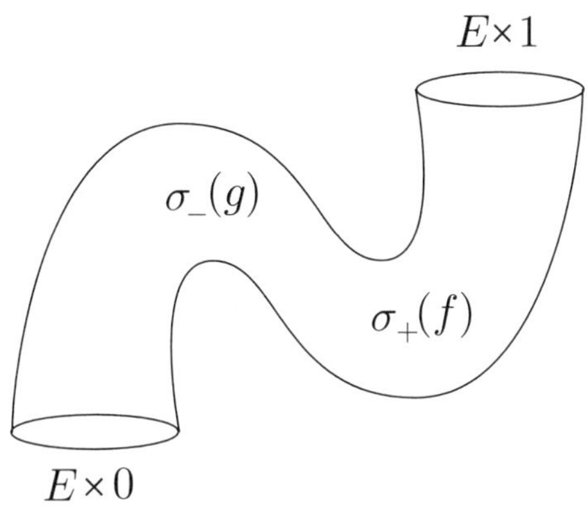


Figure 4.2: Geometric comparison map

**Proposition 4.3.5.** *Let $D = S_*(M_\sigma, \partial_0 M_\sigma; \mathcal{F})$ be the total relative singular complex of $(M_\sigma, \partial_0 M_\sigma)$ with coefficients in $\mathcal{F}$. Then $D$ has the structure of a $\Lambda\Sigma_\sigma^{min}$-filtered $R$-complex given by:*

1. *$D^A = S_*(M_\sigma^A, \partial_0 M_\sigma; \mathcal{F})$ for all $A \in \Lambda\Sigma_\sigma^{min}$.*
2. *$\lambda_A(x) \in H^i(D^{A\cup\{x\}}/D^A; R)$ is the cohomology class dual to the generator $\gamma_{\sigma x}^*[x] \in H_i(D^{A\cup\{x\}}/D^A) = H_i(M_\sigma^{A\cup\{x\}}, M_\sigma^A; \mathcal{F}) \cong R$ given above.*

**Remark 4.3.6 (Comparison case).** If we have two oriented fiberwise Morse functions $f, g$ on $E$ we get two different filtered chain complex structures, say $D_f, D_g$, on the same chain complex $D = S_*(M_\sigma, \partial_0 M_\sigma; \mathcal{F})$ for every small simplex $\sigma$ in $B$ . Consequently we can take the alternating cone of the identity $C'(D_f, D_g) = C'(id : D_g \to D_f)$ and obtain a family of acyclic filtered chain complexes indexed by the small simplices $\sigma$ of $B$. In the case when $M$ is a closed manifold, this can be visualized as the filtered chain complex of an oriented fiberwise Morse function on $(E \times I, E \times 0)$ as shown in Figure 4.2.

Let $D(\sigma) = (D_\sigma, \Sigma_\sigma^{min}, \{\lambda_A^\sigma(x)\})$ denote the above filtered $R$-complex and let $\xi^{min}(\sigma) \in \mathcal{M}_0(R, \Sigma_\sigma^{min})$ be a fixed choice of a minimal subcomplex of $D(\sigma)$ for every small simplex $\sigma$ in $B$. (The $^{min}$ in $\xi^{min}$ refers to the minimal partial ordering of $\Sigma_\sigma$.)

For any face $\tau$ of $\sigma$, $(M_\tau, \partial_0 M_\tau) \subseteq (M_\sigma, \partial_0 M_\sigma)$ so $S_*(M_\tau, \partial_0 M_\tau; \mathcal{F})$ is a subcomplex of $S_*(M_\sigma, \partial_0 M_\sigma; \mathcal{F})$. The cohomology classes $\lambda_A^\tau(x)$ for $\tau$ may not agree with the restrictions of the corresponding classes $\lambda_A^\sigma(x)$ for $\sigma$ and the poset $\Sigma_\tau^{min}$ may not be isomorphic to $\Sigma_\sigma^{min}$. However if we use the same poset $\Sigma_\sigma^{min}$ and if we use the restricted classes $\lambda_A^\sigma(x)|\tau$ we get a morphism of filtered chain complexes

$$D(\sigma, \tau) = (D_\tau, \Sigma_\sigma^{min}, \{\lambda_A^\sigma(x)|\tau\}) \to D(\sigma) = (D_\sigma, \Sigma_\sigma^{min}, \{\lambda_A^\sigma(x)\})$$

The filtered chain complex $D(\sigma, \tau)$ has the same underlying chain complex $D_\tau$ as $D(\tau)$ with a subcollection of its set of subcomplexes since $\Lambda\Sigma_\sigma^{min} \subseteq \Lambda\Sigma_\tau^{min}$

and with its cohomology classes $\lambda_A(x)$ altered by the action of $G$. Consequently the minimal subcomplex $\xi^{min}(\tau)$ of $D(\tau)$ also serves as a minimal subcomplex of $D(\sigma,\tau)$ by forgetting part of its structure and by a $G$-monomial change of basis. Let $\xi^{min}(\sigma,\tau) \in \mathcal{M}_0(R,\Sigma_\sigma^{min})$ denote this minimal subcomplex of $D(\sigma,\tau)$. Then $\xi^{min}(\sigma,\tau)$ is natural in $\sigma$ in the sense that if $\sigma \subseteq \rho$ then the identity map $I : \xi^{min}(\rho,\tau) \to \xi^{min}(\sigma,\tau)$ is a $G$-monomial morphism subordinate to the poset morphism $\Sigma_\rho^{min} \to \Sigma_\sigma^{min}$. Also note that $\xi^{min}(\tau,\tau) = \xi^{min}(\tau)$.

If $\tau_1$ is a face of $\tau_0$ then the filtered mapping $D(\sigma,\tau_1) \to D(\sigma,\tau_0)$ induces an upper triangular morphism

$$\xi^{min}(\sigma,\tau_1) \to \xi^{min}(\sigma,\tau_0)$$

subordinate to the identity morphism on $\Sigma_\sigma^{min}$ which is unique up to homotopy. Choosing such a morphism would give an element $\xi^{min}(\sigma,\tau_0,\tau_1)$ of $\mathcal{M}_1(R,\Sigma_\sigma^{min})$ with vertices $\xi^{min}(\sigma,\tau_0), \xi^{min}(\sigma,\tau_1)$. If we chose one such morphism in the universal case when $\sigma = \tau_0$ (taking the identity if $\tau_1 = \tau_0$) then the induced choices for all $\sigma \supseteq \tau_0$ will be natural in $\sigma$ in the sense that the identity mapping $I$ gives a $G$-monomial morphism

$$I : \xi^{min}(\rho,\tau_0,\tau_1) \to \xi^{min}(\sigma,\tau_0,\tau_1)$$

for all $\rho \supseteq \sigma \supseteq \tau_0 \supseteq \tau_1$.

If $\tau_2 \subseteq \tau_1 \subseteq \tau_0$ are faces of $\sigma$ then we have a diagram

$$\begin{array}{ccc} \xi^{min}(\sigma,\tau_2) & \longrightarrow & \xi^{min}(\sigma,\tau_0) \\ & \searrow \quad \nearrow & \\ & \xi^{min}(\sigma,\tau_1) & \end{array}$$

which commutes if $\tau_0,\tau_1,\tau_2$ are not distinct. In general, by Lemma 4.1.7, this diagram commutes up to a homotopy subordinate to the identity on $\Sigma_\sigma^{min}$. Choose one such homotopy in the universal case $\sigma = \tau_0$ (taking the zero homotopy in the degenerate case) and we get an element $\xi^{min}(\sigma,\tau_0,\tau_1,\tau_2) \in \mathcal{M}_2(R,\Sigma_\sigma^{min})$ for all $\sigma \supseteq \tau_0$ which is natural in $\sigma$. Proceeding in this way we get an element

$$\xi^{min}(\sigma,\tau_0,\cdots,\tau_k) \in \mathcal{M}_k(R,\Sigma_\sigma^{min})$$

for every sequence of faces $\tau_0 \supseteq \tau_1 \supseteq \cdots \supseteq \tau_k$ of $\sigma$ so that if $\sigma$ is a face of $\rho$ then the identity chain map

$$I : \xi^{min}(\rho,\tau_0,\cdots,\tau_k) \to \xi^{min}(\sigma,\tau_0,\cdots,\tau_k)$$

is a $G$-monomial (simple) morphism subordinate to the poset morphism $\Sigma_\rho^{min} \to \Sigma_\sigma^{min}$.

**Proposition 4.3.7.** *A $G$-monomial functor $\xi$ on SdB subordinate to the poset functor $\Sigma^{sd}$ (the standard poset on the last vertex) is given by taking $\xi(\sigma_0,\cdots,\sigma_k)$ to be the image of $\xi^{min}(\sigma_0,\sigma_0,\cdots,\sigma_k)$ under the simplicial forgetful mapping*

$$F : \mathcal{M}_\bullet(R,\Sigma_\sigma^{min}) \to \mathcal{M}_\bullet(R,\Sigma_\sigma)$$

*and by taking inclusion morphisms $a^*(\sigma_*) \to \sigma_*$ to the $G$-monomial morphisms $I : \xi(a^*(\sigma_*)) \to a^*\xi(\sigma_*)$ given by the identity chain map.*

*Proof.* Reducing the structure of $\Sigma_\sigma^{min}$ to the standard poset $\Sigma_\sigma$ reverses the direction of the $G$-monomial morphisms as required by the definition of a $G$-monomial functor. For example:

$$\xi(\sigma_1, \cdots, \sigma_k) = F(\xi^{min}(\sigma_1, \sigma_1, \cdots, \sigma_k))$$

$$\xrightarrow{I} F(\xi^{min}(\sigma_0, \sigma_1, \cdots, \sigma_k)) = \partial_0 \xi(\sigma_0, \cdots, \sigma_k))$$

is a $G$-monomial morphism since $\sigma_1 \subseteq \sigma_0$. □

We summarize the results of this section in the following theorem.

**Theorem 4.3.8.** *Let $M \to E \to B$ be a smooth bundle and let $f : E \to I$ be an oriented fiberwise Morse function. Let $\mathcal{F}$ be a coefficient sheaf of right $R$-modules over $E$ with stalk $\mathcal{F}_* = R$ over the base point $* \in E$. Let $G$ be the image of the holonomy $\pi_1(E) \to Aut(\mathcal{F}_*) = R^\times$ Choose any triangulation of $B$ by smooth small simplices and let $SdB$ be the first barycentric subdivision of $B$.*

*Then there exists a $G$-monomial functor*

$$\xi_f : \text{simp}\, SdB \to \text{simp}\, \mathcal{W}h_\bullet(R, G)$$

*which is well-defined up to $G$-monomial simplicial homotopy so that the homology of the total complex of $\xi_f$ is isomorphic to $H_*(E, E_0; \mathcal{F})$.*

*Furthermore the induced map on realizations*

$$|\xi_f| : B \simeq |SdB| \to |\mathcal{W}h_\bullet(R, G)|$$

*is, up to homotopy, independent of the triangulation of $B$ and the metric on $E$.*

*If $g$ is another oriented fiberwise Morse function on $E$ then the alternating mapping cone of the identity gives a $G$-monomial functor*

$$C'(\xi_f, \xi_g) : \text{simp}\, SdB \to \text{simp}\, \mathcal{W}h^h_\bullet(R, G)$$

*with subfunctor $\xi_f$ and quotient functor $C'(\xi_f, \xi_g)/\xi_f \cong \Sigma'(\xi_g)$ the alternating suspension of $\xi_g$.*

*Proof.* If we change the paths to the base point, the systems of higher homotopies $\xi(\sigma_*)$ will change by $G$-monomial isomorphisms and thus $\xi$ will change by a $G$-monomial simplicial homotopy. If we change the metric on $E$ and choose a new triangulation of $B$ which is small with respect to the conditions given by the new metric then we can find a metric on $E \times I$ and a triangulation of $B \times I$ which restricts to the two choices on the two ends. Thus $|\xi|$ is unique up to homotopy.

The fact that $H_*(E(Z(\xi))) \simeq H_*(E, E_0; \phi)$ follows from the fact that

$$Z_k(\xi(\sigma_*)) \simeq S_*(M_\sigma, \partial_0 M_\sigma; \mathcal{F})$$

and the fact that $Z$ and $S_*$ commute with pushout.

The comparison case follows from Remark 4.3.6. □

## 4.4 $G$-expansion functor of a family of GMF's

Let $E, B, M$ are as before but with $\dim M = n+1$ and $\dim B = m$. Suppose that $f : E \to I$ is a smooth function with $f^{-1}(i) = E_i$ for $i = 0, 1$ but suppose this time that the restriction $f_t = f|M_t$ of $f$ to the fiber $M_t = p^{-1}(t)$ is a GMF (*generalized Morse function*) for all $t \in B$. Thus the critical points of $f_t$ are either nondegenerate or birth-death points.

Let $\Sigma^{(1,i)}(f_t) \subseteq \Sigma(f_t)$ be the set of all birth-death points of $f_t$ of index $i$ and let

$$\Sigma^{(1,i)}(f) = \bigcup_{t \in B} \Sigma^{(1,i)}(f_t) \times t \subseteq \Sigma(f).$$

Let $\Sigma^{(1)}(f_t), \Sigma^{(1)}(f)$ be the union of all $\Sigma^{(1,i)}(f_t), \Sigma^{(1,i)}(f)$, respectively. Similarly, let $\Sigma^{(0,i)}$ and $\Sigma^{(0)}$ denote sets of nondegenerate points of index $i$ and of all indices, respectively.

We may, by a small perturbation of $f$, assume that the birth-death points of $f$ are all transversely unfolded. This means that at each $x \in \Sigma^{(1,i)}(f)$ there is a connected neighborhood $W(x)$ of $p(x)$ in $B$ and a system of local coordinates $t_1, \cdots, t_m, u_0, \cdots, u_n$ defined in a neighborhood $U_{W(x)}$ of $x$ in $p^{-1}(W(x))$ so that $t_1, \cdots, t_m : U_{W(x)} \to \mathbb{R}$ factor through $W(x)$ inducing a smooth embedding $W(x) \to \mathbb{R}^m$,

$$(p, u_1, \cdots, u_n) : U_{W(x)} \to W(x) \times \mathbb{R}^n$$

is a smooth embedding with image $W(x) \times B_\epsilon(0)$ and

$$f|U_{W(x)} = c(t_1, \cdots, t_m) - 3t_1 u_0 + u_0^3 - u_1^2 - \cdots - u_i^2 + u_{i+1}^2 + \cdots + u_n^2$$

where $c : \mathbb{R}^m \to \mathbb{R}$ is a smooth function.

For $t_1 > 0$ the function $f_t$ has two nondegenerate critical points $x^+, x^-$ with coordinates $(-\sqrt{t_1}, 0, \cdots, 0), (\sqrt{t_1}, 0, \cdots, 0)$ having indices $i+1$ and $i$ respectively. Their critical values are: $f_t(x^+) = c(t) + 2t_1^{3/2}$ and $f_t(x^-) = c(t) - 2t_1^{3/2}$. If $U_t = p^{-1}(t) \cap U_{W(x)}$ and

$$\phi_t = (u_0, \cdots, u_n) : U_t \xrightarrow{\approx} B_\epsilon(0)$$

is the diffeomorphism given by our local coordinates then we would like $\phi_t$ to be an isometry with respect to some Riemannian metric on $E$. However this would violate the parametrized Morse lemma (4.3.1) so we need a new statement of that lemma.

**Lemma 4.4.1.** *There are two neighborhoods $U(i) \subseteq V(i)$ of $\Sigma^{(1,i)}(f)$ in $E$ for each $i$ and smooth embeddings*

$$(p, \phi_\alpha) = (p, u_0, u_1, \cdots, u_n) : V_\alpha \to B \times \mathbb{R}^{n+1}$$

*defined separately on open sets $V_\alpha$ covering the entire singular set $\Sigma(f)$ and $V = \cup V(i)$ so that:*

1. *On* $V_\alpha \cap V_\beta$, $\phi_\alpha$ *and* $\phi_\beta$ *differ by a rotation* $g_{\alpha\beta} : p(V_\alpha \cap V_\beta) \to O(n+1)$

2. $f|(V_\alpha \cap U) = c \circ p + 3t_1 u_0 - \|(u_1, \cdots, u_i)\|^2 + \|(u_{i+1}, \cdots, u_n)\|^2$ *where* $t_1$ *is a nonsingular smooth function on* $U$ *which locally factors through* $p$. *(I.e., its restriction to each* $V_\alpha$ *factor through* $p$.*)*

3. $f|(V_\alpha \cap V) = h(p, u_0) - \|(u_1, \cdots, u_i)\|^2 + \|(u_{i+1}, \cdots, u_n)\|^2$ *where* $h(p, u_0) = h \circ (p, u_0)$ *for some smooth* $h : (p, u_0)(V) \to \mathbb{R}$.

4. $f|(V_\alpha - V) = c \circ p + \sum_{j=0}^{n} \pm u_j^2$

*Proof.* See, e.g., the appendix to [Igu88]. □

Now we can choose a Riemannian metric on $E$ making all fibers $M_t$ totally geodesic which is *standard* in a neighborhood of the critical set in the sense that the coordinate charts $\phi_\alpha$ in the lemma above are isometries on each fiber $V_\alpha \cap M_t$. Define *multiple incidence* using the *vertical gradient* $\nabla_{\mathsf{v}} f$.

Multiple incidence is, as always, a closed relation since the cancellation of critical points (at $t_1 = 0$) does not create any new incidences (when $t_1 < 0$). A typical event is that a critical point $z$ is incident to $x^-$ and another critical point $y$ is incident under $x^+$ but $z$ is not incident to $y$ on the open set given by $t_1 > 0$. At the moment of cancellation $z$ becomes multiply incident to $y$ by way of the birth-death point and for $t_1 < 0$ they become simply incident.

This kind of event will be discussed later. At this point, however, we will assume that birth-death points are *independent* in the sense that they are not incident to any other critical points. We will later show how this condition can be achieved, both geometrically and algebraically.

Since birth-death points are, for the moment, independent by assumption and multiple incidence is closed, the cancelling pair of critical points converging to the birth-death point will also be independent from all other critical points. (They will of course be incident to each other.)

This means there exists a fixed $\delta > 0$ so that whenever $x$ is incident to $y$ in $M_t$ then either

1. $f_t(x) > f_t(y) + 2\delta$ or

2. $(x, y) = (x^+, x^-)$ is a *cancelling pair* of critical points in a standard neighborhood of the birth-death set.

Furthermore, in the second case when $(x, y) = (x^+, x^-)$ and $f_t(x) \leq f_t(y) + 2\delta$, we are assuming that $x^+, x^-$ are independent from all other critical points.

We can now define the "small simplices." Any connected subset $D$ of $B$ will be called *small* if the difference between any pair of critical values changes by less than $\delta$ on the set $D$. In particular this means that if $x, y \in \Sigma(f_t)$ for some $t \in D$ with $f_t(x) > f_t(y) + \delta$ then $x, y$ cannot come together and cancel at a birth-death point over some other point in $D$.

Any smooth mapping $\sigma : \Delta^k \to B$ with small image will be called a (singular) *small simplex* in $B$. If $\sigma$ is such a simplex we use the notation $f_\sigma : M_\sigma \to I$

and $f_t : M_t \to I$ for $f_{\sigma(t)} : M_{\sigma(t)} \to I$ as in the last section. However we use a different definition of $\Sigma_\sigma$.

We let $\Sigma_\sigma$ be the set $\Sigma_b \subseteq M_b$ over the barycenter $b \in \Delta^k$ with a few points deleted. First, delete any critical point which lies in the same component in $\Sigma(f_\sigma)$ as a birth-death point. Then the remaining points give an embedding

$$\psi : \Sigma_\sigma \times \Delta^k \to \Sigma(f_\sigma) \subseteq M_\sigma$$

over $\Delta^k$ whose image is the union of all components of $\Sigma(f_\sigma)$ not containing any birth-death points. Now delete any pair of cancelling critical points $x^+, x^- \in \Sigma_b$ with the property that $f(\psi(x^+,t)) - f(\psi(x^-,t)) < \delta$ for all $t \in \Delta^k$. We say that $x^+$ and $x^-$ are *within* $\delta$ in that case. These points should be deleted because $\sigma$ could be a face of another small simplex $\rho$ for which the cancelling pair might be in the same component of $\Sigma_\rho$ as a birth-death point. The sets $\Sigma_\sigma$ have the property that whenever $\tau$ is a face of a small simplex $\sigma$ the inclusion $\Sigma(f_\tau) \subseteq \Sigma(f_\sigma)$ induces a monomorphism of sets $\pi_{\sigma\tau} : \Sigma_\tau \to \Sigma_\sigma$.

As before there are two partial orderings on $\Sigma_\sigma$ that we will use. One is a modified standard ordering and the other is the minimal ordering. Recall that the *standard ordering* is given by $x > y$ if $f(\psi(x,t)) > f(\psi(y,t))$ for all $t \in \Delta^k$. We modify this ordering by taking any cancelling pair $x^+, x^-$ so that $f(\psi(x^+,t)) - f(\psi(x^-,t)) = \delta$ at some $t \in \Delta^k$ and making it independent in the partial ordering from the other critical points. We will call these the *(independent) expansion pairs* of $\sigma$. Since they are, by assumption, independent in the multiple incidence ordering, the modified standard ordering is a refinement of the minimal ordering which is defined as before by making all cancelling pairs independent in the standard ordering. As before we write $\Sigma_\sigma^{min}$ for the set $\Sigma_\sigma$ with the minimal ordering.

If $\tau$ is a face of $\sigma$ the induced map $\pi_{\sigma\tau} : \Sigma_\tau \to \Sigma_\sigma$ is a closed monomorphism and the complement $S$ of $\pi_{\sigma\tau}(\Sigma_\tau)$ in $\Sigma_\sigma$ is a disjoint union of independent expansion pairs. We also have the following diagram of closed poset bijections.

$$\begin{array}{ccc}
\Sigma_\tau \coprod S & \longrightarrow & \Sigma_\sigma \\
\downarrow & & \downarrow \\
\Sigma_\tau^{min} \coprod S & \longleftarrow & \Sigma_\sigma^{min}
\end{array}$$

**Example 4.4.2.** This situation is illustrated in Figure 4.3 where two small simplices $\rho, \sigma$ meet at a common face $\tau$. Along $t_1 = 0$ there is a line of birth-death points (i.e., a line of parameter values over which the function has a birth-death point). When $t_1 > 0$ there is a cancelling pair $x^+, x^-$ which by definition are not part of the sets $\Sigma_\rho, \Sigma_\tau$ since $t_1 < (\delta/4)^{2/3}$ (making $f(\psi(x^+,t)) - f(\psi(x^-,t)) < \delta$) on these sets. For the simplex $\sigma$, $t_1 < (\delta/2)^{2/3}$ but $t_1 > (\delta/4)^{2/3}$ at the vertex $v$ so $\{x^+, x^-\}$ form an independent expansion pair in $\Sigma_\sigma$ and $\Sigma_\sigma^{min}$ but not in $\Sigma_v$ where the standard ordering is not modified.

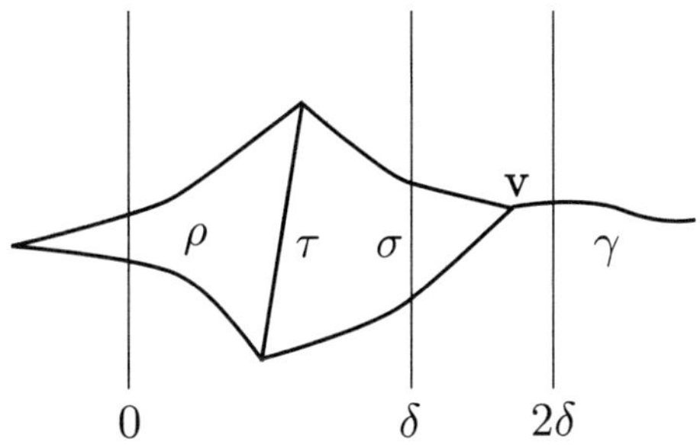


Figure 4.3: Simplices near a birth-death line

If $A$ is a closed subset of $\Sigma_\sigma^{min}$ then let $M_\sigma^A$ be defined as before as the set of all elements of $M_\sigma$ which are not multiply incident to any point in $\psi((\Sigma_\sigma - A) \times \Delta^k)$.

**Lemma 4.4.3.** *1.* $M_\sigma^{\Sigma_\sigma} = M_\sigma$

*2.* $M_\sigma^{A\cap B} = M_\sigma^A \cap M_\sigma^B$ *for all* $A, B \in \Lambda\Sigma_\sigma^{min}$.

*3.* $M_\sigma^\emptyset \simeq \partial_0 M_\sigma$

*4. If* $S = \{x^+, x^-\}$ *is an independent expansion pair in* $\sigma$ *and* $A \subseteq \Sigma_\sigma$ *is a closed set disjoint from* $S$ *then*

$$M^{A\cup S} \simeq M^A$$

*5. If* $A \subseteq \Sigma_\sigma^{min}$ *is closed and* $x$ *is a minimal element of* $\Sigma_\sigma^{min} - A$ *then for any coefficient sheaf* $\mathcal{F}$ *over* $E$ *we have*

$$H_i(M_\sigma^{A\cup\{x\}}, M_\sigma^A; \mathcal{F}) \cong \mathcal{F}_x$$

*for* $i = \deg x$ *and* $H_j(M_\sigma^{A\cup\{x\}}, M_\sigma^A; \mathcal{F}) = 0$ *for* $j \neq i$.

*Proof.* (1) and (2) are obvious. To prove (3) suppose first that $M_\sigma^\emptyset$ contains no critical points of $f_\sigma$. Then the smooth flow generated by $-\nabla_{\mathsf{v}} f$ will give a deformation retraction of $M_\sigma^\emptyset$ to $\partial_0 M_\sigma$.

If $M_\sigma^\emptyset$ has any critical points they will be cancelling pairs and birth-death points inside a standard coordinate chart for the unfolding of the birth-death point. In that case, using the local coordinates $u_0, \cdots, u_n$, we see that

$$\nabla_{\mathsf{v}} f_t = (3t_1 - 3u_0^2, -2u_1, \cdots, -2u_{i+1}, 2u_{i+2}, \cdots, 2u_n)$$

where $t_1$ is bounded by a constant $b = (\delta/2)^{2/3}$. Note that $\nabla f_t$ is parallel to the $u_0$-axis only on the $u_0$-axis. Also note that $\partial f/\partial u_0$ is negative for $|u_0| < \sqrt{b}$ and its value is bounded above by $3b$. Consequently, if $\phi : \mathbb{R}^n \to I$ is any smooth

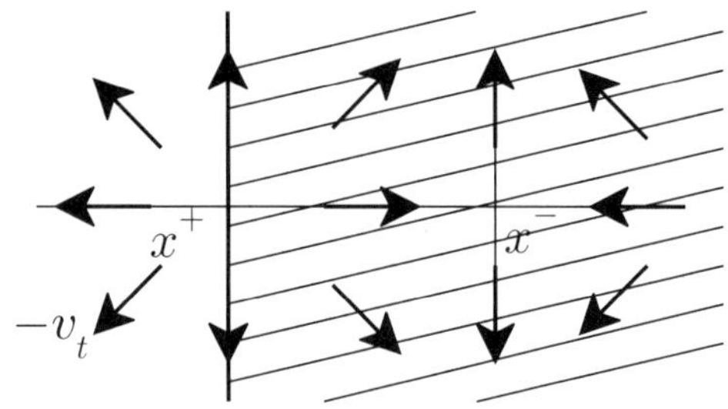


Figure 4.4: Points incident over $x^+$ and $x^-$

function which is equal to $3b$ along the line segment connecting $(-\sqrt{b}, 0 \cdots, 0)$ to $(-\sqrt{b}, 0 \cdots, 0)$ and has support in a small neighborhood of this set then

$$v_t = \phi \partial / \partial u_0 + (1 - \phi) \nabla_{\mathsf{v}} f_t$$

is a nowhere zero vector field equal to $\nabla_{\mathsf{v}} f_t$ outside a small neighborhood of the same line segment. Also the vector field $v_t$ is tangent to the set $C(x^-)$ which is given by $u_0 \geq -\sqrt{t_1}$, $u_1 = \cdots = u_{i+1} = 0$ and $u_{i+2}, \cdots, u_n \geq 0$ and points inward along its boundary $C(x^+)$. This implies that the smooth flow of $-v_t$ will preserve the complement of $C(x^+) \cup C(x^-)$ (Fig. 4.4) and give a deformation retraction of $M_\sigma^\emptyset$ to $\partial_0 M_\sigma$.

The proof of (4) is similar and (5) is proved in the same way as before (4.3.4) □

In order to get a filtered $R$-complex we need to assume that $f$ is oriented.

**Definition 4.4.4.** An *orientation* of a fiberwise GMF $f$ is defined to be a continuous orientation $[x]$ for the nondegenerate critical points so that the incidence number at birth-death points is $+1$ over the integers.

Suppose as before that $\mathcal{F}$ is a coefficient sheaf of right $R$-modules with stalk $\mathcal{F}_* = R$ over the base point $* \in E$ and $G$ is the image of $\pi_1(E)$ in $Aut(\mathcal{F}_*) = R^\times$. Choose a smooth triangulation of $B$ by small simplices and for each simplex $\sigma$ choose a homotopy class of paths $\gamma_{\sigma x}$ from $x \in \Sigma(f_{\sigma(b)}) \subseteq M_{\sigma(b)}$ to $*$ in $E$. This gives a generator

$$\gamma_{\sigma x}^*[x] \in H_i(M_\sigma^{A \cup \{x\}}, M_\sigma^A; \mathcal{F}) \cong R$$

Let $D(\sigma) = (D_\sigma, \Sigma_\sigma^{min}, \{\lambda_A^\sigma(x)\})$ denote the $\Lambda\Sigma_\sigma^{min}$-filtered $R$-complex given as before by:

1. $D_\sigma^A = S_*(M_\sigma^A, \partial_0 M_\sigma; \mathcal{F})$ for all $A \in \Lambda\Sigma_\sigma^{min}$.

2. $\lambda_A(x) \in H^i(D^{A \cup \{x\}}/D^A; R)$ is the cohomology class dual to the generator $\gamma_{\sigma x}^*[x] \in H_i(D^{A \cup \{x\}}/D^A) = H_i(M_\sigma^{A \cup \{x\}}, M_\sigma^A; \mathcal{F}) \cong R$.

Let $\xi^{min}(\sigma) \in \mathcal{M}_0(R, \Sigma_\sigma^{min})$ be a fixed choice of a minimal subcomplex of $D(\sigma)$ for every small simplex $\sigma$ in $B$. For every sequence of faces $\tau_0 \supseteq \tau_1 \supseteq \cdots \supseteq \tau_k$ of $\sigma$ we get an element

$$\xi^{min}(\sigma, \tau_0, \cdots, \tau_k) \in \mathcal{M}_k(R, \Sigma_\sigma^{min})$$

just like before.

We note that independent expansion pairs have the algebraic property that they admit no homotopies or higher homotopies. Consequently, they give a constant direct summand and the inclusion morphism

$$\xi^{min}(\sigma, \tau_0, \cdots, \tau_k)|A \to \xi^{min}(\sigma, \tau_0, \cdots, \tau_k)$$

of the restriction of $\xi^{min}(\sigma, \tau_0, \cdots, \tau_k)$ to the complement $A$ of the set of independent expansion pairs in $\Sigma_s^{min}$ is a $G$-expansion morphism.

**Theorem 4.4.5.** *A $G$-expansion functor*

$$\xi_f : \operatorname{simp} SdB \to \operatorname{simp} \mathcal{W}h_\bullet(R, G)$$

*is given by taking $\xi_f(\sigma_0, \cdots, \sigma_k)$ to be the object $\xi^{min}(\sigma_0, \sigma_0, \cdots, \sigma_k)$ with poset structure reduced to the modified standard poset $\Sigma_\sigma$ and by taking inclusion morphisms $a^*(\sigma_*) \to \sigma_*$ to the induced $G$-expansion morphisms $I : \xi_f(a^*(\sigma_*)) \to a^*\xi_f(\sigma_*)$.*

*Furthermore, $\xi_f$ is well-defined up to $G$-monomial simplicial homotopy and the homology of the total complex of $\xi_f$ is isomorphic to $H_*(E, E_0; \mathcal{F})$.*

*Also, the induced map on realizations*

$$|\xi_f| : B \simeq |SdB| \to |\mathcal{W}h_\bullet(R, G)|$$

*is, up to homotopy, independent of the triangulation of $B$ and the metric on $E$.*

*Finally, given any other oriented fiberwise GMF $g$ on $E$, the alternating mapping cone of the identity gives a $G$-expansion functor*

$$\eta : \operatorname{simp} SdB \to \operatorname{simp} \mathcal{W}h_\bullet^h(R, G)$$

*which contains $\xi_f$ as a subfunctor with quotient functor $\eta/\xi_f \cong \Sigma'(\xi_g)$.*

# 4.5 Independence of birth-death points

Given an oriented fiberwise GMF $f : E \to I$ a Riemannian metric which is standard near the critical set, we need to arrange for the birth-death set $\Sigma^{(1)}(f)$ to be independent, i.e., not incident to any other critical points. There are two ways to do this: geometrically and algebraically.

The geometric construction involves deforming the vertical gradient of $f$. Thus $\nabla_v f$ will be replaced with a *gradient-like vector field* $v$ which is defined to be any smooth vertical vector field on $E$ so that

1. $v$ agrees with the standard vertical gradient vector field of $f$ in a standard neighborhood of each critical point as specified in Lemma 4.4.1.

2. $Df_t(v_t) = \langle \nabla_{\mathsf{v}} f, v\rangle > 0$ at all nonsingular points of $f_t$.

These conditions above are obviously convex so the space of all gradient-like vector field is contractible. Thus we may use any gradient-like vector field $v$ to define $M_\sigma^A, D(\sigma), \xi$.

**Lemma 4.5.1.** *Suppose that*

$$\dim B = m \leq \operatorname{ind}(f_t, x) \leq n + 1 - m - j = \dim M - \dim B - j$$

*for all $x \in \Sigma^{(j)}(f_t)$, i.e., Morse points have index $\leq n+1-m$ and birth-death points have index $\leq n-m$. Then, there is a gradient-like vector field $v$ for $f$ equal to the vertical gradient $\nabla_{\mathsf{v}} f$ outside an arbitrarily small neighborhood of the birth-death set $\Sigma^{(1)}(f)$, so that $\Sigma^{(1)}(f)$ is independent from all other critical points with respect to the smooth flow of $v$.*

**Remark 4.5.2.** Note that the restriction on the index of $f_t$ is symmetrical since

$$\operatorname{ind}(-f_t, x) = n + 1 - j - \operatorname{ind}(f_t, x)$$

so that $\operatorname{ind}(f_t, x) \leq n+1-m-j$ is equivalent to $m \leq \operatorname{ind}(-f_t, x)$. We call the index of $-f_t$ the *coindex* of $f_t$. The number $j$ is called the *corank* of $f$.

Note also that $v - \nabla_{\mathsf{v}} f$ is not arbitrarily small, only its support (the closure of the set on which $v \neq \nabla_{\mathsf{v}} f$) is arbitrarily small.

*Proof.* By symmetry it suffices to prove that, by a deformation of $v_t$ in an arbitrarily small neighborhood of the birth-death set, the birth-death points will not be incident under any other critical point assuming that all critical points of $f_t$ have coindex at least $m$.

The point is that the set $C(x)$ of points incident to a birth-death point $x \in \Sigma^{(1,i)}(f_t)$ is an open cone on a disk of dimension $n+1-i$. Thus, if $b_t = f_t(x) + \delta$, then $C(x) \cap f_t^{-1}(b_t)$ is an $n+1-i$ disk $D_x^{n+1-i}$ which varies smoothly with $x \in \Sigma^{(1,i)}(f)$. By a deformation of $v_t$ in a smooth neighborhood of $\Sigma^{(1)}(f)$ we can deform $C(x)$ so that $D_x^{n+1-i}$ shrinks to a small subdisk of the original disk. We can also push it a small distance in a normal direction. Thus it suffices to choose a point on or near the disk $D_x^{n+1-i}$ which is not incident under any critical point of $f_t$. In other words we need a smooth function $w : \Sigma^{(1)}(f) \to E$ so that

1. $w(x) \in f_t^{-1}(b_t)$ for all $x \in \Sigma^{(1)}(f)$,

2. $w(x)$ lies arbitrarily close to $D_x^{n+1-i}$ and

3. $w(x)$ is not incident under any critical point of $f_t$.

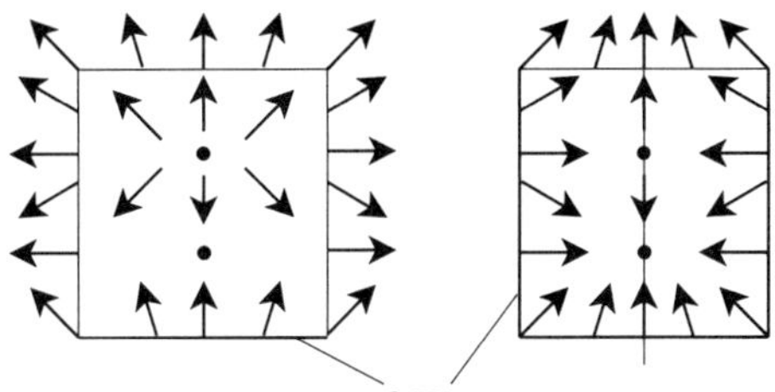


Figure 4.5: Gradient of $\pm$ suspension of $f$

However, any generic choice of such a function $w$ satisfying (1) and (2) will also satisfy (3) by transversality since the set of points in $f^{-1}(b_t)$ incident under critical points of coindex $i$ has codimension $i$. The corank is at least $m$ so in an $(m-1)$-parameter family ($\Sigma^{(1)}(f)$ being an $(m-1)$-manifold) this set will be disjoint from a generic $w(x)$. □

In general, the indices of the critical points of a fiberwise GMF $f : E \to I$ go from 0 to $n+1$. To adjust this to satisfy the lemma we "suspend" the function $f$.

**Definition 4.5.3 (Multiple suspension).** Given any smooth function $f : E \to \mathbb{R}$ we define the *$k$-fold positive/negative suspension*

$$\sigma_\pm^k f : E \times D^k \to \mathbb{R}$$

by the formula

$$\sigma_\pm^k f(x, y) = f(x) \pm \|y\|^2$$

We call

$$\begin{aligned} \sigma^N f = \sigma_+^N \sigma_-^N f : E \times D^N \times D^N &\to \mathbb{R} \\ (x, y, z) &\mapsto f(x) - \|y\|^2 + \|z\|^2 \end{aligned}$$

the *$N$-fold double suspension* of $f$.

The $N$-fold double suspension $\sigma^N f$ of $f$ has critical set $\Sigma(\sigma^N f) = \Sigma(f) \times 0 \times 0$ with indices ranging from $N$ to $N+m$. Consequently, the conditions of the lemma hold for $\sigma^N f$ if $N \geq m$. Therefore, there is a gradient-like vector field $v$ for $\sigma^N f$ with respect to which all birth-death points of $\sigma^N$ are independent.

The restriction $v_t = v|W_t$ of $v$ to the fiber $W_t = M_t \times D^N \times D^N$ agrees with $\nabla(\sigma^N f_t)$ near the boundary of $W_t$. Therefore, it points inward along

$$\partial_0 W_t = \partial_0 M_t \times D^N \times D^N \cup M_t \times \partial D^N \times D^N$$

and outward along

$$\partial_1 W_t = \partial_1 M_t \times D^N \times D^N \cup M_t \times D^N \times \partial D^N$$

(Both conditions hold on the intersection $\partial_0 W_t \cap \partial_1 W_t$ which is a corner set.) This means that the smooth flow of $-v$ sends every point in $W_t$ either to $\partial_0 W_t$ or to a critical point of $\sigma^N f_t$. Thus

$$W_t^{\emptyset} = W_t - \bigcup_{x \in \Sigma(\sigma^N f)} C(x, v_t) \simeq \partial_0 W_t$$

Here $C(x, v_t)$ denotes the set of all points in $W_t$ multiply incident to $x$ with respect to the smooth flow of $v_t$.

We proceed as before: choose $\delta > 0$, take small smooth simplices $\sigma : \Delta^k \to B$, define $\Sigma_\sigma$ and $\psi : \Sigma_\sigma \times \Delta^k \to \Sigma(\sigma^N f_\sigma)$. We note that $\Sigma_\sigma^{min}$ has a partial ordering defined in terms of multiple incidence with respect to $v_\sigma$ (the pull back of $v$ to $W_\sigma$).

For every closed subset $A$ of $\Sigma_\sigma^{min}$ let $W_\sigma^A$ be given as before by

$$W_\sigma^A = W_\sigma - \bigcup_{x \in \Sigma_\sigma - A} C(\psi(x \times \Delta^k), v_\sigma)$$

Then everything works as before with $(W_\sigma, \partial_0 W_\sigma)$ in place of $(M_\sigma, \partial_0 M_\sigma)$ and we have:

**Lemma 4.5.4.** *1.* $W_\sigma^{\Sigma_\sigma} = W_\sigma$

*2.* $W_\sigma^{A \cap B} = W_\sigma^A \cap W_\sigma^B$ *for all* $A, B \in \Lambda \Sigma_\sigma^{min}$.

*3.* $W_\sigma^{\emptyset} \simeq \partial_0 W_\sigma$

*4. If* $S = \{x^+, x^-\}$ *is an independent expansion pair in* $\sigma$ *and* $A \subseteq \Sigma_\sigma$ *is a closed set disjoint from* $S$ *then*

$$W^{A \cup S} \simeq W^A$$

*5. If* $A \subseteq \Sigma_\sigma^{min}$ *is closed and* $x$ *is a minimal element of* $\Sigma_\sigma^{min} - A$ *then for any coefficient sheaf* $\mathcal{F}$ *over* $E$ *we have*

$$H_i(W_\sigma^{A \cup \{x\}}, W_\sigma^A; \mathcal{F}) \cong \mathcal{F}_x$$

*for* $i = \deg x$ *and* $H_j(W_\sigma^{A \cup \{x\}}, W_\sigma^A; \mathcal{F}) = 0$ *for* $j \neq i$.

When $\mathcal{F}_* = R$ this means that $D_\sigma^A = S_*(W_\sigma^A, \partial_0 W_\sigma; \mathcal{F})$ has the structure of a filtered chain complex. A system of minimal subcomplexes $\xi^{min}(\sigma)$ can be used to construct a $G$-expansion functor

$$\xi : \operatorname{simp} SdB \to \operatorname{simp} \mathcal{W}h_\bullet(R, G)$$

on the first barycentric subdivision $SdB$ of $B$ (assuming we have a smooth triangulation of $B$ with small simplices).

Since the graded posets $\Sigma_\sigma$ have no elements of degree less than $N$, the $G$-expansion functor $\xi$ can be desuspended $N$ times using the alternating suspension (3.6.15).

**Theorem 4.5.5.** *The functor $\xi$ satisfies all the conditions of Theorem 4.4.5. Furthermore, if original function $f$ satisfies the condition that birth-death points are independent with respect to some metric on $E$, then the $N$-fold alternating desuspension of $\xi$ is $G$-expansion simplicially homotopic to the functor constructed in Theorem 4.4.5*

*Proof.* In the special case when the birth-death points are already independent, we can taking $v = \nabla_{\mathsf{v}}(\sigma^N f)$ without modification and we get a homotopy equivalence:

$$j : (M_\sigma^A, \partial_0 M_\sigma) \times (D^N, \partial D^N) \simeq (W_\sigma, \partial_) W_\sigma)$$

given by $j(x, y) = (x, y, 0)$ for all $A \in \Lambda\Sigma_\sigma^{min}$. The induced isomorphism

$$H_i(M_\sigma^{A\cup\{x\}}, M_\sigma^A; \mathcal{F}) \cong H_{N+i}(W_\sigma^{A\cup\{x\}}, W_\sigma^A; \mathcal{F})$$

sends $\lambda_A^x$ to $\lambda_A^x$. Therefore the $N$-fold alternating suspension of the minimal subcomplexes for the original function are minimal subcomplexes for the new function. □

We now describe an algebraic method for making birth-death points independent. It is based on a functor that makes general $G$-expansion pairs free.

**Definition 4.5.6.** A (general) *$G$-expansion pair* for $(e, P) \in \mathcal{M}_k(R)$ is a pair of elements $S = \{x^+, x^-\} \subseteq P$ so that

1. $x^- < x^+$ and $\deg x^+ = 1 + \deg x^-$
2. $x^+, x^-$ act in unison in the partial ordering, i.e., for all $z \in P$ we have
$$z > x^- \quad \Leftrightarrow \quad z \geq x^+$$
$$z \leq x^- \quad \Leftrightarrow \quad z < x^+$$
3. For every vertex $v$ the coefficient of $x^-$ in $e(v)(x^+)$ lies in $G$, i.e.,
$$e(v)(x^+) = x^- g + b(v)$$
where $b(v)$ is an $R$-linear combination of elements less than $x^-$..

Since $g \in G$ is easily seen to be independent of $v$ we call it the *incidence coefficient of $S$*. It is also easy to see that $G$-expansion pairs are disjoint. What we need to show is that there is a canonical way to make every expansion pair free.

If $S = \{x^-, x^+\}$ is a $G$-expansion pair in $(e, P)$ then we define

$$\Theta_S(e, P) = (d, S) \oplus (e/S, P - S)$$

where $d$ is the constant system on $S$ given by $dx^+ = x^- g$ and

$$e/S = p(e - eh * e)j$$

where $j : R^{P-S} \to R^P$ is the inclusion homomorphism, $p : R^P \to R^{P-S}$ is the projection and $h : R^P \to R^P$ is given by $h(x^-) = x^+ g^{-1}$ (and $h(z) = 0$ if $z \neq x^-$). The reader can verify that $\delta(e/S) = (e/S)' * (e/S)$ making $e/S$ into a system of higher homotopies.

**Lemma 4.5.7.** (1) *$\Theta_S$ is natural with respect to face and degeneracy operations and with respect to $G$-monomial morphisms sending $S$ to other $G$-expansion pairs. Furthermore, if $T$ is another $G$-expansion pair in $P - S$ then*

$$\Theta_S\Theta_T = \Theta_T\Theta_S$$

(2) *There is an upper triangular morphism*

$$(U_S, \alpha) : (e, P) \to (d, S) \oplus (e/S, P - S)$$

*which is natural in the above sense where $\alpha$ is the poset morphism given by $\alpha(A) = (A \cap S, A - S)$. Furthermore, if $S, T$ are disjoint $G$-expansion pairs which are not comparable in the partial ordering then $U_S U_T = U_T U_S$ and*

$$HU_S(HU_T) = HU_T(HU_S)$$

*where $H$ is the operator which constructs a simplicial homotopy out of an upper triangular morphism. If $S < T$ then we have the strictly weaker condition that*

$$HU_S(\Theta_T) = \Theta_T(HU_S)$$

(3) *There is a morphism*

$$(J_S, \beta) : (e/S, P - S) \to (e, P)$$

*natural in the above sense where $\beta(A) = \overline{A}$ is the closure of $A$ in $P$ for all $A \in \Lambda(P - S)$ so that the composition*

$$(e/S, P - S) \to (e, P) \to (d, S) \oplus (e/S, P - S) \to (e/S, P - S)$$

*is the identity morphism $(I, \iota)$. Furthermore, $J_S J_T = J_T J_S$ if $S, T$ are not comparable.*

(4) *If $S$ is already independent in $P$ and thus a free expansion pair in $(e, P)$ then $\Theta_S(e) = e$, $U_S$ is the identity mapping and $J_S$ is the inclusion map.*

**Remark 4.5.8.** For future reference the formulas are:
$e/S = p(e - eh * e)j$
$d = q(eh * e)q$
$U = (hep + q, p(I - e'h))$
$J = j - hej$
$HU = e - b * [qep + pe * e'h + peh * e] + \delta b * [pe'h - hep] + peb * pe'h$

Suppose for a moment that this is true. Then the $G$-expansion functor $\xi$ can be constructed on a space homotopy equivalent to $B$ as follows.

First, choose a triangulation of $B$ by small simplices as before. Let $\tilde{B}$ be obtained from $B$ by duplicating each simplex $\sigma$ having an expansion pair $S = \{x^+, x^-\} \subseteq \Sigma_\sigma$ so that $f(\psi(x^+, t)) - f(\psi(x^-, t)) = \delta$ at some $t \in \Delta^k$ and by connecting the two copies with a prism $\sigma \times I$.

For the partial ordering of the set $\Sigma_s$, we cannot modify the standard ordering to make $S$ independent since they are not independent in the minimal

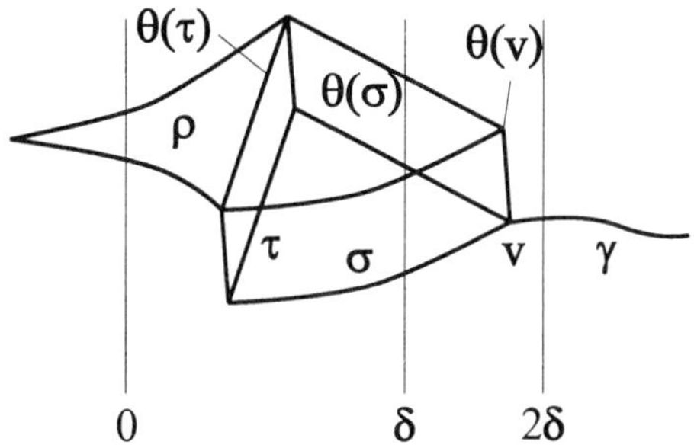


Figure 4.6: Duplicating simplices in $B$

ordering. Instead we modify both partial ordering by making $S$ act in unison in the minimal way. In other words, we add the relation $z > x^+$ if $z > x^-$ and we add $y < x^-$ if $y < x^+$ for all $z \in \Sigma_\sigma - S$. Note that this makes $y < z$ so both partial orderings of $\Sigma_s - S$ may be changed. In the duplicate simplex which we call $\theta_S(\sigma)$ we delete $S$ from the poset $\Sigma_\sigma$ and take the induced ordering:

$$\Sigma_{\theta_S(\sigma)} = \Sigma_\sigma - S, \quad \Sigma^{min}_{\theta_S(\sigma)} = \Sigma^{min}_\sigma - S$$

Simplices in $\tilde{B}$ corresponding to adjacent simplices in $B$ are adjacent in $\tilde{B}$ if and only if their posets $\Sigma_\sigma, \Sigma_\tau$ have the same size with corresponding elements deleted. Thus, in Figure 4.6, $\rho$ is adjacent to $\theta_S(\sigma)$ in $\tilde{B}$ but not to $\sigma$. Since $\tau, v$ are faces of $\sigma$ they also need to be duplicated with $\theta_S(\tau) = \rho \cap \theta_S(\sigma)$ and $\tau \subseteq \sigma$ . Figure 4.6 shows the result of this construction applied to Figure 4.3.

If $\Sigma_\sigma$ has $r$ expansion pairs as above, say $S_1, \cdots, S_r$, then we take $2^r$ copies of $\sigma$ and on each one delete a subset of these expansion pairs. We connect these with the simplicial set $\sigma \times I^r$

When expansion pairs are deleted there is no problem with the filtered chain complex. The $\Sigma_\sigma$-filtered $R$-complex with

$$D(\sigma)^A = S_*(M^A_\sigma, \partial_0 M_\sigma; \mathcal{F})$$

restricts to a $\Lambda\Sigma_{\theta_S(\sigma)}$-filtered $R$-complex by $D(\theta_S(\sigma))^A = D(\sigma)^{\overline{A}}$ where $\overline{A} = \beta(A)$ is the closure of $A$ in $\Sigma_\sigma$.

When we choose the objects $\xi^{min}(\sigma)$ and $\xi^{min}(\sigma, \tau_0, \cdots, \tau_k)$ we can choose $\xi^{min}(\theta_S(\sigma))$ and $\xi^{min}(\theta_S(\sigma), \tau_0, \cdots, \tau_k)$ in a canonical way using the operations given by Lemma 4.5.7.

First note that $\xi^{min} \in \mathcal{M}_0(R, \Sigma^{min}_\sigma)$, being a representative of a minimal subcomplex of $D(\sigma)$, comes equipped with a morphism of filtered chain complexes $Z_0(\xi^{min}(\sigma)) \to D(\sigma)$. We compose this with the morphism $Z_0(J_S)$ : $Z_0(\xi^{min}(\sigma)/S) \to Z_0(\xi^{min}(\sigma))$ to get an induced morphism $Z_0(\xi^{min}(\theta_S(\sigma)) \to D(\theta_S(\sigma))$ as indicated in the following diagram.

$$\begin{array}{ccc} Z_0(\xi^{min}(\sigma)/S) & \xrightarrow{Z_0(J_S)} & Z_0(\xi^{min}(\sigma)) \\ \downarrow & & \downarrow \\ D(\theta_S(\sigma)) & \longrightarrow & D(\sigma) \end{array}$$

This specifies $\xi^{min}(\theta_S(\sigma)) = \xi^{min}(\sigma)/S$ as a minimal subcomplex of $D(\theta_S(\sigma))$. Similarly, we let

$$\xi^{min}(\theta_S(\sigma), \theta_S(\tau_0), \cdots, \theta_S(\tau_k)) = \xi^{min}(\sigma, \tau_0, \cdots, \tau_k)/S \in \mathcal{M}_k(R, \Sigma^{min}_{\theta_S(\sigma)})$$

The upper triangular morphisms

$$U_S : (\ xi^{min}(\sigma, \tau_*), \Sigma^{min}_\sigma) \to (d, S) \oplus (\xi^{min}(\theta_S(\sigma), \theta_S(\tau_*)), \Sigma^{min}_{\theta_S(\sigma)})$$

can be used to construct a $G$-monomial simplicial homotopy

$$HU_S(\sigma) : \Delta^k \times I \to \mathcal{M}_\bullet(R, G)$$

connecting the source and target and this gives a $G$-expansion simplicial homotopy from $(\xi^{min}(\sigma, \tau_*)$ to $(\xi^{min}(\theta_S(\sigma), \theta_S(\tau_*)), \Sigma^{min}_{\theta_S(\sigma)})$. If there is more than one expansion pair being separated then we do this construction on them one at a time in increasing order. Thus the restriction of the $G$-expansion functor on $\sigma \times I^r$ is $HU_{S_r} \cdots HU_{S_1}(\sigma)$. By Lemma 4.5.7 this is well defined and has the correct value on $\sigma \times \partial I^r$.

**Theorem 4.5.9.** *Given an oriented fiberwise generalize Morse function $f$ on a smooth bundle $p : E \to B$ and a coefficient sheaf $\mathcal{F}$ with $\mathcal{F}_* = R$, the procedure outlined above defines a $G$-expansion functor*

$$\xi : \tilde{B} \to \mathcal{W}h_\bullet(R, G)$$

*on a simplicial set $\tilde{B}$ with $|\overline{B}| \simeq B$ so that*

1. $H_*(E(Z(\xi))) \cong H_*(E, E_0; \mathcal{F})$

2. *Up to homotopy the mapping*

$$|\xi| : B \simeq |\tilde{B}| \to |\mathcal{W}h_\bullet(R, G)|$$

*is independent of the choices of metric, triangulation, paths.*

3. *If the birth-death points of $f$ are already independent then $|\xi|$ agrees up to homotopy with the mapping of the same name given in Theorem 4.4.5.*

*Proof.* The projection mapping $|\tilde{B}| \to B$ is clearly a homotopy equivalence because the inverse image of each closed simplex is contractible. If the birth-death points of $f$ are already independent then $U_S : \xi(\sigma) \to \Theta(\xi(\sigma))$ is the identity map so the old and new constructions of $\xi$ form a commuting diagram:

$$\begin{array}{ccc} |\tilde{B}| & & \\ & \overset{|\xi^{new}|}{\searrow} & \\ \simeq \downarrow & & |\mathcal{W}h_\bullet(R,G)| \\ & \underset{|\xi^{old}|}{\nearrow} & \\ B & & \end{array}$$

□

*Proof of lemma.* (1) Naturality of $\Theta_S$ is obvious. To see that $\Theta_S\Theta_T = \Theta_T\Theta_S$ we just compute: if $S < T$ then

$$e/S/T = e/T/S = p(eh_S * e - eh_T * e + eh_S * eh_T * e)j$$

If $S, T$ are not comparable then the last term is zero and

$$e/S/T = e/T/S = p(eh_S * e - eh_T * e)j$$

(2) Let $U_S$ be given by

$$U_S = (hep + q, p(I - e'h))$$

where $q = I - p$ is the projection to $R^S$. Then $U_S$ is a morphism as claimed. For example, in the second coordinate, $p(I - e'h)$ is a morphism from $e$ to $e/S$ since (suppressing $*$ from the notation):

$$\begin{aligned}
&(e/S)p(I - e'h) = p(e - ehe)p(I - e'h) \\
&= pep - pehep - pepe'h + pe\underbrace{hepe'h}_{0} \\
&= pe(I - q) - pehe(I - q) - pe(I - q)e'h \\
&= pe - peq - pehe + pe\underbrace{heq}_{q^+} - pee'h + pe\underbrace{qe'h}_{q^-} \\
&= pe - pehe - pee'h \\
&= p(I - eh)e - \delta(p(I - e'h))
\end{aligned}$$

where $q^+, q^- : R^P \to R^P$ are the projections to $x^+, x^-$ with $q^+ + q^- = q$.

Furthermore, if $S, T$ are not comparable,

$$U_S U_T = (h_S ep_S + q_S, h_T ep_T + q_T, p(I - e'h_S - e'h_T))$$

which is symmetrical in $S, T$. Furthermore the double homotopy

$$HU_S(HU_T) : \Delta^k \times I \times I \to \mathcal{M}_\bullet(R)$$

has the property that all terms in which both homotopy parameters are nonzero vanish. Therefore

$$HU_S(HU_T) = e + (HU_S - e) + (HU_T - e)$$

which is also symmetrical. If $S < T$ then this is no longer true. However, $\Theta_T(HU_S) = HU_S(\Theta_T(e))$. This seems obvious from a certain point of view but the only proof we have is to compute both sides using the formula given in the proof of Lemma 3.3.5. The result is:

$$\begin{aligned}
&\Theta_T(HU_S) = d_T + p_T e p_T - p_{ST} e h_T e p_{ST} \\
&+b[-q_S e p_{ST} + q_S e h_T e p_{ST} - p_{ST} e e' h_S - p_{ST} e h_S e p_T + p_{ST} e h_S e h_T e p_{ST}] \\
&+\delta b[h_S e h_T e p_{ST} + p_{ST} e' h_S - h_S e p_{ST}] + p_{ST} e b p_{ST} e' h_S \\
&= HU_S(\Theta_T(e))
\end{aligned}$$

where $p_{ST} = p_S p_T$ is the projection away from $R^{S \cup T}$ and $b : [k] \to [1]$ is the homotopy parameter.

(3) Let $J_S : e/S \to e$ be given by

$$J_S = j - hej$$

If $S, T$ are not comparable then this gives

$$J_S J_T = (j_S - h_S e j_S)(j_T - h_T e j_T)$$

$$= j_{S \cup T} - (h_S + h_T) e j_{S \cup T} = J_T J_S$$

where $j_{S \cup T} = j_S j_T = j_T j_S$.

(4) Finally, in the case when $S$ is already independent, $e = e|S \oplus e|(P - S)$ so the cross terms are zero: $peq = qep = 0$. Consequently,

$$e = (p + q)e(p + q) = pep + qeq = p(e - ehe)p + d$$

since $d = qeq$ and $pehep = peqhqep = 0$. Thus $\Theta_S(e) = e$ and, similarly, $U_S = (q, p) = I$, $J_S = j_S$. □

**Example 4.5.10 (Morse theory on a circle).** Suppose that we have an oriented Morse function $f$ on a circle with $n$ minima and $n$ maxima and $y, x^-, x^+, z$ are four consecutive critical points as shown in Figure 4.7. Then the boundary map $\partial : R^n \to R^n$ is given by

$$\partial y = x^- u - br, \quad \partial x^+ = x^- g - zv$$

where $g, u, v, r \in G$ and $b$ is a minimum of $f$ which is not shown in the figure. Taking $S = \{x^-, x^+\}$, $e = \partial$ the construction above gives the following.

1. $h = h_S$ is given by $h(x^-) = x^+ g^{-1}$.

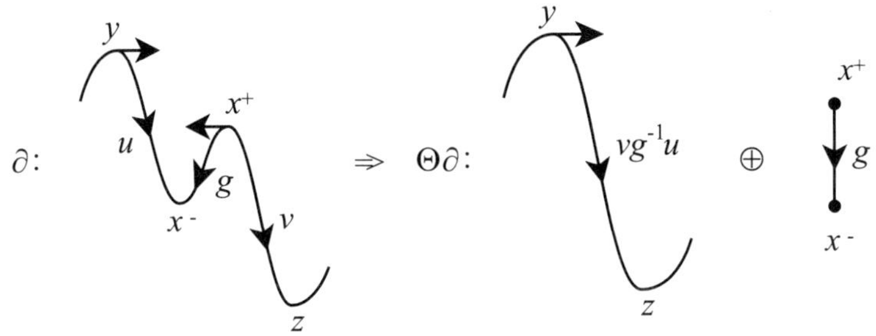


Figure 4.7: Eliminate birth-death point on a circle

2. $\Theta_S(\partial) = (d, \partial/S) = (d, p(\partial + \partial h \partial)j)$ is given by

$$dx^+ = x^- g, \quad (\partial/S)(y) = zvg^{-1}u - br$$

and $(\partial/S)(y') = \partial y'$ for all other maxima $y'$ of $f$.

3. $U_S(\partial) = I + h\partial p - p\partial h : (R^{2n}, \partial) \to (R^{2n}, \Theta_S(\partial))$ is the chain isomorphism given by $U_S = I + X$ where

$$X(y) = x^- g^{-1} u, \quad X(x^-) = -zvg^{-1}$$

and $X(y') = 0$ for all other generators.

4. $J_S = j + h\partial j : (R^{2n-2}, \partial/S) \to (R^{2n}, \partial)$ is given by

$$J_S(\partial)(y) = y - x^- g^{-1} u$$

and $J_S(\partial)(y') = y'$ for all other generators.

## 4.6 The framed function theorem

In this section we show how the framed function theorem [Igu87] gives a canonical oriented fiberwise GMF and therefore a $G$-expansion functor $\xi : \operatorname{simp} B \to \mathcal{W}h_\bullet(R, G)$.

**Definition 4.6.1.** A *framed function* on a smooth manifold $M$ is a GMF $f : M \to I$ which is nonsingular on $\partial M$ together with a framing $(v_1, \cdots, v_i)$ of the negative eigenspace of $D^2 f$ at each critical point of $f$.

The framing of $f$ can be used to give an orientation of $f$. However, we must use the reverse orientation at each Morse point given by $(v_i, \cdots, v_1)$. This will give the correct sign for the incidence of Morse points in an unfolding of a birth-death point.

**Definition 4.6.2.** Given a smooth bundle $p : E \to B$ with fibers $M_t$, a *fiberwise framed function* on $E$ is a fiberwise GMF $f : E \to \mathbb{R}$ together with a smooth family of framings. At each birth-death point of index $i$ we define the $i+1$st framing vector $v_{i+1}$ to be $\partial/\partial u_0$ in standard coordinates and we require that each vector $v_i$ be a smooth function on its domain

$$\bigcup_{j+k \geq i} \Sigma^{(j,k)}(f) \subseteq E$$

With the reverse orientation a fiberwise framed function gives an oriented fiberwise GMF. Note that the positive and negative suspensions $\sigma_+^k f, \sigma_-^k f$ of $f$ are also fiberwise framed functions in an obvious way, adding $k$ vectors $y_1, \cdots, y_k$ at the beginning of the framing in the case of $\sigma_-^k f$. We will also discuss cases in which "twisted suspensions" of framed functions are framed.

The main theorem about framed functions is given as follows.

**Theorem 4.6.3 (Framed function theorem [Igu87]).** *If* $\dim B \leq \dim M$ *then there exists a fiberwise framed function* $f : E \to I$*. If* $\dim B < \dim M$ *then* $f$ *is unique up to framed homotopy.*

Using the framed function theorem we can construct a canonical framed function on $E$ or some stabilization of $E$ and suspending further we can make all birth-death points independent. Thus we get:

**Corollary 4.6.4.** (1) *There exists a framed function*

$$f : E \times D^k \times D^k \times D^m \to \mathbb{R}$$

*so that* $f$ *has no critical points of index* $< k$ *and all birth-death points of all birth-death points are independent with respect to some gradient-like vector field for* $f$*.*

(2) *The* $G$*-expansion functor of* $f$ *is the* $k$*-fold alternating suspension of a functor*

$$\xi : \text{simp}\, SdB \to \mathcal{W}h_\bullet(R, G)$$

*which is well-defined up to* $G$*-expansion simplicial homotopy. Consequently,* $H_*(EZ(\xi)) \cong H_*(E, E_0; \mathcal{F})$*.*

(3) *If* $\pi_1 B$ *acts trivially on* $H_*(M_*, \partial M_*; \mathcal{F})$ *then we get a continuous mapping*

$$|C(\xi)| : B \to |\mathcal{W}h_\bullet^h(R, G)|$$

*well defined up to homotopy.*

*Proof.* First choose any smooth function $g : E \to I$ so that each $g_t = g|M_t$ is nonsingular on $\partial M$ and $g^{-1}(i) = E_i$ for $i = 0, 1$. Then take a large positive suspension say $\sigma_+^m g : E \times D^m \to \mathbb{R}$ where $m \geq \dim B - \dim M$. Then $\dim(E \times D^m) \geq 2 \dim B$ so there exists a framed function $f$ on $E \times D^m$ which is equal to $\sigma_+^m g$ on $\partial(E \times D^m)$

If the birth-death points of $f$ are not independent with respect to some admissible metric on $E \times D^m$ then take a multiple double suspension

$$\sigma^k f : E \times D^k \times D^k \times D^m \to \mathbb{R}$$

and proceed as in the last section.

(3) follows from (2) according to the results of Sections 3.5 and 3.6. □

Now we need to consider "twisted suspensions" of framed functions where the suspension disk is an oriented disk bundle over $E$. Suppose that $D$ is the unit disk bundle of a smooth $k$-plane bundle over $E$ with structure group $SO(k)$. Then we have a well-defined norm function $\|\cdot\| : D \to I$ whose square is smooth. Given an fiberwise framed function $f : E \to \mathbb{R}$ on the total space of a smooth bundle $E \to B$ we can define the *twisted suspensions*

$$\sigma_\pm^D(f) : D \to \mathbb{R}$$

on $D$ considered as a bundle over $B$ by

$$\sigma_\pm^D(f)(x) = f(\pi(x)) \pm \|x\|^2$$

where $\pi : D \to E$ is the projection map.

It is clear that the positive suspension $\sigma_+^D(f)$ is always framed. The negative suspension is not in general framable. However it is framable in some cases.

**Proposition 4.6.5.** *If $E \to B$ is an oriented circle bundle then the framed function $f$ on $E$ can be chosen so that the twisted negative suspension $\sigma_-^D(f)$ is framable. Furthermore the $G$-expansion functor of $\sigma_-^D(f)$ is the $k$-fold alternating suspension of the $G$-expansion functor of $f$.*

*Proof.* A positively framed function on an oriented $S^1$-bundle can be given by a system of disjoint local sections over $B$. (See [IK93] or the end of Chapter II.) If we take sections $s_\alpha : U_\alpha \to D$ of $p : E \to B$ over overlapping contractible regions $U_\alpha$ in $B$ then we can choose trivializations of the vector bundle over the image $s_\alpha(U_\alpha)$ of each $U_\alpha$ and use these to add $k$ vectors to the framing of the function $\sigma_-^D(f)$ over the singular component corresponding to $U_\alpha$. The other statement is obvious from the fact that $D$ is an oriented disk bundle. □

# Chapter 5

# Homotopy Type of the Whitehead Space

## Introduction

We prove two main theorems in this chapter.

- $\mathcal{W}h^h_\bullet(R,G) \simeq FC^h_\bullet(BG,\xi_R)$: The Whitehead space is homotopy equivalent to the simplicial category of filtered chain complexes.

- $\mathcal{W}h^h_\bullet(R,G) \simeq \mathcal{W}_\bullet(R,G)$: The Whitehead space is also homotopy equivalent to the space of matrices of Chapter II. This is called the "two-index theorem."

The first theorem gives us the homotopy type of the Whitehead space as the title promises since the homotopy type of $FC^h_\bullet(X,\xi)$ is known by a result of John Klein and myself ([IK95], Corollary 5.6.2 below). This work is summarized in section 5.6.

In the previous chapter we proved the equivalence between the Whitehead space and the space of filtered chain complexes on the level of objects and this was enough to do Morse theory. In this chapter we need to extend this to morphisms. This is done in sections 5.1, 5.2 and 5.3. The first main theorem (5.3.4, 5.3.5) says that the two simplicial categories are homotopy equivalent. We conclude that the results of [IK95] also apply to the Whitehead space.

The "two-index theorem" tells us that the formula for the Borel regulator derived in Chapter II extends in some canonical way to systems of higher homotopies (the Whitehead space). It does not tell us what the extended formula is but its existence allows us to compute the higher Reidemeister torsion in many cases. Section 5.9 contains the proof of the two index theorem. The precise statement is that the simplicial full subcategory of $\mathcal{W}h^h_\bullet(R,G)$ consisting of acyclic chain complexes which are nonzero in only two (fixed) consecutive indices is a deformation retract. This simplicial category is equivalent to the category of invertible matrices discussed in Chapter II.

Section 5.7 and Chapter VI describe the consequences of the 2-index theorem. We can define the higher FR-torsion more generally, list many of its properties and compute it in several cases.

Sections 5.4, 5.5 and 5.8 deal with the conjugate transpose involution on the Whitehead space. §5.4 explains the definition. §5.5 gives a very detailed proof of the obvious fact that the families of chain complexes given by an oriented fiberwise GMF $f$ and its "negative" $1-f$ are related by conjugate transfer. §5.8 proves that the involution preserves the higher FR-torsion up to the sign $(-1)^{n-1}$ where $n$ is the dimension of the fiber.

## 5.1 Filtered endomorphisms

We return to an analysis of the correspondence between filtered chain complexes and systems of higher homotopies. We recall that the multiple mapping cylinder construction gives a homotopy equivalence

$$Z : \operatorname{simp} \mathcal{M}_\bullet(R, P) \to \mathcal{F}(R, P)$$

for a fixed finite graded poset $P$ and for each object $D \in \mathcal{F}(R, P)$ any minimal subcomplex of $D$ represents a vertex $e_0 \in \mathcal{M}_0(R, P)$. Let $\mathcal{M}_\bullet(D)$ denote the connected component in $\mathcal{M}_\bullet(R, P)$ of the vertex $e_0$. Since the fiber $Z_\bullet/D$ of $Z$ over $D$ is contractible by 4.1.7, $\mathcal{M}_\bullet(D)$ is independent of the choice of $e_0$.

In some cases the component $\mathcal{M}_\bullet(D)$ is contractible. This means that any two complexes are filtered chain isomorphic, any two filtered chain isomorphisms are filtered chain homotopic, etc. This will be very useful, allowing us to avoid the topology of parametrized Morse theory. We start with the easier case when $D$ is minimal, i.e., $D \cong R^P$ and $e_0 : D \to D$ is the boundary map.

**Definition 5.1.1.** If $(D, e_0), (E, e_1) \in \mathcal{M}_0(R, P)$, so that $E \cong D \cong R^P$ as graded $R$-modules, then we define the *filtered homomorphism complex Fhom* $(E, D)$ to be the chain complex defined as follows.

1. $Fhom_k(E, D)$ is the additive group of all $R$-module homomorphisms $f : E \to D$ which are homogeneous of degree $k$ and *filtered* in the sense that $f(x)$ is an $R$-linear combination of elements $y < x \in P$ for all $x \in P$.

2. The boundary map $\partial : Fhom_{k+1}(E, D) \to Fhom_k(E, D)$ is given by $\partial(f) = e_0 \circ f + (-1)^k f \circ e_1$.

In the special case when $E = D$ we call $Fhom(D, D)$ the *filtered endomorphism complex* of $D$ and denote it by $Fend(D)$. We say that $Fhom(E, D)$ is *acyclic* if it has no homology in nonnegative degrees and if there exists an $f \in Fhom_0(E, D)$ with boundary $\partial(f) = e_1 - e_0$. In that case $1 + f$ will be a *filtered chain isomorphism* $(E, e_1) \to (D, e_0)$.

**Example 5.1.2.** Let $P$ be the linearly ordered set

$$y_1 > y_2 > x_2 > y_3 > x_3 > \cdots > y_n > x_n > x_1$$

with the $x_i$ of degree 0 and the $y_i$ of degree 1. Let $u$ be an invertible element of a ring $R$. Let $(E, e_0)$ be the $\Lambda P$-filtered chain complex $0 \to R^n \to R^n \to 0$

with boundary map

$$e_0 = \begin{pmatrix} 1 & & & & & -u \\ -1 & 1 & & & & \\ & -1 & \cdots & & & \\ & & & \cdots & & \\ & & & -1 & 1 & \\ & & & & -1 & 1 \end{pmatrix}$$

Then $Fend(E)$ is acyclic. This is true for all incidence matrices coming from Morse functions on circles. (See Proposition 5.1.8.)

**Lemma 5.1.3.** *a) If $D, E \in \mathcal{M}_0(R, P)$ are filtered chain isomorphic then $Fend(D)$, $Fend(E)$, $Fhom(E, D)$ are chain isomorphic.*

*b) If $Fend(D)$ is acyclic then the connected component $\mathcal{M}_\bullet(D)$ of $D$ in $\mathcal{M}_\bullet(R, P)$ is contractible.*

*Proof.* (a) If $F = 1 + f : D \to E$ is a filtered chain isomorphism then left composition with $F$ gives a chain isomorphism $Fhom(E, D) \cong Fend(E)$ and right composition with $F$ gives a chain isomorphism $Fhom(E, D) \cong Fend(D)$.

(b) Any simplicial map $\partial\Delta^k \to \mathcal{M}_\bullet(D)$ is equivalent to a family of endomorphisms $e(v_0, v_1, \cdots, v_n)$ of $R^P$ for $k \geq v_0 > \cdots > v_n \geq 0$ and $n < k$ satisfying the condition that $e' * e = \delta e$ and $(R^P, e(i))$ is filtered chain isomorphic to $D$. To extend this simplicial map to $\Delta^k$ it suffices to find the single endomorphism $f = e(k, k-1, \cdots, 1, 0)$ of $R^P$ so that

$$e(k) \circ f + (-1)^k f \circ e(0) = \delta e(k, k-1, \cdots, 1, 0)$$

$$+ \sum_{i=1}^{k-1} (-1)^{i-1} e(k, \cdots, k-i) e(k-i, \cdots, 0)$$

As in the proof of 3.4.6 the RHS is a $(k-1)$-cycle in $Fhom(E(0), E(k))$ therefore it is equal to $\partial(f) = e(k) \circ f + (-1)^k f \circ e(0)$ for some $f \in Fhom_k(E(0), E(k))$. $\square$

The above lemma immediately implies the following.

**Lemma 5.1.4.** *Suppose that $D$ is a $\Lambda P$-filtered $R$-complex and $(e, \phi) \in Z_\bullet/D$. Then*

*a) $Fend(e)$ is unique up to chain homotopy. (It does not depend on the choice of $(e, \phi)$, so we write $Fend(D) = Fend(e)$.)*

*b) If $Fend(D) = Fend(e)$ is acyclic then $\mathcal{M}_\bullet(D)$ is contractible.*

**Proposition 5.1.5.** *Suppose that $K$ is a simplicial set and $A$ is a simplicial subset so that $A_0 = K_0$. Suppose that $F : A \to \mathcal{M}_\bullet(R, P)$ is a simplicial map with the property that the $Fend(F_0(v))$ is acyclic for all $v \in A_0 = K_0$ and so that $F(v), F(w)$ are filtered isomorphic for any two vertices of $A$ which form the endpoints of an edge in $K$. Then $F$ extends to a simplicial map $\overline{F} : K \to \mathcal{M}_\bullet(R, P)$. Furthermore, $F$ is unique up to simplicial homotopy rel $A$.*

**Remark 5.1.6.** We are interested in the particular case when $A = K_0$. In that case the proposition says that $\overline{F} : K \to \mathcal{M}_\bullet(R, P)$ is uniquely determined up to simplicial homotopy by its values on vertices.

*Proof.* Let $n$ be the smallest degree in which $A_n \neq K_n$. Then the elements of $K_n - A_n$ are all nondegenerate and $F$ extends to each of these $n$-simplices by Lemma 5.1.3. For the uniqueness up to homotopy, replace $(K, A)$ by $(K \times I, K \times 0 \cup K \times 1 \cup A \times I)$. □

**Theorem 5.1.7.** *A fiberwise framed function on an $S^1$-bundle gives a family of chain complexes satisfying the above condition, i.e., $Fend(D_v)$ is acyclic at all vertices $v$. Thus we do not need to find the actual geometrically defined higher mapping cylinders since they are uniquely determined up to homotopy.*

*Proof.* If $f : E \to \mathbb{R}$ is a fiberwise framed function on a smooth $S^1$-bundle $p : E \to B$ then there is a positive real valued function $\delta$ on $B$ so that for any three consecutive critical points $x, y, z \in S^1_t = p^{-1}(t)$ of $f|S^1_t$ either $|f(x) - f(y)| > \delta(t)$ or $|f(z) - f(y)| > \delta(t)$. Furthermore, $\delta$ can be chosen so that whenever two consecutive critical points $x, y$ of $f|S^1_t$ have critical values within $\delta(t)$ of each other then $x, y$ are nondegenerate critical points in *cancelling position* in the sense that the framing vector on one of them (the maximum) points towards the other (the minimum).

Given such an $\delta$ and a generic $t \in B$ (so that $f|S^1_t$ is Morse), the critical set $\Sigma_t$ of $f|S^1_t$ has a partial ordering given by $x < y$ if $f(x) < f(y)$ and one of the following conditions is satisfied.

1. $x, y$ are adjacent critical points.
2. $x, y$ lie in a sequence of adjacent critical points $z_0, z_1, z_2, \cdots, z_{2n}, z_{2n+1}$ where $(z_1, z_2), (z_3, z_4), \cdots, (z_{2n-1}, z_{2n})$ are cancelling pairs (with critical values differing by at most $\delta(t)$).

(This is the modified minimal ordering as described in section 10.) These conditions are enough to show that $Fend(D_t)$ is acyclic as explained in the lemma below. □

**Lemma 5.1.8.** *Let $R$ be a ring and let $\pi_1 E \to R^\times$ be any one-dimensional representation of $\pi_1 E$. Let $D_t$ be the $\Lambda\Sigma_t(f)$-filtered $R$-complex given by the Morse function $f|S^1_t$ with the modified minimal partial ordering of $\Sigma_t(f)$ described above. Then $Fend(D_t)$ is acyclic.*

*Proof.* By induction on the number of cancelling pairs in $\Sigma_t$. If this number is zero then $x < y$ is possible only when $x$ has degree 0 and $y$ has degree 1 making $Fend(D_t)$ zero in all nonnegative degrees.

If there is at least one cancelling pair $S = \{x^-, x^+\}$ then we can cancel the pair $x^-, x^+$ giving a new complex $E_t$ and a filtered chain isomorphism $U_S : D_t \cong E_t \oplus F$ where $F = (R^2, d)$ with $dx^+ = x^- g$ as given in Example 4.5.10. By Lemma 5.1.3, $Fend(D_t)$ is chain isomorphic to $Fend(E_t \oplus F)$ so it suffices to show that $Fend(E_t \oplus F)$ is homotopy equivalent to $Fend(E_t)$ which is acyclic by induction. But

$$Fend(E_t \oplus F) \cong Fend(E_t) \oplus Fend(F) \oplus Fhom(F, E_t) \oplus Fhom(E_t, F)$$

and the last three terms are acyclic since $Fend(F)$ is zero in nonnegative degrees,

$$Fhom(F, E_t) \cong \operatorname{Hom}(F, X) \simeq 0$$

($F$ being contractible) where $X$ is the subcomplex of $E_t$ generated by all elements of $\Sigma_t$ less than $x$ and

$$Fhom(E_t, F) \cong \operatorname{Hom}(E_t/Y, F) \simeq 0$$

where $Y$ is the subcomplex of $E_t$ generated by all elements of $\Sigma_t$ not greater than $x^+$. □

## 5.2 The $G$-monomial category

In this section we will look at the components of the category of $G$-monomial morphisms in $\mathcal{W}h_\bullet(R, G)$. Since $G$-monomial morphisms preserve the number of elements in $P$ of degree $i$ for each $i \geq 0$, the sequence of numbers $|P| = (|P_0|, |P_1|, \cdots)$ is constant on each component.

**Definition 5.2.1.** If $G$ is a subgroup of the group of units of a ring $R$ and $n_* = (n_0, n_1, n_2, \cdots)$ is a finite sequence of nonnegative integers then we define the *$G$-monomial category* $\mathcal{W}h_\bullet(R, G, n_*)$ to be the simplicial subcategory of $\mathcal{W}h_\bullet(R, G)$ consisting of the $G$-monomial morphisms between objects $(e, P)$ where $P$ is a finite graded posets (whose underlying set lies in some fixed infinite set $\Omega$) with $|P| = n_*$. We represent $G$-monomial morphisms by

$$(\alpha, \gamma)_* : (e_0, P) \to (e_1, P')$$

where

1. $\alpha : P \to P'$ is a closed poset bijection (i.e. a bijection whose inverse is order preserving).
2. $\gamma : P \to G$ is a set map.
3. $(\alpha, \gamma)_* z_0(v_*)(x) = z_1(v_*)(\alpha(x))\gamma(\alpha(x))$ for all $x \in P$ and admissible $v_* = (v_0, \cdots, v_n)$.

In order to compare $\mathcal{W}h_\bullet(R, G, n_*)$ with an analogous category of filtered chain complexes we will decompose both into simplicial bicategories.

Let $\mathcal{A} = \mathcal{W}h_\bullet(R, 1, n_*)$ be the simplicial subcategory of $\mathcal{W}h_\bullet(R, G, n_*)$ with the same simplicial set of objects but with morphisms $(\alpha, 1)_*$ allowing only the trivial element of $G$. For each finite graded poset $P$ with $|P| = n_*$ let $\mathcal{B}(P)$ be the simplicial subcategory of $\mathcal{W}h_\bullet(R, G, n_*)$ with simplicial object set $\mathcal{M}_\bullet(R, P)$ and morphisms $(id_P, \gamma)_*$ allowing only the identity map on $P$. Let $\mathcal{B} = \coprod \mathcal{B}(P)$.

Clearly any morphism $(\alpha, \gamma)_* : (e_1, P') \to (e_0, P)$ in $\mathcal{W}h_\bullet(R, G, n_*)$ factors uniquely as compositions

$$(\alpha, \gamma) = (\alpha, 1)(id_{P'}, \gamma) = (id_P, \gamma \circ \alpha^{-1})(\alpha, 1)$$

forming a commuting square

$$\begin{array}{ccc} (e_1, P') & \xrightarrow{(id,\gamma)} & A \\ \downarrow{\scriptstyle \alpha} & & \downarrow{\scriptstyle \alpha} \\ B & \xrightarrow[(id,\gamma\circ\alpha^{-1})]{} & (e_0, P) \end{array}$$

where $A, B$ are uniquely determined by $(e_0, P), P', \alpha, \gamma$. Therefore the geometric realization $|\mathcal{W}h_\bullet(R, G, n_*)|$ of our simplicial category is naturally homotopy equivalent to the realization of the simplicial bicategory $\mathcal{AB}$ of all commuting squares as indicated above.

Now we construct the unstable simplicial category $FC(BG_\bullet, \xi_R, n_*)$ where $BG_\bullet$ is the nerve of $G$ considered as a groupoid with one object. Thus $BG_n = G^n$ with face and degeneracy operators given by multiplication of adjacent entries and by inserting 1. And $|BG_\bullet|$ is a $K(G, 1)$.

The projective bundle $\xi_R$ over the simplicial set $BG_\bullet$ is given by the representation $G \to R^\times = Aut_R(R_R)$. Thus the fiber over any $g = (g_k, \cdots, g_1) \in BG_k$ is $\xi_R(g) = R_R$ with simplicial operators $a^* : R \to R$, for $a : [j] \to [k]$, given by left multiplication by $g_m g_{m-1} \cdots g_1$ where $m \in [k] = \{0, \cdots, k\}$ is the smallest element in the image of $a$.

We recall that a filtered $R$-complex over $(BG_k, \xi_R)$ is a triple $(E, P, \gamma)$ where $\gamma : P \to BG_k = G^k$ is a set map and $E$ is a $\Lambda P$-filtered chain complex with cohomology classes $\lambda_A(x)$ giving isomorphisms

$$H_{deg(x)}(E^{A\cup\{x\}}, E^A) \cong \xi_R(\gamma(x)) = R$$

for all admissible pairs $(A, x)$ which commute with simplicial operators $a^*$.

**Definition 5.2.2.** Let $FC(BG_\bullet, \xi_R, n_*)$ be the simplicial category of objects $(E, P, \gamma)$ as above with $|P| = n_*$ satisfying the additional condition $E^\emptyset = 0$ and with morphisms $(f, \alpha) : (E', P', \gamma') \to (E, P, \gamma)$ where $\alpha : P' \to P$ is a closed bijection and $f : E' \to E$ is a morphism of $R$-complexes s.t.

1. $\gamma' = \gamma \circ \alpha$
2. $f(E'^A) \subseteq E^{\alpha(A)}$ for all $A \in \Lambda P'$
3. $\lambda_A(x) = f^* \lambda_{\alpha(A)}(\alpha(x)) = \lambda_{\alpha(A)}(\alpha(x)) \circ f_*$ for all admissible pairs $(A, x)$ in $P'$. In other words the following diagram commutes.

$$\begin{array}{ccc} H_{deg(x)}(E'^{A\cup\{x\}}, E'^{A}) & \xrightarrow{f_*} & H_{deg(x)}(E^{A\cup\{x\}}, E^{A}) \\ & {\scriptstyle \lambda_A(x)} \searrow \quad \swarrow {\scriptstyle \lambda_{\alpha(A)}(\alpha(x))} & \\ & R & \end{array}$$

We note that the definition of $FC(X, \xi_R)$ generalizes in an obvious way to the case when $BG_\bullet$ is replaced by any simplicial set $X$ and $\xi_R$ is replaced by a functor $\xi : \operatorname{simp} X \to \text{Mod-}R$ so that $\xi(x) \cong R_R$ for all $x \in X$.

The simplicial category $FC(BG_\bullet, \xi_R, n_*)$ has two simplicial subcategories $\mathcal{C}, \mathcal{D}$ having all the objects of $FC(BG_\bullet, \xi_R, n_*)$ but $\mathcal{C}$ has morphisms $(f, id_P)$ and $\mathcal{D}$ has morphisms $(id_E, \alpha) : (E, P \to (E, Q)$ so that $E^A = E^{\alpha(A)}$ for all $A \in \Lambda P$. Any morphism $(f, \alpha) : (E, P) \to (D, Q)$ in $FC(BG_k, \xi_R, n_*)$ has a natural factorization:

$$(E, P) \xrightarrow{(f, id_P)} (D, P) \xrightarrow{(id_D, \alpha)} (D, Q)$$

where $(D, P)$ is uniquely determined by the condition $D^A = D^{\alpha(A)}$. Therefore, by Lemma 5.2.4 below, we have:

$$|FC(BG_\bullet, \xi_R, n_*)| \simeq |\mathcal{C}\mathcal{D}|$$

Let $\mathcal{N}_\bullet\mathcal{C}\mathcal{D}$ be the nerve of $\mathcal{C}\mathcal{D}$ in the $\mathcal{D}$-direction. This is a bisimplicial category where the objects of $\mathcal{N}_m\mathcal{C}\mathcal{D}_k$ are sequences

$$(E, P^0) \xrightarrow{\alpha_1} (E, P^1) \xrightarrow{\alpha_2} \cdots \xrightarrow{\alpha_m} (E, P^m)$$

over $BG_k$. These are uniquely determined by $(E, P^m), \gamma : P^m \to BG_k$ and the sequences of refinements $P^0 \to P^1 \to \cdots \to P^m$ of the poset $P^m$. Morphisms in this category will fix $P^0 \to P^1 \to \cdots \to P^m$ and $\gamma : P^m \to BG_k$. They are given by chain maps $f : E \to D$ so that $(f, id_{P^m}) : (E, P^m) \to (D, P^m)$ is a morphism in $\mathcal{C}_k$, the induced morphism $(E, P^i) \to (D, P^i)$ being uniquely determined.

Let $\operatorname{simp} \mathcal{N}_{\bullet\bullet}\mathcal{A}\mathcal{B}$ be the bisimplicial category given by taking the bisimplicial nerve of $\mathcal{A}\mathcal{B}$ (a trisimplicial set) and considering the category of simplices in the original simplicial direction. Then the objects of $\operatorname{simp} \mathcal{N}_{mk}\mathcal{A}\mathcal{B}$ are given by $([q], *)$ where $*$ is a rectangular commuting diagram of the following shape:

$$\begin{array}{ccccccccc}
\cdot & \xrightarrow{\alpha_1} & \cdot & \xrightarrow{\alpha_2} & \cdots & \xrightarrow{\alpha_{m-1}} & \cdot & \xrightarrow{\alpha_m} & \cdot \\
{\scriptstyle \gamma_1\alpha_m\cdots\alpha_1}\downarrow & & \downarrow & & & & \downarrow & & \downarrow{\scriptstyle \gamma_1} \\
\cdot & \longrightarrow & \cdot & \longrightarrow & \cdots & \longrightarrow & \cdot & \longrightarrow & \cdot \\
{\scriptstyle \gamma_2\alpha_m\cdots\alpha_1}\downarrow & & \downarrow & & & & \downarrow & & \downarrow{\scriptstyle \gamma_2} \\
\vdots & & \vdots & & \cdots & & \vdots & & \vdots \\
{\scriptstyle \gamma_k\alpha_m\cdots\alpha_1}\downarrow & & \downarrow & & & & \downarrow & & \downarrow{\scriptstyle \gamma_k} \\
\cdot & \xrightarrow{\alpha_1} & \cdot & \xrightarrow{\alpha_2} & \cdots & \xrightarrow{\alpha_{m-1}} & \cdot & \xrightarrow{\alpha_m} & \cdot
\end{array}$$

where the horizontal arrows $(\alpha_i, 1) : (e, P^{i-1}) \to (e, P^i)$ are in $\mathcal{A}$ and the vertical arrows are in $\mathcal{B}$. Morphisms are given by simplicial operators at each vertex in the above diagram.

**Proposition 5.2.3.** *The multiple mapping cylinder construction gives a bisimplicial functor*

$$Z_\bullet : \operatorname{simp} \mathcal{N}_{\bullet\bullet}\mathcal{AB} \to \mathcal{N}_\bullet \mathcal{CD}$$

*sending* $([q], \{(e, P^i), \alpha_i, \gamma_j\})$ *to* $(Z_q(e), P^0 \to P^1 \to \cdots \to P^m, \gamma = (\gamma_1, \cdots, \gamma_k) : P^m \to G^k)$*. This simplicial functor satisfies Quillen's theorem A in every bidegree and therefore*

$$|\mathcal{W}h_\bullet(R, G, n_*)| \simeq |FC(BG_\bullet, \xi_R, n_*)|$$

*Proof.* The additional data given by $P^0, \cdots, P^{m-1}, \alpha_i, \gamma_j$ is the same in both domain and range and therefore the fiber category is the same as the category of simplices in the contractible fiber $Z_\bullet/(D, P^m)$ of Lemma 4.1.7. The restriction to the subcategory satisfying $E^\emptyset = 0$ changes nothing since every $Z_k(e)$ satisfies that condition. □

We used the following lemma which is equivalent to Waldhausen's "swallowing lemma" [Wal85] and has a similar proof.

**Lemma 5.2.4 (Modified swallowing lemma).** *Suppose that* $\mathcal{A}, \mathcal{B}$ *are subcategories of a category* $\mathcal{C}$ *and suppose that every morphism* $f : X \to Y$ *in* $\mathcal{C}$ *factors naturally with respect to* $\mathcal{B}$ *as a composition*

$$X \xrightarrow{f_\mathcal{A}} W(f) \xrightarrow{f_\mathcal{B}} Y$$

*where* $f_\mathcal{A}$ *is in* $\mathcal{A}$ *and* $f_\mathcal{B}$ *is in* $\mathcal{B}$*. Then* $|\mathcal{AB}| \simeq |\mathcal{C}|$*.*

**Remark 5.2.5.** By *natural with respect to* $\mathcal{B}$ we mean that, given $f : X \to Y, f' : X' \to Y'$ in $\mathcal{C}$ and morphisms $g_X : X \to X', g_Y : Y \to Y'$ in $\mathcal{B}$ so that $g_Y \circ f = f' \circ g_X$, there is an induced morphism $g_W : W(f) \to W(f')$ in $\mathcal{B}$ so that the following diagram commutes.

$$\begin{array}{ccccc} X & \xrightarrow{f_\mathcal{A}} & W(f) & \xrightarrow{f_\mathcal{B}} & Y \\ \downarrow{\scriptstyle g_X} & & \downarrow{\scriptstyle g_W} & & \downarrow{\scriptstyle g_Y} \\ X' & \xrightarrow{f'_\mathcal{A}} & W(f') & \xrightarrow{f'_\mathcal{B}} & Y' \end{array}$$

Note also that by symmetry we could just as well assume naturality with respect to $\mathcal{A}$.

*Proof.* Waldhausen's swallowing lemma ([Wal85]) says that $|\mathcal{AC}| \simeq |\mathcal{C}|$ if $\mathcal{I} \subseteq \mathcal{A} \subseteq \mathcal{C}$ where $\mathcal{I}$ is the subcategory of all identity morphisms in $\mathcal{C}$. This is a special case of the above lemma since every morphism $f : X \to Y$ has the *trivial factorization* $X \to_{id} X \to_f Y$ where the identity on $X$ is in $\mathcal{A}$ by assumption and the trivial factorization is obviously natural with respect to any subcategory of $\mathcal{C}$.

The swallowing lemma also implies our lemma because of the following.

*Claim:* $|\mathcal{AB}| \simeq |\mathcal{CB}|$ (under the assumptions of the lemma).

This will prove the lemma since, by the swallowing lemma with $\mathcal{A}, \mathcal{B}$ reversed, $|\mathcal{CB}| \simeq |\mathcal{C}|$. On the other hand, the claim also implies the swallowing lemma by taking $\mathcal{B} = \mathcal{C}$. In that case, using the trivial factorization, the claim gives us $|\mathcal{AC}| \simeq |\mathcal{CC}|$ for any $\mathcal{I} \subseteq \mathcal{A} \subseteq \mathcal{C}$. Consequently, $|\mathcal{AC}| \simeq |\mathcal{CC}| \simeq |\mathcal{IC}| = |\mathcal{C}|$.

*Proof of claim:* Take the nerve of both bicategories in the first coordinate. Then we have an inclusion functor of simplicial categories $j : \mathcal{N}_\bullet\mathcal{AB} \to \mathcal{N}_\bullet\mathcal{CB}$. The assumed natural factorization above gives a simplicial functor $r : \mathcal{N}_\bullet\mathcal{CB} \to \mathcal{N}_\bullet\mathcal{AB}$ by

$$r\left(X_0 \xrightarrow{f^1} X_1 \xrightarrow{f^2} \cdots \xrightarrow{f^n} X_n\right) = \left(W(h_0) \xrightarrow{h^1_{\mathcal{A}}} W(h_1) \xrightarrow{h^2_{\mathcal{A}}} \cdots \xrightarrow{h^n_{\mathcal{A}}} W(h_n)\right)$$

where $h_0 = id_{X_0}$ and $h^i = f^i \circ h^{i-1}_{\mathcal{B}} : W(h^{i-1}) \to X_i$. The morphisms $h^0_{\mathcal{B}}, \cdots, h^n_{\mathcal{B}}$ give a natural transformation $j \circ r \to id_{\mathcal{N}_\bullet\mathcal{AC}}$ and also $r \circ j \to id_{\mathcal{N}_\bullet\mathcal{AB}}$. Consequently, $|\mathcal{AB}|$ is a deformation retract of $|\mathcal{CB}|$. □

## 5.3 Expansions in filtered chain complexes

We will stabilize the space of filtered chain complexes $FC(BG_\bullet, \xi_R, n_*)$ in a way analogous to the insertion of expansion morphisms in $\mathcal{W}h_\bullet(R,G)$. We will then get $|FC(BG_\bullet, \xi_R)| \simeq |\mathcal{W}h_\bullet(R,G)|$. We will also see exactly how the previous version of the Whitehead space $\mathcal{W}_\bullet(R,G)$ is related to $\mathcal{W}h_\bullet(R,G)$.

The morphisms in the simplicial category $\mathcal{W}h_\bullet(R,G)$ factor, by definition, as the composition of an expansion morphism

$$(e,P) \to (e,P) \oplus (d,S)$$

and a $G$-monomial morphism $(e,P) \oplus (d,S) \to (e_1,P_1)$. Since this factorization is natural up to isomorphism, the modified swallowing lemma (5.2.4) gives us a homotopy equivalence

$$|\mathcal{W}h_\bullet(R,G)| \simeq |me\mathcal{W}h_\bullet(R,G)|$$

where

$me\mathcal{W}\boldsymbol{h}_\bullet(R,G)$ is the bicategory $me\mathcal{W}h_\bullet(R,G) = m\mathcal{W}h_\bullet(R,G)e\mathcal{W}h_\bullet(R,G)$,

$e\mathcal{W}\boldsymbol{h}_\bullet(R,G)$ is the simplicial subcategory of expansion morphisms and

$m\mathcal{W}\boldsymbol{h}_\bullet(R,G) = \coprod \mathcal{W}\boldsymbol{h}_\bullet(R,G,n_*)$ is the category of $G$-monomial morphisms.

By comparing definitions we see that this simplicial bicategory contains the simplicial bicategory $\mathcal{W}_\bullet(R,G)$ of Chapter II. In fact it contains several copies.

For any $i \geq 0$ let $\mathcal{W}h_\bullet^{[i,i+1]h}(R,G)$ denote the simplicial full subcategory of $\mathcal{W}h_\bullet^h(R,G)$ of objects $(e,P)$ where $P = P_i \cup P_{i+1}$ has elements only in degrees $i, i+1$ and $e$ is acyclic. If $me\mathcal{W}h_\bullet^{[i,i+1]h}(R,G)$ denotes the corresponding simplicial full subbicategory of $me\mathcal{W}h_\bullet(R,G)$ then we have the following.

**Proposition 5.3.1.** *$me\mathcal{W}h_\bullet^{[i,i+1]h}(R,G)$ is isomorphic to $\mathcal{W}_\bullet(R,G)$ for all $i \geq 0$ and consequently,*

$$|\mathcal{W}h_\bullet^{[i,i+1]h}(R,G)| \simeq |me\mathcal{W}h_\bullet^{[i,i+1]h}(R,G)| \cong |\mathcal{M}_\bullet(R,G)|$$

**Remark 5.3.2.** The two-index theorem (5.7.1) which is proved below tells us that $\mathcal{W}h_\bullet^{[i,i+1]h}(R,G)$ is a deformation retract of $\mathcal{W}h_\bullet^h(R,G)$. Therefore we will be able to conclude that

$$|\mathcal{W}h_\bullet^h(R,G)| \simeq |\mathcal{M}_\bullet(R,G)|.$$

Now consider the bisimplicial category $me_\bullet\mathcal{W}h_\bullet(R,G)$ given by taking the nerve of $me\mathcal{W}h_\bullet(R,G)$ in the $e$-direction. An *object* in $me_k\mathcal{W}h_j(R,G)$ is an element $(e,P)$ of $\mathcal{W}h_j(R,G)$ together with a sequence of subobjects

$$(e,P) = (e^k,P^k) \supseteq (e^{k-1},P^{k-1}) \supseteq \cdots \supseteq (e^0,P^0)$$

so that $(e^{i-1},P^{i-1}) \to (e^i,P^i)$ is an expansion morphism, i.e., $P - P^i$ is a union of free expansion pairs for $(e,P)$ for all $i$. Since $e^i$ is uniquely determined by $P^i$ and $e$ we can write this object as

$$((e,P^k),P^{k-1},\cdots,P^0)$$

A *morphism*

$$((e,P),P^{k-1},\cdots,P^0) \to ((d,Q),Q^{k-1},\cdots,Q^0)$$

is then a $G$-monomial morphism

$$(\alpha,\gamma)_* : (e,P) \to (d,Q)$$

so that $\alpha(P^i) = Q^i$ for all $i$. It necessarily follows that $(\alpha,\gamma)$ induces $G$-monomial morphisms $(e^i,P^i) \to (d^i,Q^i)$ for all $i$. Also, $\alpha$ necessarily takes free expansion pairs in $P - P^0$ to free expansion pairs in $Q - Q^0$ and it also follows that $\gamma(x^+) = \gamma(x^-)$ for each free expansion pair in $P - P^0$.

An analogous definition in the category of filtered chain complexes is given as follows.

**Definition 5.3.3.** (a) If $(E,P) \in FC(BG_j,\xi_R,n_*)$ then $S = \{x^-,x^+\}$ is called a *free expansion pair* in $(E,P)$ if:

1. $S$ is a free expansion pair in the graded poset $P$, i.e., $\deg x^+ = 1 + \deg x^-$, $x^- < x^+$ and $S, P - S$ are both closed.

2. $\gamma(x^-) = \gamma(x^+) \in G^j$.

3. $E = E^S \oplus E^{P-S}$. (This is where we need $E^\emptyset = 0$.)

4. $\lambda(x^+) = \delta\lambda(x^-) = \lambda(x^-) \circ \partial$, i.e., the following diagram commutes where $d = \deg x^-$.

$$\begin{array}{ccc} H_{d+1}(E^S, E^{\{x^-\}}) & \xrightarrow{\partial} & H_d(E^{\{x^-\}}, E^\emptyset) \\ & {\scriptstyle\lambda(x^+)} \searrow \quad \swarrow {\scriptstyle\lambda(x^-)} & \\ & R & \end{array}$$

(b) The stabilized version $FC(BG_\bullet, \xi_R)$ of $FC(BG_\bullet, \xi_R, n_*)$ is defined to be the simplicial category with simplicial object set

$$\coprod_{n_*} FC(BG_\bullet, \xi_R, n_*)$$

and whose morphisms are compositions of *expansions*

$$(E, P) \to (E, P) \oplus (D, S)$$

where $S = \coprod S_i$ is a union of free expansion pairs $S_i$ in $(D, S)$ with morphisms $(f, \alpha) : (E, P) \oplus (D, S) \to (E', P')$ in $FC(BG_\bullet, \xi_R, n_*)$.

The morphisms of $FC(BG_\bullet, \xi_R)$ factor naturally. (As in (b) above with $(D, S) = \bigoplus (E'^{\alpha(S_i)}, \alpha(S_i))$.) Therefore, following a discussion similar to the one above, we get:

$$|FC(BG_\bullet, \xi_R)| \simeq |weFC(BG_\bullet, \xi_R)| \simeq |we_\bullet FC(BG_\bullet, \xi_R)|$$

where $we_\bullet FC(BG_\bullet, \xi_R)$ is a bisimplicial category with objects in bidegree $(k, \ell)$ given by sequences $((E, P), P^{k-1}, \cdots, P^0)$ where $(E, P) \in FC(BG_\ell, \xi_R, n_*)$ and $P = P^k \supseteq P^{k-1} \supseteq \cdots \supseteq P^0$ is a sequence of subposets so that $P - P^i$ is a union of free expansion pairs in $(E, P)$ for all $i$.

**Theorem 5.3.4.** *The multiple mapping cylinder construction gives a homotopy equivalence*

$$|\mathcal{W}h_\bullet(R, G)| \simeq |FC(BG_\bullet, \xi_R)|$$

*Proof.* In $me_\bullet \mathcal{W}h_\bullet(R, G)$ we have the bisimplicial subcategories $\mathcal{A}, \mathcal{B}$ defined as before (in section 13) and $|\mathcal{W}h_\bullet(R, G)| \simeq |me_\bullet \mathcal{W}h_\bullet(R, G)| \simeq |\mathcal{AB}|$. By taking bisimplicial nerves and the category of simplices in the original simplicial direction we get a trisimplicial category simp $\mathcal{N}_{\bullet\bullet}\mathcal{AB}$ which is just like the bisimplicial category with the same name of the previous section with an additional simplicial direction given by the filtration $P = P^k \supseteq P^{k-1} \supseteq \cdots \supseteq P^0$.

Similarly the bisimplicial category $we_\bullet FC(BG_\bullet, \xi_R)$ has bisimplicial subcategories $\mathcal{C}, \mathcal{D}$ having all the objects $((E, P), P^{k-1}, \cdots, P^0)$ of $we_\bullet FC(BG_\bullet, \xi_R)$ but with morphisms in $\mathcal{C}$ being identity maps on the posets $P = P^k \supseteq P^{k-1} \supseteq \cdots \supseteq P^0$ and morphisms in $\mathcal{D}$ being the identity on $E$. As before we let $\mathcal{N}_\bullet \mathcal{CD}$ be the nerve in the $\mathcal{D}$-direction. This is a trisimplicial category where the objects in tridegree $(j, k, \ell)$ are given by filtered chain complexes $(E, P) \in FC(BG_\ell, \xi_R)$ together with a filtration

$$P = P^k \supseteq P^{k-1} \supseteq \cdots \supseteq P^0$$

so that $P - P^i$ is a union of free expansion pairs in $(E, P)$ and with sequences of closed bijections

$$P^{i0} \to P^{i1} \to \cdots \to P^{ij} = P^i$$

which up to isomorphism are the same as refinements of the partial ordering of the terminal object $P^i = P^{ij}$. Since the partial ordering on free expansion pairs is not allowed to be refined this data is redundant up to isomorphism for $i > 0$ and is essentially given by the sequence of refinements $P^{00} \to P^{01} \to \cdots \to P^{0j} = P^0$ of $P^0$. Thus an object of $\mathcal{N}_\bullet\mathcal{CD}$ in tridegree $(j, k, \ell)$ is given up to canonical isomorphism by $(E, P)$ plus additional data:

$$((E, P), P^{00} \to \cdots \to P^{0j} = P^0 \subseteq \cdots \subseteq P^k = P, \gamma : P \to G^\ell).$$

This additional data is the same as in simp$\mathcal{N}_{\bullet\bullet}\mathcal{AB}$ so the multiple mapping cylinder construction gives a trisimplicial functor simp$\mathcal{N}_{\bullet\bullet}\mathcal{AB} \to \mathcal{N}_\bullet\mathcal{CD}$ which is a homotopy equivalence in every tridegree just as in the proof of Proposition 5.2.3. □

**Corollary 5.3.5.** *The multiple mapping cylinder construction gives a homotopy equivalence*

$$|\mathcal{W}h^h_\bullet(R, G)| \simeq |FC^h(BG_\bullet, \xi_R)|$$

*where the $^h$ denoted the simplicial full subcategories of acyclic complexes in both cases. In particular we have isomorphisms of groups:*

$$\pi_0|FC^h(BG_\bullet, \xi_R)| \cong \pi_0|\mathcal{W}h^h_\bullet(R, G)| \cong Wh_1(R, G)$$

*(using direct sum as the group structure on $FC^h(BG_\bullet, \xi_R)$ and $\mathcal{W}h_\bullet(R, G)$) and*

$$|FC^h(BG_\bullet, \xi_R)| \simeq |FC^s(BG_\bullet, \xi_R)| \times Wh_1(R, G)$$

*where $FC^s(BG_\bullet, \xi_R)$ is the null component of $FC^h(BG_\bullet, \xi_R)$.*

One consequence of this is that we can define the *Whitehead torsion* $\tau(E, P) \in Wh_1(R, G)$ of every acyclic complex in $FC(BG_\bullet, \xi_R)$ to be the torsion of the corresponding element (its minimal subcomplex) in $\mathcal{W}h^h_\bullet(R, G)$.

## 5.4 The involution on Whitehead space

It is well-known that contravariant functors induce continuous maps on geometric realizations. But the geometric realization of a (small) category is actually the geometric realization of the simplicial nerve of the category. A contravariant functor $f : \mathcal{A} \to \mathcal{B}$ induces an "antisimplicial map" $f_* : \mathcal{N}_\bullet \mathcal{A} \to \mathcal{N}_\bullet \mathcal{B}$ which induces a continuous mapping

$$|\mathcal{A}| = |\mathcal{N}_\bullet \mathcal{A}| \xrightarrow{|f_*|} |\mathcal{N}_\bullet \mathcal{B}| = |\mathcal{B}|.$$

Since the involution on $\mathcal{W}h_\bullet^{[0,m]}(R,G)$ is antisimplicial we will explore this briefly.

## Antisimplicial maps

**Definition 5.4.1.** If $X_\bullet, Y_\bullet$ are simplicial objects in a category $\mathcal{C}$ then an *antisimplicial map* $F : X_\bullet \to Y_\bullet$ is a sequence of morphisms $F_k : X_k \to Y_k$, $k \geq 0$, with the property that for all $a : [j] \to [k]$ in the category $\Delta$, the following diagram commutes in $\mathcal{C}$ where $\overline{a} = op_k \circ a \circ op_j$, $op_k : [k] \to [k]$ being the unique order reversing bijection, (i.e., $\overline{a}(j-i) = k - a(i)$).

$$\begin{array}{ccc} X_k & \xrightarrow{F_k} & Y_k \\ {\scriptstyle a^*}\downarrow & & \downarrow{\scriptstyle \overline{a}^*} \\ X_j & \xrightarrow{F_j} & Y_j \end{array}$$

In the case of simplicial sets or simplicial spaces, the antisimplicial map $F$ induces a continuous map of realizations

$$|F| : |X_\bullet| \to |Y_\bullet|$$

given by reversing the order of the barycentric coordinates of $\Delta^k$. Thus, on $X_k \times \Delta^k$, $|F|(x,t) = (F_k(x), t^{op})$ where $(t_0, \cdots, t_k)^{op} = (t_k, \cdots, t_0)$.

Another way to describe the same map (up to homeomorphism) is to take the category of simplices. The antisimplicial map $F$ induces a (covariant) functor

$$F_* : \operatorname{simp} X_\bullet \to \operatorname{simp} Y_\bullet$$

by $F_*([k], x) = ([k], F_k(x))$ and $F_*(a^*) = \overline{a}^*$.

Thus, given a contravariant functor $f : \mathcal{A} \to \mathcal{B}$ we get an antisimplicial map of nerves $\mathcal{N}_\bullet f : \mathcal{N}_\bullet \mathcal{A} \to \mathcal{N}_\bullet \mathcal{B}$ which induces a covariant functor on the category of simplices: $\operatorname{simp} \mathcal{N}_\bullet f : \operatorname{simp} \mathcal{N}_\bullet \mathcal{A} \to \operatorname{simp} \mathcal{N}_\bullet \mathcal{B}$.

**Example 5.4.2.** Take any group $G$ and consider the antiautomorphism $\mathcal{I}(g) = g^{-1}$. Then $\mathcal{I}$ is a contravariant automorphism of $G$ considered as a category with one object. The induced map of nerves, given by $\mathcal{N}_k \mathcal{I}(g_0, \cdots, g_k) = (g_k^{-1}, \cdots, g_0^{-1})$ is antisimplicial and induces a mapping $|\mathcal{N}_\bullet \mathcal{I}| : BG \to BG$. This mapping is not the identity but it is homotopic to the identity since it induces the identity map on $\pi_1 = G$. The mapping $|\mathcal{N}_\bullet \mathcal{I}|$ takes the loop corresponding to $g \in G$ and maps it to the loop corresponding to $g^{-1}$ but running backwards.

This example can be generalized to any *symmetric simplicial set*, which we define to be any simplicial set with the property that every set map $a : [j] \to [k]$ induces a map $a^* : X_k \to X_j$. (But these are not part of the structure of the simplicial set. )

**Proposition 5.4.3.** *Let $X_\bullet$ be a symmetric simplicial set and let $J : X_\bullet \to X_\bullet$ be the antisimplicial involution given by $J(x) = x \circ op$. Then the induced involution $|J|$ on the realization $X_\bullet$ is homotopic to the identity.*

*Proof.* (First we verify that $J$ is antisimplicial: $J(a)(x) = \overline{a}^* J(x) = J(x)\overline{a} = (x \circ op) \circ (op \circ a \circ op) = x \circ a \circ op = J(a^* x)$.)

To show that $|J|$ is homotopic to the identity we go to the category of simplices. Then there are natural transformations from simp $J$ and from the identity functor on simp $X_\bullet$ to a third functor which is just the disjoint union of the two: $\Phi([k], x) = ([2k+1], \phi_k^*(x))$ where $\phi_k : [2k+1] \to [k]$ is given by $\phi_k(i) = \min(i, 2k+1-i)$. □

We leave it to the reader to show that the nerve of a groupoid with one object is symmetric and that Example 5.4.2 is thus a special case of Proposition 5.4.3.

**Remark 5.4.4.** The simplicial categories $\mathcal{W}_\bullet(R, G)$, $\mathcal{W}_\bullet^{diff}(\mathbb{C}^m, G)$, $\mathcal{W}_\bullet^{cont}(\mathbb{C}^m, G)$, $\mathcal{W}_\bullet^{iso\text{-}d}(\mathbb{C}^m)$, $\mathcal{W}_\bullet^{iso\text{-}d}(\mathbb{C}^m)/_{conj}$ are all symmetric so the anti-involution given by reversing the order of the vertices (in Definition 2.1.11) is homotopic to the identity.

## Conjugate transpose on $\mathcal{W}h_\bullet^{[0,m]}(R, G)$

We are now ready to describe the involution on the Whitehead space. First we fix a positive integer $m$ and consider only chain complexes which are zero above degree $m$. This defines a simplicial full subcategory $\mathcal{W}h_\bullet^{[0,m]}(R, G)$ of $\mathcal{W}h_\bullet(R, G)$.

Given a conjugation on $(R, G)$ the involution on $\mathcal{W}h_\bullet^{[0,m]}(R, G)$ is given by turning the objects "upside-down" replacing endomorphisms of $R^P$ with their conjugate transpose and by inverse conjugate transpose on morphisms. The definition is basically a combination of Definitions 2.1.11 and 2.5.6.

**Definition 5.4.5.** The involution $ct_m$ on $\mathcal{W}h_k^{[0,m]}(R, G)$ is given as follows.

*(a)* On objects $ct_m$ is the antisimplicial involution given by $ct_m(e, P) = (e^{op}, P^{op})$ where

1. $P^{op} = \{\overline{x} : x \in P\}$ with both degree and partial ordering reversed. ($\deg \overline{x} = m - \deg x$ and $\overline{x} < \overline{y} \Leftrightarrow x > y$.)

2. $e^{op}(v_0, \cdots, v_k) = e(v_k, \cdots, v_0)^*$ where $(\ )^*$ denotes conjugate transpose.

*(b)* On $G$-monomial morphisms $(f, \alpha) : (e_0, P_0) \to (e_1, P_1)$ we get $ct_m(f, \alpha) = (f^{op}, \alpha^{op})$ where

1. $\alpha^{op} : P_0^{op} \to P_1^{op}$ is given by $\alpha^{op}(\overline{x}) = \overline{\alpha(x)}$.

2. $f^{op}(j) = (f(j)^*)^{-1}$ $(f(v_0, \cdots, v_n) = 0$ for $n \neq 0)$.

*(c)* On $G$-expansions $(e, P) \to (e, P) \oplus (d, S)$ we take the induced $G$-expansion $(e^{op}, P^{op}) \to (e^{op}, P^{op}) \oplus (d^{op}, S^{op})$. (Note that the involution preserves direct sum.)

**Proposition 5.4.6.** *When $m = 1$, the natural isomorphism*

$$\phi : \mathcal{W}_\bullet(R, G) \to me\mathcal{W}h_\bullet^{h[0,1]}(R, G)$$

*is, up to homotopy, compatible with the conjugate transpose involution (i.e., $\phi \circ ct \simeq ct_1 \circ \phi$).*

*Proof.* The conjugate transpose involution is defined using the same formula. The reason that $\phi$ does not strictly commute with the involution is because we used the symmetry of $\mathcal{W}_\bullet(R, G)$ to make $ct$ into a true involution (instead of an anti-involution). However, by Proposition 5.4.3 this changes the involution only up to homotopy. □

## 5.5 Poincaré duality and conjugate transpose

In this section we go over the definition of the Poincaré duality mapping

$$\mathrm{D} : H^k(M, \partial_0 M; \mathbb{Z}[\pi_1 M]) \xrightarrow{\approx} H_{n-k}(M, \partial_1 M; \mathbb{Z}[\pi_1 M])$$

and verify the obvious fact that the family of cellular chain complexes associated to an oriented fiberwise GMF is the conjugate transpose of the one given by its negative. The statement is as follows.

Let $p : X \to B$ be a smooth manifold bundle which is *vertically oriented* in the sense that the vertical tangent bundle of $X$ is oriented. In other words, the fiber $M$ is oriented and the action of $\pi_1 B$ on $M$ preserves the orientation. Let $f : X \to I$ be an oriented fiberwise GMF. Then:

**Theorem 5.5.1.** *The function $1 - f : X \to I$ has the structure of an oriented fiberwise GMF and the associated families of cellular chain complexes*

$$\xi_f, \xi_{1-f} : \operatorname{simp} B \to \mathcal{W}h^{[0,n]}(\mathbb{Z}[\pi_1 X], \pi_1 X)$$

*can be chosen so that they are related by conjugate transpose:*

$$\xi_{1-f} = ct_n(\xi_f).$$

Note that we only need to consider the universal coefficient system $(R, G) = (\mathbb{Z}[\pi_1 X], \pi_1 X)$. Any other pair is induced from this universal case. Since we have already shown that $\xi_f$ (and thus $\xi_{1-f}$) is uniquely determined up to homotopy, we only need to consider the existence question in this section. I.e., we do not need to discuss how the results depend on any choices that we make.

## Elementary example

First consider the well-known single manifold case. Suppose that $M$ is a compact smooth oriented manifold with $\partial M = \partial_0 M \coprod \partial_1 M$. Let $\mathcal{F}$ be the $R = \mathbb{Z}[\pi_1 M]$-coefficient sheaf over $M$ so that the complex $S_*(M; \mathcal{F})$ of singular chains in $M$ with coefficients in $\mathcal{F}$ is the total singular complex (with $\mathbb{Z}$-coefficients) of the universal covering $\widetilde{M}$ of $M$.

Suppose that $f : M \to I$ is a Morse function (with the property $f^{-1}(i) = \partial_i M$ for $i = 0, 1$). Take as an example the case when $f$ has only two critical values $\frac{1}{2}, \frac{3}{4}$ and 8 critical point $\Sigma(f) = \{x_1, x_2, x_3, y_1, \cdots, y_5\}$ where each $x_i$ has index $k$ and critical value $\frac{1}{4}$ and each $y_j$ has index $k+1$ and critical value $\frac{3}{4}$. Let $W_- = f^{-1}[0, \frac{1}{2}]$, $W_+ = f^{-1}[\frac{1}{2}, 1]$ and $L = W_- \cap W_+ = f^{-1}(\frac{1}{2})$. Then $H_k(W_-, \partial_0 M; \mathcal{F})$ and $H_{k+1}(W_+, L; \mathcal{F})$ are free right $R$-modules for which bases can be given by choosing paths from each $z \in \Sigma(f)$ to the base point in $M$ and orientations for the negative eigenspace of $D^2 f$ at each $z \in \Sigma(f)$. The connecting homomorphism

$$g : H_{k+1}(W_+, L; \mathcal{F}) \cong H_{k+1}(M, W_-; \mathcal{F}) \to H_k(W_-, \partial_0 M; \mathcal{F})$$

is a morphism of free right $R$-modules given by left multiplication by a $3 \times 5$ matrix $A$ with coefficients in $R$. It is the boundary map in the cellular chain complex associated to the Morse function $f$.

Passing to cohomology, we get a morphism of free left $R$-modules

$$g^* : H^k(W_-, \partial_0 M; R) \to H^{k+1}(M, W_-; R) \cong H^{k+1}(W_+, L; R)$$

given by right multiplication by the same matrix $A$. Using the (unique) conjugation on $(R, \pi_1 M)$ we convert the left action of $R$ into a right action and the new matrix for the same mapping $g^*$ is the conjugate transpose $A^* = \overline{A}^t$ of $A$.

Poincaré duality gives isomorphisms D of right $R$-modules in the following diagram which commutes up to sign ($\overline{g} \circ \mathrm{D} = (-1)^{k+1} \mathrm{D} \circ g^*$) by the well-known properties of the Poincaré duality map.

$$\begin{array}{ccccc}
H^k(W_-, \partial_0 M; R) & \xrightarrow{\mathrm{D}} & H_{n-k}(W_-, L; \mathcal{F}) & \xrightarrow{\cong} & H_{n-k}(M, W_+; \mathcal{F}) \\
\Big\downarrow{\scriptstyle g^*} & & & & \Big\downarrow{\scriptstyle \overline{g}} \\
H^{k+1}(M, W_-; R) & \xrightarrow{\cong} & H^{k+1}(W_+, L; R) & \xrightarrow{\mathrm{D}} & H_{n-k-1}(W_+, \partial_0 M; \mathcal{F})
\end{array}$$

Here, $\overline{g}$ is the connecting homomorphism. Since $\overline{g}$ is also the boundary map for the cellular chain complex of $1 - f$ we have verified Theorem 5.5.1 in this example.

## Chain and cochain formulas

in order to generalize this example we need to write down formulas for the maps at the chain level. The first step is to restrict the size of the singular simplices and make them transverse to small unstable disks for each critical point of $f$. To do this we first must choose a gradient-like vector field for $f$ which is standard near each critical point. Choose an $\epsilon > 0$ which is much smaller than the difference between critical values of $f$. ($\epsilon < \frac{1}{16}$ is good enough in the above example.) For each $c \in \Sigma(f)$ let $D_-(c)$, $D_+(c)$ be the stable and unstable $2\epsilon$-disks for $c$ given by

$$D_-(c) = \{x \in M \;:\; x \text{ is incident under } c \text{ and } f(x) \geq f(c) - 2\epsilon\}$$
$$D_+(c) = \{x \in M \;:\; x \text{ is incident over } c \text{ and } f(x) \leq f(c) + 2\epsilon\}$$

An orientation for the negative eigenspace of $D^2 f$ at $c$ is the same as a tangential orientation of $D_-(c)$ or a normal orientation of $D_+(c)$. As before we choose such an orientation and we choose a path from each $c$ to the base point.

Let $S_*^\epsilon(M;\mathcal{F})$ denote the subcomplex of $S_*(M;\mathcal{F})$ generated by all singular simplices $\sigma : \Delta^k \to M$ which are transverse to every ascending disk $D_+(c)$ and to its boundary $\partial D_+(c)$ and whose image under $f$ has length less than $\epsilon$. (We say that $\sigma : \Delta^k \to M$ is *transverse* to a submanifold $K$ of the interior of $M$ of codimension $d$ if $\sigma(\tau)$ is disjoint from $K$ for every face $\tau$ of $\Delta^k$ of dimension less than $d$. If $K$ is normally oriented then every $d$-simplex which is transverse to both $K$ and $\partial K$ has a well-defined intersection number with $K$.) By the usual arguments, $S_*^\epsilon(M;\mathcal{F})$ is a deformation retract of $S_*(M;\mathcal{F})$ and the analogous statement holds for nice subcomplexes of $S_*(M;\mathcal{F})$. For example, $S_*^\epsilon(L;\mathcal{F}) \simeq S_*(L;\mathcal{F})$ if $L$ is transverse to each $D_+(c)$ and $\partial D_+(c)$.

Let $W_-(c) = f^{-1}[0, f(c)+\epsilon]$, $W_+(c) = f^{-1}[f(c) - 2\epsilon, 1]$. Then $\partial W_- = \partial_0 M \coprod L_+$ and $\partial W_+ = \partial_1 M \coprod L_-$ where $L_-(c) = f^{-1}(f(c) - 2\epsilon)$ and $L_+(c) = f^{-1}(f(c)+\epsilon)$. If the index of $c \in \Sigma(f)$ is $k$ then we have a well defined $k$-cocycle

$$c_+^* \in S_\epsilon^k(W_-(c), \partial_0 M; R) := \mathrm{Hom}_R(S_k^\epsilon(W_-(c), \partial_0 M; \mathcal{F}), R)$$

given by the transverse intersection number (of each singular simplex of $\widetilde{M}$) with the liftings of $D_+(c)$ in $\widetilde{M}$. This is dual to the relative cycle $c_- \in S_k^\epsilon(W_+(c), L_-; \mathcal{F})$ given by triangulating the descending disk $D_-(c)$. (So that $c_+^*(c_-) = 1$.) We first need to extend both $c_+^*$ and $c_-$ to all of $M$ so that they form dual bases for a minimal subcomplex of the filtered chain complex $E = S_*^\epsilon(M, \partial_0 M; \mathcal{F})$.

**Lemma 5.5.2.** *For each $c \in \Sigma(f)$ there exists a chain $z(c) \in S_k^\epsilon(M, \partial_0 M; \mathcal{F})$ and a cochain $c^* \in S_\epsilon^k(M, \partial_0; R)$ where $k = \mathrm{ind}(c)$ with the following properties.*

1. *$z(c) - c_-$ has support in $f^{-1}[0, f(c) - 2\epsilon]$ where $c_-$ is as above ($c_-$ is a triangulation of $D_-(c)$).*

2. *$\{z(c) : c \in \Sigma(f)\}$ generates an $R$-subcomplex of $E$ (namely, the minimal subcomplex).*

3. *The restriction of $c^*$ to $W_-(c)$ is equal to the cochain $c_+^*$ described above.*

4. *$\{c^* : c \in \Sigma(f)\}$ generates an $R$-subcomplex of $S_\epsilon^*(M, \partial_0 M; R)$.*

5. $c_i^*(z(c_j)) = \delta_{ij}$, *i.e.*, $\{c^*\}$ *is dual to* $\{z(c)\}$ *on the minimal subcomplex.*

**Remark 5.5.3.** (a) If $z(c)$, $c^*$ exist as claimed, then the intersections of the kernels of $c^*$ necessarily forms a strongly acyclic subcomplex $Q$ of $E$ which is complementary to the minimal subcomplex $K$ generated by the chains $z(c)$. ($Q$ is complementary to $K$ by (5) and it is a subcomplex by (4).)

(b) Since $\{c^*\}$ is dual to $\{z(c)\}$, the incidence matrix for the coboundary operator on the $c^*$'s is the same as the incidence matrix for the boundary operator for the $z(c)$'s, i.e., if $\partial c_j = \sum c_i a_{ij}$ then $\delta c_i^* = \sum a_{ij} c_j^*$. Note that the matrix $A = (a_{ij})$ is strictly upper triangular with respect to the standard partial ordering of $\Sigma(f)$ by critical value.

*Proof.* The proof of the lemma will be by induction on the number of critical values of $f$. If this number is 1 then the lemma is easy. We can just extend the disks $D_-(c)$, $D_+(c)$ to the boundary of $M$ and this gives the desired extensions $z(c)$ and $c^*$ of $c_-$ and $c_+^*$, respectively.

Now suppose that $f$ has at least two critical points and let $s$ be the maximum critical value of $f$. Let $\Sigma_+(f) = \Sigma(f) \cap f^{-1}(s)$ and let $\Sigma_-(f) = \Sigma(f) - \Sigma_+(f)$. Let $M_- = f^{-1}[0, s-2\epsilon]$. Then, by induction, we have a minimal subcomplex $K$ of $E_- = S_*^\epsilon(M_-, \partial_0 M; \mathcal{F})$ generated by chains $z(b_i)$ where $b_i \in \Sigma_-(f)$ and complementary strongly acyclic subcomplex $Q$ which is the intersection of the kernels of the cochains

$$b_i^* \in S_\epsilon^*(M_-, \partial_0 M; R) = \mathrm{Hom}_R(E_-, R)$$

which are dual to the $z(b_i)$'s.

On $M_+ = f^{-1}[s-2\epsilon, 1]$ we have relative cocycles

$$c_k \in E_+ = S_*^\epsilon(M_+, f^{-1}(s-2\epsilon); \mathcal{F})$$

generating a minimal subcomplex $K_+$ of $E_+$ and cochains which can be extended to cocycles $c_k^*$ on $E_+$ so that the intersections of these cocycles is an acyclic subcomplex $Q_+$ of $E_+$ complementary to $K_+$. We need to extend each $c_k$ to a relative chain in $E = S_*^\epsilon(M, \partial_0 M; \mathcal{F})$ so that its boundary lies in $K$. To do this note that $\partial c_k$ has support in $M_-$ so it decomposes as a sum

$$\partial c_k = \partial_K c_k + \partial_Q c_k$$

where $\partial_K c_k$, $\partial_Q c_k$ are cycles in $K$, $Q$. Since $Q$ is acyclic, $\partial_Q c_k$ is the boundary of some $c_k' \in Q$ and $z(c_k) = c_k + c_k'$ is the desired extension of $c_k$.

It remains to extend the cochains $b_i^* : E_- \to R$ to all of $E$. First we try the trivial extension which is zero on all singular simplices not contained in $M_-$. Let $a_{ij} \in R$ be the incidence coefficients. Then

$$d_i = \delta b_i^* - \sum_{i<j} a_{ij} b_j^*$$

is equal to zero on $E_-$. If $i$ is maximal then $d_i$ is a cocycle on $E$ and therefore on $E_+$ so, taking the decomposition $E_+ = K_+ \oplus Q_+$, we have that $d_i | K_+$, $d_i | Q_+$

are both cocycles. Since $Q_+$ is acyclic, $d_i|Q_+ = \delta d_i'$ where $d_i' \in \mathrm{Hom}_R(Q_+, R)$. Since $K_+$ is generated by the $c_k$'s, $d_i|K_+ = \sum r_{ik}c_k^*$ for some $r_{ik} \in R$.

We replace $b_i^*$ with $b_i^* - d_i'$ (letting $d_i' = 0$ on $K_+$). Then $b_i^*(z(c_k)) = 0$ and

$$\delta b_i^* = \sum a_{ij}b_j^* + \sum r_{ik}c_k^*.$$

Proceeding by downward induction on $i$, let $i$ be maximal so that $\delta b_i$ is not of the above form. Then

$$\delta d_i = \sum_j a_{ij}\delta b_j = \sum_{j,k} a_{ij}r_{jk}c_k^* = 0.$$

$(\delta d_i(z(c_k)) = d_i\partial z(c_k) = 0$ since $\partial z(c_k)$ lies in $E_-$ on which $d_i = 0$.) Consequently, $d_i$ is a cycle and we proceed as before. The resulting extended cochains $b_i^*$ together with the cocycles $c_k^*$ and the relative chains $z(b_i)$ and $z(c_k)$ fulfill the conditions of the lemma. □

Poincaré duality comes from an orientation on the manifold $M$. Thus we need $x \in S_n^\epsilon(M, \partial M)$ a relative integral $n$-cycle representing the orientation class $[x] = [M]$ of $M$. (For example choose a smooth triangulation of $M$ by small smooth simplices transverse to the disks $D_\pm(c)$ and their boundaries and let $x$ be the sum of the $n$-simplices in this triangulation.) If $x = \sum x_i m_i$, $m_i \in \mathbb{Z}$ then let

$$\widetilde{x} = \sum \widetilde{x}_i m_i \in S_n^\epsilon(M, \partial M; \mathcal{F})$$

be an $n$-chain in $\widetilde{M}$ which maps to $x$. Then the standard diagonal

$$\Delta\widetilde{x} \in \bigoplus_{p+q=n} S_p^\epsilon(M, \partial M; \mathcal{F}) \otimes S_q^\epsilon(M, \partial M; \mathcal{F})$$

is given by

$$\Delta\widetilde{x} = \sum_i \sum_{p+q=n} f_p(\widetilde{x}_i) \otimes b_q(\widetilde{x}_i)m_i$$

where $f_p, b_q$ denote front $p$-face and back $q$-face. We will use the standard Sweedler abbreviation [Swe69] $\Delta\widetilde{x} = \sum_{(\widetilde{x})} \widetilde{x}_{(1)} \otimes \widetilde{x}_{(2)}$. Then the Poincaré duality isomorphism

$$\mathrm{D} : H^k(M, \partial_0 M; R) \xrightarrow{\approx} H_{n-k}(M, \partial_1 M; \mathcal{F})$$

is given at the chain level by

$$\mathrm{D}(c^*) = \sum_{(\widetilde{x})} \widetilde{x}_{(2)}\overline{c^*(\widetilde{x}_{(1)})}.$$

This can also be written as $\mathrm{D}(c^*) = \mu(c^* \otimes 1)\Delta(x)$ where conjugation is used to convert the left action of $R$ into a right action. This is clearly independent of the choice of liftings $\widetilde{x}$ of $x$.

If $c \in \Sigma(f)$ then $c^*$, as given by Lemma 5.5.2, has support in $f^{-1}[f(c) - \epsilon, 1]$. Consequently, $\mathrm{D}(c^*)$ has support in $f^{-1}[f(c) - 2\epsilon, 1]$ since the front face and back face of any simplex share one vertex. Therefore, $\mathrm{D}(c^*)$ is subordinate to the filtration of $\overline{E}$ given by $1-f$ and Theorem 5.5.1, in the case $B=*$, follows from the following computation.

**Lemma 5.5.4.** *As a relative chain in $\overline{E} = S_*^{\epsilon}(M, \partial_1 M; \mathcal{F})$ we have:*

$$\partial\,\mathrm{D}(c_i^*) = (-1)^{k+1} \sum \mathrm{D}(c_j^*)\overline{a_{ij}}$$

*when $c_i \in \Sigma(f)$ has index $k$, i.e., $\mathrm{D}'(c_i) = (-1)^{\binom{k+1}{2}}\,\mathrm{D}(c_i^*)$ generate a subcomplex of $\overline{E}$ which is the conjugate transpose of the minimal subcomplex of $E$ generated by the chains $z(c_i)$.*

*Proof.* Since $x$ is a relative $n$-cycle, $\partial x$ has support in $\partial M$. This implies that $\mu(1\otimes\iota)(1\otimes c^*)\Delta(\partial\widetilde{x})$ has support in $\partial_1 M$ for each $c \in \Sigma(f)$. Since $\Delta$ is a chain map this says (when $\deg c = k$) that

$$\sum_{(\widetilde{x})} \partial\widetilde{x}_{(2)}\overline{c^*(\widetilde{x}_{(1)})} + (-1)^k \sum_{(\widetilde{x})} \widetilde{x}_{(2)}\overline{c^*(\partial\widetilde{x}_{(1)})} = 0.$$

In other words,

$$\partial\,\mathrm{D}(c_i^*) = (-1)^{k+1} \sum_{(\widetilde{x})} \widetilde{x}_{(2)}\overline{(\delta c^*)(\widetilde{x}_{(1)})} = (-1)^{k+1} \sum \mathrm{D}(c_j^*)\overline{a_{ij}}$$

since $\overline{(\delta c^*)(\widetilde{x}_{(1)})} = \sum \overline{a_{ij} c_j^*(\widetilde{x}_{(1)})} = \sum c_j^*(\widetilde{x}_{(1)})\overline{a_{ij}}$. □

## The parametrized case

We now go to the parametrized case. We have a smooth vertically oriented compact manifold bundle $p: X \to B$ with fiber $M$ and subbundles $\partial_0 X, \partial_1 X$ with fibers $\partial_0 M, \partial_1 M$ forming the boundary of $M$. We have $f: X \to I$ an oriented fiberwise GMF. Following the procedure given in Chapter IV, we choose a triangulation of $B$ by small generic smoothly embedded simplices $\sigma : \Delta^k \to B$. We will construct a family of chains, cochains and dual chains given by a family of orientation classes, all over the ring $R = \mathbb{Z}[\pi_1 X]$. We will construct these by induction on the dimension $k$ of $\sigma$, extending the given choices on $\partial\sigma$. Then it will be clear that we have a compatible family over all simplices of our triangulation of $B$ and we will get functors

$$\xi_f, \xi_{1-f} : \mathrm{simp}\, B \to \mathcal{W}h^{[0,n]}(\mathbb{Z}[\pi_1 X], \pi_1 X).$$

It will then be sufficient to show that $\xi_f$ and $\xi_{1-f}$ are related by conjugate transpose on each simplex $\sigma$.

We restrict ourselves to one $k$-simplex $\sigma$ which we regard as a subset of $B$ and we write $X_\sigma = p^{-1}(\sigma)$. Let $\Sigma(\sigma)$ be the set of singularities of $f$ over the barycenter of $\sigma$, deleting any expansion pairs and let

$$\psi : \Sigma(\sigma) \times \sigma \to X_\sigma$$

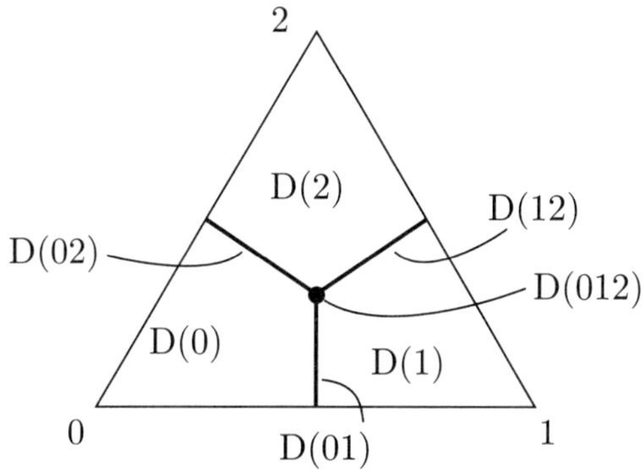


Figure 5.1: Dual cells in $\Delta^2$

be the embedding whose image is the parametrized singular set of $f$ over $\sigma$. We take the standard partial ordering on $\Sigma(\sigma)$ given by $c_i < c_j$ if $f(\psi(c_i, t)) < f(\psi(c_j, t))$ for all $t \in \sigma$.

Assume we have an $\epsilon > 0$ (chosen before triangulating $B$) with the property that the image $f(\psi(c \times \sigma))$ has length $< \epsilon/N$, where $N$ is the maximum number of critical values of $f$ on any fiber of $X$, and so that, if $c_i$ is incident under $c_j$ then $f(\psi(c_j, t)) - f(\psi(c_i, t)) \geq 7\epsilon$ for all $t \in \sigma$ . We also assume that all expansion pairs are independent (and all pairs of incident critical points with critical values separated by less than $7\epsilon$ are expansion pairs) and we ignore them.

For each $c \in \Sigma(\sigma)$ we take embedded families of stable and unstable $3\epsilon$-disks $D_-(c) \times \sigma$ and $D_+(c) \times \sigma$ defined as before by:

$$D_-(c) \times \sigma = \{(x, t) \text{ incident under } \psi(c, t) \,:\, f(\psi(x, t)) \geq f(\psi(c, t)) - 3\epsilon\}$$
$$D_+(c) \times \sigma = \{(x, t) \text{ incident over } \psi(c, t) \,:\, f(\psi(x, t)) \leq f(\psi(c, t)) + 3\epsilon\}$$

To write a formula for the Poincaré duality mapping at the chain level we need a triangulation of $X_\sigma = p^{-1}(\sigma)$. This triangulation should have three properties.

- It should consist of "$\epsilon$-small" simplices in the same sense as before.
- It should agree with the given triangulation of $X_\tau$ for each proper face $\tau$ of $\sigma$.
- It should be transverse to $D_+(c) \times \mathrm{D}(\tau)$ and $\partial D_+(c) \times \mathrm{D}(\tau)$ for all $c \in \Sigma(\sigma)$ and $\tau \subseteq \sigma$, i.e., any simplex of dimension less than $\operatorname{ind} c + \dim \tau$ should be disjoint from $D_+(c) \times \mathrm{D}(\tau)$ and similarly for $\partial D_+(c) \times \sigma$.

Here $\mathrm{D}(\tau)$ is the dual cell of $\tau$ in $\sigma$ consisting of all points in $\sigma$ whose barycentric coordinates $t_i$ corresponding to vertices in $\tau$ are maximal. (See Figure 5.1.) Suppose we have such a triangulation. Then, for every face $\tau$ of $\sigma$ we get an

$(n+p)$-chain $(p = \dim \tau)$

$$x(\tau) \in S^{\epsilon}_{n+p}(X_\tau, \partial_0 X_\tau \coprod \partial_1 X_\tau; \mathcal{F})$$

having the following property.

$$\partial x(\tau) = \sum_{i=0}^{p} (-1)^i x(\partial_i \tau) \tag{5.1}$$

We then have the following parametrized version of Lemma 5.5.2 in which we use the following notation for any real number $a$.

$$W_{\leq a}(c, \tau) = \{(x, t) \,:\, t \in \tau \text{ and } f(x,t) - f(\psi(c,t)) \leq a\}$$

and similarly $W_{\geq a}(c,\tau)$, $W_{<a}(c,\tau)$, etc. are given by imposing the stated conditions on $f(x,t) - f(\psi(c,t))$.

**Lemma 5.5.5.** *There exist chains $c(\tau) \in S^{\epsilon}_{p+d}(X_\sigma, \partial_0 X_\sigma; \mathcal{F})$ and cochains $c(\tau)^* \in S^{p+d}_{\epsilon}(X_\tau, \partial_0 X_\tau; R)$ where $p = \operatorname{ind} c$, $d = \dim \tau$ with the following properties.*

1. *$c(\tau) = c(\tau)_- - c(\tau)'$ where $c(\tau)_-$ is given by a triangulation of $D_-(c)$ $\times \tau$ and $c(\tau)'$ has support in $W_{\leq -3\epsilon}(c, \tau)$.*
2. *The boundary of $c_\beta(\tau)$ is given by*
$$\partial c_\beta(\tau) = -\sum_{i=0}^{d} (-1)^i c_\beta(\partial_i \tau) + \sum_{\alpha<\beta} \sum_{j=0}^{d} (-1)^j c_\alpha(f_j(\tau)) e_{\alpha\beta}(b_{d-j}(\tau)) \tag{5.2}$$
*for some $e_{\alpha\beta}(\rho) \in R$.*
3. *The restriction of $c(\tau)^*$ to $W_{\leq\epsilon}(c, \sigma)$ is given by the $R$-intersection number with $D_+(c) \times \mathrm{D}(\tau)$ in $\widetilde{X}_\sigma$.*
4. *$c(\tau)^* = 0$ on $X_\rho$ if $\tau$ is not contained in $\rho$.*
5. *$\delta c_i(\tau)^*$ is an $R$-linear combination of $c_j(\rho)^*$ where $c_j \geq c_i$ and $\rho \supseteq \tau$.*
6. *$c_i(\tau)^*(c_j(\rho)) = \delta_{ij}\delta_{\tau\rho}$, i.e., $c_i(\tau)^*$ is dual to $c_i(\tau)$.*

**Remark 5.5.6.** The boundary formula in (5.2) is supposed to be the same as the boundary map for $Z_k$ in Remark 3.1.4. The reason it looks different is because we reversed the order of the vertices in $\tau$. This means that the boundary map for the dual chains $\bar{c}(\tau)$ should be in the correct order and should satisfy the following boundary formula in agreement with Remark 3.1.4.

$$\partial \bar{c}_\alpha(\tau) = -\sum_{i=0}^{d} (-1)^{d-i} \bar{c}_\alpha(\partial_i \tau) + \sum_{\beta>\alpha} \sum_{j=0}^{d} (-1)^j \bar{c}_\beta(b_j(\tau)) \overline{e_{\alpha\beta}(f_{d-j}(\tau))} \tag{5.3}$$

The chains $c(\tau)$ generate the multiple mapping cylinder for the system of higher homotopies $e(\rho)$ given by

$$e(\rho)(c_\beta) = \sum_{\alpha<\beta} c_\alpha e_{\alpha\beta}(\rho).$$

We also note that since the cochains $c(\tau)^*$ are dual to the chains $c(\tau)$, the boundary formula (5.2) gives the coboundary formula:

$$\delta c_\alpha(\tau)^* = \sum_v (-1)^{i+1} c_\alpha(\tau * v)^* + \sum_\rho (-1)^d e_{\alpha\beta}(b_0(\tau) * \rho) c_\beta(\tau * \rho)^* \tag{5.4}$$

where the first sum is over all vertices $v$ of $\sigma$ which are not in $\tau$, $\tau * x$ denotes the join of $\tau$ and $x$ (the smallest face of $\sigma$ containing both $\tau$ and $x$) and $i$ is the number of vertices of $\tau$ which are less than $v$ (so that $\partial_i(\tau * v) = \tau$). The second sum is over all faces $\rho$ of $\sigma$ (including the empty face) which are disjoint from $\tau$ and have vertices greater than the last vertex $b_0(\tau)$ of $\tau$ (so that $\tau = f_d(\tau * \rho)$). In particular, $\tau$ contains the last vertex of $\sigma$ if and only if the second sum consists of the single term $(-1)^d e_{\alpha\beta}(b_0(\tau)) c_\beta(\tau)^*$.

*Proof.* The proof is analogous to the proof of Lemma 5.5.2. As before, the intersection of the kernels of the cochains $c(\tau)^*$ generate a strongly acyclic subcomplex $Q$ of the filtered chain complex $S_*^\epsilon(X_\sigma, \partial_0 X_\sigma; \mathcal{F})$ which is complementary to the subcomplex $K$ generated by the chains $c(\tau)$ for all $c \in \Sigma(\sigma)$ and all (nonempty) faces $\tau$ of $\sigma$. For each vertex $v$ of $\sigma$, the chains $c(v)$ generate a minimal subcomplex $K(v)$ which is homotopy equivalent to $S_*^\epsilon(X_\sigma, \partial_0 X_\sigma; \mathcal{F})$.

The proof will be by induction, first on the dimension $k$ of $\sigma$. So we may assume that $c(\tau)$, $c(\tau)^*$ already exist for proper faces $\tau$ of $\sigma$ and we need only construct $c(\sigma)$ and $c(\sigma)^*$ for all $c \in \Sigma(\sigma)$. Next we do induction on the number of lavers of the poset $\Sigma(\sigma)$ where a *layer* is a component of the union $\cup f(\psi(c\times\sigma))$ of images of the critical set components $\psi(c \times \sigma)$. Since each image $f\psi(c \times \sigma)$ has length $< \epsilon/N$, each layer (for $\sigma$) has length $< \epsilon$. Since $f\psi(c\times\tau) \subseteq f\psi(c\times\sigma)$ each face $\tau$ of $\sigma$ has a larger (or equal) number of smaller layers.

Suppose first that there is only one layer for $\sigma$. Then for each $c \in \Sigma(\sigma)$, the set $W_{\leq -\epsilon}(c, \sigma)$ has no critical points and is therefore homotopy equivalent to $\partial_0 X_\sigma$. Consequently every relative cycle is a boundary and we can complete the chains $c(\sigma)_-$, given by triangulating the stable disk bundle $D_- \times \sigma$, to a relative cycle as claimed in 1. The cochains $c(\tau)^*$ can also be completed in an analogous way so that 3 and 4 are satisfied.

In detail: We triangulate $D_-(c) \times \sigma$ making sure that the triangulation extends the given triangulation over $\partial\sigma$. Then the difference between the boundary of $c(\sigma)_-$ and the alternating sum of the given chains $c(\partial_i\sigma)$ is a relative cycle with support in $W_{\leq -3\epsilon}(c, \sigma)$. Consequently this relative cycle is the relative boundary of some chain $c(\sigma)'$ which has support in $W_{\leq -3\epsilon}(c, \sigma)$ so that

$$\partial c_- + \sum (-1)^i \partial c(\partial_i \sigma) = \partial c(\sigma)'.$$

Then $c(\sigma) = c(\sigma)_- - c(\sigma)'$ satisfies Condition 1 with $W_{\leq -3\epsilon}(c, \sigma)$ instead of $W_{\leq -\epsilon}(c, \sigma)$. We can therefore strengthen Condition 1 as follows.

1. Each $c(\tau)'$ in Condition 1 has support in $W_{\leq -3\epsilon}(c,\tau)$ provided that the set $W_{=-3\epsilon}(c,\tau)$ has no critical points.

2. If $W_{=-3\epsilon}(c,\tau)$ contains critical points of $f$ then we assume that $c(\tau)'$ has support in a set $W(c,\tau) = f^{-1}[0,s] \cap X_\tau$ for some $s = s(c,\tau)$ having the property that

   (a) $s(c,\tau)$ is not in any layer of $\tau$ (i.e., is not a critical value).

   (b) $s(c,\tau) \subseteq s(c,\rho)$ if $\tau \subseteq \rho$.

   (c) $W_{\leq -3\epsilon}(c,\tau) \subseteq W(c,\tau) \subseteq W_{\leq -\epsilon}(c,\tau)$

($s(c,\tau)$ can be chosen to be a real number slightly larger than the supremum of the highest layer of $\tau$ which meets the set $W_{=-3\epsilon}(c,\tau)$.)

Now suppose that $\sigma$ has more than one layer. Then $\Sigma(\sigma) = \Sigma_+(\sigma) \coprod \Sigma_-(\sigma)$ where $\Sigma_+(\sigma)$ is the set of all critical points in the top layer of $\sigma$. By induction on the number of layers we may assume that we have the chains $c_\alpha(\sigma)$ for all $c_\alpha \in \Sigma_-(\sigma)$ and that the cochains $c_\alpha(\sigma)^*$ are defined up to some regular value $s$ below the top layer of $\sigma$. For each $c_\beta \in \Sigma(\sigma)$ we take $c_\beta(\sigma)_-$ as before. Then the chain

$$b = \partial c_\beta(\sigma)_- + \sum_{i=0}^{k} c_\beta(\partial_i\sigma) - \sum_{\alpha<\beta}\sum_{j=1}^{k}(-1)^j c_\alpha(f_j(\sigma))e_{\alpha\beta}(b_{k-j}(\sigma))$$

is defined and has support in $W(c_\beta,\sigma)$. However, $W(c_\beta,\sigma)$ is homotopy equivalent to its intersection with $X_v$ where $v$ is the last vertex of $\sigma$. Since the chains $c_\alpha(v)$ generate a minimal subcomplex of $S_*^\epsilon(W(c_\beta,\sigma)\cap X_v;\mathcal{F})$, the chain $b$ above is homotopic to a linear combination of $c_\alpha(v)$'s:

$$b = \partial c_\beta(\sigma)' + \sum_\alpha c_\alpha(v) r_\alpha.$$

Letting $c_\beta(\sigma) = c_\beta(\sigma) - c_\beta(\sigma)'$ and $e_{\alpha\beta}(\sigma) = r_\alpha$, we get (5.4) as required.

For $c_\beta \in \Sigma_+(\sigma)$, the cochains $c_\beta(\sigma)^*$ is easily constructed. So, take $c_\alpha \in \Sigma_-(\sigma)$ maximal so that $c_\alpha(\sigma)^*$ has not yet been extended to all of $X_\sigma$. ($c_\alpha(\sigma)^*$ is already defined on some subset $W_-$ of $X_\sigma$ given by $f \leq s$ for some regular value $s$ between the top two levels of $\sigma$.) Let $E_+ = S_*^\epsilon(X_\sigma, W_-;\mathcal{F})$. Then $E_+ = K_+ \oplus Q_+$ where $K_+$ is generated by the chains $c_\beta(\tau)$ for all $c_\beta \in \Sigma_+(\sigma)$ and $\tau \subseteq \sigma$ and $Q_+$ is the intersection of the kernels of the cochains $c_\beta(\tau)^*$.

Let $d_\alpha = \delta c_\alpha(\sigma)^* - b_\alpha$ where $b_\alpha$ is the righthand side of (5.4). These terms have already been extended to $X_\sigma$ by maximality of $\alpha$ and they are zero on $X_\rho$ for all proper faces $\rho$ of $\sigma$ by downward induction on $\alpha$. Consequently, the cochain $d_\alpha$ is equal to zero on the subcomplex $K_+$ and on the sum $Q_0 = \sum_{\rho\neq\sigma} Q_\rho$ where $Q_\rho$ is the intersection of $Q_+$ with the total singular complex of $X_\rho$. Since the complement $Q_+$ of $K_+$ is acyclic and $Q_\tau$ is acyclic for all $\tau$, $b_\alpha = \delta d'_\alpha$ for some $d'_\alpha \in \mathrm{Hom}_R(Q_+/Q_0, R)$ and $c_\alpha(\sigma) - d'_a$ is the desired extension of $c_\alpha(\sigma)$ to all of $X_\sigma$. □

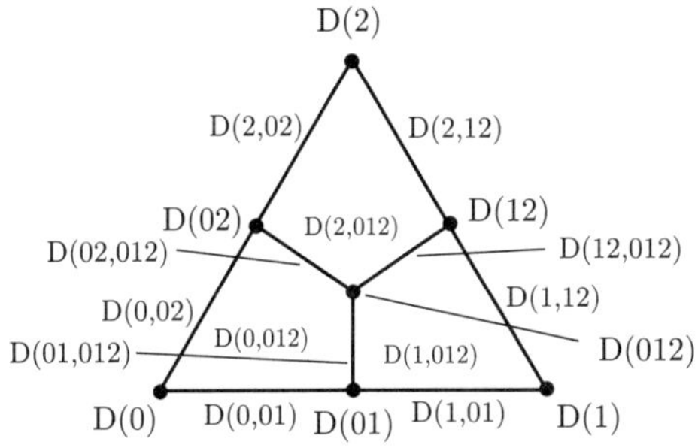


Figure 5.2: $\Delta^2$ has $3^{k+1} - 2^{k+1} = 19$ dual chains for each $c \in \Sigma(\sigma)$

We can now define the *dual chains* $\mathrm{D}\, c(\tau, \lambda)$ for all $\tau \subseteq \lambda \subseteq \sigma$ by

$$\mathrm{D}\, c(\tau, \lambda) = \sum_{(\widetilde{x})} \widetilde{x}_{(2)}(\lambda)\overline{c(\tau)^* \widetilde{x}_{(1)}(\lambda)}. \tag{5.5}$$

There are too many dual chains. Figure 5.2 shows the dual chains for $\Delta^2$ as given by (5.5). The conjugate transpose complex has only $2^{k+1} - 1 = 7$ generators for each element of $\Sigma(\sigma)$. To fix this we will give a natural upper triangular morphism from the complex generated by the dual chains $\mathrm{D}\, c(\rho, \tau)$ to the (abstract) complex generated by the symbols $\overline{c}(\tau)$ with boundary map given by (5.3). But first we need the following.

**Lemma 5.5.7.** *The dual chain* $\mathrm{D}\, c(\tau, \lambda) \in S_*^\epsilon(X_\lambda, \partial_1 X_\lambda; \mathcal{F})$ *has support in* $W_{\geq -2\epsilon}(c, \tau)$ *and its boundary is given as follows where we use the notation of Equation (5.4) above and* $\ell = \dim \lambda$.

$$\partial\, \mathrm{D}\, c_\alpha(\tau, \lambda) = \sum_{i=0}^{\ell} (-1)^{p+d+i}\, \mathrm{D}\, c_\alpha(\tau, \partial_i \lambda) \tag{5.6}$$

$$+ \sum_{v} (-1)^{p+d+i}\, \mathrm{D}\, c_\alpha(\tau * v, \lambda) \tag{5.7}$$

$$+ \sum_{\beta > \alpha} \sum_{\rho} (-1)^{p+1}\, \mathrm{D}\, c_\beta(\tau * \rho, \lambda)\overline{e_{\alpha\beta}(b_0(\tau) * \rho)} \tag{5.8}$$

*Proof.* Condition 1 follows from Lemma 5.5.5.3 which implies that the support of $c(\tau)^*$ lies in $W_{\geq -3\epsilon}(c, \sigma)$ and from (5.5) which tells us that $\mathrm{D}\, c(\tau, \lambda)$ is a sum of $\epsilon$-small simplices having at least one vertex in $W_{> -\epsilon}(c, \lambda)$. To prove (2) we lift Equation (5.1) to the covering $\widetilde{X}_\lambda$ and apply the diagonal operator $\Delta$. Then we get the following equivalence modulo the diagonal action of $\pi_1 X$ where $p = \deg c$, $q = n - p$, $d = \dim \tau$, $\ell = \dim \lambda$, $s = \ell - d$ and superscripts denote degree before boundary is taken.

$$\sum_{(\widetilde{x})} \partial\widetilde{x}_{(1)}(\lambda)^{p+d+1} \otimes \widetilde{x}_{(2)}(\lambda)^{q+s-1} + \sum_{(\widetilde{x})}(-1)^{p+d}\widetilde{x}_{(1)}(\lambda)^{p+d} \otimes \partial\widetilde{x}_{(2)}(\lambda)^{q+s}$$
$$\cong \sum_{i=0}^{\ell}\sum_{(\widetilde{x})}(-1)^{i}\widetilde{x}_{(1)}(\partial_i\lambda) \otimes \widetilde{x}_{(2)}(\partial_i\lambda).$$

Applying $\mu(c_\alpha(\tau)^* \otimes 1)$ to this we get

$$\partial\,\mathrm{D}\,c_\alpha(\tau,\lambda) = \sum_{i=0}^{\ell}(-1)^{p+d+i}\,\mathrm{D}\,c_\alpha(\tau,\partial_i\lambda) + \sum_{(\widetilde{x})}(-1)^{p+d+1}\widetilde{x}_{(2)}(\lambda)\overline{(\delta c_\alpha(\tau)^*)\widetilde{x}_{(1)}(\lambda)}$$

The first sum is equal to (5.6) and the second sum can be expanded using (5.4) to get:

$$\sum_{(\widetilde{x})}\sum_{v}(-1)^{p+d+i}\widetilde{x}_{(2)}(\lambda)\overline{c_\alpha(\tau * v)^*\widetilde{x}_{(1)}(\lambda)}$$
$$+\sum_{(\widetilde{x})}\sum_{\beta>\alpha}\sum_{\rho}(-1)^{p+1}\widetilde{x}_{(2)}(\lambda)\overline{e_{\alpha\beta}(b_0(\tau) * \rho)c_\beta(\tau * \rho)^*\widetilde{x}_{(1)}(\lambda)}$$

These sums are equal to (5.7) and (5.8). □

*Proof of Theorem 5.5.1.* We need only one more step to prove the theorem. We need to show that the free $R$-complex D generated by the dual chains $\mathrm{D}\,c(\tau,\lambda)$ contains a copy of the free $R$-complex $\overline{C}$ generated by the symbols $\overline{c}(\tau)$ with boundary map given by (5.3). We also need to check that the inclusion map $\overline{C} \to \mathrm{D}$ is both natural and filtered and that it gives a minimal subcomplex of $S_*^\epsilon(X_v, \partial_1 X_v; \mathcal{F})$ over each vertex $v$ of $\sigma$. (Naturality implies that it induces a functor on the category of small generic simplices in $B$. If it is filtered and correct over the vertices it must be correct up to homotopy over every simplex by the uniqueness of minimal subcomplexes.)

For each face $\rho$ of $\sigma$ let $\mathrm{D}(\rho)$ denote the subcomplex of D generated by all $\mathrm{D}\,c(\tau,\lambda)$ where $\lambda \subseteq \rho$ and let $\overline{C}(\rho)$ denote the subcomplex of $\overline{C}$ generated by all $\overline{c}(\tau)$ where $\tau \subseteq \rho$. Then we need to construct a natural chain map $\phi : D \to \overline{C}$ which induces a chain homotopy equivalence $D(\rho) \simeq \overline{C}(\rho)$ for all $\rho \subseteq \sigma$. Such a map is given as follows where we use the notation $p = \deg c_\alpha$, $d = \dim\tau$, $\ell = \dim\lambda$ and $s = \ell - d$.

$$\phi\,\mathrm{D}\,c_a(\tau,\lambda) = \begin{cases}(-1)^{\binom{p-s+1}{2}+d}c_\alpha(b_s(\lambda)) & \text{if } \tau = f_d(\lambda),\\ 0 & \text{otherwise.}\end{cases} \tag{5.9}$$

It is a easy calculation to show that this is chain map. (It took me only a week to do it.)

It remains to show that $\phi$ induces a homotopy equivalence $\mathrm{D}(\rho) \simeq \overline{C}(\rho)$ for each $\rho \subseteq \sigma$. We do this by induction on the dimension $k$ of $\sigma$. If $k = 0$ then the statement reduces to the one manifold case discussed earlier. So suppose that $\phi$ induces a homotopy equivalence $\mathrm{D}(\rho) \simeq \overline{C}(\rho)$ for all proper faces $\rho$ of $\sigma$. Then we have the following commuting diagram of free $R$-complexes with exact rows where the left hand vertical arrow is a homotopy equivalence by induction on $k$.

$$\begin{array}{ccccccccc} 0 & \longrightarrow & \sum \mathrm{D}(\partial_i\sigma) & \longrightarrow & \mathrm{D}(\sigma) & \longrightarrow & \mathrm{D}(\sigma)/\mathrm{D}(\partial\sigma) & \longrightarrow & 0 \\ & & \downarrow \phi & & \downarrow \phi & & \downarrow \overline{\phi} & & \\ 0 & \longrightarrow & \sum \overline{C}(\partial_i\sigma) & \longrightarrow & \overline{C}(\sigma) & \longrightarrow & \overline{C}(\sigma)/\overline{C}(\partial\sigma) & \longrightarrow & 0 \end{array}$$

The generators of the quotient complex $\mathrm{D}(\sigma)/\mathrm{D}(\partial\sigma)$ are given by $\mathrm{D}\,c_\alpha(\tau,\sigma)$. This generator maps to zero in $\overline{C}(\sigma)/\overline{C}(\partial\sigma)$ unless $\tau = v_0 = f_0(\sigma)$ is the first vertex of $\sigma$ in which case $\overline{\phi}(\mathrm{D}\,c_a(v,\sigma)) = \pm\overline{c}_\alpha(\sigma)$. Consequently $\overline{\phi}$ is onto and the kernel is the subcomplex $Q$ of $\mathrm{D}(\sigma)/\mathrm{D}(\partial\sigma)$ generated by all $\mathrm{D}\,c_\alpha(\tau,\sigma)$ where $\tau \neq v_0$. But a chain contraction of $Q$ can be given by

$$h\,\mathrm{D}\,c_a(\tau,\sigma) = \begin{cases} (-1)^{p+d-1}\,\mathrm{D}\,c_a(\partial_0\tau,\sigma) & \text{if } v_0 \in \tau, \\ 0 & \text{otherwise.} \end{cases}$$

We conclude that $\overline{\phi}$ in the above diagram is a homotopy equivalence and therefore $\mathrm{D}(\sigma) \simeq \overline{C}(\sigma)$ by the 5-lemma. □

## 5.6 Waldhausen $K$-theory

In this section we outline the argument of [IK95] in which Waldhausen $K$-theory is used to show that the space of acyclic filtered chain complexes has the correct homotopy type. Since $|\mathcal{W}h^h(R,G)| \simeq |FC^h(BG,\xi_R)|$, the same will be true for the Whitehead space. For more details see [IK95] and [Wal85]. The theorem we will prove is:

**Theorem 5.6.1.** *[IK95] Given any simplicial set $X_\bullet$ and any functor $\xi : \mathrm{simp}\,X_\bullet \to$ Mod-$R$ so that $\xi(x) \cong R_R$ for all $x \in X_\bullet$ we have a natural homotopy fiber sequence*

$$|FC^h(X_\bullet,\xi)| \to Q(|X_\bullet|_+) \to \overline{\mathbb{Z}} \times BGL(R)^+$$

*where $Q = \Omega^\infty S^\infty$, $BGL(R)^+$ is the Quillen plus construction and $\overline{\mathbb{Z}}$ is the image of $\mathbb{Z} = K_0\mathbb{Z}$ in $K_0R$. Furthermore the map $|FC^h(X_\bullet,\xi)| \to Q(|X_\bullet|_+)$ is given by sending $(E,P,\gamma)$ to the poset $P$ considered as a configuration of positive and negative particles over $|X_\bullet|$ by the map $\gamma$ (the sign being given by $(-1)^{deg}$).*

In the special case $(X_\bullet,\xi) = (BG_\bullet,\xi_R)$, Corollary 5.3.5 gives us the following.

**Corollary 5.6.2.** *For any subgroup $G$ of the group of units of a ring $R$ we get a natural homotopy fiber sequence:*

$$|\mathcal{W}h^h_\bullet(R,G)| \to Q(BG_+) \to \overline{\mathbb{Z}} \times BGL(R)^+.$$

This theorem is an application of Waldhausen's fibration theorem [Wal85] which in this case says the following. (We note that by Lemma 5.2 in [Wal78] the fibration theorem implies its simplicial version and we can use a different cylinder functor in each degree.)

**Theorem 5.6.3 (Fibration theorem).** *Let $\mathcal{K}_\bullet$ be a simplicial (or multi-simplicial) category with cofibrations and two kinds of weak equivalences $w\mathcal{K} \subseteq h\mathcal{K}_\bullet$ and suppose that in each degree $\mathcal{K}_k$ has a simplicial cylinder functor $T_k$ so that the weak equivalences in $h\mathcal{K}_k$ satisfy the cylinder axiom, saturation axiom and the extension axiom. Then the following sequence of inclusion functors has a canonical null homotopy of its composition and gives a homotopy fiber sequence on realizations.*

$$w\mathcal{S}_\bullet\mathcal{K}_\bullet^h \to w\mathcal{S}_\bullet\mathcal{K}_\bullet \to h\mathcal{S}_\bullet\mathcal{K}_\bullet$$

**Remark 5.6.4.** These three axioms are obviously satisfied when $h\mathcal{K}$ is the category of quasi-isomorphisms in a category of chain complexes with extra structure and $T_k(f)$ is the mapping cylinder (with extra structure) of a chain map $f : A \to B$ since:

1. The cylinder axiom says that the morphism $T_k(f) \to B$ is in $h\mathcal{K}_k$.
2. The saturation axiom says that when a composition of morphisms $f \circ g$ lies in $h\mathcal{K}_k$ then $f$ is in $h\mathcal{K}_k$ if and only if $g$ is in $h\mathcal{K}_k$.
3. The extension axiom says that when three morphisms give a map between two extension sequences

$$(f_A, f_B, f_C) : (A \rightarrowtail B \twoheadrightarrow C) \to (A' \rightarrowtail B' \twoheadrightarrow C')$$

   and if the two outside maps $f_A, f_C$ lie in $h\mathcal{K}_k$ then so does $f_B : B \to B'$. This will of course follow from the 5-lemma.

It remains for us to define $\mathcal{K}_\bullet$ ($e_\bullet\mathcal{K}(X_\bullet, \xi)$ in the notation below) and show that the three terms in Theorem 5.6.1 are the loop spaces of the realizations of the corresponding terms in the fibration theorem. (Corollary 5.6.16, Theorem 5.6.17 and Theorem 5.6.22, respectively.)

We begin with the category of finite graded posets over a simplicial set $X_\bullet$. This definition does not agree with the definition used earlier. However, the two definitions agree on all cofibrations, quotient maps and weak equivalences.

**Definition 5.6.5.** If $X_\bullet$ is a simplicial set we define the simplicial category of *finite graded posets over* $X_\bullet$ to be the simplicial Waldhausen category ($\mathcal{Q}(X_\bullet)$, $co\mathcal{Q}(X_\bullet)$, $w\mathcal{Q}(X_\bullet)$) given as follows.

(a) An *object* of $\mathcal{Q}(X_\bullet)$ in degree $k$ is a finite graded poset $P$ together with a map $\gamma : P \to X_k$.

(b) A *morphism* $\alpha : (P, \gamma) \to (Q, \gamma')$ in $\mathcal{Q}_k(X_\bullet)$ is a pointed mapping $\alpha : P_+ \to Q_+$ (where $P_+ = P \coprod \{*\}$ with the new element $*$ being less than all other elements of $P$) which takes closed sets to closed sets and so that $\gamma(x) = \gamma'(\alpha(x))$

whenever $\alpha(x) \neq *$. The set of points in $P$ which go to $*$ will be called the *kernel* of $\alpha$. (So $\alpha^{-1}(*) = (\ker \alpha)_+$.)

(c) A morphism $\alpha : P \to Q$ is a *cofibration* if it is an order preserving monomorphism.

(d) A morphism $\alpha : P \to Q$ is a *weak equivalence* if it is a closed bijection.

**Lemma 5.6.6.** $\Omega|w\mathcal{S}_\bullet\mathcal{Q}(X_\bullet)| \simeq \coprod_{n\geq 0} Q(|X_\bullet|_+)$

*Proof.* Let $\mathcal{C}(X_\bullet) \subseteq \mathcal{Q}(X_\bullet)$ be the simplicial full subcategory of posets over $X_\bullet$ with null ordering. Then there is an obvious retraction $r : \mathcal{Q}(X_\bullet) \to \mathcal{C}(X_\bullet)$ given by forgetting the partial ordering and a natural weak equivalence $(P,\gamma) \to r(P,\gamma)$. Consequently, $|w\mathcal{S}_\bullet\mathcal{Q}(X_\bullet)| \simeq |w\mathcal{S}_\bullet\mathcal{C}(X_\bullet)|$. Since the points of different degrees do not interact at all in $\mathcal{C}(X_\bullet)$ we have that

$$\mathcal{C}(X_\bullet) = \prod_{n\geq 0} \mathcal{C}_n(X_\bullet)$$

where $\mathcal{C}_n(X_\bullet)$ is the simplicial full subcategory of $\mathcal{C}(X_\bullet)$ consisting of sets over $X_\bullet$ all of whose elements have degree $n$. In each degree, $w\mathcal{S}_\bullet\mathcal{C}_n(X_\bullet) = i\mathcal{S}_\bullet\mathcal{C}_n(X_\bullet)$ is identical to $BB\Sigma_X(\mathbf{1})$ of [Seg73]. The lemma follows. □

In order to coalesce the different copies of $Q(|X_\bullet|_+)$ in $w\mathcal{S}_\bullet\mathcal{Q}(X_\bullet)$ we need to use expansion pairs and introduce another simplicial direction analogous to the construction in the last section.

A morphism in $\mathcal{Q}(X_\bullet)$ will be called an *expansion* if it is a split cofibration $P \rightarrowtail P \vee S$ where $S$ is a union of expansion pairs, an *expansion pair* being defined to be a pair $T = \{x^-, x^+\} \subseteq S$ so that $\deg x^+ = 1 + \deg x^-$, $x^- < x^+$, $T$ and $S - T$ are both closed and $\gamma(x^-) = \gamma(x^+)$. (The same as (1),(2) in Definition 5.3.3.) We let $e\mathcal{Q}(X_\bullet)$ denote the category of all expansions.

**Definition 5.6.7.** Let $e_\bullet\mathcal{Q}(X_\bullet)$ denote the bisimplicial nerve of $e\mathcal{Q}(X_\bullet)$ considered as a bisimplicial Waldhausen category as follows:

(a) The objects of $e_k\mathcal{Q}(X_\ell)$ are sequences of split cofibrations

$$P \rightarrowtail P \vee S_1 \rightarrowtail \cdots \rightarrowtail P \vee S_k$$

of posets over $X_\ell$ where each $S_i$ is a union of expansion pairs. We sometimes write this as $\{P \vee S_i\}_{i=0\cdots k}$ where $S_0 = \emptyset$.

(b) A morphisms

$$(P \rightarrowtail P \vee S_1 \rightarrowtail \cdots \rightarrowtail P \vee S_k) \to (Q \rightarrowtail Q \vee T_1 \rightarrowtail \cdots \rightarrowtail Q \vee T_k)$$

in $e_k\mathcal{Q}(X_\ell)$ is a morphism $\alpha : P \vee S_k \to Q \vee T_k$ in $\mathcal{Q}(X_\ell)$ so that $\alpha^{-1}(Q) = P - \ker \alpha$ and the inverse image of each expansion pair in $T_i - T_{i-1}$ is a union of expansion pairs in $S_i - S_{i-1}$. [Thus, the expansion pairs act as a unit and we have an isomorphism of categories:

$$e_k\mathcal{Q}(X_\ell) \cong \mathcal{Q}(X_\ell) \times \underbrace{\mathcal{C}(X_\ell) \times \cdots \times \mathcal{C}(X_\ell)}_{k}.\ ]$$

(c) Cofibrations are order preserving monomorphisms and weak equivalences are closed bijections just like before.

The expansion pairs of degree $d, d+1$ will coalesce the $d, d+1$-st copies of $Q(|X_\bullet|_+)$ and we get:

**Theorem 5.6.8.** $\Omega|wS_\bullet e_\bullet \mathcal{Q}(X_\bullet)| \simeq Q(|X_\bullet|_+)$.

**Definition 5.6.9.** The *mapping cylinder* $T(\alpha)$ of a morphism $\alpha : P \to Q$ in $\mathcal{Q}(X_\bullet)$ is defined by $T(\alpha) = P \coprod Q \coprod \Sigma P$ with elements of $P, Q$ ordered as before, $\Sigma P = \{\sigma x : x \in P\}$ with $\gamma(\sigma x) = \gamma(x)$ and $\deg \sigma x = 1 + \deg x$ being the suspension of $P$ with partial ordering given by:

1. $\sigma x > y \in P$ iff $x \geq y$ in $P$

2. $\sigma x > z \in Q$ iff $\alpha(y) \geq z$ in $Q$ for some $y \leq x$ in $P$

Equivalently, $A \cup B \cup \Sigma C$ is closed in $T(\alpha)$ if $C \subseteq A$ are closed in $P$ and $B$ is a closed subset of $Q$ containing $\alpha(C)$. Let $\pi : T(\alpha) \to Q$ be given by $\pi = (\alpha, id, *)$, i.e., $\pi(x) = \alpha(x), \pi(\sigma x) = *$ if $x \in P$ and $\pi(y) = y$ if $y \in Q$.

Then $T(\alpha)$ together with the inclusion morphisms $P \vee Q \to T(\alpha)$ and projection $\pi : T(\alpha) \to Q$ forms a simplicial cylinder functor for $\mathcal{Q}(X_\bullet)$. There is no bisimplicial cylinder functor for $e_\bullet \mathcal{Q}(X_\bullet)$. However, we will work with the simplicial cylinder functors $T_k$ for $e_k \mathcal{Q}(X_\bullet)$ given on each morphism $\alpha : \{P \vee S_i\} \to \{Q \vee T_i\}$ by

$$T_k(\alpha) = \{T(\alpha_0 : P \to Q) \vee S_i \vee T_i\}_{i=0,\cdots,k}$$

with projection map $\pi_k : T_k(\alpha) \to \{Q \vee T_i\}$ given by $\pi_k = (\pi, \alpha|S_i, id_{T_i})$.

Now suppose that $X_\bullet$ is a simplicial set and $\xi : \operatorname{simp} X_\bullet \to$ Mod-$R$ is a functor taking each element of $X_\bullet$ to a free right $R$-module of rank 1. Then the bisimplicial category of filtered chain complexes $we_\bullet FC(X_\bullet, \xi)$ will become the category of weak equivalence in a bisimplicial Waldhausen category $e_\bullet \mathcal{K}(X_\bullet, \xi)$ defined as follows.

**Definition 5.6.10.** (a) The bisimplicial category $e_\bullet \mathcal{K}(X_\bullet, \xi)$ has the same bisimplicial object set as $we_\bullet FC(X_\bullet, \xi)$ (5.3.3). Thus, in bidegree $(k, \ell)$, the objects are sequences of expansions:

$$(E, P) \rightarrowtail (E, P) \vee (D_1, S_1) \rightarrowtail \cdots \rightarrowtail (E, P) \vee (D_k, S_k)$$

in $FC(X_\ell, \xi)$ where each $S_i - S_{i-1}$ is a union of expansion pairs in the filtered chain complex

$$(E, P) \vee (D_k, S_k) = (E \oplus D_k, P \vee S_k).$$

We write these objects as $\{(E, P) \vee (D_i, S_i)\}_{i=0\cdots k}$ where $(D_0, S_0) = (0, \emptyset)$.

(b) A morphism $\{(E, P) \vee (D_i, S_i)\} \to \{(F, Q) \vee (C_i, T_i)\}$ in $e_k \mathcal{K}(X_\ell, \xi)$ is a pair $(f, \alpha)$ where $\alpha : \{P \vee S_i\} \to \{Q \vee T_i\}$ is a morphism in $e_k \mathcal{Q}(X_\ell)$ and $f : E \oplus D_k \to F \oplus C_k$ is a filtered morphism with respect to $\alpha : P \vee S_k \to Q \vee T_k$, i.e., satisfying the three conditions in Definition 5.2.2. (Except that condition (3) only makes sense if $x$ is not in the kernel of $\alpha$.)

(c) $(f, \alpha)$ is a *cofibration* in $e_k \mathcal{K}(X_\ell, \xi)$ if $\alpha$ is a cofibration in $e_k \mathcal{Q}(X_\ell)$ and $f$ maps each $E^A \oplus D_k^B$ isomorphically onto $F^{\alpha(A)} \oplus C_k^{\alpha(B)}$.

(d) $(f, \alpha)$ is a *w-equivalence* in $e_k\mathcal{K}(X_\ell, \xi)$ if $\alpha$ is a weak equivalence in $e_k\mathcal{Q}(X_\ell)$, i.e., a closed bijection. Thus:

$$we_\bullet \mathcal{K}(X_\bullet, \xi) = we_\bullet FC(X_\bullet, \xi).$$

(e) $(f, \alpha)$ is an *h-equivalence* in $e_k\mathcal{K}(X_\ell, \xi)$ if $f : E \oplus D_k \to F \oplus C_k$ is a chain homotopy equivalence.

**Definition 5.6.11.** The *mapping cylinder* $T_k(f, \alpha)$ of a morphism $(f, \alpha)$ : $\{(E, P) \vee (D_i, S_i)\} \to \{(F, Q) \vee (C_i, T_i)\}$ in $e_k\mathcal{K}(X_\bullet, \xi)$ is given by $T_k(f, \alpha) = (T_k(f), T_k(\alpha))$ where $T_k(f) = (E \oplus D_k) \oplus (F \oplus C_k) \oplus \Sigma E$ with boundary map given by $\partial(x, y, \sigma z) = (\partial x - z, \partial y + f(z), -\sigma\partial^E z)$. The $\Lambda T_k(\alpha)$-filtration of $T_k(f)$ is given by

$$T_k(f)^{(A_1 \cup A_2) \cup (B_1 \cup B_2) \cup \Sigma C} = (E^{A_1} \oplus D_k^{A_2}) \oplus (F^{B_1} \oplus C_k^{B_2}) \oplus \Sigma(E^C)$$

The cohomology classes are given in the obvious way for $P, S_k, Q, T_k$ and for $\sigma x \in \Sigma P$ they are given by $\lambda(\sigma x) = -\delta\lambda(x)$. Finally the projection $(p, \pi_k)$ : $(T_k(f), T_k(\alpha)) \to (F, Q) \vee (C_k, T_k)$ is given by $p = (f, id, 0)$.

The projection map is an $h$-equivalence. In fact it is a simple homotopy equivalence by the following definition.

**Definition 5.6.12.** The *Whitehead group* of $(X_\bullet, \xi)$ is defined by

$$Wh_1(X_\bullet, \xi) = \pi_0|we_\bullet \mathcal{K}^h(X_\bullet, \xi)|$$

with addition given by direct sum: $[A] + [B] = [A \vee B]$. The additive inverse is given by $-[A] = [\Sigma(A)]$ where $\Sigma(A) = T_k(id_A)/(A \vee A)$. The *torsion* of an $h$-equivalence $(f, \alpha)$ is defined to be the Whitehead torsion of its mapping cone $C_k(f, \alpha) = T_k(f, \alpha)/\{(E, P) \vee (D_i, S_i)\}$.

$$\tau(f, \alpha) = \tau C_k(f, \alpha) \in Wh_1(X_\bullet, \xi)$$

(Although $C_k$ does not commute with simplicial operators, its torsion is invariant.) An $h$-equivalence with zero torsion is called a *simple homotopy equivalence* or simple $s$-equivalence and the bisimplicial category of $s$-equivalences is denoted $se_\bullet \mathcal{K}(X_\bullet, \xi)$.

**Lemma 5.6.13.** $|we_\bullet \mathcal{K}^h(X_\bullet, \xi)| \simeq |we_\bullet \mathcal{K}^s(X_\bullet, \xi)| \times Wh_1(X_\bullet, \xi)$

**Theorem 5.6.14.** *Given any simplicial set $X_\bullet$ and any functor $\xi : \operatorname{simp} X_\bullet \to$ Mod-R so that $\xi(x) \cong R_R$ for all $x \in X_\bullet$ we have natural homotopy fiber sequences*

$$w\mathcal{S}_\bullet e_\bullet \mathcal{K}^h(X_\bullet, \xi) \to w\mathcal{S}_\bullet e_\bullet \mathcal{K}(X_\bullet, \xi) \to h\mathcal{S}_\bullet e_\bullet \mathcal{K}(X_\bullet, \xi)$$

$$w\mathcal{S}_\bullet e_\bullet \mathcal{K}^s(X_\bullet, \xi) \to w\mathcal{S}_\bullet e_\bullet \mathcal{K}(X_\bullet, \xi) \to s\mathcal{S}_\bullet e_\bullet \mathcal{K}(X_\bullet, \xi)$$

Now we will identify the terms in these sequences.

**Theorem 5.6.15.**

$$\Omega(|w\mathcal{S}_\bullet e_\bullet \mathcal{K}^s(X_\bullet,\xi)|) \simeq |we_\bullet \mathcal{K}^s(X_\bullet,\xi)| \simeq |FC^s(X_\bullet,\xi)|$$

*Proof.* [IK95] Since the simplicial category $w\mathcal{F}_\bullet e_k \mathcal{K}^s(X_\ell,\xi)$ is contractible for each $(k,\ell)$ , $|w\mathcal{F}_\bullet e_\bullet \mathcal{K}^s(X_\bullet,\xi)| \simeq *$ ([Wal85]). It therefore suffices to show that the natural sequence:

$$we_\bullet \mathcal{K}^s(X_\bullet,\xi) \to w\mathcal{F}_\bullet e_\bullet \mathcal{K}^s(X_\bullet,\xi) \to w\mathcal{S}_\bullet e_\bullet \mathcal{K}^s(X_\bullet,\xi)$$

is a homotopy fibration. Since $we_\bullet \mathcal{K}^s(X_\bullet,\xi)$ is connected by definition of $\mathcal{K}^s$, so is $w\mathcal{S}_k e_\bullet \mathcal{K}^s(X_\bullet,\xi)$ for all $k \geq 0$. Thus it suffices to show that for each $k \geq 0$ the sequence

$$we_\bullet \mathcal{K}^s(X_\bullet,\xi) \to w\mathcal{F}_k e_\bullet \mathcal{K}^s(X_\bullet,\xi) \to w\mathcal{S}_k e_\bullet \mathcal{K}^s(X_\bullet,\xi)$$

is a homotopy fibration, or equivalently that we have a homotopy equivalence:

$$w\mathcal{F}_k e_\bullet \mathcal{K}^s(X_\bullet,\xi) \to we_\bullet \mathcal{K}^s(X_\bullet,\xi) \times w\mathcal{S}_k e_\bullet \mathcal{K}^s(X_\bullet,\xi)$$

given by sending the object $A_0 \rightarrowtail \cdots \rightarrowtail A_k$ in $w\mathcal{F}_k e_\bullet \mathcal{K}^s(X_\bullet,\xi)$ to the pair

$$(A_0, A_1/A_0 \rightarrowtail \cdots \rightarrowtail A_k/A_0)$$

where each $A_i$ is a sequence

$$A_i = \{(E_i, P_i) \vee (D_{ij}, S_{ij})\}.$$

By mapping this back to $w\mathcal{F}_k e_\bullet \mathcal{K}^s(X_\bullet,\xi)$ in the obvious way we get a retraction

$$w\mathcal{F}_k e_\bullet \mathcal{K}^s(X_\bullet,\xi) \xrightarrow{r_k} w\mathcal{F}_k e_\bullet \mathcal{K}^s(X_\bullet,\xi)$$
$$(A_0 \rightarrowtail \cdots \rightarrowtail A_k) \mapsto (A_0 \rightarrowtail A_1/A_0 \vee A_0 \rightarrowtail \cdots \rightarrowtail A_k/A_0 \vee A_0)$$

and it suffices to show that this retraction is a homotopy equivalence for each $k \geq 0$. For this it suffices to prove $r_k$ induces a homotopy equivalence on $w\mathcal{F}_k e_\ell \mathcal{K}^s(X_\bullet,\xi)$ in each bidegree $(k,\ell)$.

The objects of $w\mathcal{F}_k e_\ell \mathcal{K}^s(X_\bullet,\xi)$ are given by:

$$\begin{array}{ccccc}
(E_0,P_0) & \rightarrowtail & \cdots & \rightarrowtail & (E_k,P_k) \\
\downarrow & & & & \downarrow \\
(E_0,P_0)\vee(D_{01},S_{01}) & \rightarrowtail & \cdots & \rightarrowtail & (E_k,P_k)\vee(D_{k1},S_{k1}) \\
\downarrow & & & & \downarrow \\
\cdots & & & & \cdots \\
\downarrow & & & & \downarrow \\
(E_0,P_0)\vee(D_{0\ell},S_{0\ell}) & \rightarrowtail & \cdots & \rightarrowtail & (E_k,P_k)\vee(D_{k\ell},S_{k\ell})
\end{array}$$

which we denote by $\{(E_i,P_i)\vee(D_{ij},S_{ij})\}$ with $(D_{i0},S_{i0}) = (0,\emptyset)$ and with $k,\ell$ understood and fixed. We want to show that the simplicial functor

$$\begin{array}{ccc}
w\mathcal{F}_k e_\ell \mathcal{K}^s(X_\bullet,\xi) & \xrightarrow{r_{k\ell}} & \mathcal{F}_k e_\ell \mathcal{K}^s(X_\bullet,\xi) \\
\{(E_i,P_i)\vee(D_{ij},S_{ij})\} & \mapsto & \{(E_0,P_0)\vee(E_i/E_0,P_i/P_0)\vee(D_{ij},S_{ij})\}
\end{array}$$

is homotopic to the identity. To do this we will construct two other simplicial functors $f, g$ on $w\mathcal{F}_k e_\ell \mathcal{K}^s(X_\bullet, \xi)$ and three natural transformations:

$$id \leftarrow f \rightarrow g \leftarrow r_{k\ell}.$$

The functor $f$ will change only the posets $P_i$ for $i > 0$, replacing them with the "maximal extension" $Q_i$ of $P_0$ by $P_i/P_0$. Thus $Q_i$ has the same underlying graded set over $X_\bullet$ as $P_i$ but with closed subsets either closed in $P_0$ or $P_0$ union a closed subset of $P_i/P_0$. Since $\Lambda Q_i \subseteq \Lambda P_i$ we get an induced $\Lambda Q_i$-filtration of $E_i$ and we can define an object

$$f(\{(E_i, P_i) \vee (D_{ij}, S_{ij})\}) = \{(E_i, Q_i) \vee (D_{ij}, S_{ij})\}$$

giving a functor in a unique way so that the identity mapping gives a natural transformation $f \rightarrow id$.

To construct the functor $g$ we need a new filtration of $E_k$. Let $F$ be the chain complex $E_k$ with the $\Lambda P_0$-filtration given by $F^A = E_k^A = E_0^A$ if $A \subsetneqq P_0$ and $F^{P_0} = E_k$. Since $E_k/E_0$ is acyclic, $(F, P_0)$ is an object of $\mathcal{K}(X_\bullet, \xi)$ and we can let $g$ be the functor on $w\mathcal{F}_k e_\ell \mathcal{K}^s(X_\bullet, \xi)$ taking $\{(E_i, P_i) \vee (D_{ij}, S_{ij})\}$ to

$$\{(F, P_0) \vee (E_i/E_0, P_i/P_0) \vee (D_{ij}, S_{ij})\}.$$

Then the inclusion map $E_0 \rightarrow F$ gives a morphism $(E_0, P_0) \rightarrow (F, P_0)$ and thus a natural transformation $r_{k\ell} \rightarrow g$.

For each $i \geq 0$ let

$$(j, \beta) : (E_i, Q_i) \rightarrow (F, P_0)$$

be given by the inclusion map $j : E_i \rightarrow E_k$ and the unique poset retraction $\beta : Q_i \rightarrow P_0$ with kernel $Q_i - P_0$, i.e., $\beta(x) = x$ if $x \in P_0$ and $\beta(y) = *$ if not. Together with the quotient map $(E_i, Q_i) \rightarrow (E_i/E_0, Q_i/P_0) = (E_i/E_0, P_i/P_0)$ we get a morphism

$$(E_i, P_i) \rightarrow (F, P_0) \vee (E_i/E_0, P_i/P_0)$$

in $w\mathcal{K}(X_\bullet, \xi)$ defining a natural transformation $f \rightarrow g$ completing the proof. □

Since $\pi_0|we_\bullet \mathcal{K}^h(X_\bullet, \xi)| \cong Wh_1(X_\bullet, \xi)$ is an abelian group it follows easily that:

**Corollary 5.6.16.**

$$\Omega(|w\mathcal{S}_\bullet e_\bullet \mathcal{K}^h(X_\bullet, \xi)|) \simeq |we_\bullet \mathcal{K}^h(X_\bullet, \xi)| \simeq |FC^h(X_\bullet, \xi)|$$

The next step is to identify $w\mathcal{S}_\bullet e_\bullet \mathcal{K}(X_\bullet, \xi)$.

**Theorem 5.6.17.**

$$\Omega|w\mathcal{S}_\bullet e_\bullet \mathcal{K}(X_\bullet, \xi)| \simeq \Omega|w\mathcal{S}_\bullet e_\bullet \mathcal{Q}(X_\bullet)| \simeq Q(|X_\bullet|_+)$$

*Proof.* We will show that the obvious forgetful functor $F : e_\bullet\mathcal{K}(X_\bullet, \xi) \to e_\bullet\mathcal{Q}(X_\bullet)$ induces an equivalence in $K$-theory:

$$\Omega|w\mathcal{S}_\bullet e_\bullet\mathcal{K}(X_\bullet, \xi)| \simeq \Omega|w\mathcal{S}_\bullet e_\bullet\mathcal{Q}(X_\bullet)|.$$

It suffices to prove this in each bidegree $(k, \ell)$. So, we will show that the exact forgetful functor

$$e_k\mathcal{K}(X_\ell, \xi) \to e_k\mathcal{Q}(X_\ell)$$

induces an equivalence in $K$-theory for every $(k, \ell)$.

The first step is to reduce to the case $k = 0$. As we noted earlier, the Waldhausen category $e_k\mathcal{Q}(X_\ell)$ is a product of $k+1$ Waldhausen categories:

$$e_k\mathcal{Q}(X_\ell) \cong \mathcal{Q}(X_\ell) \times \underbrace{\mathcal{C}(X_\ell) \times \cdots \times \mathcal{C}(X_\ell)}_{k}$$

since the $k+1$ components are independent, being mixed only the simplicial operators in the $e$-direction and the points in each expansion pair lie over the same point in $X_\ell$. Similarly $e_k\mathcal{K}(X_\ell, \xi)$ is a product of $k+1$ Waldhausen categories:

$$e_k\mathcal{K}(X_\ell, \xi) \cong e_0\mathcal{K}(X_\ell, \xi) \times \underbrace{\mathcal{D}(X_\ell, \xi) \times \cdots \times \mathcal{D}(X_\ell, \xi)}_{k}$$

where $\mathcal{D}(X_\ell, \xi)$ is the category of filtered chain complexes which are unions of expansion pairs.

Claim 1: The forgetful functor $\mathcal{D}(X_\ell, \xi) \to \mathcal{C}(X_\ell)$ induces an equivalence of K-theories.

*Proof of Claim 1:* Each object $P$ of $\mathcal{C}(X_\ell)$ corresponds to a unique system of higher homotopies up to canonical isomorphism. By Lemma 4.1.7 this implies (in the current notation) that $w\mathcal{D}(W_\ell, \xi)/P$ is contractible. Thus $w\mathcal{D}(W_\ell, \xi) \simeq i\mathcal{C}(X_\ell) = w\mathcal{C}(X_\ell)$. Similarly,

$$w\mathcal{S}_j\mathcal{D}(W_\ell, \xi) \simeq i\mathcal{S}_j\mathcal{C}(X_\ell) = w\mathcal{S}_j\mathcal{C}(X_\ell)$$

for every $j \geq 0$ proving our claim.

It remains to show that the forgetful functor $e_0\mathcal{K}(X_\ell, \xi) \to \mathcal{Q}(X_\ell)$ induces an equivalence of K-theories. For this step we need a *height function* on each poset $P$. By this we mean a positive real valued function $h$ on $P$ with the property that $h(x) > 1 + h(y)$ wherever $x > y$ in $P$. Since these are convex conditions, the space of all functions $h$ having this property is contractible for all $P$. Therefore $\mathcal{Q}(X_\ell)$ has the same $K$-theory as the simplicial Waldhausen category $\mathcal{H}_\bullet\mathcal{Q}(X_\ell)$ where $\mathcal{H}_k\mathcal{Q}(X_\ell)$ is the category of finite graded posets over $X_\ell$ together with continuous families of height functions $h_t : P \to (0, \infty)$, $t \in \Delta^k$ with morphisms respecting the height function ($h_t(\alpha(x)) = h_t(x)$ when $\alpha(x) \neq *$). Let $\mathcal{H}_\bullet e_0\mathcal{K}(X_\ell, \xi)$ be the corresponding construction for $e_0\mathcal{K}(X_\ell, \xi)$. Then it suffices to show that the simplicial forgetful functor

$$F_\bullet : \mathcal{H}_\bullet e_0\mathcal{K}(X_\ell, \xi) \to \mathcal{H}_\bullet\mathcal{Q}(X_\ell)$$

induces an equivalence in $K$-theory. We will show that this holds in every degree $k$.

For each $0 \leq n < m$ let $\mathcal{H}_k^{(n,m]}\mathcal{Q}(X_\ell)$ denote the full Waldhausen subcategory of $\mathcal{H}_k\mathcal{Q}(X_\ell)$ consisting of posets $P$ over $X_\ell$ together with $\Delta^k$-families of height functions $h_t$ so that

$$n < \max\{h_t(x) \,:\, t \in \Delta^k\} \leq m$$

for all $x \in P$. Let $\mathcal{H}_k^{(n,m]}e_0\mathcal{K}(X_\ell, \xi)$ be defined similarly. Then

$$\mathcal{H}_k\mathcal{Q}(X_\ell) = \bigcup \mathcal{H}_k^{(n,m]}\mathcal{Q}(X_\ell)$$

and similarly for $e_0\mathcal{K}^{deg}(X_\ell, \xi)$ so it suffices to show the following.

Claim 2. For each $0 \leq n < m$ the forgetful functor

$$F_k^{(n,m]} : \mathcal{H}_k^{(n,m]}e_0\mathcal{K}(X_\ell, \xi) \to \mathcal{H}_k^{(n,m]}\mathcal{Q}(X_\ell)$$

induces an equivalence in $K$-theory.

*Proof of Claim 2:* This is true for $n = m-1$ by the same argument as in the proof of the claim since the posets in $\mathcal{H}_k^{(m-1,m]}\mathcal{Q}(X_\ell)$ are completely unordered and thus the cellular chain complex is uniquely determined up to isomorphism. So suppose that $n < m-1$ and the statement holds for $(n, m-1]$. In that case each object of $\mathcal{H}_k^{(n,m]}\mathcal{Q}(X_\ell)$ has a canonical subobject in $\mathcal{H}_k^{(n,m-1]}\mathcal{Q}(X_\ell)$ and the quotient object lies in $\mathcal{H}_k^{(m-1,m]}\mathcal{Q}(X_\ell)$. Therefore, by Waldhausen's additivity theorem, we have

$$w\mathcal{S}_\bullet\mathcal{H}_k^{(n,m]}\mathcal{Q}(X_\ell) \simeq w\mathcal{S}_\bullet\mathcal{H}_k^{(n,m-1]}\mathcal{Q}(X_\ell) \times w\mathcal{S}_\bullet\mathcal{H}_k^{(m-1,m]}\mathcal{Q}(X_\ell).$$

A similar decomposition holds for $\mathcal{H}_k^{(n,m]}e_0\mathcal{K}(X_\ell, \xi)$ and the forgetful functor respects the decomposition. Therefore $F_k^{(n,m]}$ gives an equivalence in $K$-theory by induction on $m - n$ proving our second claim and completing the proof of the theorem. □

Finally we want to identify $h\mathcal{S}_\bullet e_\bullet\mathcal{K}(X_\bullet, \xi)$. This will be easy since the homotopy equivalences satisfy the cylinder axiom and we can use Waldhausen's approximation theorem [Wal85]. First we have to restrict to finitely generated chain complexes.

**Lemma 5.6.18.** *Let $e_\bullet\mathcal{K}^{fg}(X_\bullet, \xi)$ be the bisimplicial full Waldhausen subcategory of $e_\bullet\mathcal{K}(X_\bullet, \xi)$ consisting of sequences of filtered chain complexes $\{(E, P) \vee (D_i, S_i)\}_{i=0\cdots k}$ where the total space $\bigoplus_j (E \oplus D_k)_j$ is finitely generated. Then the inclusion functor*

$$h\mathcal{S}_\bullet e_\bullet\mathcal{K}^{fg}(X_\bullet, \xi) \to h\mathcal{S}_\bullet e_\bullet\mathcal{K}(X_\bullet, \xi)$$

*is a homotopy equivalence on realizations.*

*Proof.* Since homotopy equivalences in each $e_k\mathcal{K}^{fg}(X_\bullet, \xi)$ satisfy the cylinder axiom, this follows easily from Waldhausen's approximation theorem using the fact that every filtered chain complex has a minimal subcomplex. □

Now we compare $e_\bullet\mathcal{K}^{fg}(X_\bullet, \xi)$ with one of the models for the $K$-theory of $R$.

**Definition 5.6.19.** For any ring $R$ let $(\mathcal{K}(R), co\mathcal{K}(R), h\mathcal{K}(R))$ be the Waldhausen category of bounded complexes of finitely generated projective $R$-modules, monomorphism of such complexes with projective quotients and homotopy equivalences. Let $\mathcal{K}^{sf}(R)$ be the full subcategory whose objects $E$ have the property that $\sum(-1)^n[E_n]$ is stably free, i.e., lies in the image $\overline{\mathbb{Z}}$ of $K_0\mathbb{Z} = \mathbb{Z}$ in $K_0R$.

**Lemma 5.6.20.** $\Omega|h\mathcal{S}_\bullet\mathcal{K}(R)| \simeq K_0R \times BGL(R)^+$ *and*

$$\Omega|h\mathcal{S}_\bullet\mathcal{K}^{sf}(R)| \simeq \overline{\mathbb{Z}} \times BGL(R)^+.$$

**Remark 5.6.21.** For future reference we note that $\mathcal{K}^{sf}(R)$ contains as a Waldhausen subcategory (taking complexes which are nonzero only in degree 0) the category $\mathcal{K}_0^{sf}(R)$ of all stably free f.g. projective $R$-modules, with split monomorphisms as cofibrations and isomorphism as weak equivalences and that the exact inclusion functor $\mathcal{K}_0^{sf}(R) \to \mathcal{K}^{sf}(R)$ induced a homotopy equivalence of K-theories $i\mathcal{S}_\bullet\mathcal{K}_0^{sf}(R) \simeq h\mathcal{S}_\bullet\mathcal{K}^{sf}(R)$. Furthermore the composition $|h\mathcal{K}^{sf}(R)| \to \Omega|h\mathcal{S}_\bullet\mathcal{K}^{sf}(R)| \simeq \Omega|i\mathcal{S}_\bullet\mathcal{K}_0^{sf}(R)|$ is the alternating sum of the maps

$$|h\mathcal{K}^{sf}(R)| \xrightarrow{H_i} |i\mathcal{K}_0^{sf}(R)| \to \Omega|i\mathcal{S}_\bullet\mathcal{K}^{sf}(R)|.$$

where $H_i : \mathcal{K}^{sf}(R) \to \mathcal{K}_0^{sf}(R)$ is the $i$-th homology functor.

**Theorem 5.6.22.**

$$\Omega|h\mathcal{S}_\bullet e_\bullet\mathcal{K}(X_\bullet, \xi)| \simeq \overline{\mathbb{Z}} \times BGL(R)^+$$

*Proof.* Let $h_\bullet\mathcal{K}^{sf}(R)$ be the simplicial full Waldhausen subcategory of $\mathcal{F}_\bullet\mathcal{K}^{sf}(R)$ of all cofibration sequences which are also homotopy equivalences. Then the constant functor $\mathcal{K}^{sf}(R) \to h_\bullet\mathcal{K}^{sf}(R)$ is an exact equivalence and therefore induces an equivalence in $K$-theory:

$$h\mathcal{S}_\bullet\mathcal{K}^{sf}(R) \simeq h\mathcal{S}_\bullet h_\bullet\mathcal{K}^{sf}(R)$$

The forgetful functor $F : e_\bullet\mathcal{K}^{fg}(X_k, \xi) \to h_\bullet\mathcal{K}^{sf}(R)$ given by $F\{(E,P) \vee (D_i, S_i)\} = \{E \vee D_i\}$ is exact and also induces an equivalence of K-theories:

$$h\mathcal{S}_\bullet e_\bullet\mathcal{K}^{sf}(X_k, \xi) \simeq h\mathcal{S}_\bullet h_\bullet\mathcal{K}^{sf}(R)$$

for each $k \geq 0$ by the approximation theorem. Since these forgetful functors commute with the simplicial operators on $X_\bullet$ they induce a homotopy equivalence

$$\Omega|h\mathcal{S}_\bullet\mathcal{K}(X_\bullet, \xi)| \simeq \Omega|h\mathcal{S}_\bullet\mathcal{K}^{sf}(R)| \simeq \overline{\mathbb{Z}} \times BGL(R)^+.$$

□

## 5.7 Higher FR-torsion and the 2-index theorem

The fact that the Whitehead space $\mathcal{W}h^h_\bullet(R,G)$ has the correct homotopy type (Corollary 5.6.2) means that, theoretically, one should be able to define higher FR-torsion invariants such as

$$\tau_k \in H^{2k}(\mathcal{W}h^h_\bullet(\mathbb{C},U(1))).$$

However, we prefer a more direct approach using the 2-index theorem which says:

**Theorem 5.7.1 (2-index theorem).** *$\mathcal{W}h^h_\bullet(R,G)$ contains as a deformation retract the simplicial full subcategory $\mathcal{W}h^{[i,j]h}_\bullet(R,G)$ of acyclic complexes which are nonzero only in degrees $i, i+1, \cdots, j$ provided that $0 \le i < j$.*

Since $|\mathcal{W}h^{[i,i+1]h}_\bullet(R,G)| \simeq |\mathcal{W}_\bullet(R,G)|$ by Prop. 5.3.1 we have the following corollary.

**Corollary 5.7.2.** *The realizations of the simplicial bicategories $\mathcal{W}_\bullet(\mathbb{C}^m,G)$, $\mathcal{W}^{cont}_\bullet(\mathbb{C}^m,G)$, $\mathcal{W}^{diff}_\bullet(\mathbb{C}^m,G)$, $\mathcal{W}^{iso-c}_\bullet(\mathbb{C}^m,G)$ and $\mathcal{W}^{iso-d}_\bullet(\mathbb{C}^m,G)$ of Chapter II are all homotopy equivalent to $|\mathcal{W}h^h_\bullet(M_m(\mathbb{C}),G)|$ and consequently form homotopy fiber sequences:*

$$|\mathcal{W}_\bullet(\mathbb{C}^m,G)| \to Q(BG_*) \to \mathbb{Z} \times BGL(\mathbb{C})^+$$

*where $BG$ and $BGL(\mathbb{C})$ are the classifying spaces of $G$ and $GL(\mathbb{C})$ as discrete groups.*

### 5.7.1 The suspension theorem

The 2-index theorem will be proved in Section 5.9. This section will deal with the preliminaries and consequences of the two index theorem. One of the preliminary issues we need to address is how the space of chain complexes changes with the choice of the index $i$. This is answered in the expected way by the suspension theorem below.

**Theorem 5.7.3 (Suspension theorem).** *The alternating suspension*

$$\Sigma' : \mathcal{W}h^h_\bullet(R,G) \to \mathcal{W}h^h_\bullet(R,G)$$

*acts as $-1$, i.e., $\Sigma' + id \simeq 0$ where addition is given by direct sum. Furthermore, the deformation will not involve poset elements in new degrees. In particular, the double suspension*

$$(\Sigma')^2 = \Sigma^2 : \mathcal{W}h^{[i,j]h}_\bullet(R,G) \to \mathcal{W}h^{[i,j+2]h}_\bullet(R,G)$$

*is homotopic to the inclusion functor for all $0 \le i \le j$.*

**Remark 5.7.4.** (a) In order to describe mappings of $\mathcal{W}h^h_\bullet(R,G)$ to itself we will use the *last simplex functor* $\xi$ on the diagonal of the nerve of $\mathcal{W}h^h_\bullet(R,G)$ which assigns to each sequence of morphisms

$$\sigma_0 \xrightarrow{f_1} \sigma_1 \xrightarrow{f_2} \cdots \xrightarrow{f_k} \sigma_k$$

in $\mathcal{W}h_k(R,G)$ the value $\xi(f_1,\cdots,f_k) = \sigma_k$. This is a canonical $G$-expansion functor representing the identity mapping on $\mathcal{W}h^h_\bullet(R,G)$.

(b) We will use a generalized homotopy of $\xi$ given by

1. embedding $B$ as a deformation retract in another simplicial set $Z(B)$,
2. extending $\xi$ to a $G$-expansion functor $Z(\xi)$ on $Z(B)$ and
3. restricting $Z(\xi)$ to another deformation retract $\widetilde{B}$ of $Z(B)$.

The resulting $G$-expansion functor $\tilde{\xi}$ over $\widetilde{B}$ will be called a *deformation* of $\xi$ and $Z(\xi)$ will be called the *trace* of the deformation. We note that a deformation of the restriction of $\xi$ to a simplicial subset $A \subseteq B$ extends to $B$ by taking $Z(B) = B \cup Z(A)$ and $\widetilde{B} = (B - A) \cup \widetilde{A} \cup Z(A \cap B)$.

(c) Finally, we note that the proof of the suspension theorem given below also proves that the alternating cone $C'(id_\xi)$ of the identity morphism on any $G$-expansion functor

$$\xi : B \to \mathcal{W}h^{[i,j]}_\bullet(R,G)$$

is null-homotopic in $\mathcal{W}h^{[i,j+1]h}_\bullet(R,G)$.

*Proof.* We will show that for any $G$-expansion functor $\xi$ with chain contractible fibers on a simplicial base $B$, the functor $\xi \oplus \Sigma'(\xi)$ has a null deformation, i.e., a deformation (as described above) to the zero functor. By taking $B$ and $\xi$ to be the universal example as described above, the theorem will follow. We will prove this first in case where $\xi$ is a $G$-monomial functor (with no $G$-expansion pairs). In this case the standard suspension $\Sigma(\xi)$ is defined and we will show first that $\xi \oplus \Sigma(\xi)$ is $G$-expansion simplicially null homotopic.

Over each simplex $\sigma$ in $B$ we take $\xi(\sigma) = (e, P)$. Then

$$\xi(\sigma) \oplus \Sigma(\xi(\sigma)) = (e \oplus \Sigma(e), P \coprod \Sigma P)$$

and we want to construct a $G$-expansion null homotopy of this. First, we change the partial ordering of $P \coprod \Sigma P$ so that every element of $P$ is less than every element of $\Sigma P$. This gives the cone $C(\omega)$ of the "maximal morphism" $\omega : P \to P$ and the functor becomes equal to the fiberwise mapping cone of the zero morphism $\xi \to \xi$. The identity map gives a $G$-monomial natural transformation:

$$C(0,\omega) = (e \oplus \Sigma'(e), C(\omega)) \to (e \oplus \Sigma'(e), P \coprod \Sigma P)$$

Next we note that the identity morphism $\xi \to \xi$ is fiberwise null-homotopic by Lemma 3.4.6. By Theorem 3.4.4, this gives us a $G$-monomial simplicial homotopy from $C(0)$ to $C(id_\xi)$.

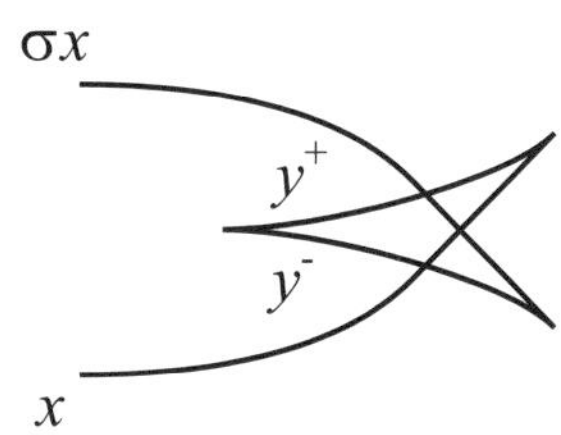


Figure 5.3: Swallowtail deformation

The next step is to weaken the partial ordering so that the poset becomes the cone on the identity morphism on $P$. Then strengthen it so that each pair $\{x, \sigma x\}$ acts in unison, making it into a (general) expansion pair. By Lemma 4.5.7 we can make these pairs simultaneously independent (using the simplicial homotopies $HU$ of 4.5.7) and then cancel them with expansion morphisms.

Since no new poset points were introduced, this proves that

$$j + \Sigma : \mathcal{W}h_\bullet^{[i,k]h}(R, G, n_*) \to \mathcal{W}h_\bullet^{[i,k+1]h}(R, G)$$

is null homotopic for all $n_*$ where $j$ is the inclusion functor. The same argument also shows that the cone on the identity morphism on any functor to $\mathcal{W}h_\bullet^{[i,k]}(R, G, n_*)$ is null homotopic in $\mathcal{W}h_\bullet^{[i,k+1]}(R, G)$. To prove that $j + \Sigma'$ is null-homotopic, or equivalently, that the alternating cone of the identity on any $G$-expansion functor is null-homotopic, we need to modify the above argument as follows.

The alternating suspension $\Sigma'$ differs from the usual suspension only by a sign in every even degree. This changes little except for the last step. When $x$ has odd degree, the pair $\{x, \sigma x\}$ becomes an expansion pair with incidence $-1$. These cannot be cancelled (unless $-1 \in G$). So we do the following deformation.

First create another true expansion pair $\{y^-, y^+\}$ with incidence number 1. Then change the ordering to $y^- < x < \sigma x < y^+$, subtract $\sigma x$ from $y^+$ and $x$ from $y^-$ changing the incidence matrix to:

$$\begin{pmatrix} 1 & 0 \\ -1 & 1 \end{pmatrix} \begin{pmatrix} -1 & 0 \\ 0 & 1 \end{pmatrix} \begin{pmatrix} 1 & -1 \\ 0 & 1 \end{pmatrix} = \begin{pmatrix} -1 & 1 \\ 1 & 0 \end{pmatrix}.$$

Since $y^-, y^+$ are no longer incident we can change the partial ordering to $x < y^+ < y^- < \sigma x$. This gives two expansion pairs $\{x, y^+\}$ and $\{y^-, \sigma x\}$ with incidence number 1 which we are allowed to cancel. The deformation can be visualized by the "swallowtail" graphic show in Figure 5.3. The vertical symmetry in the picture reflects the algebraic symmetry. (The matrix equation above is equal to its transpose.) The Morse theoretic origins of this $A_4$ unfolding are not important except to justify the name: "swallowtail deformation."

This gives a null deformation of $j + \Sigma' : \mathcal{W}h_\bullet^{[i,k]h}(R, G, n_*) \to \mathcal{W}h_\bullet^{[i,k]h}(R, G)$ for all $n_*$. To extend this null deformation to all of $\mathcal{W}h_\bullet^{[i,k]h}(R, G)$ we need to deal with the expansion pairs in the original functor $\xi$.

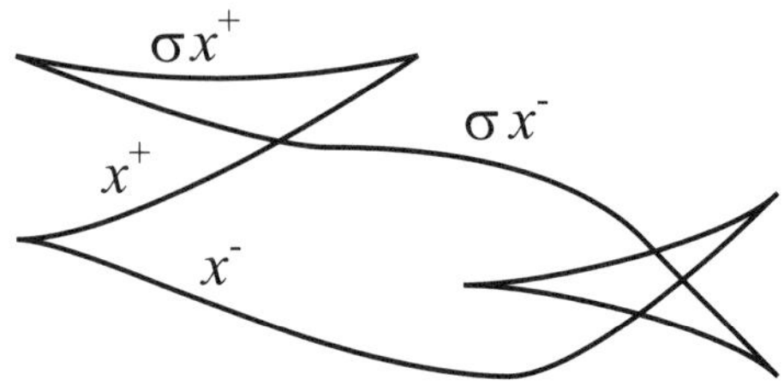


Figure 5.4: Result of cancellation at an expansion pair

If we start with $\xi$ being a $G$-expansion functor over $B$ then the deformation described above will work to eliminate all points in $P \cup \Sigma P$ except along the original expansion pair set $C$ of $\xi$. Over an odd-even expansion pair $\{x^-, x^+\}$ of $\xi$ we will have an even-odd expansion pair $\{\sigma x^-, \sigma x^+\}$ of $\Sigma'\xi$. The deformation described above will make $\{x^+, \sigma x^+\}$ into a $+1$ expansion pair and the other pair $\{x^-, \sigma x^-\}$ into a $-1$ expansion pair. Thus the cancellation procedure creates the one dimensional $G$-expansion functor indicated in Figure 5.4 times the simplicial set $C$.

Since the coefficients in this one parameter deformation are all integers, this is a path in $\mathcal{W}h^h(\mathbb{Z}, 1)$ representing an element of the trivial group

$$\pi_1 \mathcal{W}h^h(\mathbb{Z}, 1) = Wh_1(\mathbb{Z}, 1) = 0.$$

Consequently, there exists a null deformation having only integer coefficients. One such deformation is indicated in Figure 5.5. The case of even-odd expansion pairs in $\xi$ is similar. (Just turn all the pictures upside-down!) Multiply this by $C$ to get a null deformation of the entire functor. □

### 5.7.2 Properties of higher FR-torsion

Using the 2-index theorem, the suspension theorem and the formulas for higher FR-torsion given in other chapters we can define the higher Franz-Reidemeister torsion invariants and compute them in some cases. We first go over the construction that defines the higher FR-torsion and more generally the mapping into $|\mathcal{W}h^h_\bullet(R, G)|$.

Let $p : E \to B$ be a smooth fibration with fiber $M$. Let $\rho : \pi_1 E \to U(n)$ be a unitary representation of $\pi_1 E$ and let $\mathcal{F}$ be the associated Hermitian coefficient sheaf over $E$ with stalk $\cong \mathbb{C}^n$. Suppose that $\pi_1 B$ acts trivially on $H_*(M; \mathcal{F})$. Then we have a well defined $G$-expansion functor $C(\xi)$ (the canonical cone of the functor obtained from a framed function) on the category of generic small simplices of $B$ giving a continuous mapping

$$|C(\xi)| : B \to |\mathcal{W}h^h_\bullet(M_n(\mathbb{C}), U(n))|$$

well defined up to homotopy.

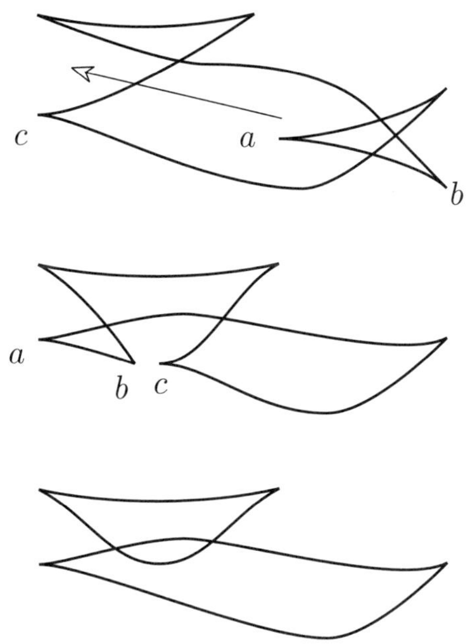


Figure 5.5: Slide bottom of swallowtail to left and cancel

**Theorem 5.7.5.** *Under these conditions there are well defined higher FR-torsion invariants*

$$\tau_k(E;\mathcal{F}) \in H^{2k}(B;\mathbb{R})$$

*for* $k \geq 1$ *given by* $\tau_k(E;\mathcal{F}) = |C(\xi)|^*(\tau_k(n))$ *where*

$$\tau_k(n) \in H^{2k}(\mathcal{W}h^h_\bullet(M_n(\mathbb{C}), U(n));\mathbb{R})$$

*are universal classes having the following properties.*

1. (naturality) $\tau_k(E;\mathcal{F})$ *is natural, i.e., given any smooth mapping of* $M$*-bundles* $(f,g) : (B',E') \to (B,E)$ *we have*

$$f^*\tau_k(E;\mathcal{F}) = \tau_k(E', g^*\mathcal{F}).$$

2. (additivity lemma) *If* $p_i : M_d(\mathbb{C})^m \to M_d(\mathbb{C})$ *is the projection to the* $i$*-th factor and* $s : M_d(\mathbb{C})^m \to M_{dm}(\mathbb{C})$ *is the ring homomorphism given by* $s(A_1, \cdots, A_m) = A_1 \oplus \cdots \oplus A_m$ *then*

$$s^*(\tau_k(dm)) = \sum p_i^*(\tau_k(d))$$

$\in H^{2k}(\mathcal{W}h^h_\bullet(M_d(\mathbb{C})^m, U(d)^m);\mathbb{R}).$

3. (1st transfer lemma)

$$J^*(\tau_k(dm)) = \mathrm{N}^*(\tau_k(d))$$

*in* $H^{2k}(\mathcal{W}h^h_\bullet(M_m(M_d(\mathbb{C})), S_m(U(d))); \mathbb{R})$ *where*

$$J : \mathcal{W}h^h_\bullet(M_m(M_d(\mathbb{C})), S_m(U(d))) \to \mathcal{W}h^h_\bullet(M_{md}(\mathbb{C}), U(md))$$

*is the simplicial inclusion functor with* $S_m(U(d))$ *being the group of* $m \times m$ *monomial matrices with coefficients in* $U(d)$ *and*

$$\mathrm{N} : \mathcal{W}h^h_\bullet(M_m(M_d(\mathbb{C})), S_m(U(d))) \to \mathcal{W}h^h_\bullet(M_d(\mathbb{C}), U(d))$$

*is the simplicial transfer functor given in Def. 2.5.15 above and in the definition below.*

*Proof.* By the two-index theorem the cohomology classes $\tau_k(n)$ defined on

$$|\mathcal{W}_\bullet(\mathbb{C}^n, U(n))| \cong |\mathcal{W}h^{[0,1]h}_\bullet(M_n(\mathbb{C}), U(n))|$$

extend to $|\mathcal{W}h^h_\bullet(M_n(\mathbb{C}), U(n))|$ and the last two properties follow from the corresponding statements for $\mathcal{W}_\bullet(\mathbb{C}^n, U(n))$. Naturality of $\tau_k$ follows from the naturality of $C(\xi)$. □

The functor N is defined on the level of $\mathcal{W}h_\bullet(R, G)$ as follows.

**Definition 5.7.6.** If $G$ is a subgroup of $R^\times$ and $S_n(G)$ is the group of $G$-monomial matrices in $M_n(R)$ then let

$$\mathrm{N} : \mathcal{W}h_\bullet(M_n(R), S_n(G)) \to \mathcal{W}h_\bullet(R, G)$$

be the simplicial functor sending the object $(e, P)$ to $(\mathrm{N}_n\, e, (nP)')$ where $(nP)' = P \times \{1, 2, \cdots, n\}$ with the partial ordering $(x, i) < (y, j)$ if either

1. $x < y$ in $P$ and $\{x, y\}$ is not a free $G$-expansion pair in $(e, P)$ or
2. $\{x, y\}$ is a free $G$-expansion pair with $e(v)(y) = xg$ and $i = \sigma(j)$ where $\sigma$ is the underlying permutation of $g$.

And $\mathrm{N}_n\, e$ is the system of endomorphisms $\mathrm{N}_n\, e(v_*)$ of $R^{(nP)'} = (R^n)^P$ with coefficients

$$\mathrm{N}_n\, e(v_*)_{(x,i)(y,j)} = (e(v_*)_{xy})_{ij}\,.$$

We need two more properties of $\mathcal{W}h_\bullet(R, G)$.

**Lemma 5.7.7 (Conjugation lemma).** *The ring automorphism* $\phi^g$ *of* $R$ *given by conjugation by an element* $g \in G$ *induces simplicial automorphisms of* $\mathcal{W}h_\bullet$ $(R, G)$ *and* $\mathcal{W}h^h_\bullet(R, G)$ *which are simplicially homotopic to the identity.*

*Proof.* Multiplication by $g$ is a $G$-monomial natural transformation from the identity functor to the functor $\phi^g_*$ induced by $\phi^g$. □

**Lemma 5.7.8 (Splitting lemma).** *Let* $\xi$ *be a* $G$*-expansion functor on a simplicial set* $B$ *and suppose that* $\xi$ *has a subfunctor* $\eta$ *so that either*

1. *the homology bundles of* $\xi$, $\eta$ *and* $\xi/\eta$ *are constant and projective or*
2. $R$ *is a field and* $\xi, \eta, \xi/\eta$ *have upper triangular homology.*

*Then*

$$C(\xi) \simeq C(\eta) \oplus C(\xi/\eta),$$

*i.e., the sequence* $C(\eta) \to C(\xi) \to C(\xi/\eta)$ *is a split extension sequence up to homotopy.*

**Remark 5.7.9.** We note that in the case when $R = K$ is a field, whenever two of the functors $\eta, \xi, \xi/\eta$ have upper triangular homology, so does the third. This follows from the long exact homology sequence (5.10) at any vertex $v$ of $B$ and the fact that finite dimensional upper triangular $K[\pi_1 B]$-modules form a Serre subcategory of the category of all $K[\pi_1 B]$-modules (by uniqueness of composition factors).

*Proof.* Case 0. We first do the easy case when either $\eta$ or $\xi/\eta$ is fiberwise acyclic. By Proposition 3.3.2, $\xi$ is the fiberwise mapping cone of some morphism

$$g : \Sigma^{-1}(\xi/\eta) \to \eta.$$

(We note that the desuspension $\Sigma^{-1}(\xi/\eta)$ of $\xi/\eta$ is not a $G$-expansion functor. It is a $\pm G$-expansion functor since all the incidence numbers change sign. Also the desuspension may have elements of degree $-1$. However these flaws cause no problems.) By Proposition 3.4.15, $g$ will be null homotopic if either its domain or range is acyclic. By Lemma 3.6.3 homotopic morphisms give simplicially homotopic mapping cones so the extension $\eta \to \xi \to \xi/\eta$ is simplicially homotopic to a split extension.

Next suppose that $\eta$, $\xi$ and $\xi/\eta$ have constant free homology (resp. upper triangular homology in the field case) and the kernel, image and cokernel of the map $j_* : H_*(\eta(\sigma)) \to H_*(\xi(\sigma))$ induced by the inclusion $j : \eta \to \xi$ give constant free (resp. upper triangular) bundles over $B$. Look at the long exact homology sequence at any vertex $v$ of $B$:

$$0 \to H_n(\eta(v)) \to H_n(\xi(v)) \to H_n((\xi/\eta)(v)) \to H_{n-1}(\eta(v)) \to \cdots \qquad (5.10)$$

We note that in all cases the $\pi_1 B$-fixed submodule of the first nonzero term in this sequence maps split monomorphically into the $\pi_1 B$-fixed submodule of the next term. We use the notation:

$$sM = M^{\pi_1 B} = \pi_1 B\text{-}fixed\ submodule\ of\ M$$

(i.e., the set of all elements of $M$ left fixed by the action of $\pi_1 B$) for any $R[\pi_1 B]$-module $M$.

There are three different cases depending on where the first nonzero group in this sequence lies.

Case 1. Suppose that $H_n(\eta(v))$ is the first nonzero group in this sequence. Then choose a basis $Q$ for $sH_n(\eta(v))$ and consider $Q$ as a graded poset (all in degree $n$) with empty ordering. Let $\epsilon_n(Q)$ be the unique monomial functor on $B$ over the ring $R$ subordinate to the constant poset $Q$. Let $f : \epsilon_n(Q) \to \eta$ be any morphism which induces the inclusion map in degree $n$ homology. Then the composition $j \circ f : \epsilon_n(Q) \to \eta \to \xi$ induces a split monomorphism on $\pi_1 B$-fixed degree $n$ homology groups.

The fiberwise cone of $j \circ f$ contains the fiberwise cone of $f$. (In fact $C(j \circ f) = C(f) \cup \xi$.) Consequently $(C(j \circ f), C(f))$ is a new pair satisfying the condition of the lemma. However the long exact homology sequence of this pair either starts lower or has a smaller first term. Therefore, by induction on the length of the long exact homology sequence and the rank of the first nonzero term, the lemma holds for $(C(j \circ f), C(f))$. However, $C(f)$ has the same canonical cone as $\eta$ since it is the first step in the construction of this cone. (Recall that the homology must be killed from the top down by an iterated mapping cone starting with the $\pi_1 B$-fixed submodule of the maximal homology.) Similarly, $C(j \circ f)$ has the same canonical cone as $\xi$. The quotients are the same: $C(j \circ f)/C(f) \cong \xi/\eta$. Consequently,

$$C(\xi) \cong C(C(j \circ f)) \simeq C(C(f)) \oplus C(C(j \circ f)/C(f)) \cong C(\eta) \oplus C(\xi/\eta).$$

Case 2. Suppose that $H_n(\xi(v))$ is the first nonzero group in the long exact homology sequence (5.10). Choose a basis $Q$ for $sH_n(\xi(v))$. Since the mapping $q_* : sH_n(\xi(v)) \to sH_n((\xi/\eta)(v))$ induced by $q : \xi \to \xi/\eta$ is a split monomorphism, $Q$ also forms part of a basis for $sH_n((\xi/\eta)(v))$.

Let $f : \epsilon_n(Q) \to \xi$ be any morphism which induces the inclusion map in degree $n$ homology. Then $q \circ f : e_n(Q) \to \xi/\eta$ induces a monomorphism to the top homology of $\xi/\eta$. The fiberwise cone $C(f)$ contains $\eta$ and the quotient is

$C(q \circ f)$. The lemma holds for $(C(f), \xi)$ by induction so, as in Case 1, we have:

$$C(\xi) \cong C(C(f)) \simeq C(C(q \circ f)) \oplus C(\eta) \cong C(\xi/\eta) \oplus C(\eta).$$

Case 3. Suppose that $H_n((\xi/\eta)(v))$ is the first nonzero group in the long exact sequence. Let $Q$ be a basis for $H_n((\xi/\eta)(v))$. Then $Q$ forms part of a basis for $H_{n-1}(\eta(v))$. Let $f : \epsilon_{n-1}(Q) \to \eta$ be a morphism which induces this monomorphism in degree $n-1$ homology. Then, as before, the fiberwise cone of $f$ has the same canonical cone as $\eta$.

Now comes something new. Let $S$ be a sum of $(-1)$-expansion pairs, one for each element of $Q$, so that $\epsilon_{n-1}(Q) \subseteq S$, the quotient is $\epsilon_n(Q)$ and the incidence numbers are all $-1$ so that the suspension of $S$ is a union of ordinary expansion pairs. Since $\xi$ has no fiber homology above degree $n-1$, and the morphism $j \circ f : \epsilon_{n-1}(Q) \to \eta \to \xi$ induces zero in homology, it extends to a morphism $h : S \to \xi$. Since $S$ is acyclic, $h$ is null homotopic and the fiberwise cone $C(h)$ is simplicially homotopic to a $G$-expansion of $\xi$. Consequently, its canonical cone is equivalent to the canonical cone of $\xi$. (In fact we can take $C(C(h)) = C(h) \cup C(\xi)$.) The cone of $h$ contains the cone of $f$ and the pair $(C(h), C(f))$ satisfies the lemma by induction on the length of the long exact sequence and the rank of the first nonzero term. Also the quotient $C(h)/C(f)$

is equal to the fiberwise cone of the morphism $\overline{h} : \epsilon_n(Q) \to \xi/\eta$ which induces an isomorphism in $\pi_1 B$-fixed degree $n$ homology. Thus

$$C(\xi) \simeq C(C(h)) \simeq C(C(f)) \oplus C(C(\overline{h})) \cong C(\eta) \oplus C(\xi/\eta).$$

This settles all case when the homologies are all free, e.g., when $R$ is a field.

Case 4. In the general case of constant projective homology, we note first that the long exact homology sequence (5.10) must split, being an exact sequence of projective $R$-modules. Consequently, the kernel, image and cokernel of $j_* : H_*(\eta(v)) \to H_*(\xi(v))$ are projective. To make these all free we simply add a fixed $R$-complex $\theta$ containing a subcomplex $\theta_0$ so that

(a) both $\theta_0$ and $\theta/\theta_0$ are free,

(b) $\theta_0$, $\theta$ and $\theta/\theta_0$ have projective homology complementary to the homology of $\eta$, $\xi$ and $\xi/\eta$ respectively and

(c) the kernel, image and cokernel of the map $H_*(\theta_0) \to H_*(\theta)$ are complementary to the kernel, image and cokernel of $j_* : H_*(\eta(v)) \to H_*(\xi(v))$.

(Note that (c) implies (b).) Then the pair $(\xi \oplus \theta, \eta \oplus \theta_0)$ has free homology in all respects and the argument above applies. Also, by definition of the canonical cone, this new pair gives the same three canonical cones as the original pair.

It is easy to construct the pair $(\theta, \theta_0)$. Choose fixed projective complexes $C_1, \cdots, C_6$ as follows.

1. Let $C_1, C_2, C_3$ be f.g. projective $R$-complexes with homology complementary to the homology kernel, image and cokernel bundles.
2. Let $C_4$ be the mapping cone of the identity on $C_1$. (Thus $C_4$ is acyclic, $C_1 \subseteq C_4$ and $C_4/C_1 \cong \Sigma C_1$.)
3. Since $C_1 \oplus C_2 \oplus \eta$ has free homology, there is an acyclic projective complex $C_5$ so that $C_1 \oplus C_2 \oplus C_5$ is a free $R$-complex.
4. Since $C_3 \oplus \Sigma C_1 \oplus \xi/\eta$ has free homology there is another acyclic projective complex $C_6$ so that $C_3 \oplus \Sigma C_1 \oplus C_6$ is free.

Then $\theta_0 = C_1 \oplus C_2 \oplus C_5$ and $\theta = C_2 \oplus \cdots \oplus C_6$ satisfy the required conditions. □

The easy case of the Splitting Lemma (Case 0) can be used to prove the additivity of the cone of the comparison map with respect to composition:

**Proposition 5.7.10.** *Suppose that $f : \xi_0 \to \xi_1$ and $g : \xi_1 \to \xi_2$ are morphisms of $G$-expansion functors over a simplicial set $B$ so that both $f$ and $g$ are fiberwise homotopy equivalences. Then the alternating cones of $f, g, g \circ f$ are related by the formula:*

$$C'(g \circ f) \simeq C'(f) \oplus C'(g).$$

**Remark 5.7.11.** If the higher FR-torsion of a fiberwise homotopy equivalence is defined to be the higher FR-torsion of its alternating cone, this implies that

$$\tau_k(g \circ f) = \tau_k(f) + \tau_k(g).$$

*Proof.* By the Splitting Lemma, $C'(f) \oplus C'(g)$ is simplicially homotopic to the alternating cone of

$$\begin{pmatrix} f & id_{\xi_1} \\ 0 & g \end{pmatrix} : \xi_0 \oplus \xi_1 \to \xi_1 \oplus \xi_2.$$

This in turn is isomorphic to the alt. cone of $j_1 \circ f$ where $j_i$ is the inclusion of $\xi_i$ into the alt. cone $C'(id, g)$ of $(id, g) : \xi_1 \to \xi_1 \oplus \xi_2$. However, the mapping $f$ of $\xi_0$ into the suspended copy of $\xi_1$ in $C'(id, g)$ gives a homotopy between $j_1 \circ f$ and $-j_2 \circ g \circ f$.

The alt. cone of $-j_2 \circ g \circ f$ contains $C'(-g \circ f)$ with quotient $C'(id_{\xi_1}) \simeq 0$. (See the proof of the suspension theorem.) This shows that

$$C'(f) \oplus C'(g) \simeq C'(-g \circ f).$$

Replacing $(f, g)$ with $(id_{\xi_0}, g \circ f)$ we get

$$C'(id_{\xi_0}) \oplus C'(g \circ f) \simeq C'(-g \circ f).$$

Since $C'(id_{\xi_0}) \simeq 0$, the proposition follows. □

### 5.7.3 Higher FR-torsion of oriented sphere bundles

We are now in a position to compute the higher Franz-Reidemeister torsion invariants for oriented linear sphere bundles and their associated lens-space bundles. Except for the case of even dimensional spheres these calculations seem to agree with the computations for the corresponding analytic torsion forms by Bunke, Bismut and Goette ([Bun], [BG00]).

Let $S^{2n-1} \to S(\gamma) \to B$ be the sphere bundle of a complex $n$-plane bundle $\gamma$ and let $m \geq 2$. Then the cyclic group $Z_m = \mathbb{Z}/m\mathbb{Z}$ acts freely on $S(\gamma)$ by multiplication by a primitive $m$-th root of unity $\zeta$ and the resulting lens-space bundle

$$S^{2n-1}/Z_m \to S(\gamma)/Z_m \to B$$

has trivial fiberwise homology with respect to the twisted one dimensional coefficient system $\mathcal{F}_\zeta$ having holonomy $\zeta$ on each fiber. This gives a well-defined mapping

$$f_\gamma : B \to |\mathcal{W}h_\bullet^h(\mathbb{C}, Z_m)|$$

obtained from the canonical fiberwise framed function on $S(\gamma)/Z_m$. Consequently, there are associated higher FR-torsion invariants

$$\tau_k(S(\gamma)/Z_m, \mathcal{F}_\zeta) \in H^{2k}(B; \mathbb{R})$$

**Theorem 5.7.12.** *The higher FR-torsion invariant $\tau_k(S(\gamma)/Z_m, \mathcal{F}_\zeta)$ is related to the Chern character $ch(\gamma)$ of $\gamma$ by the formula:*

$$\tau_k(S(\gamma)/Z_m, \mathcal{F}_\zeta) = -m^k \Re\left(\frac{1}{i^k}\mathcal{L}_{k+1}(\zeta)\right) ch_{2k}(\gamma)$$

*where $ch_{2k}(\gamma)$ is the degree $2k$ term of the Chern character of $\gamma$ and*

$$\mathcal{L}_{k+1}(\zeta) = \sum_{n=1}^{\infty} \frac{\zeta^n}{n^{k+1}}.$$

*Proof.* By the splitting principle we may assume that $\gamma$ splits as a sum of line bundles $\gamma \cong \sum_{1 \le i \le n} \lambda_i$ and that the structure group of $\gamma$ is $U(1)^n$. Then $S(\gamma)$ is the fiberwise join of the $n$ circle bundles $S(\lambda_i)$. This means that if we choose fiberwise framed functions $f_i : S(\lambda_i)/Z_m \to \mathbb{R}$, with images in the open interval $(0, 1)$, then, considering these as $m$ periodic functions on $S(\lambda_i)$, we can combine them to construct a fiberwise framed function $f$ on $S(\gamma)/Z_m$, for example by the formula:

$$f(z_1, \cdots, z_n) = \sum_{i=1}^{n} i|z_i|^2 + \exp(-|z_i|^{-2}) f_i\left(\frac{z_i}{|z_i|}\right). \tag{5.11}$$

Since the derivative of $f$ can only be nonzero when all but one of the coordinates $z_1, \cdots, z_n$ is zero, we see that the image of $f$ lies in $[1, n+1]$ and the restriction of $f$ to $f^{-1}[i, i+1]$ is a twisted suspension of $f_i$ (plus the constant $i$). By Proposition 4.6.5, $f$ is framable and its $G$-expansion functor $\xi$ has a filtration with associated graded functor $\oplus \Sigma^{2i-2}\xi_i$ where $\xi_i$ is the $G$-expansion functor of $f_i$. By the Splitting Lemma 5.7.8 and the suspension theorem 5.7.3 this implies that

$$C(\xi) \simeq \bigoplus \Sigma^{2i-2} C(\xi_i) \simeq \bigoplus C(\xi_i).$$

Consequently, by Theorem 7.10.2 we have:

$$\tau_k(S(\gamma)/Z_m, \mathcal{F}_\zeta) = \sum_{i=1}^{n} \tau_k(S(\lambda_i)/Z_m, \mathcal{F}_{\lambda_i})$$

$$= \sum_{i=1}^{n} -(-1)^{[k/2]} m^k (\Re \text{ or } \Im) \mathcal{L}_{k+1}(\zeta) \frac{c_1(\lambda_i)^k}{k!}$$

$$= -m^k \Re\left(\frac{1}{i^k}\mathcal{L}_{k+1}(\zeta)\right) ch_{2k}(\gamma)$$

□

Now we proceed to the exceptional case when $m = 1$. This is the case of the sphere bundle itself without any action of any group. The fiberwise homology of the sphere bundle $S^{2n-1} \to S(\gamma) \to B$ associated to the complex $n$-plane bundle $\gamma$ over $B$ is nontrivial. However, it is constant, i.e., the action of $\pi_1(B)$ on $H_*(S^{2n-1})$ is trivial. Consequently we get a canonical mapping

$$B \to |\mathcal{W}h_\bullet^h(\mathbb{Z}, 1)|.$$

well defined up to homotopy. Since higher FR-torsion classes pull back to $\mathcal{W}h_\bullet^h(\mathbb{Z}, 1)$ along the natural map to $\mathcal{W}h_\bullet^h(\mathbb{C}, 1)$ we get higher FR-torsion invariants $\tau_k(S(\gamma)) \in H^{2k}(B; \mathbb{R})$.

**Theorem 5.7.13.** *$\tau_{2k+1}(S(\gamma)) = 0$ and*

$$\tau_{2k}(S(\gamma)) = (-1)^{k+1}\zeta(2k+1)ch_{4k}(\gamma)$$

*where $ch_{4k}(\gamma)$ is the degree $4k$ term of the Chern character of $\gamma$ and*

$$\zeta(2k+1) = \mathcal{L}_{2k+1}(1) = \sum_{n=1}^{\infty} \frac{1}{n^{2k+1}}.$$

**Remark 5.7.14.** This theorem includes the computation of the higher FR-torsion for oriented circle bundles with untwisted coefficients. This is the exceptional case where the fibers $(S^1)$ are not acyclic.

*Proof.* First we reduce to the case when $\gamma$ is a circle bundle. Following the proof of Theorem 5.7.12, we use the splitting principle to make $\gamma$ into a sum of line bundles $\gamma \cong \sum \lambda_i$, choose fiberwise framed functions $f_i : S(\lambda_i) \to I$ and combine them to form a framed function $f : S(\gamma) \to [1, n+1]$.

As before the expansion functor $\xi_f$ of $f$ (the $G$-expansion functor with $G = 1$) has a filtration with successive quotients $\xi_1, \Sigma^2\xi_2, \cdots, \Sigma^{2i-2}\xi_n$. Since these all have free homology and all extensions have free homology, the Splitting Lemma 5.7.8 applies. Combined with the suspension theorem we get

$$C(\xi_f) \simeq \bigoplus_{i=1}^{n} C(\xi_i).$$

This reduces us to the case when $n = 1$ and $S(\gamma)$ is a circle bundle.

We now assume that $S(\gamma) \to B$ is the circle bundle associated to a complex line bundle $\gamma$ and we compute the torsion $\tau_k(S(\gamma))$ using the transfer argument of Chapter II. Since we will be taking torsion of the same bundle with different coefficients we write $\tau_k(S(\gamma)) = \tau_k(S(\gamma), 1)$ to represent trivial coefficients.

Take any integer $m \geq 2$ and let $f_m : S(\gamma)/Z_m \to I$ be a fiberwise framed function. This gives a functor

$$\xi_m : \operatorname{simp} B \to \operatorname{simp} \mathcal{W}h_\bullet(\mathbb{Q}[Z_m], Z_m).$$

Since $S(\gamma)/Z_m$ is an oriented circle bundle and $\mathbb{Q}[Z_m]$ is semisimple, the homology bundle is constant and projective. Therefore we can kill the homology and construct another functor

$$C(\xi_m) : \operatorname{simp} B \to \operatorname{simp} \mathcal{W}h_\bullet^h(\mathbb{Q}[Z_m], Z_m)$$

which is an extension of $\xi_m$ by a constant functor.

Let $\zeta$ be a primitive $m$-th root of unity and let $G = \langle\zeta\rangle$ be the subgroup of $\mathbb{Q}(\zeta)$ generated by $\zeta$. We will construct two simplicial functors

$$\phi : \mathcal{W}h^h_\bullet(\mathbb{Q}[Z_m], Z_m) \to \mathcal{W}h^h_\bullet(\mathbb{Q}(\zeta), G)$$

$$\psi : \mathcal{W}h^h_\bullet(\mathbb{Q}[Z_m], Z_m) \to \mathcal{W}h^h_\bullet(\mathbb{Q}, 1)$$

satisfying the following.

1. $\tau_k(\phi(C(\xi_m))) = \sum_{j=0}^{m-1} \tau_k(S(\gamma)/Z_m, \zeta^j)$ where
   (a) $\tau_k(S(\gamma)/Z_m, 1) = m^k \tau_k(S(\gamma), 1)$ and
   (b) $\tau_k(S(\gamma)/Z_m, \zeta^j) = -m^k \Re\left(\frac{1}{i^k}\mathcal{L}_{k+1}(\zeta^j)\right) ch_{2k}(\gamma)$ if $\zeta^j \neq 1$.
2. $\tau_k(\psi(C(\xi_m))) = \tau_k(S(\gamma), 1)$.
3. $\tau_k(\phi(C(\xi_m))) = \tau_k(\psi(C(\xi_m)))$

Theorem 5.7.12 gives us (1)(b). If we prove the other four statements then we get:

$$\tau_k(S(\gamma), 1) = m^k \tau_k(S(\gamma), 1) - m^k \sum_{j=1}^{m-1} \Re\left(\frac{1}{i^k}\mathcal{L}_{k+1}(\zeta^j)\right) ch_{2k}(\gamma)$$

and we can conclude that:

$$\begin{aligned}\tau_k(S(\gamma), 1) &= \frac{m^k}{m^k - 1} \sum_{j=1}^{m-1} \Re\left(\frac{1}{i^k}\mathcal{L}_{k+1}(\zeta^j)\right) ch_{2k}(\gamma) \\ &= -m^k \Re\left(\frac{1}{i^k}\mathcal{L}_{k+1}(1)\right) ch_{2k}(\gamma)\end{aligned}$$

proving the theorem.

We begin by defining

$$\phi : \mathcal{W}h^h_\bullet(\mathbb{Q}[Z_m], Z_m) \to \mathcal{W}h^h_\bullet(\mathbb{Q}(\zeta), G)$$

to be the direct sum of the simplicial functors $\phi^j_*$ induced by the ring homomorphisms $\phi^j : \mathbb{Q}[Z_m] \to \mathbb{Q}(\zeta)$ given by $\phi^j(t) = \zeta^j$ for $j = 0, \cdots, m-1$. If $\zeta^j \neq 1$ then $\phi^j_*(\xi_m)$ is a fiberwise acyclic subfunctor of $\phi^j_*(C(\xi_m))$ and the quotient is acyclic and constant. Therefore $\phi^j_*(C(\xi_m))$ and $\phi^j_*(\xi_m)$ have the same higher FR-torsion. If $j = 0$ then $\phi^0_*(C(\xi_m))$ is by definition one choice of the fiberwise cone of $\phi^0_*(\xi_m)$. Therefore we have

$$\tau_k(\phi^j_*(C(\xi_m))) = \tau_k(S(\gamma)/Z_m, \zeta^j)$$

for all $j$, proving (1).

If we take $\gamma$ to be the universal circle bundle over $(S^2)^{2k}$ then $S(\gamma)/Z_m$ is the circle bundle of the pull back of $\gamma$ along a map $h^{2m} : (S^2)^{2k} \to (S^2)^{2k}$ of degree $m^{2k}$. Therefore we see that

$$\tau_k(S(\gamma)/Z_m, 1) = m^{2k}\tau_{2k}(S(\gamma), 1)$$

proving (1)(a).

Now we define $\psi$ to be the composition of two simplicial functors:

$$\mathcal{W}h^h_\bullet(\mathbb{Q}[Z_m], Z_m) \xrightarrow{\lambda_*} \mathcal{W}h^h_\bullet(M_m(\mathbb{Q}), \Sigma_m) \xrightarrow{\mathrm{N}} \mathcal{W}h^h_\bullet(\mathbb{Q}, 1)$$

Here $\lambda_*$ is the simplicial functor induced by the ring homomorphism $\lambda : \mathbb{Q}[Z_m] \to M_m(\mathbb{Q})$ given by sending $t$ to an $m \times m$ cyclic permutation matrix $T$ with $\Sigma_m = \langle T \rangle$ being the subgroup of the $m \times m$ matrix ring $M_m(\mathbb{Q})$ generated by $T$. N was defined in 5.7.6

Let $f : S(\gamma) \to I$ be the composition of $f_m : S(\gamma)/Z_m \to I$ with the covering map $S(\gamma) \to S(\gamma)/Z_m$. Then $f$ is an $m$-periodic fiberwise framed function and the resulting expansion functor $\xi$ is related to $\xi_m$ by the transfer homomorphism $\mathrm{N} \circ \lambda_* : \mathcal{W}h(\mathbb{Q}[Z_m], Z_m) \to \mathcal{W}h(\mathbb{Q}, 1)$ described above. The canonical cone of $\xi_m$ is an extension of $\xi_m$ by a trivially filtered $G$-monomial functor, i.e., with a constant poset. Since we are now assuming that the base is $(S^2)^k$ which is simply connected, the functor $\mathrm{N}(\lambda_*(C(\xi_m)))$ is, likewise, an extension of $\xi = \mathrm{N}(\lambda_*(\xi_m))$ by a trivially filtered monomial functor. Since $\mathbb{Q}$ is a field and the base is 1-connected, all homology bundles are projective and constant. Consequently, the Splitting Lemma applies, telling us that $\mathrm{N}(\lambda_*(C(\xi_m)))$ which is its own canonical cone is the sum of $C(\xi)$ and the canonical cones of the successive quotients which all have the form $\epsilon_i$. But $C(\epsilon_i)$ is a sum of expansion pairs and therefore null homotopic. Thus $C(\xi) \simeq \mathrm{N}(\lambda_*(C(\xi_m)))$, proving (2).

It remains to show that $\tau_k(\phi(C(\xi_m))) = \tau_k(\psi(C(\xi_m)))$. The proof of this is embodied in the following diagram.

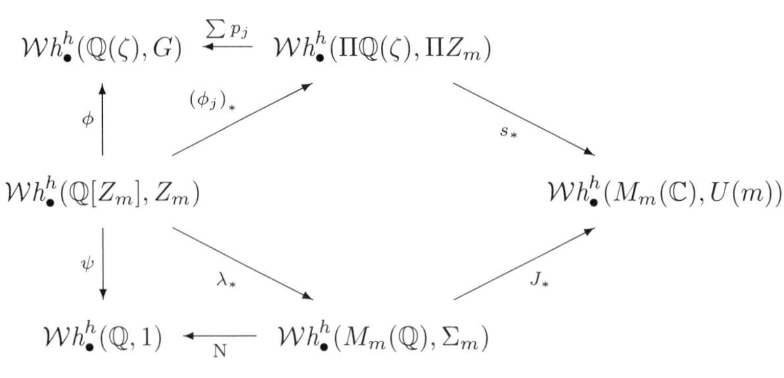


This diagram commutes. The two triangles commute by definition of $\phi$ and $\psi$. The quadrilateral commutes by the conjugation lemma (5.7.7) since the ring homomorphisms $s \circ (\phi_j), J \circ \lambda : \mathbb{Q}[Z_m] \to M_m(\mathbb{C})$ are conjugate by the unitary matrix $\left(\frac{1}{\sqrt{n}}\zeta^{ij}\right)$. The additivity lemma (5.7.5(2)) says that

$$\tau_k(\phi(C(\xi))) = \tau_k((\Sigma p_j)_*(\phi_j)_* C(\xi)) = \tau_k(s_*(\phi_j)_* C(\xi)).$$

The 1-st transfer lemma (5.7.5(3)) says that

$$\tau_k(J_*\lambda_* C(\xi)) = \tau_k(\mathrm{N}\,\lambda_* C(\xi)) = \tau_k(\psi(C(\xi))).$$

Consequently, $\tau_k(\psi(C(\xi))) = \tau_k(\phi(C(\xi)))$ proving (3) and completing the proof of the theorem. □

Theorem 5.7.13 above assumes that the sphere bundle $S^{2n-1} \to S(\gamma) \to B$ is the unit sphere bundle of a complex $n$-plane bundle $\gamma$. To extend this computation to arbitrary oriented linear sphere bundles we need the following formula.

**Lemma 5.7.15.** *Suppose that $\xi, \eta$ are oriented real vector bundles over a space $B$. Then the higher FR-torsion invariants of the unit sphere bundle of $\xi \oplus \eta$ are given by*

$$\tau_{2k}(S(\xi \oplus \eta)) = (-1)^m \tau_{2k}(S(\xi)) + (-1)^n \tau_{2k}(S(\eta))$$

*where $m = \dim \eta$, $n = \dim \xi$.*

*Proof.* The sphere bundle $S(\xi \oplus \eta)$ is the fiberwise join of the sphere bundles $S(\xi)$ and $S(\eta)$. Consequently, the lemma holds in the following two cases.

1. If either $\xi$ or $\eta$ is a trivial bundle.
2. If both $\xi$ or $\eta$ are complex vector bundles.

The general case follows from the fact that the natural map

$$BU(n) \to BSO(2n)$$

induces a monomorphism in rational cohomology. We give the details below.

First suppose that $\eta$ is trivial. Let $f : S(\xi) \to \mathbb{R}$, $g : S(\eta) \to \mathbb{R}$ be fiberwise framed functions. (We may need to take a positive suspension of $S(\xi)$. We can take $g$ to be the composition: $S(\eta) \to B \times S^m \to S^m \to \mathbb{R}$.) Construct an oriented fiberwise generalized Morse function on $S(\xi \oplus \eta)$ by putting $S(\eta)$ on the bottom and $S(\xi)$ on the top (as in (5.11)). Since $\eta$ is trivial, the result is $g$ plus an untwisted $m$-fold suspension of $f$. Thus, by the suspension theorem,

$$\tau_{2k}(S(\xi \oplus \eta)) = (-1)^m \tau_{2k}(S(\xi))$$

which is as required since $\tau_{2k}(S(\eta)) = 0$.

In the case when $\xi$ and $\eta$ are both complex vector bundles the formula follows from Theorem 5.7.13. In the general case we first make both $\xi$ and $\eta$ even dimensional by adding a trivial line bundle to each if necessary. If we add a trivial (real) line bundle to $\eta$ then both $\tau_{2k}(S(\xi \oplus \eta))$ and $\tau_{2k}(S(\eta))$

change sign and $m$ increases by 1. Consequently all three terms in our formula change sign and so we may assume that $\xi$ and $\eta$ have dimension, say, $2n$ and $2m$. Since the natural map $BU(n) \to BSO(2n)$ induces a monomorphism in rational cohomology, so does the map:

$$BU(n) \times BU(m) \to BSO(2n) \times BSO(2m).$$

By naturality of the higher FR-torsion invariants, the difference

$$\tau_{2k}(S(\xi \oplus \eta)) - \tau_{2k}(S(\xi)) - \tau_{2k}(S(\eta)) \in H^{4k}(B; \mathbb{R})$$

in the universal even dimensional case $B = BSO(2n) \times BSO(2m)$ goes to zero in $H^{4k}(BU(n) \times BU(m); \mathbb{R})$ so it must already be zero. □

**Theorem 5.7.16.** *Suppose that $\xi$ is an oriented real $n$-plane bundle over $B$. Then*

$$\tau_{2k}(S(\xi)) = (-1)^{n+k+1}\zeta(2k+1)ch_{4k}(\xi)$$

*where the Chern character of a real bundle is defined to be half the Chern character of its complexification.*

*Proof.* As a real bundle $\xi \otimes \mathbb{C} \cong \xi \oplus \xi$, which by the lemma above has torsion:

$$\tau_{2k}(S(\xi \otimes \mathbb{C})) = \tau_{2k}(S(\xi \oplus \xi)) = (-1)^n 2\tau_{2k}(S(\xi)).$$

Compare this with the formula from Theorem 5.7.13:

$$\tau_{2k}(S(\xi \otimes \mathbb{C})) = (-1)^{k+1}\zeta(2k+1)ch_{4k}(\xi \otimes \mathbb{C}).$$

The theorem follows. □

## 5.8 The involution again

In this section we examine once more how the conjugate transpose involution acts on the higher FR-torsion invariants. The theorem on the level of the Whitehead space is as follows.

**Theorem 5.8.1.** *Given any group $G$, a subring $K$ of $\mathbb{C}$ which is invariant under complex conjugation and any unitary representation $\rho : G \to U(d)$, the higher FR-torsion invariants*

$$\tau_k(\rho) = \rho^*(\tau_k(d)) \in H^{2k}(\mathcal{W}h_\bullet^{h[0,n]}(K[G], G); \mathbb{R})$$

*are anti-invariant with respect to the conjugate transpose involution, i.e.,*

$$ct_n(\tau_k(\rho)) = (-1)^{n-1}\tau_k(\rho). \tag{5.12}$$

**Remark 5.8.2.** We need $K$ to be invariant under conjugation in order for the conjugate transpose involution on $\mathcal{W}h_\bullet^{h[0,n]}(K[G], G)$ to be defined.

If we combine this with Theorem 5.5.1 we get the following. Suppose that $M, X, B$ are as given in §5.5, i.e., $(X, \partial_0 X, \partial_1 X) \to B$ is a vertically oriented smooth manifold bundle with fiber $(M, \partial_0 M, \partial_1 M)$ where $\partial M = \partial_0 M \coprod \partial_1 M$. Let $\mathcal{F}$ be a $\mathbb{C}^m$-coefficient bundle over $X$ given by a unitary representation $\rho : \pi_1 X \to U(m)$. Let $G = \rho(\pi_1 X)$ be the image of $\rho$.

**Corollary 5.8.3.** *Suppose that the fiberwise homology $H_*(M, \partial_0 M; \mathbb{C}[G])$ is constant and projective (as a $\mathbb{C}[G]$-bundle). Let $f : X \to I$ be an oriented fiberwise GMF with $f^{-1}(i) = \partial_i X$ for $i = 0, 1$. Then the higher FR-torsion invariant $\tau_k(C(\xi_f))$ associated with $f$ differs from the invariant associated with $1 - f$ by a factor of $(-1)^{n-1}$ where $n = \dim M$:*

$$\tau_k(C(\xi_{1-f})) = (-1)^{n-1}\tau_k(C(\xi_f))$$

**Remark 5.8.4.** A heuristically important special case of this corollary is given by taking $f$ to be a framed function. If $n$ is odd then the torsions of $f$ and $1-f$ agree. If $n$ is even then their average is zero:

$$\frac{\tau_k(C(\xi_f) + \tau_k(C(\xi_{1-f}))}{2} = 0. \tag{5.13}$$

Since the negative $-f$ (or $1 - f$) of a framed function is just as canonical as $f$, one could argue that this average is the "correct" intrinsic higher FR-torsion invariant. Also this average agrees with what happens for the analytic torsion form [BL95] adding weight to the speculation that they are always proportional.

*Proof.* Note first that $\xi_f$ is a composition of two functors $\xi_f = \phi \circ \eta_f$:

$$\operatorname{simp} B \xrightarrow{\eta_f} \mathcal{W}h_\bullet^{[0,n]}(\mathbb{C}[G], G) \xrightarrow{\phi} \mathcal{W}h_\bullet^{[0,n]}(\mathbb{C}^m, U(m))$$

where $\phi$ is induced by the inclusion $G \subseteq U(m)$. Similarly, $\xi_{1-f} = \phi \circ \eta_{1-f}$. Also the canonical cones $C(\xi_f)$, $C(\xi_{1-f})$ are induced from $\mathcal{W}h_\bullet^{h[0,n]}(\mathbb{C}[G], G)$ by $C(\xi_f) = \phi_0 \circ C(\eta_f)$ and $C(\xi_{1-f}) = \phi_0 \circ C(\eta_{1-f})$ where $\phi_0$ is the restriction of $\phi$ to $\mathcal{W}h_\bullet^{h[0,n]}(\mathbb{C}[G], G)$.

Theorem 5.5.1 says that $\eta_{1-f} \simeq ct_n(\eta_f)$. By the Splitting Lemma this implies that $C(\eta_{1-f}) \simeq ct_n C(\eta_f)$. To see this note that $C(\eta_f)$ is an extension of $\eta_f$ by a filtered functor $\theta$ whose associated graded functor is trivial. $ct_n C(\eta_f)$ is therefore an extension of $ct_n(\theta)$ by $ct_n(\eta_f) \simeq \eta_{1-f}$. Since $ct_n(\theta)$ is filtered with trivial layers, $C(ct_n(\theta)) \simeq 0$. The splitting lemma therefore gives us:

$$ct_n C(\eta_f) = C(ct_n C(\eta_f)) \simeq C(ct_n(\theta)) \oplus C(ct_n(\eta_f)) \simeq C(\eta_{1-f}).$$

By the theorem above we conclude that:

$$\begin{aligned}
\tau_k(C(\xi_{1-f})) &= C(\eta_{1-f})^*(\phi_0^*\tau_k(m)) \\
&= ct_n C(\eta_f)^*(\phi_0^*\tau_k(m)) \\
&= (-1)^{n-1} C(\eta_f)^*(\phi_0^*\tau_k(m)) \\
&= (-1)^{n-1}\tau_k C(\xi_f)
\end{aligned}$$

□

Theorem 5.8.1 follow from the following theorem due to Keith Dennis.

**Theorem 5.8.5 (R.K.Dennis).** *If $G$ is any group and $K$ is any commutative ring then the inclusion map*

$$G \to GL_k(K[G])$$

*induces a split monomorphism in homology with coefficients in $K$ for all $k \geq 1$.*

*Proof.* Pass to Hochschild homology and use Morita equivalence: (See [Igu84] for details.)

$$\begin{array}{ccc}
H_*(G;K) & \longrightarrow & H_*(GL_k(K[G]);K) \\
\downarrow \cong & & \downarrow \cong \\
HH_*(K[G];K) & \longrightarrow & HH_*(K[GL_k(K[G])];K) \\
\downarrow \cong & & \downarrow \\
HH_*(M_k(K[G]);M_k(K)) & \longleftarrow & HH_*(K[GL_k(K[G])];M_k(K))
\end{array}$$

The natural map $K[GL_k(K[G])] \to M_k(K[G])$ is a ring homomorphism since $K$ is commutative. □

Since the homotopy groups of $BMon(G)^+ \times \mathbb{Z} \simeq Q(BG_+)$ are rationally the same as the homology as $G$, (i.e., $\pi_k Q(BG_+) \otimes \mathbb{Q} \cong H_k(G, \mathbb{Q})$ for $k \geq 0$) we get:

**Corollary 5.8.6.** *The mapping*

$$Q(BG_+) = BMon(G)^+ \times \mathbb{Z} \to BGL(K[G])^+ \times \mathbb{Z}$$

*induced by the inclusion $Mon(G) \hookrightarrow GL(K[G])$ induces a monomorphism on rational homotopy groups if $K$ is a subring of $\mathbb{C}$.*

**Remark 5.8.7.** We note that in the case $R = K[G]$ when $\mathbb{Z} \subseteq K$, $K_0\mathbb{Z}$ maps monomorphically into $K_0 R$ (and thus $\overline{\mathbb{Z}} = \mathbb{Z}$) since the rank of a free $R$-module is detected by the Hattori-Stallings trace

$$tr : K_0 R \to HH_0(R;R)$$

which sends a projective module $P$ to the trace of the projection to $P$ when $P$ is embedded as a split summand of a free $R$-module. The augmentation $\epsilon : K[G] \to K$ induces a map

$$HH_0(R;R) \xrightarrow{\epsilon_*} HH_0(R;K) = K$$

and $\epsilon_* tr(R^n) = n \in K$.

Now we need the following theorem.

**Theorem 5.8.8.** *Suppose that $R = K[G]$ where $K$ is a subring of $\mathbb{C}$. Then we have the following commuting square. (Terms are defined in Section 5.6.)*

$$\begin{array}{ccc} BMon(G)^+ \times \mathbb{Z} & \longrightarrow & BGL(K[G])^+ \times \overline{\mathbb{Z}} \\ \downarrow \simeq & & \downarrow \simeq \\ \Omega|w\mathcal{S}_\bullet \mathcal{K}(BG_\bullet, \xi_R)| & \longrightarrow & \Omega|h\mathcal{S}_\bullet \mathcal{K}(BG_\bullet, \xi_R)| \end{array}$$

Suppose that this is true. Then we get the following corollary.

**Corollary 5.8.9.** *The simplicial inclusion functor of the stable Volodin category $\mathcal{V}_\bullet(K[G])$ into the Whitehead category $\mathcal{W}h_\bullet^h(K[G], G)$ induces an epimorphism on rational homotopy groups and therefore a monomorphism in rational cohomology.*

Theorem 5.8.1 now follows from the two-index theorem (which implies that $\mathcal{W}h_\bullet^{h[0,n]}(R, G) \simeq \mathcal{W}h_\bullet^h(R, G)$ for all $n \geq 1$) and the suspension theorem (Theorem 5.7.3).

*Proof of Theorem 5.8.1.* If $n = 1$ we know from Proposition 2.5.17 that $ct_1 = 1$ on the image of $H_*(\mathcal{V}_\bullet(K[G]); \mathbb{R})$. By Corollary 5.8.9 (and the two-index theorem) the image is all of $H_*(\mathcal{W}h_\bullet^{h[0,1]}(K[G], G); \mathbb{R})$.

When $n > 1$ the two-index theorem tells us that $\mathcal{W}h_\bullet^{h[0,1]}(K[G], G)$ is a deformation retract of $\mathcal{W}h_\bullet^{h[0,n]}(K[G], G)$. The conjugate transpose functor $ct_n$ then flips this up to $\mathcal{W}h_\bullet^{h[n-1,n]}(K[G], G)$. Consequently, $ct_n$ is $ct_1$ followed by an $(n-1)$-fold (alternating) suspension:

$$ct_n \simeq (\Sigma')^{n-1} \circ ct_1.$$

We then invoke the suspension theorem which says that (alternating) suspension acts as $-1$ on additive cohomology classes. ($\tau_k$ is additive by Theorem 5.7.5.2.) □

We return to the proof of Theorem 5.8.8 and its corollary.

*Proof of Corollary 5.8.9.* Let $\mathcal{K} = \mathcal{K}(BG_\bullet, \xi_R)$ where $R = K[G]$. Then by Theorem 5.6.1 there is a homotopy fiber sequence

$$w\mathcal{S}_\bullet e_\bullet \mathcal{K}^h \to w\mathcal{S}_\bullet e_\bullet \mathcal{K} \to h\mathcal{S}_\bullet e_\bullet \mathcal{K}.$$

By Corollary 5.8.6 and Theorem 5.8.8, the map $\Omega|w\mathcal{S}_\bullet e_\bullet \mathcal{K}^h| \to \Omega|w\mathcal{S}_\bullet e_\bullet \mathcal{K}|$ is trivial on rational homotopy groups. Consequently, the next mapping

$$X = \Omega^2|h\mathcal{S}_\bullet e_\bullet \mathcal{K}| \to \Omega|w\mathcal{S}_\bullet e_\bullet \mathcal{K}^h| = Y \tag{5.14}$$

is an epimorphism on rational homotopy groups. However, by the results of [IK95] as explained in §5.6, the spaces $X, Y$ above can be identified with $\mathcal{V}_\bullet(K[G])$ and $\mathcal{W}h^h_\bullet(K[G], G)$. This proves the first statement. It remains to show that the map $X \to Y$ above induces an epimorphism in rational homology.

Since $\pi_0 \mathcal{W}h^h_\bullet(K[G], G) = Wh_1(K[G], G)$ is a quotient of $\pi_0\mathcal{V}_\bullet(K[G])$, the mapping (5.14) is surjective on components. Since the components of $X$ (and those of $Y$) are homotopy equivalent to each other, it now suffices to show that the restriction of (5.14) to the null components $X_0$, $Y_0$ induces an epimorphism in rational homology. This follows from the fact that the mapping $X \to Y$ is a morphism of infinite loop spaces. ($\mathcal{S}_\bullet$ is the delooping operator [Wal85].)

Choose a basis for each $\mathbb{Q}$-vector space $\pi_i Y_0 \otimes \mathbb{Q}$ which lifts to $\pi_i Y_0$. These can be used to construct a wedge of spheres $W$ with a map $f : W \to Y_0$ so that the induced map

$$\Omega^\infty S^\infty(f) : \Omega^\infty S^\infty(W_+) \to Y_0$$

is a rational homotopy equivalence. Since the cokernels of the maps $\pi_i X_0 \to \pi_i Y_0$ are torsion groups for all $i \geq 1$ we may choose the elements of $\pi_i Y_0$ so that they lift to $\pi_i X_0$. Then the mapping $f : W \to Y_0$ lifts to $X_0$ giving a map $\widetilde{f} : W \to X_0$ which induces a lifting $\Omega^\infty S^\infty(\widetilde{f})$ of $\Omega^\infty S^\infty(f)$ to $X_0$. This gives a rational section for the map $X_0 \to Y_0$ and thus for the mapping in (5.14). □

*Proof of Theorem 5.8.8.* The group $Mon_n(G)$ is equivalent to a category with one object. We view this object as the set $\overline{n} = \{1, 2, \cdots, n\}$ with morphisms $\overline{n} \to \overline{n}$ being permutations together with maps $\overline{n} \to G$. This category is equivalent to the larger category whose objects are all $n$ element sets $S$ (in some fixed universe) with morphisms being bijections $S \to T$ together with maps $\gamma : S \to G$. Similarly, $GL_n(R)$ is equivalent to the category with the same set of objects but with morphisms being $R$-linear isomorphisms $R^S \to R^T$. The first category naturally embeds in the second. Therefore, by the swallowing lemma, $BGL_n(R)$ is homotopy equivalent to the bicategory which is $G$-monomial in one direction and $R$-isomorphic in the other.

Take the nerve in the $G$-direction as we did in §5.2. Then we get, on one side, the simplicial category of all $n$ point sets over $BG_\bullet$ and, on the other side, the simplicial category with the same set of objects but whose morphisms $S \to T$ are isomorphisms $R^S \to R^T$.

Let $\mathcal{A}_\bullet$ be the simplicial Waldhausen category of finite sets over $BG_\bullet$. ($\mathcal{A}_\bullet = \mathcal{C}_0(BG_\bullet)$ in the proof of Lemma 5.6.6.) Let $\mathcal{B}_\bullet$ be the simplicial Waldhausen category of finite sets $(S, \gamma)$ over $BG_\bullet$ and morphisms $(f, \alpha) : (S, \gamma) \to (T, \gamma')$ where $\alpha : (S, \gamma) \to (T, \gamma')$ is a morphism in $\mathcal{A}_\bullet$ and $f : R^S \to R^T$ is a homomorphism of right $R$-modules so that $f(x)$ is in the $R$-submodule generated by $\alpha(x) \subseteq T$ for all $x \in S$ and where $R^S$ is more precisely given by

$$R^S = \bigoplus_{x \in S} \xi_R(\gamma(x)).$$

The cofibrations of $\mathcal{B}_\bullet$ will be morphisms $(f, \alpha)$ where $\alpha : S \rightarrowtail T$ is a cofibration of sets and $f$ is an isomorphism of $R^S$ onto $R^{\alpha(S)}$. The isomorphisms of $\mathcal{B}_\bullet$ will be morphisms $(f, \alpha)$ where $f : R^S \to R^T$ is an $R$-module isomorphism. Then we have the following commuting diagram.

$$\begin{array}{ccccc}
\coprod BMon_n(G) & \xrightarrow{\simeq} & |i\mathcal{A}_\bullet| & \longrightarrow & \Omega|iS_\bullet\mathcal{A}_\bullet| \\
\downarrow & & \downarrow & & \downarrow \\
\coprod BGL_n(K[G]) & \xrightarrow{\simeq} & |w\mathcal{B}_\bullet| & \longrightarrow & \Omega|wS_\bullet\mathcal{B}_\bullet|
\end{array}$$

The rest of the argument is contained in the following commuting diagram with terms defined in Section 5.6.

$$\begin{array}{ccccc}
iS_\bullet\mathcal{A}_\bullet & \xrightarrow{\simeq} & wS_\bullet e_\bullet\mathcal{K}(BG_\bullet, \xi_R) & \longrightarrow & hS_\bullet e_\bullet\mathcal{K}(BG_\bullet, \xi_R) \\
\downarrow & & \downarrow & & \downarrow \simeq \\
wS_\bullet\mathcal{B}_\bullet & \longrightarrow & hS_\bullet h_\bullet\mathcal{K}^{sf}(R) & \xleftarrow{\simeq} & hS_\bullet e_\bullet\mathcal{K}^{fg}(BG_\bullet, \xi_R)
\end{array}$$

□

## 5.9 Proof of the 2-index theorem

This section is devoted to the proof of the 2-index theorem. In light of Theorem 5.7.3 and the fact that

$$\mathcal{W}h^h_\bullet(R, G) = \lim_{\rightarrow} \mathcal{W}h^{[0,j]h}_\bullet(R, G)$$

it suffices to prove the following lemma.

**Lemma 5.9.1 (Handle exchange lemma).** *$\mathcal{W}h^{[i+1,j]h}_\bullet(R, G)$ is a deformation retract of $\mathcal{W}h^{[i,j]h}_\bullet(R, G)$ when $0 \le i \le j - 2$.*

*Proof.* The proof of this is basically a linearization of the proof of the smooth handle exchange lemma given in [Igu88] which in turn was basically a smooth version of the proof in the PL category given by Hatcher [Hat75]. The outline of the argument is analogous to the outline of the smooth handle exchange lemma given in [Igu88], page 227.

Step 0. (*preparation*) Let $\xi$ be any $G$-expansion functor on a simplicial base $B$ so that the poset $P(\tau)$ of $\xi(\tau)$ has elements only in degrees $i$ through $j$. Then we want to show that the set of elements $P_i(\tau)$ of this poset of degree $i$ can be eliminated for all $\tau \in B$ by a deformation of $\xi$ (in the sense of 5.7.4) without creating any new points with degrees outside this range. By applying this to the universal example the lemma will be proved.

We begin our preparation by replacing $B$ with a *simplicial complex* $K$, i.e., a simplicial set in which each nondegenerate simplex $\sigma \in K^{nd}_k$ has $k+1$ distinct vertices which determine it uniquely. This can be done by taking $K$ to be the second barycentric subdivision of the nondegenerate realization of $B$.

$$K = \coprod_{k\geq 0} B_k \times Sd^2\Delta^k/_{(face\ identifications)}$$

This is a simplicial complex whose vertices are sequences of face operators

$$(\sigma_0, [n_0]) \xrightarrow{a_1} (\sigma_1, [n_1]) \xrightarrow{a_2} \cdots \xrightarrow{a_k} (\sigma_k, [n_k])$$

where $n_0 < n_1 < \cdots < n_k$ and each $a_i : [n_{i-1}] \to [n_i]$ is an order preserving monomorphism with $\sigma_{i-1} = a_i^*\sigma_i$. A finite sequence of such sequences $(\sigma^i, [k_i])_{0\leq i\leq m}$ span an $m$-simplex in $K$ if they are related by non-identity face operators $[k_0] \to [k_1] \to \cdots \to [k_m]$, i.e., if each sequence $\sigma^i$ is a subsequence of $\sigma^{i+1}$.

As a simplicial set $K$ is a simplicial subset of $\mathcal{N}\operatorname{simp}\mathcal{N}\operatorname{simp} B$ and consequently we have a simplicial map $K \to B$ given by the last vertex map (([Wal85],p.359) which is a homotopy equivalence by the realization lemma ([Seg73]). We take the pull back of $\xi$ to $K$ and it suffices for us to deform $\xi$ over $K$ to eliminate the poset points of degree $i$.

Suppose by induction on $k \geq 0$ that we have constructed a deformation of $\xi$ over a homotopy cylinder $Z(K^{k-1})$ of the $(k-1)$-skeleton $K^{k-1}$ of $K$. ($K^{k-1}$ is the simplicial subset of $K$ generated by nondegenerate simplices of dimension $< k$.) Suppose also that $Z(K^{k-1})$ is a simplicial complex containing subcomplexes $K^{k-1}$ and $\widetilde{K^{k-1}}$ as deformation retracts. Suppose also that the projection map $p : Z(K^{k-1}) \to K^{k-1}$ has the property that the inverse image in both $Z(K^{k-1})$ and $\widetilde{K^{k-1}}$ of each simplex in $K^{k-1}$ is contractible. Finally suppose that this deformation has the desired property that $\xi|\widetilde{K^{k-1}}$ has no poset elements of degree $i$ and that furthermore the total set of degree $i$ points

$$\Sigma^i(\xi) = \bigcup_{j\geq 0}\bigcup_{\tau} P_i(\tau) \times \Delta^j$$

over $p^{-1}(\sigma)$ is a disjoint union of contractible sets meeting $\sigma$.

Assuming all of this, then for every $k$-simplex $\sigma$ of $K$ we are reduce to the relative case where we must eliminate the degree $i$ points of $P$ over the contractible simplicial set $X(\sigma) = \sigma \cup p^{-1}(\partial\sigma)$ without changing $\xi$ over the simplicial subset $\partial X = \widetilde{\partial\sigma} \simeq \partial\Delta^k$. In other words at the end of the deformation $Z(X)$ we want $\widetilde{X} \cap X$ to be equal to $\partial X$. Then the union of these deformations over all $k$-simplices will produces the desired deformation on $K^k$.

We now restrict attention to $\xi$ over the contractible simplicial set $X$ and we will work relative to the simplicial subset $\partial X$. We may assume that the total degree $i$ set $\Sigma^i(\xi)$ is a disjoint union of contractible sets mapping homeomorphically onto various contractible subsets of $X$. We will eliminate the components of $\Sigma^i(\xi)$ over $X$ one at a time. So take one component. Let $Y$ be the image of this set in $X$. Then we have one element $x_0 \in P_i(\tau)$ for every simplex $\tau$ in $Y$.

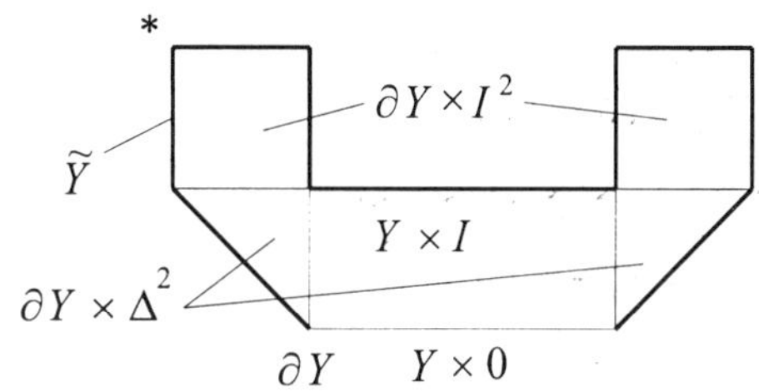


Figure 5.6: Preparing the expansion pairs

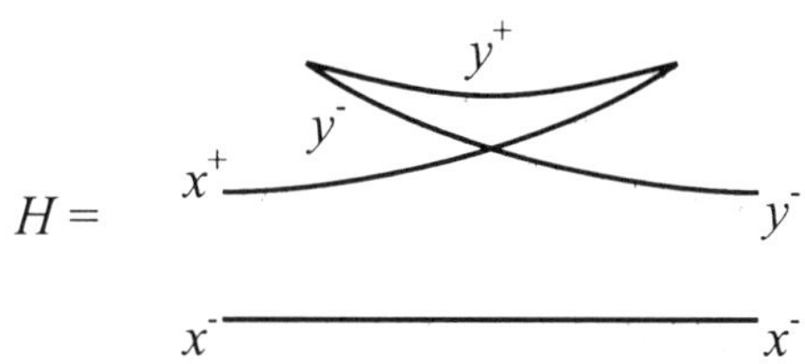


Figure 5.7: The deformation $H$

This element is cancelled along expansion morphisms over s simplicial subset $\partial Y$ of $Y$. We need to eliminate $x_0$ over $Y$ in a nice way so that all the induction hypotheses are preserved.

Step 1. (*preparation of the expansion pairs*) We will eliminate $x_0$ by making it part of an expansion pair. However, along $\partial Y$ we are assuming that this point already forms part of an expansion pair, say $\{x_0, y_0\}$. So, we need to deform $\xi$ near $\partial Y$ to prepare this existing expansion pair so that it will match the new one which we are creating. The trace of the deformation that does this is indicated in Figure 5.6.

First, we take the simplicial homotopy $\overline{\xi}$ of $\xi$ to $\xi_1$ given by simply expanding the expansion pairs along $\partial Y$. Then, along $\partial Y \times I \subseteq Y \times I$ we attach $\partial Y \times \Delta^2$. We extend $\overline{\xi}|(\partial Y \times I)$ to $\partial Y \times \Delta^2$ in a trivial way, then we take the homotopy $H \times id_{I'}$ of $\xi_1|(\partial Y \times I')$ where $I'$ is the top edge of the triangle (not attached to the original $\partial Y$) and $H$ is the homotopy indicated in Fig. 5.7 and described algebraically below. Figure 5.6 shows $Z(Y) = Y \times I \cup \partial Y \times \Delta^2 \cup \partial Y \times I' \times I$ with $\widetilde{Y}$ given by heavy lines. The restriction of the functor to $\widetilde{Y}$ is indicated in Fig. 5.8.

Algebraically, the deformation $H$ is given as follows. Let $C$ be the free $R$-complex generated by $x_0, y_0, y^-, y^+$ with degrees $i, i+1, i+1, i+2$ respectively and boundary map given by $\partial y_0 = x_0, \partial y^- = -x_0$ and $\partial y^+ = y^- + y_0$. Then $S_0 = \{x_0, y_0\}, T = \{y^-, y^+\}$ and $U = \{y_0, y^+\}$ all form general expansion pairs in $C$. The formulas in Lemma 4.5.7 tell us how to make these pairs independent. To make $S_0$ and $T$ independent we make $y_0 < y^-$ in the partial ordering and subtract $y_0$ from $y^-$ and to make $U$ independent we reverse the ordering of these two elements and subtract $y^-$ from $y_0$. The homotopy $H$ is now given by:

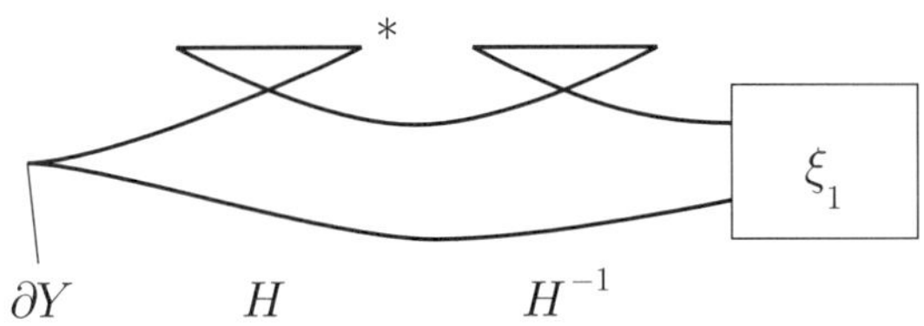


Figure 5.8: $\xi$ on $\widetilde{Y}$

1. Starting with the expansion pair $S_0 = \{x_0, y_0\}$ over $\partial Y$ we take the expansion morphism creating the pair $T = \{y^-, y^+\}$.

2. Change the partial ordering to $x_0 < y_0 < y^- < y^+$ and add $y_0$ to $y^-$ making the complex $C$ (plus the other elements of the poset $P$).

3. Change the partial ordering again to $x_0 < y^- < y_0 < y^+$ and subtract $y^-$ from $y_0$ making $U = \{y_0, y^+\}$ into a general expansion pair.

4. Finally, cancel $U$ with an expansion morphism going the other way. What remains is $y^-, x_0$ with incidence $-1$ (plus the other elements of $P$).

The point $*$ in Figures 5.6 and 5.8 is a new copy of $\partial Y$ bounding a new copy of $Y$ in $\widetilde{Y}$. We call these $\partial D$ and $D$. The restriction of the extended functor to $D$ will be called $\xi|D$.

Step 2. (*the morphism* $f : \eta \to \xi|D$) Let $\eta$ be the constant functor on $D$ given by $\eta(\tau) = (d, S)$ where $S = \{x^-, x^+\}$ with degrees $i, i+1$ and $dx^+ = -x^-$. Suppose for a moment that there exists a morphism $f : \eta \to \xi|D$ which is subordinate to a poset morphism $\alpha : S \to P$ satisfying the following.

1. $W(\tau) = \alpha_\tau(x^-)$ is a closed subset of $P(\tau)$ for every $\tau \in D$ so that $W(\tau) \supseteq W(\sigma)$ whenever $\tau \subseteq \sigma$. (This is part of the definition of $\alpha$ being a morphism.)

2. $W(\tau)$ containing $x_0$ as a maximal element for every $\tau \in D$.

3. $W(\tau) = \{x_0\}$ if $\tau \in \partial D$.

4. $\alpha_\tau(S) = P(\tau)$ for every $\tau \in D$.

5. $f_\tau(x^-) = x_0 + \textit{lower terms}$ for all $\tau$.

6. $f_\tau(x^+) = y^-$ for all $\tau \in \partial D$ where $y^- \in P_{i+1}(\tau)$ is the new point created in Step 1 with boundary $-x_0$.

Step 3. (*sufficiency of the existence of* $(f, \alpha)$) If $(f, \alpha)$ exists satisfying the above six conditions then we can use it to eliminate $x_0$ from $\xi|D$ as follows. By an expansion morphism we can deform $\xi|D$ into the fiberwise cone $C(f, \alpha)$ of $(f, \alpha)$. (By definition, the suspensions of $x^-, x^+$ in this cone have incidence $+1$ so they form an expansion pair $\Sigma S$. The formulas in Lemma 4.5.7 tell us that

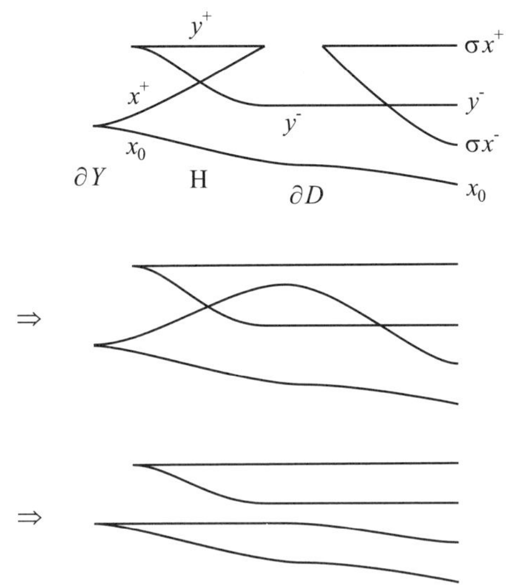


Figure 5.9: Extending the expansion pair to $\widetilde{Y}$

$C(f,\alpha)/\Sigma S = \xi|D$.) Then $x_0, \sigma x^-$ form a new expansion pair by (5) and we are in a position to cancel them. However, before we do, we extend this expansion pair to all of $\widetilde{D}$, the end of the first deformation. The situation is indicated at the top of Figure 5.9. (The $H^{-1}$ portion of Fig 5.8 is pushed to the right, out of the frame of the picture.) Along $\partial D$ we have expansion morphisms going both ways, inward into $D$ and outward into the rest of $\widetilde{Y}$ (right and left in the picture, respectively). By construction of the homotopy $H$, these are identical and therefore can be cancelled. Going to the right, $y^-$ is added to $\sigma x^-$. This cancels step (3) in the homotopy $H$. The end result is that $\sigma x^-$ becomes the continuation of $y_0$ forming an expansion pair with $x_0$ throughout $\widetilde{D}$ and allowing us to eliminate $x_0$ entirely. (See Fig. 5.9.) Thus, by induction, the proof would be complete.

Step 4. (*construction of* $(f,\alpha)$) For each $\tau \in D$ let $\alpha_\tau(x^-) = W(\tau)$ be the closure of $x_0$ in $P(\tau)$ and let $\alpha_\tau(S) = P(\tau)$. Then conditions (1)-(4) will be satisfied. To construct $f_\tau$ we first need to reverse the containment relation $W(\tau) \supseteq W(\sigma)$ for $\tau \subseteq \sigma$.

Let $V(\tau) \subseteq W(\tau)$ be the subset (not closed) of all elements $y$ which are multiply incident under $x_0$ in the system of higher homotopies $\xi(\tau)$, i.e., $V(\tau)$ is the closure of $x_0$ in the minimal partial ordering of $P(\tau)$ allowed by the system $\xi(\tau)$. Then $V$ satisfies the "radial condition" that $V(\tau) \subseteq V(\sigma)$ whenever $\tau \subseteq \sigma$. Also $\xi(\tau)|V(\tau)$ is a system of higher homotopies, i.e., gives a subcomplex $Z(\xi(\tau)|V(\tau))$ of $Z(\xi(\tau))$. We will attempt to construct the morphism $f : \eta \to \xi|D$ inductively on simplices of $D$ so that it satisfies the following additional condition.

(7) $f_\tau(x^-)$ is an $R$-linear combination of elements of $V(\tau)$ for all $\tau$.

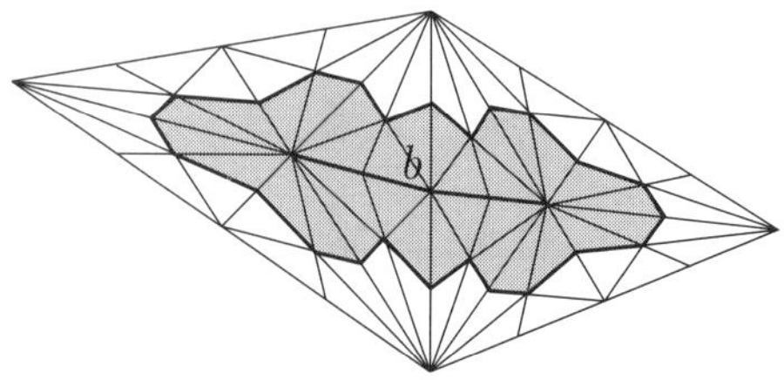


Figure 5.10: $K(b) = \overline{St}(SdB(b))$

We say that $f$ is *subordinate* to $V$ is it satisfies this condition.

Over each vertex $v \in D_0$ we take $f_v(x^-) = x_0$. Since $\xi(v)$ is contractible, $x_0$ must be the boundary of some $i+1$ chain $z(v)$. We let $f_v(x^+) = z(v)$. According to Proposition 3.4.15 the morphism $f$ can be extended to $x^-$ over the $k$-simplex $\tau$ satisfying condition (7) if and only if a certain obstruction

$$o(f) \in H^{i+k-1}(Z(\xi(\tau)|V(\tau)))$$

vanishes. Since $\xi(\tau)$ is always contractible, there is never any obstruction to extending $f$ to $x^+$.

Step 5. (*killing the obstruction*) In this step we will deform the functor so that the obstructions to the construction of $(f, \alpha)$ vanish. In this proof we will use the face that $D$ is a simplicial complex as arranged in Step 1 and use geometric arguments.

For each edge $e$ of $D$ take the barycenter $b$ and let $B(b)$ be the subcomplex of the first barycentric subdivision $SdD$ of $D$ consisting of the closure of all simplices in $SdD$ which meet $b$ but contain neither of the vertices of $e$. (This is the "dual cell" of $e$.) Let $K(b) = \overline{St}(SdB(b))$ be the closed star of $SdB(b)$ in the second barycentric subdivision of $D$. (See Figure 5.10.) Then $B(b)$ and $K(b)$ are both contractible and $K(b)$ does not meet any of the other edges of $D$.

Take the composition of the last vertex maps $Sd^2D \to SdD \to D$ and take its (triangulated) mapping cylinder. Pull back the bundle $\xi|D$ to this mapping cylinder. Then we need to deform $\xi$ over $Sd^2D$ to kill the obstruction.

Take the edges in lexicographic order according to their endpoints. This is a partial ordering but if we restrict attention to one simplex the edges are totally ordered: $e_1, e_2, \cdots$. Taking the barycenter $b_i$ corresponding to the edge $e_i$, do the following where we start with $Z(K(b_1)) = K(b_1)$.

1. On all of $Z(K(b_i))$ create an expansion pair $S_i = \{x_i^-, x_i^+\}$ in degrees $i, i+1$ so that $x_i^-$ is incident to $x_0$ (the point we are trying to eliminate). This is possible since $Z(K(b_i))$ is contractible.

2. Considering $T_i = \{x_0, x_i^-\}$ as an expansion pair, do the deformation $HU$ of Lemma 4.5.7 which makes $W_i$ independent.

3. After $T_i$ becomes independent over $Z(K(b_i))$ we change the partial ordering to make $T_i$ maximal.

4. Let $Z(K(b_{i+1}))$ be the trace of $K(b_{i+1})$ under the deformations done in the first $i$ steps.

Since there are three steps in the deformation we subdivided $I$ into three 1-simplices at $\frac{1}{3}$ and $\frac{2}{3}$. The key point is that in the deformation $HU$ of 4.5.8 points independent from an expansion pair remain independent through the homotopy. Consequently, the points at the top, $Z(K(b_i)) \times 1$, of the $i$-th homotopy that have been made independent of $x_0$ (i.e., every point of $P$ except for $x_i^-$) remains independent during the subsequent homotopies. The new thing that happens at these points is that $x_i^-$ also becomes independent of $x_0$, being replaced in turn by $x_{i+1}^-, x_{i+2}^-, \cdots$. Therefore, the sets $V(\sigma)$ will satisfy the condition that points are being added but not subtracted during the homotopies. (This is the crucial "radial condition" that our deformation needs to preserve.)

The trace $Z(D)$ of the deformation described above is the subset of $Sd^2 \times I^D$ where $I^D = \Pi I_i$ is the product of one copy $I_i$ of $I$ for each $b_i$ given by

$$Z(D) = Sd^2 D \cup \bigcup_{\sigma \in Sd^2 D} \sigma \times \prod_{\sigma \in K(b_i)} I_i.$$

The new copy $\widetilde{D}$ of $D$ inside $Z(D)$ is given as follows.

1. Let $L = \widetilde{D} \cap Sd^2 D$ be the closure of the union of all simplices of $Sd^2 D$ not contained in any $K(b_i)$.

2. If $\sigma \in L$ then
$$\widetilde{D}|\sigma = Z(D)|\sigma = \sigma \times \prod_{\sigma \in K(b_i)} I_i.$$

3. If $\sigma \notin L$ then the coordinate of $\Pi I_i$ corresponding to the first $K(b_i)$ containing $\sigma$ must be 1. Thus
$$\widetilde{D}|\sigma = Z(D)|\sigma = \sigma \times 1 \times \prod_{\sigma \in K(b_j), j > i} I_j,$$
i.e., $\widetilde{D}$ contains only the end of the first homotopy $H(b_i)$ performed on $\sigma$ and the traces of all subsequent homotopies.

Since the inverse image in $\widetilde{D}$ of every simplex in $Sd^2 D$ is a cube, $\widetilde{D} \simeq D$.

We will now show that $(f, \alpha)$ can be constructed over $\widetilde{D}$. We begin by constructing $W(\tau) = \alpha_\tau(x^-)$ and $V(\tau) \subseteq W(\tau)$ for each simplex $\tau$ in $\widetilde{D}$.

For each simplex $\tau$ in $Sd^2 D$ let $W(\tau) = W(\sigma)$ and $V(\tau) = V(\sigma)$ where $\sigma$ is the smallest simplex in $D$ containing $\tau$. During the $i$-th homotopy, we keep the same $W, V$ until the deformation parameter reaches $\frac{2}{3}$. For any simplex over $[\frac{2}{3}, 1]$ we change $W$ so that $W(\tau) = P - \{x_i^-\}$ for each $\tau \in Z(W(b_i))$. At $\frac{2}{3}$ we keep the same $V$, only changing it over simplices $\tau$ of $Z(K(b_i))$ whose interiors map to $(\frac{2}{3}, 1]$ by taking $V(\tau) := W(\tau)$ on those simplices. We keep $\alpha_\tau(S) = P(\tau)$ for all $\tau$.

We will construct $f$ on $\widetilde{D}$ one step at a time. As in Step 4 we will choose $f$ so that it is subordinate to $V$. First we note that $L$ is a disjoint union of contractible neighborhoods of each vertex of $D$. In fact each component is a cone. (Look again at Figure 5.9.) Furthermore, for each simplex $\tau$ in the base of this cone, $V(\tau)$ is equal to $V(C(\tau))$. Consequently, we can let $f_v(x^-) = x_0$ at each vertex $v$ of $D$ and we can use Corollary 3.4.17 to extend $f$ to $L$ so that it is subordinate to $V$.

Next, we take the first homotopy on $K(b_1) \cap L$. Since $V(\tau)$ is increasing in the direction of the homotopy we can extend $f$, keeping it subordinate to $V$, to $(K(b_1) \cap L) \times I$. At the end, on $(K(b_1) \cap L) \times 1$, the obstruction groups is given by

$$H_{i+k-1}(E(\xi(\tau))|V(\tau)) = \begin{cases} R & \text{if } k = 1 \\ 0 & \text{otherwise} \end{cases}$$

This means the only problem is in the extension of $f$ to the 1-simplices of $(K(b_1) \cap L) \times 1$. However, the obstruction $o(f)$ on each edge is equal to the difference between the coefficient of $x_0$ in $f_v(x^-)$ and in $f_w(x^-)$ where $v, w$ are the end points of the edge. Since this coefficient is always equal to 1, there is no obstruction.

At the $i$-th stage we extend $f$ on the trace of the homotopy of $K(b_i)$ on the region in which it is already defined. When the deformation parameter becomes 1, we again have the situation where the obstruction is equal to zero on the 1-simplices by construction and the entire obstruction group vanishes on all higher simplices. (On any new vertices we always take $f_v(x^-) = x_0$.) Therefore, $f$ can be extended as needed on all of $\widetilde{D}$.

Step 6. (*making expansion pairs independent*) The deformation we have constructed has one defect. We did not keep expansion pairs independent as required. However, there is a simple way to fix this. We deform the entire deformation using the procedure of section 10. By Lemma 4.5.7(4) this does not change the original functor $\xi$ so the result is the desired deformation. □

# Chapter 6

# The Framing Principle and Bökstedt's Theorem

In this chapter we will show that the formula for the higher FR-torsion of a linear sphere bundle (Theorem 5.7.16) applies in a more general context which we call the "framing principle." We use this principle to solve a conjecture of Allen Hatcher (namely that an exotic element in the higher homotopy groups of $Diff(D^n)$ can be constructed topologically without the use of algebraic $K$-theory). This was first proved by Marcel Bökstedt [Bök84] using homotopy theory and Waldhausen $K$-theory. We begin in the introduction with the framing principle for disk bundles and prove the general framing principle in §6.1. In Sections 6.2 and 6.3 we review pseudoisotopy theory and the methods of Jean Cerf. Then in §6.4 we give a proof of Hatcher's conjecture using the framing principle. The next section (6.5) discusses the involution in pseudoisotopy theory and its relationship to the higher FR-torsion. We use these to give some applications of Hatcher's construction. Finally, in §6.6 we use the framing principle to obtain three examples of a product formula for higher FR-torsion.

## Introduction

Smooth bundles $p : (E, E_0) \to B$ in general admit many nonhomotopic oriented fiberwise GMF's $f : E \to [0, \infty)$ with $f^{-1}(0) = E_0$. Each such $f$ gives a mapping

$$\xi_f : B \to |Wh_\bullet(\mathbb{Z}, 1)|$$

well defined up to homotopy. A canonical choice $\xi_+$ is given by taking $f$ to be a fiberwise framed function on $(E, E_0)$, or a fiberwise framed function on any positive suspension $(E \times D^N, E_0 \times 0)$ of $(E, E_0)$. (Another canonical choice $\xi_-$ is given by taking $f$ to be the negative of a fiberwise framed function.) Since $\xi_f$ and $\xi_+$ are given by taking minimal subcomplexes of filtered chain complexes which have the same total complex (the total singular complex of $p^{-1}(\sigma)$ for all small simplices in $B$), the identity map on these total complexes, when taken together for all $\sigma$, gives a well defined morphism $F_f : \xi_f \to \xi_+$ which we call the *comparison map*. (We also have a morphism $\overline{F}_f : \xi_f \to \xi_-$ which we call the *reverse comparison map*. Since $F_f$ is a fiber homotopy equivalence, its alternating cone is a map

$$C'(F_f) = C'(\xi_+, \xi_f) : B \to |Wh^h_\bullet(\mathbb{Z}, 1)|.$$

We note that if we take a unitary representation $\rho : \pi_1 E \to U(d)$ we get a mapping $C'(F_f; \rho) : B \to |\mathcal{W}h^h_\bullet(M_d(\mathbb{C}), U(d))|$. However, the resulting higher FR-torsion invariants depends only on the dimension $d$ of the representation so we do not get anything new. (Theorem 6.1.1).

The *framing principle* tells us that the difference between $\xi_f$ and $\xi_+$ is given by the extent to which $f$ is not framable, more precisely the higher FR-torsion invariants of the comparison map $F_f$ are multiples of the Chern character of the complexification of the negative eigenspace bundle of $f$. If the homology bundle of $(E, E_0)$ is untwisted and free then we can form the canonical cones

$$C(\xi_f), C(\xi_+) : B \to |Wh^h_\bullet(\mathbb{Z}, 1)|$$

which, by the Splitting Lemma and suspension theorem, are related to $F_f$ by the formula

$$C(\xi_f) \oplus C'(F_f) \simeq C(\xi_+). \tag{6.1}$$

Consequently, $\tau_*(C'(F_f))$ measures the difference between the higher FR-torsion invariants of $\xi_f$ and $\xi_+$. We call this the *comparison torsion class* and use the notations:

$$\tau_*(F_f) = \tau_*(\xi_+, \xi_f) = \tau_*(C'(F_f)).$$

The *reverse framing principle* says that $\overline{F}_{-f} = ct_n(F_f)$ so that $C'(\overline{F}_{-f}) = (-1)^{n-1} C'(F_f)$.

We first work out the framing principle in the special case of a linear disk bundle modulo its boundary. (This is in fact the universal example.) Suppose that $\gamma$ is an oriented $n$-plane bundle over a space $B$ with structure group $SO(n)$ and let $D(\gamma)$ be the unit disk bundle of $\gamma$ with boundary $\partial_0 D(\gamma) = S(\gamma)$ the $S^{n-1}$-bundle. There is an obvious fiberwise GMF $f : D(\gamma) \to \mathbb{R}$ given by $f(x) = 1 - \|x\|^2$. Then $f^{-1}(0) = \partial_0 D(\gamma)$ and each fiber has exactly one critical point given by the zero section. Since $\gamma$ is oriented, so is $f$. The corresponding map $\xi_f$ is constant and its higher FR-torsion is trivial.

A fiberwise framed function $g$ for $D(\gamma)$ can be given as follows. First, construct a fiberwise framed function for the sphere bundle $h : S(\gamma) \to [0, c]$ where the constant $c$ will be chosen later. The function $g$ will have a minimum instead of a maximum along the zero section and a negative suspension of $h$ along the sphere of radius $\frac{1}{2}$. It will be modelled on a smooth function $\phi : I \to I$ having the following properties. (See Figure 6.1.)

1. $\phi(r) = 1 - r^2$ for $r$ close to 1.

2. $\phi(r) = \frac{1}{2} - (r - \frac{1}{2})^2$ for $r$ close to $\frac{1}{2}$.

3. $\phi(r) = r^2 + \frac{1}{4}$ for $r$ close to 0.

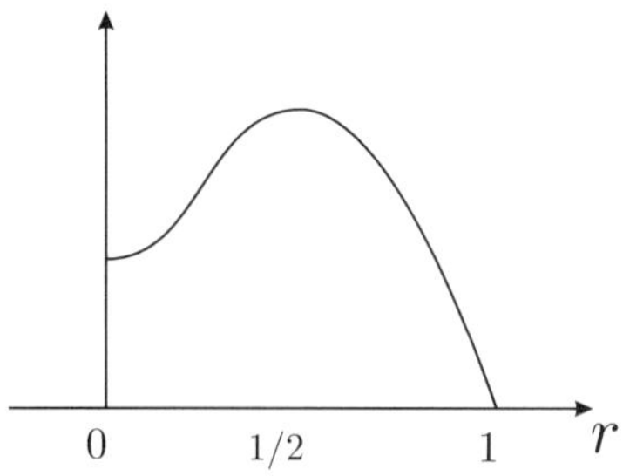


Figure 6.1: The function $\phi(r)$

4. $\phi'(r) \neq 0$ for $r \neq 0, \frac{1}{2}, 1$.

The fiberwise framed function $g : D(\gamma) \to \mathbb{R}$ can then be given by

$$g(x) = \phi(\|x\|) + \psi(\|x\|)h\left(\frac{x}{\|x\|}\right)$$

where $\psi : I \to I$ is a smooth function equal to 1 near $\frac{1}{2}$ and with support in $[\frac{1}{4}, \frac{3}{4}]$. On each fiber, $g$ has a minimum at the origin and is the negative suspension of $h$ along the sphere of radius $\frac{1}{2}$ centered at the origin. If the maximum value $c$ of $|h|$ is sufficiently small then $g$ will have no other critical points.

It follows from the suspension theorem that $\tau_*(D(\gamma)) = -\tau_*(S(\gamma))$. The higher FR-torsion of the sphere bundle is given in Theorem 5.7.16. Together with (6.1) we get:

**Lemma 6.0.1 (Framing principle for disk bundles).**

$$\tau_{2k}(F_f) = \tau_{2k}(C(\xi_+)) = (-1)^{n+k}\zeta(2k+1)ch_{4k}(\gamma)$$

*and* $\tau_{2k+1}(F_f) = 0$.

We note that $g$ can be modified so that it is equal to $f$ outside an arbitrarily small neighborhood of the zero section. To do this let $g_\epsilon : D(\gamma) \to \mathbb{R}$ be given by

$$g_\epsilon(x) = 1 - \epsilon^2(1 - g(\tfrac{1}{\epsilon}x)).$$

Then $g_\epsilon = f$ outside the $\epsilon$-disk neighborhood of the zero section.

Now change the signs of all the functions above. We can go through the same argument or we can use Corollary 5.8.3 to get the following.

**Lemma 6.0.2 (Reverse framing principle for disk bundles).**

$$\tau_{2k}(\overline{F}_{-f}) = (-1)^{n-1}\tau_{2k}(F_f) = (-1)^{k+1}\zeta(2k+1)ch_{4k}(\gamma)$$

*and* $\tau_{2k+1}(\overline{F}_{-f}) = 0$.

## 6.1 The framing principle

Now we state the general framing principle. Suppose that $p : (E, E_0) \to B$ is a smooth bundle and $f : E \to I$ is an oriented fiberwise GMF so that $f^{-1}(0) = E_0$. Suppose further that $f$ *is framed along the birth-death set.* If $\overline{\Sigma^i}(f) \subseteq E$ denotes the closure of the set of Morse critical points of $f$ of index $i$ then the negative eigenspace bundle $\gamma_f^i$ of $f$ along $\overline{\Sigma^i}(f)$ is an oriented $i$-plane bundle which is trivial over $\partial\overline{\Sigma^i}(f)$ so $ch_{4k}(\gamma_f^i) = \frac{1}{2} ch_{4k}(\gamma_f^i \otimes \mathbb{C})$ is an element of $H^{4k}(\overline{\Sigma^i}(f), \partial\overline{\Sigma^i}(f); \mathbb{Q})$.

**Theorem 6.1.1 (Framing principle).** *The higher FR-torsion invariants of the comparison map $F_f$ with respect to any $d$-dimensional Hermitian coefficient system $\mathcal{F}$ on $E$ are given by*

$$\tau_{2k}(F_f; \mathcal{F}) = d \sum_{i \geq 0} (-1)^{i+k} \zeta(2k+1) p_* ch_{4k}(\gamma_f^i) \tag{6.2}$$

*where*

$$p_* : H^{4k}(\overline{\Sigma^i}(f), \partial\overline{\Sigma^i}(f); \mathbb{R}) \to H^{4k}(B; \mathbb{R})$$

*is the push-down operator. Also, $\tau_{2k+1}(F_f; \mathcal{F}) = 0 \in H^{4k+2}(B; \mathbb{R})$. In particular, given $d$, $\tau_*(F_f; \mathcal{F})$ is independent of $\mathcal{F}$.*

**Remark 6.1.2.** Note that since $p : \overline{\Sigma^i}(f) \to B$ is a finite-to-one mapping the push-down is given at the cochain level by

$$\langle p_*(c), \sigma \rangle = \sum \langle c, \sigma_j \rangle$$

for all small simplices $\sigma$ in $B$ where $\sigma_j$ are the liftings of $\sigma$ to $\overline{\Sigma^i}(f)$. Taking $c$ to be zero in a neighborhood of $\partial\overline{\Sigma^i}(f)$, $p_*(c)$ will be a cocycle representing the push-down of $c$ ([KP72]).

*Proof.* The idea is very simple. We modify $f$ on the singular sets $\overline{\Sigma^i}(f)$ one index at a time. Then, it suffices to show that $\tau_{2k}(F_f)$ changes by the corresponding summand of (6.2). To do this we paste in the disk bundle example explained above and apply the framing principle for disk bundles.

Let $Z$ denote $\overline{\Sigma^i}(f)$ minus a small open collar neighborhood of $\partial\overline{\Sigma^i}(f)$. Let $\gamma$ be the vertical tangent bundle of $E$ along $Z$ with decomposition $\gamma = \gamma^- \oplus \gamma^+$ into the negative and positive eigenspace bundles of $D^2 f$. Let $\pi : D_\epsilon(\gamma) \to Z$ be the $\epsilon$-disk bundle of $\gamma$. There is an embedding $\phi : D_\epsilon(\gamma) \to E$ of $D_\epsilon(\gamma)$ into a tubular neighborhood of $Z$ in $E$ making the following diagram commute.

$$\begin{array}{ccc} D_\epsilon(\gamma) & \xrightarrow{\phi} & E \\ \pi \downarrow & & \downarrow p \\ Z & \xrightarrow{p|Z} & B \end{array}$$

For each $t \in Z$ let $f_t : \pi^{-1}(t) \to \mathbb{R}$ be the restriction of $f \circ \phi$ to the $\epsilon$-disk $\pi^{-1}(t) \in D_\epsilon(\gamma)$. Let $x_t^0 \in \pi^{-1}(t)$ be the zero section (so $\phi(x_t^0) = t$). Then, taking coordinates $y, z$ in $\gamma^-, \gamma^+$ we may, by the parametrized Morse lemma (4.4.1), assume that

$$f_t\phi(y,z) = f_t(x_t^0) - \|y\|^2 + \|z\|^2$$

We will construct a smooth function $g : D_\epsilon(\gamma) \to \mathbb{R}$ with the following properties.

1. $g = f$ in a neighborhood of $\partial D_\epsilon(\gamma)$.
2. $g$ is homotopic fixing $\partial D_\epsilon(\gamma)$ to a framed function.
3. $\Sigma(g) = \Sigma(f) \coprod L$ where no point of $L$ is multiply incident under any point in $\Sigma(f)$, i.e., $\Sigma_\sigma(f)$ is a closed subset of $\Sigma_\sigma^{min}(g)$ for every small simplex $\sigma$ in $B$.

*Condition 1* allows us to modify $f$ in a neighborhood of $\phi(D_\epsilon(\gamma))$ of $Z$ by replacing it with $g \circ \phi^{-1}$. Let $h$ denote this new function.

*Condition 2* says that the torsion of the comparison map $F_h$ for $h$ is given by (6.2) with term $i$ deleted. (This is by induction on the number of indices $i$ for which $\gamma_f^i$ is nontrivial.)

*Condition 3* tells us that $\xi_f$ is a subfunctor of $\xi_h$ with quotient $\xi_h/\xi_f$ being the $G$-expansion functor $p_*\xi_L$ on $B$ given by $L$. The comparison map goes the wrong way $F : \xi_h \to \xi_f$ according to the criterion for $\alpha$ in 4.2.12 but it is the identity on $\xi_f \subseteq \xi_h$. Consequently, the alternating cone $C'(F)$ of $F$ is an extension of $C'(id_{\xi_f})$ by $\Sigma'(p_*\xi_L)$. By Remark 5.7.4(b), $C'(id_{\xi_f}) \simeq 0$. By the Splitting Lemma and suspension theorem we have

$$C'(F) = C'(id_{\xi_f}) \oplus \Sigma'(p_*\xi_L) = -p_*\xi_L. \tag{6.3}$$

But $C'(F) \simeq C'(F_h) - C'(F_f)$ by Proposition 5.7.10, so it suffices to show that

4. $\tau_{2k}(\xi_L) = (-1)^{i+k}\zeta(2k+1)ch_{4k}(\gamma^-) \in H^{4k}(Z, \partial Z; \mathbb{R})$
5. $\tau_{2k}(p_*\xi_L) = p_*\tau_{2k}(\xi_L) \in H^{4k}(B; \mathbb{R})$

where $\xi_L$ is the $G$-expansion functor on $Z$ given by $L$.

*Proof that 4,5 are sufficient:* Since $\xi_L$ comes from a disk bundle over $Z$ (with contractible fibers) the restriction to $D_\epsilon(\gamma)$ of the twisted coefficient system $\mathcal{F}$ is essentially trivial and the higher FR-torsion of $\xi_L$ is independent of the choice of $\mathcal{F}$. Consequently we may assume that $\mathcal{F} = E \times \mathbb{C}^d$ and the additivity lemma (2.5.14) tells us that we get $d$ times the torsion for trivial 1-dimensional coefficients. Therefore, we take (4), multiply by $d$ and push down to $B$. The result is term $i$ of (6.2).

*Proof of 5:* This follows from the 2-index theorem and the formal characteristics of the cocycle $F_{2k} - G_{2k}$ representing the Borel regulator which defines the higher FR-torsion. By the two index theorem we may assume that the critical points of $L$ are in two adjacent indices. (We can use either the geometric 2-index theorem of [Igu88] applied to the function $g$ or the algebraic 2-index theorem 5.7.1 applied to $\xi_L$.) The cocycle $F_{2k} - G_{2k}$ is obviously additive with respect to direct sum of matrices and consequently satisfies the formula in 6.1.2.

The construction of $g$ goes as follows. First we may assume that the positive eigenspace bundle $\gamma^+$ of $D^2 f_t$ along $Z$ admits a nowhere zero section $\partial/\partial z_1$. (If it doesn't we take a positive suspension of the original function $f$ and start over.) In this direction we first modify $f$ by creating a cancelling pair $\{x_t^+, x_t^-\}$ in the positive $z_1$-direction from the zero section $x_t^0$. This can be done using the embedding $\phi$ by replacing $f \circ \phi(y, z) = f_t(x_t^0) - \|y\|^2 + \|z\|^2$ with a function which looks like:

$$f_t(x_t^0) - \|y\|^2 + \psi(z_1) + \sum_{j=2}^{n-i} z_j^2$$

where $\psi$ is a function as indicated in Figure 2. Thus $x_t^+, x_t^-$ have index $i+1, i$ respectively.

We choose the orientations of $x_t^+, x_t^-$ so that the incidence of $x_t^+$ with $x_t^0$ and $x_t^-$ are both $+1$, i.e., $x_t^+$ is in cancelling position with both. Along $\partial Z$ we cancel $x_t^+, x_t^-$ with a set of birth-death points. We also use a partition of unity function to make the function $g$ equal to $f$ near $\partial D_\epsilon(\gamma)$ without introducing any unwanted critical points.

The new critical sets $Z^+ = \{x_t^+\}$, $Z^- = \{x_t^-\}$ are equivalent to $Z$ with negative eigenspace bundles $\gamma^- \oplus \epsilon^1$ and $-\gamma^-$, respectively, where $\epsilon^1$ is the trivial line bundle over $Z$ and $-\gamma^-$ is $\gamma^-$ with the opposite orientation.

Since $x_t^+, x_t^-$ are in cancelling position for all $t \in Z$, we may assume that they are not incident under any critical points of $h$ where $h$ is $f$ with $f|\phi(D_\epsilon(\gamma))$ replaced by $g \circ \phi^{-1}$. (See Lemma 4.5.1.)

Finally, we deform $g$ in a small neighborhood of $Z^-$ so that it is fiberwise framed. This is done by pasting in a $\gamma^+$-suspension of a fiberwise framed function for the disk bundle $D(\gamma^-)$ along $Z^-$.

Now we verify that $g$ satisfies the required properties.

(1) This is routine.

(2) Since $x_t^0, x_t^+$ are in cancelling position we can cancel them eliminating all critical points of $g$ in $D_\epsilon(\gamma)$ which are not framed.

(3) $L$ consists of $Z^+$ together with the modified $Z^-$. Since the original $Z^+, Z^-$ are not incident under $\Sigma(f)$, we can do our deformation producing the new $Z^-$ away from the compact set of all points incident under $\Sigma(f)$.

(4) is the framing principle for disk bundles (6.0.1). □

Since the framing principle was proved geometrically by deforming the fiberwise GMF $f$, we can change the signs of the functions to get a framing principle for $-f$.

**Corollary 6.1.3 (Reverse framing principle).** *Let $f : E \to I$ be as in 6.1.1* *and suppose that $E$ is vertically oriented (so that $-f$ is oriented fiberwise). Then*

$$\tau_{2k}(\overline{F}_{-f};\mathcal{F}) = (-1)^{n-1}\tau_{2k}(F_f;\mathcal{F}) = d\sum_{i\geq 0}(-1)^{n-i+k+1}\zeta(2k+1)p_*ch_{4k}(\gamma_f^i) \tag{6.4}$$

*and* $\tau_{2k+1}(\overline{F}_{-f};\mathcal{F}) = 0$.

*Proof.* The geometric arguments stay the same. Only the algebra is different. The arrows are all reversed and $\xi_{-f} = ct_n(\xi_f)$. Condition 3 in the above proof of the framing principle tells us that $\xi_{-f}$ is a quotient of $\xi_{-h}$ with kernel equivalent to $p_*ct_n(\xi_L)$ by Theorem 5.5.1. The natural comparison map now goes in the correct direction $\overline{F} : \xi_{-f} \to \xi_{-h}$. However, the kernel is not suspended so we still have the same sign:

$$C'(\overline{F}) = C'(id_{\xi_{-f}}) \oplus p_*ct_n(\xi_L) = (-1)^{n-1}p_*\xi_L.$$

The corollary now follows from 4 and 5 in the proof of 6.1.1 above. □

The framing principle can be used to compute the higher FR-torsion in a number of cases. One easy case is the projective bundle associated to a complex vector bundle.

Suppose that $\xi^n$ is a complex $n$-plane bundle over a space $B$ and let $P(\xi) \to B$ be the associated $\mathbb{C}P^{n-1}$-bundle. Since $U(n)$ is connected the action of $U(n)$ on the integral homology of $\mathbb{C}P^{n-1}$ is trivial. Consequently, the homology bundle of $P(\xi)$ is constant. It is also free since $H_*(\mathbb{C}P^{n-1})$ is free of rank 1 in every even degree $\leq 2n-2$. Therefore the higher FR-torsion invariants are defined.

**Theorem 6.1.4.** *For each $k \geq 1$ we have*

$$\tau_{2k}(P(\xi)) = \frac{1}{2}(-1)^{k+1}\zeta(2k+1)ch_{4k}(End(\xi))$$

*and $\tau_{2k-1}(P(\xi)) = 0$ where $End(\xi) = \mathrm{Hom}(\xi,\xi)$ is the endomorphism bundle of $\xi$ over $B$.*

*Proof.* Using the splitting principle one may assume that $\xi$ is a sum of line bundles $\xi = \lambda_1 \oplus \cdots \oplus \lambda_m$. Then a fiberwise oriented Morse function $f : P(\xi) \to \mathbb{R}$ can be given by

$$f[z_1, z_2, \cdots, z_n] = \frac{\sum j|z_j|^2}{\sum |z_j|^2}.$$

The critical points of $f$ on each fiber are the points where one coordinate is nonzero. Call these $x_1 = [*, 0, \cdots, 0]$, $x_2 = [0, *, 0, \cdots, 0]$, etc. Then we claim that the negative eigenspace bundle of the critical point $x_i$ is isomorphic to

$$\mathrm{Hom}(\lambda_i, \lambda_1 \oplus \cdots \oplus \lambda_{i-1}) \cong \bigoplus_{j=1}^{i-1} \mathrm{Hom}(\lambda_i, \lambda_j).$$

On each fiber the isomorphism is the derivative at 0 of the map sending $\phi : \lambda_i \to \oplus\lambda_j$ to the point

$$[\phi_1(z), \phi_2(z), \cdots, \phi_{i-1}(z), z, 0, \cdots, 0]$$

which is independent of the choice of $z \neq 0$. Since the cellular chain complex bundle $C(\xi_f)$ is isomorphic to the homology bundle, its higher torsion is zero. Consequently, by (6.1) the higher torsion of $P(\xi)$ is given by the comparison map for $f$. The framing principle now says that

$$\tau_{2k}(P(\xi)) = \tau_{2k}(F_f) = \sum_{i=1}^{n}\sum_{j=1}^{i-1}(-1)^{k+1}\zeta(2k+1)ch_{4k}(\mathrm{Hom}(\lambda_i, \lambda_j))$$

(and $\tau_{2k-1}(P(\xi)) = 0$). But $\mathrm{Hom}(\lambda_i, \lambda_j)$ is a line bundle with dual $\mathrm{Hom}(\lambda_j, \lambda_i)$ so

$$ch_{4k}(\mathrm{Hom}(\lambda_i, \lambda_j)) = ch_{4k}(\mathrm{Hom}(\lambda_j, \lambda_i)).$$

Also, $\mathrm{Hom}(\lambda_i, \lambda_i)$ is trivial for each $i$. Thus

$$\sum_{i>j} ch_{4k}(\mathrm{Hom}(\lambda_i, \lambda_j)) = \frac{1}{2}\sum_{i,j} ch_{4k}(\mathrm{Hom}(\lambda_i, \lambda_j)) = \frac{1}{2}ch_{4k}(\mathrm{Hom}(\xi, \xi)).$$

The theorem follows. □

## 6.2 Pseudoisotopies

Let $M$ be a compact smooth manifold with boundary. Then a *pseudoisotopy* of $M$ is defined to be a diffeomorphism of $M \times I$ which is the identity on $M \times 0 \cup \partial M \times I$. The group of all such diffeomorphisms with the $C^\infty$ topology ([Hir76]) is called the *pseudoisotopy* or *concordance space* of $M$:

$$\mathcal{C}(M) = Diff(M \times I \text{ rel } M \times 0 \cup \partial M \times I).$$

Note that we have the following restriction map $r$ with fiber $Diff(M \times I$ rel $\partial(M \times I))$ which we denote $Diff(M \times I \text{ rel } \partial)$ for short.

$$Diff(M \times I \text{ rel } \partial) \to \mathcal{C}(M) \xrightarrow{r} Diff(M \text{ rel } \partial M). \tag{6.5}$$

Jean Cerf showed in the Appendix of [Cer68] that restriction maps of this kind are fibrations.

There are suspension maps

$$\sigma_+, \sigma_- : \mathcal{C}(M) \to \mathcal{C}(M \times I)$$

which are given by the diagrams in Figure 6.2. In these diagrams each point represents one copy of $M$. Thus each arrow represents a copy of $M \times I$. These

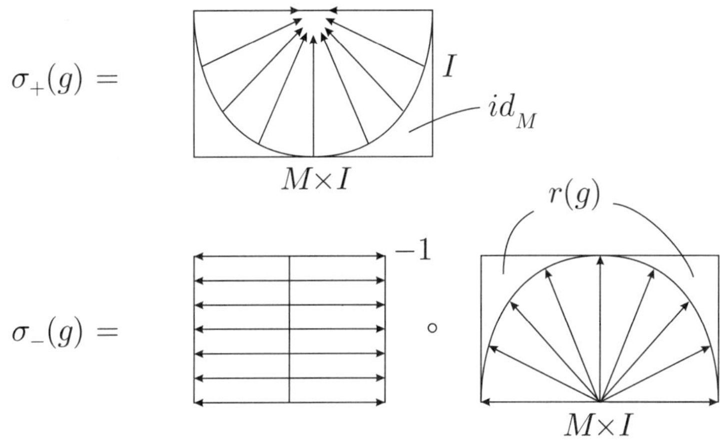


Figure 6.2: +/- suspension of pseudoisotopies

are mapped to themselves by $g \in \mathcal{C}(M)$. The other points in the two squares represent copies of $M$ which are mapped to themselves by $id_M$ or $r(g)$ as indicated. See [Igu88] for details.

Figure 6.3 shows that $\sigma_+, \sigma_-$ are homotopy inverses, i.e., we have the following.

**Proposition 6.2.1.** *The negative suspension map* $\sigma_- : \mathcal{C}(M) \to \mathcal{C}(M \times I)$ *is homotopic to the composition of* $\sigma_+$ *with group inversion* $(g \mapsto \sigma_+(g)^{-1})$.

The stability theorem ([Igu88]) says:

**Theorem 6.2.2.** *The suspension maps* $\sigma_+, \sigma_-$ *are* $k$*-connected if* $\dim M \geq \max(2k+7, 3k+4)$. *Furthermore,*

$$\sigma_+^m : \mathcal{C}(M) \to \mathcal{C}(M \times I^m)$$

*is an epimorphism on* $\pi_k$ *if* $\dim M \geq 2k+7$ *and* $m >> k$.

**Definition 6.2.3.** The *stable pseudoisotopy space* is defined to be the direct limit $\mathcal{P}(M) = \lim_\to \mathcal{C}(M \times I^m)$ with respect to positive suspension.

Since any smooth embedding $M \to N$ induces a continuous mapping $\mathcal{C}(M) \to \mathcal{C}(N)$ by generalized suspension and any two isotopic embeddings induce homotopic maps on concordance spaces, it is easy to see that:

**Proposition 6.2.4.** $\mathcal{P}$ *is a homotopy functor on finite cell complexes in the sense that* $\mathcal{P}(X)$ *is well-defined up to homotopy equivalence for any finite cell complex* $X$ *and maps* $\mathcal{P}(M) \to \mathcal{P}(N)$ *are determined up to homotopy by a homotopy class of maps* $M \to N$.

The following well-known theorem is due to Waldhausen. ([Wal87a], [Wal87b].)

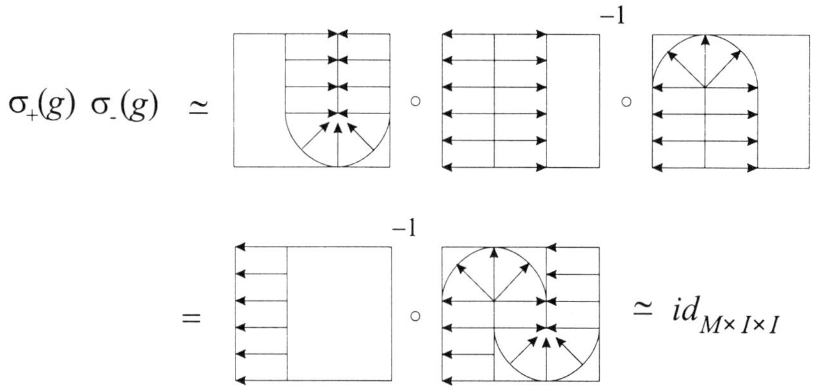


Figure 6.3: $\sigma_+, \sigma_-$ are homotopy inverses

**Theorem 6.2.5.**

$$\pi_i \mathcal{P}(*) \otimes \mathbb{Q} = \begin{cases} \mathbb{Q} & \textit{if } i \equiv 3 \mod 4 \\ 0 & \textit{if not} \end{cases}$$

## 6.3 Cerf's approach

The first theorem in smooth pseudoisotopy theory was obtained by Jean Cerf.

**Theorem 6.3.1.** *[Cer70] If* $\dim M \geq 5$ *and* $M$ *is* 1*-connected then* $\mathcal{C}(M)$ *is connected.*

To prove this Cerf used Morse theory. Let $\mathcal{F}(M)$ be the space of all smooth functions $f : M \times I \to I$ which are *admissible* in the sense that they are equal to the projection map $p_I : M \times I \to I$ near $\partial(M \times I) = \partial M \times I \cup M \times \partial I$. Let $\mathcal{E}(M)$ be the subspace of $\mathcal{F}(M)$ consisting of all admissible smooth functions on $M \times I$ without singularities. Then Cerf observed that:

**Lemma 6.3.2.** *Composition with* $p_I$ *gives a homotopy equivalence:*

$$\mathcal{C}(M) \simeq \mathcal{E}(M).$$

*Proof.* The inverse map $\mathcal{E}(M) \to \mathcal{C}(M)$ is given by integrating the gradient vector field. □

Consequently we have:

**Proposition 6.3.3.** $\pi_k \mathcal{C}(M) \cong \pi_{k+1}(\mathcal{F}(M), \mathcal{E}(M))$.

In other words, an element of $\pi_k \mathcal{C}(M)$ is given by a smooth family of admissible functions

$$f_t : M \times I \to I, \quad t \in D^{k+1}$$

whose singular set $\Sigma(f)$ is contained in the interior of $M \times I \times D^{k+1}$.

Note that $[g_t] \in \pi_k\mathcal{C}(M)$ corresponds to a smooth bundle $E \to S^{k+1}$ given by pasting together two copies of $M \times I \times D^{k+1}$ along $M \times I \times \partial D^{k+1}$ using the diffeomorphism $g(x,t) = (g_t(x), t)$. Thus the fibers of $E$ are diffeomorphic to $M \times I$ and the subbundle corresponding to $M \times 0 \cup \partial M \times I$ is trivial. Let $f : E \to I$ be a smooth function which is admissible over each fiber and which is equal to the projection $p_I : M \times I \to I$ over the southern hemisphere $S^{k+1}_-$ of $S^{k+1}$. Then over the boundary of the northern hemisphere $S^{k+1}_+$ we have $p_I \circ g_t$ which extends by $f_t$ to all of $S^{k+1}_+$.

## 6.4 Hatcher's construction

Allen Hatcher gave the following construction for an element of $\pi_3\mathcal{C}(D^N) \cong \pi_4(\mathcal{F}(D^N), \mathcal{E}(D^N))$ for large $N$ and asked if it gave a nontrivial element.

Choose an element in the kernel of the J-homomorphism

$$J : \pi_3(O(n)) \to \pi_{n+3}(S^n).$$

This is the homomorphism on $\pi_3$ induced by the map on spaces

$$O(n) \to G_{n-1} \to \Omega^n S^n$$

given by sending a rotation of $\mathbb{R}^n$ to a self-map of $S^{n-1}$ ($G_{n-1}$ is the space of self-homotopy equivalences of $S^{n-1}$) and suspending to get a pointed self map of $S^n$. Both groups are stable in the sense that $\pi_3 O(n) \cong \pi_3 O$ and $\pi_3\Omega^n S^n \cong \pi_3\Omega^\infty S^\infty = \pi_3^s(S^0)$ for $n \geq 5$. For $O(n)$ this is obvious from the fibration sequence

$$O(n) \to O(n+1) \to S^n$$

which implies that $\pi_k O(n) \cong \pi_k O$ for $n \geq k+2$. For $S^n$ this follows from the Freudenthal suspension theorem which implies that $\pi_{n+k}S^n$ is also stable for $n \geq k+2$.

So let $n \geq 5$ and take a generator of the kernel of $J : \pi_3 O(n) \to \pi_{n+3}S^n$. This is represented by a continuous mapping of triples

$$g : (D^3 \times I, \partial(D^3 \times I), \partial D^3 \times I \cup D^3 \times 0) \to (\Omega^n S^n, O(n), 1)$$

which we may assume is smooth. In other words we have a smooth family of smooth mappings $g_{tu} : S^n \to S^n$, $(t,u) \in D^3 \times I$ so that

1. $g_{tu}$ is the identity map for $(t,u) \in D^3 \times 0 \cup \partial D^3 \times I$.

2. $g_{t1}$ is a rotation for each $t \in D^3$.

By transversality and the Whitney trick we can assume that $g_{tu}$ lifts to a family of smooth embeddings

$$\tilde{g}_{tu} : S^n \to S^n \times D^{n+4}.$$

We view $S^n \times D^{n+4}$ as a subspace of $S^n \times S^{n+4}$. (See Figure 6.5.)

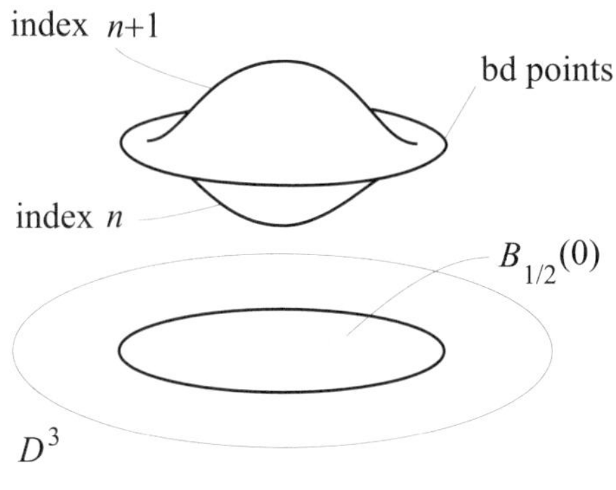


Figure 6.4: Graphic of a 3-lens

With this family of embeddings, Hatcher constructs a 4-parameter family of oriented GMF's on $D^{2n+4} \times I$ for $n \geq 5$. First he creates a "3-lens." This is a 3 parameter family of functions $f_t : D^{2n+4} \times I \to I$, $t \in D^3$ so that for $t$ in the open ball $B_{\frac{1}{2}}(0)$ of radius $\frac{1}{2}$ centered at the origin. $f_t$ is a Morse function with exactly two critical points $x^-, x^+$ with indices $n$, $n+1$ which are in cancelling position and which cancel along a 2-sphere of birth-death points over the boundary of $B_{\frac{1}{2}}(0)$. Although the singular set $\Sigma(f) \subset (D^{2n+1} \times I) \times D^3$ is an embedded 3-sphere, its *graphic* (the parametrized set of critical values $\{(t, f_t(x)) \in D^3 \times I | (x,t) \in \Sigma(f)\}$) is shaped like a "lens" as shown in Figure 6.4.

Hatcher's idea was to eliminate the critical points in this 3-parameter family of functions in two different ways resulting in a "4-lens." This would represent an element of $\pi_4(\mathcal{F}(D^{2n+4}), \mathcal{E}(D^{2n+4}))$ which by Cerf is isomorphic to $\pi_3\mathcal{C}(D^{2n+4})$.

For each $t$ in the 3-dimensional ball $B_{\frac{1}{2}}(0)$ let $L_t$ be the level surface of $f_t$ between the two critical levels. Then $L_t$ is diffeomorphic to $S^n \times S^{n+4}$ minus a disk. The ascending sphere of $x_t^-$ is $* \times S^{n+4}$ and the descending sphere of $x_t^+$ is $S^n \times *$. Everyone know that since these spheres intersect transversely at one point for each $t \in B_{\frac{1}{2}}(0)$ the critical points are in cancelling position ([Mil65]). Thus we can cancel them by a deformation $f_{tu}$. This forms the first half of the 4-lens.

We use the family of embeddings $g_{tu}$ constructed above to create a second null-deformation of $f_t$. Since any isotopy of the descending spheres of the critical points $x_t^+$ can be realized by a deformation of the gradient vector field of $f_t$, we can deform the descending spheres of $x_t^+$ by $g_{tu}$. During the deformation $(0 < u < 1)$ these descending spheres are not always in cancelling position with the ascending sphere of $x_t^-$. However, at the end, when $u = 1$ they are again in cancelling position so we can cancel the critical points eliminating the 3-lens again. This second null-deformation of the 3-lens is known as a "rolling collapse" because the descending spheres are rotating (by $g_{t1}$).

The resulting 4-parameter family of functions gives an element

$$\eta_1 \in \pi_4(\mathcal{F}(D^{2n+4}), \mathcal{E}(D^{2n+4})) \cong \pi_3\mathcal{C}(D^{2n+4})$$

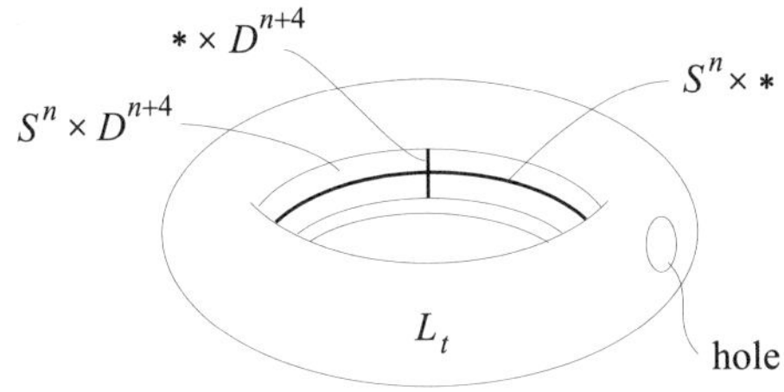


Figure 6.5: $L_t$ is a level surface of $f_t$.

for any $n \geq 5$. (By the Stability Theorem 6.2.2 this element is realized in $\pi_3\mathcal{C}(D^{13})$.) Hatcher suggested that this element was nonzero.

This was first proved (in the stable case $n = \infty$) by Marcel Bökstedt [Bök84] using different terminology. To understand this we need to reformulate the problem at the level of spaces. Instead of taking an element of the kernel of $J$ we can take any homotopy group of the fiber $\Omega(G/O)$ of the mapping $J : O \to G$ where $G$ is the direct limit of $G_n$'s. Then we note that Hatcher's construction actually gives a mapping $\Omega(G/O) \to \mathcal{P}(*)$. Passing to classifying spaces we get a map

$$G/O \to Wh^{diff}(*)$$

where $Wh^{diff}(*)$ is a model for the classifying space of the stable pseudoisotopy space $\mathcal{P}(*)$. Therefore, Hatcher's conjecture follows from the following theorem of Bökstedt.

**Theorem 6.4.1.** *[Bök84] The natural mapping $G/O \to Wh^{diff}(*)$ is a rational homotopy equivalence.*

We will now see that the framing principle gives another logically independent proof of Hatcher's conjecture by giving us the value of the higher FR-torsion on Hatcher's pseudoisotopy element. In the case of the generator of the kernel of $J : \pi_3(O) = \mathbb{Z} \to \mathbb{Z}/24 = \pi_3^s(S^0)$ this is $\pm 24\zeta(3)$ times the generator of $H^4(S^4;\mathbb{Z})$. The theorem is as follows.

**Theorem 6.4.2.** *Hatcher's construction gives an element $\eta_1$ of $\pi_3(\mathcal{C}(D^n))$ for any $n \geq 13$ with higher FR-torsion invariant $\tau_2(\eta_1) \in H^4(S^4;\mathbb{R})$ equal to $\pm 24\zeta(5)$ times the generator of $H^4(S^4;\mathbb{Z})$. More generally his construction gives $\eta_k \in \pi_{4k-1}(\mathcal{C}(D^n))$ for any $n \geq 8k+5$ whose torsion invariant $\tau_{2k}(\eta_k)$ is equal to $\pm b_k\zeta(2k+1)$ times the generator of $H^{4k}(S^{4k};\mathbb{Z})$ when $k$ is odd and $\pm\frac{1}{2}b_k\zeta(2k+1)$ times the generator of $H^{4k}(S^{4k};\mathbb{Z})$ when $k$ is even where $b_k$ is the order of the image of $J_{4k-1} : \pi_{4k-1}O \to \pi_{4k-1}^s(S^0)$.*

**Remark 6.4.3.** It follows from the Adam's Conjecture that $b_k$ is the denominator of $\frac{B_{2k}}{4k}$ where $B_{2k}$ is the $k$-th Bernoulli number ([Ada66], [Sul74], [Qui71], [Qui73], [Ada78], Chap. 5).

*Proof.* Since the incidence matrix for Hatcher's family of functions is either zero or the identity over all points in $D^{4k}$, the torsion of the comparison map is equal to the torsion of the bundle. The framing principle gives the torsion of the comparison map. The value of the Chern character of the bundle representing a generator of the kernel the $J$-homomorphism is equal to $b_k$ times the Chern character on the generator of $\pi_{4k}BO$ which is equal to either the generator of $H^{4k}(BO;\mathbb{Z})$ for $k$ odd or half the generator of $H^{4k}(BO;\mathbb{Z})$ for $k$ even by the following well-known results.

First, Raoul Bott's integrality theorem says that if $\xi$ generates $K(S^{2k})$ then the Chern character of $\xi$ is the generator of $H^{2k}(S^{2k};\mathbb{Z})$ ([Bot69], p.28, [Hus66], p.278). Combine this with the fact that the mapping $U \to O$ induced by the inclusion $U(n) \to O(2n)$ is an isomorphism on $\pi_{8k+3}$ and is multiplication by 2 on $\pi_{8k-1}$. This in turn follows from the fact that $\pi_i(O/U) = \pi_{i+1}O = 0$ for $i = 8k+3$ and has order 2 for $i = 8k-1$ ([Bot59], reprinted in [Bott:K(X)], p.181, or [Mil63], p.136). □

By Hatcher's construction and Theorem 6.2.5, the framing principle implies Bökstedt's theorem. And the computation of the torsion of lens space bundles (5.7.12) can be viewed as a generalization of Bökstedt's theorem to the complex case.

## 6.5 The involution on $\mathcal{C}(M)$

Finally we note that there are involutions on $Diff(M \times I \text{ rel } \partial)$, $\mathcal{C}(M)$ and $\mathcal{E}(M)$, all of which we will denote by $\iota(f) = \overline{f}$, given as follows.

$$\begin{array}{ll} \overline{g} = \phi \circ g \circ \phi & \text{for } g \in Diff(M \times I \text{ rel } \partial), \\ \overline{g} = \phi \circ (r(g) \times id_I)^{-1} \circ g \circ \phi & \text{for } g \in \mathcal{C}(M), \\ \overline{f} = \gamma \circ f \circ \phi & \text{for } g \in \mathcal{E}(M) \end{array}$$

where $\gamma, \phi$ are the involutions on $I, M \times I$ given by $\gamma(t) = 1-t$, $\phi(x,t) = (x, 1-t)$ (so that $\gamma \circ p = p \circ \phi$). Note that the inclusion map $j : Diff(M \times I \text{ rel } \partial) \to \mathcal{C}(M)$ and projection $p : \mathcal{C}(M) \to \mathcal{E}(M)$ commute with the involution. In particular this implies that on homotopy groups with the prime 2 inverted, $j$ induces maps

$$j_*^+ : \pi_i^+ Diff(M \times I \text{ rel } \partial) \to \pi_i^+ \mathcal{C}(M)$$

$$j_*^- : \pi_i^- Diff(M \times I \text{ rel } \partial) \to \pi_i^- \mathcal{C}(M)$$

where $\pi_i^+, \pi_i^-$ indicate the subgroups of $\pi_i$ on which the involution acts as $+1, -1$ respectively.

Take the fibration sequence (6.5) for both $M$ and $M \times I'$ where we use $I'$ to denote the second copy of the interval $I$:

$$\begin{array}{ccccc} Diff(M \times I \text{ rel } \partial) & \xrightarrow{j} & \mathcal{C}(M) & \xrightarrow{r} & Diff(M \text{ rel } \partial M) \\ & & \big\downarrow \sigma_+ & & \\ Diff(M \times I' \times I \text{ rel } \partial) & \xrightarrow{j'} & \mathcal{C}(M \times I') & \xrightarrow{r'} & Diff(M \times I' \text{ rel } \partial) \end{array}$$

Then we have the following elementary observations.

**Lemma 6.5.1.** *1.* $r(\overline{g}) = r(g)^{-1}$ *for all* $g \in \mathcal{C}(M)$. *(So* $\pi_i^+\mathcal{C}(M)$ *goes to zero in* $\pi_i Diff(M \text{ rel } \partial M) \otimes \mathbb{Z}[\frac{1}{2}]$ *and the connecting homomorphism maps* $\pi_i Diff(M \text{ rel } \partial M)$ *into* $\pi_{i-1}^- Diff(M \times I \text{ rel } \partial)$ *and therefore* $j$ *induces an isomorphism*

$$\pi_i^+ Diff(M \times I \text{ rel } \partial) \otimes \mathbb{Z}[\tfrac{1}{2}] \cong \pi_i^+\mathcal{C}(M) \otimes \mathbb{Z}[\tfrac{1}{2}].)$$

*2.* $\sigma_- \circ \iota = \iota \circ \sigma_+$ *on* $\mathcal{C}(M)$, *i.e.,* $\sigma_-(\overline{g}) = \overline{\sigma_+(g)}$ *for all* $g \in \mathcal{C}(M)$. *(So suspension anticommutes with involution.)*

*3.* $j'$ *and* $r'$ *are equivariant with respect to the "lateral" involution* $s$ *which flips the* $I'$ *factor (*$s(g) = \lambda \circ g \circ \lambda$ *where* $\lambda(x,t,u) = (x, 1-t, u)$*). Furthermore:*

a. $s$ *is the same as* $\iota$ *on* $Diff(M \times I' \text{ rel } \partial)$ *if we identify* $I'$ *with* $I$,

b. $s \simeq \iota$ *on* $Diff(M \times I' \times I \text{ rel } \partial)$ *and*

c. $s$ *is the identity on* $\sigma_+(\mathcal{C}(M)) \subseteq \mathcal{C}(M \times I')$.

*(So* $\pi_i^- Diff(M \times I' \text{ rel } \partial) \otimes \mathbb{Z}[\frac{1}{2}]$ *goes isomorphically to* $\pi_{i-1}^- Diff(M \times I^2 \text{ rel } \partial) \otimes \mathbb{Z}[\frac{1}{2}]$ *and* $\pi_i^+ Diff(M \times I^2 \text{ rel } \partial) \otimes \mathbb{Z}[\frac{1}{2}]$ *goes isomorphically onto* $\pi_i^+\mathcal{C}(M \times I') \otimes \mathbb{Z}][\frac{1}{2}]$ *in the surjective stability range.)*

*4. The composition*

$$r' \circ \sigma_+ \circ j : Diff(M \times I \text{ rel } \partial) \to Diff(M \times I' \text{ rel } \partial) \cong Diff(M \times I \text{ rel } \partial)$$

*is homotopic to the mapping* $g \mapsto g\overline{g}$. *(So* $r'$ *induces an isomorphism* $\pi_i^- \mathcal{C}(M \times I') \otimes \mathbb{Z}[\frac{1}{2}] \cong \pi_i^+ Diff(M \times I' \text{ rel } \partial) \otimes \mathbb{Z}[\frac{1}{2}]$ *in the surjective stability range.)*

*Proof.* The homotopy in (4) is given by squeezing the support of $g$ down to $M \times [0, \frac{1}{2}]$ (and consequently, that of $\overline{g}$ down to $M \times [\frac{1}{2}, 1]$) . (The *support* of a diffeomorphism $g$ is the closure of the set of points on which it is not the identity.)

The homotopy in (3b) is given by first squeezing the support down to $M \times D^2$ where $D^2$ is the disk of radius $\frac{1}{2}$ centered at $(\frac{1}{2}, \frac{1}{2})$ and then rotating the flipping axis by $90°$. □

Since suspension anticommutes with the involution by Proposition 6.2.1 and 2 above we define the involution on $\mathcal{P}(M)$ to be the one compatible with the involution $(-1)^{\dim M}\iota$ on $\mathcal{C}(M)$. Then we get the following.

**Theorem 6.5.2.** *In the surjective stable range (*$\dim M \geq \max(2i+7, 3i+4)$*) we have:*

*1.* $\pi_i^+ Diff(M \times I \text{ rel } \partial) \otimes \mathbb{Z}[\frac{1}{2}] \cong \pi_i^+\mathcal{P}(M) \otimes \mathbb{Z}[\frac{1}{2}]$ *when* $\dim M$ *is even.*

2. $\pi_i^+ Diff(M \times I \text{ rel } \partial) \otimes \mathbb{Z}[\frac{1}{2}] \cong \pi_i^- \mathcal{P}(M) \otimes \mathbb{Z}[\frac{1}{2}]$ *when* $\dim M$ *is odd.*

3. $\pi_i^- (Diff(M \times I \text{ rel } \partial) \otimes \mathbb{Z}[\frac{1}{2}] \cong \pi_0^- Diff(M \times I^{i+1} \text{ rel } \partial) \otimes \mathbb{Z}[\frac{1}{2}]$.

In the special where $M$ is a disk, $\pi_0 Diff(D^{n+i} \text{ rel } \partial)$ is a finite group by [KM63] and [Cer68]. Consequently we get:

**Corollary 6.5.3.**

$$\pi_i^- Diff(D^n \text{ rel } \partial D^n) \otimes \mathbb{Q} = 0$$

*when* $n \geq \max(2i+8, 3i+5)$.

To compute $\pi_i^+ Diff(D^n \text{ rel } \partial D^n) \otimes \mathbb{Q}$ we need to determine how the rational generator of $\pi_{4k-1}\mathcal{P}(*)$, given by Hatcher's example, behaves with respect to the involution $\iota$.

**Lemma 6.5.4.** *When the manifold $M$ is parallelizable then the involution on $\mathcal{C}(M)$ acts as multiplication by $(-1)^{\dim M}$ on the higher FR-torsion $\tau_k(\mathcal{C}(M); \mathcal{F})$ $\in H^{2k}(B\mathcal{C}(M); \mathbb{R})$ with respect to a Hermitian coefficient bundle $\mathcal{F}$ given by any unitary representation $\rho$ of $\pi_1 M$.*

*Proof.* If $f_t$ is a family of framed functions on $M \times I$ then $\overline{f_t}$ will also be framable, at least stably. The lemma now follows from Corollary 5.8.3. □

Since the rational generator of $\pi_{4k-1}\mathcal{P}(*)$ is given by Hatcher's example which is detected by $\tau_{2k}$, Lemma 6.5.4 and Theorem 6.2.5 gives us the old result that

$$\pi_{4k-1}\mathcal{P}(*) \otimes \mathbb{Q} = \pi_{4k-1}^+ \mathcal{P}(*) \otimes \mathbb{Q} = \mathbb{Q}. \tag{6.6}$$

However, a more direct application of Lemma 6.5.4 tells us that the element $\eta_k \overline{\eta_k}$ of $\pi_{4k-1} Diff(D^{2n+1} \text{ rel } \partial)$ obtained from Hatcher's example $\eta_k \in \pi_{4k-1}\mathcal{C}(D^{2n})$ has twice the torsion. Thus:

**Theorem 6.5.5.** *Hatcher's construction gives an element*

$$\eta_k \overline{\eta_k} \in \pi_{4k-1} Diff(D^{2n+1} \text{ rel } \partial D^{2n+1})$$

*with higher FR-torsion $\tau_{2k}(\eta_k \overline{\eta_k}) = \pm a_k b_k \zeta(2k+1)$ times the generator of $H^{4k}(S^{4k}; \mathbb{Z})$ for all $n \geq 4k+3$ where $a_k = 1, 2$ depending on whether $k$ is even or odd and*

$$b_k = denom \frac{B_{2k}}{4k}$$

*is the order of the image of $J_{4k-1} : \pi_{4k-1} O \to \pi_{4k-1}^s(S^0)$, $B_{2k}$ being the $k$-th Bernoulli number.*

**Remark 6.5.6.** (a) Note that $Homeo(D^N \text{ rel } \partial D^N)$ is contractible for all $N$ by the Alexander trick (shrinking the support of $g$ to the center point of $D^N$). Consequently, $\eta_k$ represents an element of $\pi_{4k}$ of the fiber of the natural mapping

$$BDiff(D^{2n+1} \text{ rel } \partial D^{2n+1}) \to BHomeo(D^{2n+1} \text{ rel } \partial D^{2n+1}).$$

(b) It was already known by Waldhausen's theory (6.6) and the elementary involution argument of 6.5.2 that

$$\pi_{4k-1} Diff(D^{2n+1} \text{ rel } \partial D^{2n+1}) \otimes \mathbb{Q} = \mathbb{Q}$$

in the stable range (see [FH78], [Hsi79]) and that the generator corresponds to the rationally nontrivial element of $K_{4k+1}\mathbb{Z}$ which also has rank 1 by Borel and thus the higher FR-torsion, properly defined, must be a nonzero rational multiple of the value of the Riemann zeta function at $2k+1$.

**Proposition 6.5.7.** *Let $M$ be a compact smooth manifold and let $\partial M = \partial_0 M \coprod \partial_1 M$. Then the higher FR-torsion invariants give well defined cohomology classes $\tau_{2k} \in H^{4k}(B;\mathbb{R})$ where*

$$B = fiber(BDiff(M \text{ rel } \partial_0 M) \to BHomeo(M \text{ rel } \partial_0 M)).$$

*Proof.* Let $M \to E \to B$ be the canonical smooth bundle over $B$. Then $E$ contains a subbundle $\partial_0 M \times B$ and there is a fiberwise homeomorphism $\phi : E \to M \times B$. Let $\xi, \xi'$ be the expansion functors on $B$ given by $E$ and $M \times B$. Since these are given by filtrations on the total singular complex of the restrictions of $E, M \times B$ to small simplices in $B$, $\phi$ induces a homotopy equivalence $\phi_* : \xi \to \xi'$ and we can define $\tau_{2k}$ to be the negative of the higher FR-torsion of the fiberwise alternating cone $C'(\phi_*)$ of $\phi_*$:

$$\tau_{2k} = -\tau_{2k}(C'(\phi_*)) \in H^{4k}(B;\mathbb{R})$$

(We take the minus sign since $C'(\phi_*)$ is an extension of the constant functor $\xi'$ by the alternating suspension of $\xi$.) □

**Corollary 6.5.8.** *Let $N$ be a closed manifold of dimension $2n+1 \geq 8k+5$ and let $D \subseteq M$ be any embedded $(2n+1)$-disk. Then the mapping*

$$Diff(D^{2n+1} \text{ rel } \partial D^{2n+1}) \to fiber(Diff(M) \to Homeo(M))$$

*which factors through $Diff(M \text{ rel } (M - intD))$ is rationally nontrivial on $\pi_{4k-1}$.*

*Proof.* We claim that the mapping in real cohomology induced by the mapping

$$BDiff(D^{2n+1} \text{ rel } \partial D^{2n+1}) \to B = fiber(BDiff(M) \to BHomeo(M))$$

preserves the higher FR-torsion invariant $\tau_{2k}$ (whose restriction to $H^{4k}(S^{4k};\mathbb{R})$ is nonzero). To see this note that the canonical bundle $M \to E \to B$ when restricted to $B' = BDiff(M \text{ rel } (M - intD))$ differs from $M \times B'$ only inside $D \times B'$. Consequently, $\xi|B', \xi'|B'$ also differ only inside this set. Following the proof of the suspension theorem we can cancel $\xi|B', \xi'|B'$ outside of $D \times B'$ so

$$C'(\phi_*)|B' \simeq C'(\phi_*|D \times B') \simeq C'(\xi|D \times B').$$

□

## 6.6 FR-torsion for fiber products

In this section we will look at the problem of computing the higher FR-torsion of a fiber product of smooth bundles. We obtain formulas in some special cases and speculate on what the general formula might be.

Suppose that $E \to B$ is a smooth manifold bundle with a closed manifold fiber $M$. Let $\mathcal{F}$ be a flat Hermitian coefficient system on $E$ so that the fiberwise homology bundle $H_*(M;\mathcal{F})$ is constant over $B$. Suppose that $N$ is another compact manifold with $\partial N = \partial_0 N \coprod \partial_1 N$. Then $E \times N$ is a smooth bundle over $B$ with fiber $M \times N$ and $\mathcal{F}$ pulls back to a local coefficient system $\mathcal{F}'$ on $E \times N$ so that

$$H_*(M \times N, M \times \partial_0 N; \mathcal{F}') \cong H_*(M;\mathcal{F}) \otimes H_*(N, \partial_0 N).$$

Consequently the fiberwise homology of $E \times N$ is untwisted and $\tau_k(E \times N; \mathcal{F}')$ is defined.

**Theorem 6.6.1.** *The higher FR-torsion of $E \times N$ is given by*

$$\tau_k(E \times N; \mathcal{F}') = \chi(N, \partial_0 N)\tau_k(E;\mathcal{F})$$

*where $\chi(N, \partial_0 N) = \chi(N) - \chi(\partial_0 N)$ is the relative Euler characteristic of $(N, \partial_0 N)$.*

*Proof.* Choose a fixed Morse function $g : N \to [0, m]$ with distinct integer critical values and with $\partial_0 N = g^{-1}(0)$, $\partial_1 N = g^{-1}(m)$ for some $m \in \mathbb{Z}$. Let $f : E \to (0,1)$ be a fiberwise framed function. Then

$$h(x, y) = f(x) + g(y) \tag{6.7}$$

is a fiberwise framed function on $E \times N$ whose critical values never take integer values and $\{h^{-1}[0, n]\}$ is a filtration of the critical set.

If $y \in N$ is a critical point of $g$ of index $i$ and critical value $n$ then $f^{-1}[n, n+1]$ is a negative $i$-fold suspension of $g$ so the torsion of $h^{-1}[n, n+1]$ is given by

$$\tau_k(h^{-1}[n, n+1]; \mathcal{F}') = (-1)^i \tau_k(E;\mathcal{F}).$$

The theorem now follows from the Splitting Lemma (5.7.8). □

In the next example we consider the fiber product of a lens-space bundle $E_m$ over $B$ and another bundle $M \to B$ having the property that there exists an oriented fiberwise Morse function $g_M$ on $M$ having distinct critical values on every fiber. We call such bundles *Morse bundles*. For example, oriented linear sphere bundles are Morse bundles and the fiber product of any two Morse bundles is a Morse bundle. We note that the fiber $S$ of $M$ need not be an oriented manifold. For example, any trivial bundle $S \times B$ is a Morse bundle.

The cellular chain complex associated to a fiberwise oriented Morse function $g_M$ with distinct critical values is evidently upper triangular. Since upper triangular $\pi_1 B$-modules over any field form a Serre class (Remark 5.7.9) it follows that the fiberwise rational homology bundle is upper triangular. Consequently, it follows from the splitting lemma that the higher FR-torsion given by $g_M$ is trivial. So the higher FR-torsion of $M$ with trivial coefficients is defined and is given by the framing principle:

**Theorem 6.6.2 (Higher FR-torsion for Morse bundles).**

$$\tau_{2k}(M) = \sum_{i\geq 0} \sum_{x\in\Sigma^i(g_M)} (-1)^{i+k}\zeta(2k+1)ch_{4k}(\gamma_x^i)$$

*where $\Sigma^i(g_M)$ is the set of index $i$ critical points of $g_M$ on one fiber of $M$ and $\gamma_x^i$ is the negative eigenspace bundle over $B$ of the critical point set corresponding to $x \in \Sigma^i(g_M)$.*

Suppose that $\gamma$ is a complex vector bundle over $B$ of any dimension. Let $S(\gamma)$ be the associated sphere bundle. For any $m \geq 2$ and any nontrivial $m$-th root of unity $\zeta$ let $E_m = S(\gamma)/(\mathbb{Z}/m)$ with coefficient bundle $\mathcal{F}_\zeta$. Then $\mathcal{F}_\zeta$ pulls back to a one-dimensional coefficient system $\mathcal{F}'_\zeta$ on the fiber product $E_m \oplus M$ with respect to which the fibers (over $B$) are acyclic (by the Künneth formula).

**Theorem 6.6.3.** $\tau_k(E_m \oplus M, \mathcal{F}'_\zeta) = \chi(S)\tau_k(E_m, \mathcal{F}_\zeta)$

$$= -\chi(S)m^k\Re\left(\frac{1}{i^k}\mathcal{L}_{k+1}(\zeta)\right)ch_{2k}(\gamma)$$

*where $\chi(S)$ is the Euler characteristic of the fiber $S$ of the Morse bundle $M$.*

*Proof.* We put together a fiberwise framed function for $E_m$ and the fiberwise Morse function $g_M$ as in (6.7) above. In the proof of the computation of $\tau_k(E/Z_m)$ (Theorem 5.7.12) we saw that it was an iterated extension of circle bundles. We get an $i$-fold twisted suspension of this for every critical point of $g_M$ of index $i$. By Proposition 4.6.5, we get $(-1)^i\tau_k(E/Z_m)$ for each critical point of $g_M$ of index $i$. The theorem follows. □

In the final example we take the fiber product of two Morse bundles $M_1, M_2$ with closed manifold fibers $S_1, S_2$. Since $M_1 \oplus M_2$ is also a Morse bundle the higher FR-torsions of $M_1, M_2, M_1 \oplus M_2$ with trivial coefficients is given by Theorem 6.6.2 and we get the following comparison.

**Corollary 6.6.4.**

$$\tau_{2k}(M_1 \oplus M_2) = \chi(S_1)\tau_{2k}(M_2) + \chi(S_2)\tau_{2k}(M_1).$$

**Remark 6.6.5.** This formula suggests that the higher FR-torsion is the "derivative" of the Euler characteristic of the fiber.

*Proof.* By Theorem 6.6.2 we have

$$
\begin{aligned}
\tau_{2k}(M_1) &= \sum_{(i,x)} (-1)^{i+k}\zeta(2k+1)ch_{4k}(\gamma_x^i) \\
\tau_{2k}(M_2) &= \sum_{(j,y)} (-1)^{j+k}\zeta(2k+1)ch_{4k}(\gamma_y^j) \\
\tau_{2k}(M_1 \oplus M_1) &= \sum_{(i,x)}\sum_{(j,y)} (-1)^{i+j+k}\zeta(2k+1)ch_{4k}(\gamma_x^i \oplus \gamma_y^j) \\
&= \sum_{(j,y)} (-1)^j \sum_{(i,x)} (-1)^{i+k}\zeta(2k+1)ch_{4k}(\gamma_x^i) \\
&+ \sum_{(i,x)} (-1)^i \sum_{(j,y)} (-1)^{j+k}\zeta(2k+1)ch_{4k}(\gamma_y^j) \\
&= \chi(S_2)\tau_{2k}(M_1) + \chi(S_1)\tau_{2k}(M_2).
\end{aligned}
$$

□

Applying this to the case when $M_i$ are even dimensional linear sphere bundles we get the following example.

**Corollary 6.6.6.** *Suppose that $\gamma_1, \cdots, \gamma_n$ are odd dimensional oriented real vector bundles over $B$. Then the higher FR-torsion of the fiber product of their associated sphere bundles $S(\gamma_i)$ is given by*

$$\tau_{2k}(S(\gamma_1) \oplus \cdots \oplus S(\gamma_n)) = 2^{n-1}(-1)^k\zeta(2k+1)ch_{4k}(\gamma_1 \oplus \cdots \oplus \gamma_n).$$

These three examples suggest the following general product formula.

**Conjecture 6.6.7.** Suppose that $\tau_k(E_i, \mathcal{F}_i)$ are defined for $i = 1, 2$. Then

$$\tau_k(E_1 \oplus E_2, \mathcal{F}_1 \otimes \mathcal{F}_2) = \chi(M_1, \mathcal{F}_1)\tau_k(E_2, \mathcal{F}_2) + \chi(M_2, \mathcal{F}_2)\tau_k(E_1, \mathcal{F}_1)$$

where $\chi(M_i, \mathcal{F}_i) = \chi(M_i) \dim \mathcal{F}_i$ is the Euler characteristic of the fiber $M_i$ of $E_i$ with coefficients in $\mathcal{F}_i$ and $E_1 \oplus E_2$ is the fiber product of $E_1$, $E_2$ over $B$.

# Chapter 7

# Proof of Complexified Bökstedt Theorem

This chapter contains the calculation of the higher Franz-Reidemeister torsion invariant for principal $S^1$-bundles. This is the fundamental calculation upon which all the other computations in this book are based either by the transfer argument of Chapter II or the framing principle of 6.1.1. Only the statement of the theorem (in §7.10) is needed. The rest of the chapter can safely be skipped. This chapter logically comes after Chapter II and before Chapter V and is independent of Chapters III and IV. An outline of the argument goes as follows.

## Outline of the argument

We begin in Section 7.1 by constructing what we call systems of local sections for principal $S^1$-bundles $E \to B$ which we assume are smooth without loss of generality. For each $b \in B$ this gives a finite subset $\Sigma_b$ of the fiber over $b$ which varies smoothly with $b$.

The points in $\Sigma_b$ are cyclically ordered and separated by positive distances $\alpha_j$ adding up to $2\pi$. In Section 7.2 we take a fixed positive number $c$ $(0 < c < 1)$ and define the *continuous incidence matrix* $G_b$ to be

$$G_b = \begin{pmatrix} 1 & & & & -e^{ic\alpha_n} \\ -e^{ic\alpha_1} & 1 & & & \\ & -e^{ic\alpha_2} & 1 & & \\ & & \cdots & 1 & \\ & & & -e^{ic\alpha_{n-1}} & 1 \end{pmatrix}$$

where $n$ is the cardinality of $\Sigma_b$.

The continuous incidence matrices are conjugate to the discrete incidence matrices $g_b$:

$$G_b = D_b(x) g_b(x) D_b(x)^{-1}$$

where $x = e^{ic/m}$, $g_b(x)$ is a matrix whose entries are integer polynomials in $x$ and $D_b(x)$ is a diagonal matrix with diagonal entries being powers of $x$ with variable exponents.

Using the continuous incidence matrices we obtain a new version of the integral invariant in Section 7.3. The new version basically replaces the contraction $h^u$ of $h = g_b(x)g_b(x)^*$ with a new contraction

$$h_r = (D_b(r)g_b(rx)D_b(r)^{-1})(D_b(r)g_b(rx)D_b(r)^{-1})^*$$

of $h_1 = h$. It follows from Stokes that over any $2k$-cycle in the base $B$ the integral of $\mathrm{Tr}((h^{-u}dh^u)^{2k+1})$ is equal to the integral of $\mathrm{Tr}((h_r^{-1}dh_r)^{2k+1})$.

Since the invariant $\tau_k(\lambda)$ is a natural cohomology class of degree $2k$ it suffices to compute it for any $U(1)$-bundle $\lambda$ with $c_1(\lambda)^k \neq 0$. We may therefore restrict our attention to the $S^1$-bundle $\gamma^{\otimes k}$ over $(S^2)^k$. We begin in Section 7.4 by constructing an explicit system of local sections for $\gamma^{\otimes k}$.

In Section 7.5 we show that it is enough to examine what happens over a finite set of isolated points in $(S^2)^k$. With respect to local coordinates $(t_1, \cdots, t_k, s_1, \cdots, s_k)$ what is happening at these points is:

1. The number of local sections is equal to one plus the number of nonnegative coordinates $t_i$. (Labelling the local sections $0, 1, 2, \cdots, k$ the $i$-th section is defined only for $t_i \geq 0$.)

2. As $s_i$ increases the $i$-th section moves vertically with respect to the 0-th section.

In Section 7.6 we write down a family of matrices $K(t, s, r, x)$ which smoothes out the discontinuity of the continuous incidence matrix as the coordinates $t_i$ change sign. This reduces the problem to computing an explicit integral of the form

$$\int_{(t,s,r)\in I^{2k}\times I} \mathrm{Tr}((K^{-1}dK + (K^{-1}dK)^*)^{2k+1})$$

In Section 7.7 we present some elementary formulas expressing the terms in the above integral as linear combinations of polylogarithms paying particular attention to the $(k+1)$-polylogarithm component.

In Section 7.8 we use these formulas to show that the $(k+1)$-polylogarithm component of the integral above is equal to that of another integral of the form

$$\int_{I^{2k}\times I} \mathrm{Tr}(T^{2k+1} + (T^{2k+1})^*)$$

where $T$ is a matrix 1-form on $I^{2k} \times I$ with only one term in each entry.

In Section 7.9 of this chapter we formulate and solve an elementary combinatorial problem (counting certain types of necklaces) and show that the answer can be expressed as the integral of $\mathrm{Tr}(T^{2k+1})$.

We finish in Section 7.10 with a final statement of the theorem.

## 7.1 Systems of local sections for $S^1$-bundles

Suppose that $p : E \to B$ is an oriented principal $U(1)$-bundle. Then by a *system of sections* for $E$ we mean a locally finite covering of the base $B$ by closed sets

$C_\alpha$ whose interiors cover $B$ together with sections $s_\alpha : C_\alpha \to E$ of $p$ whose images $s_\alpha(C_\alpha)$ are disjoint. The union $\Sigma$ of these images will be called the *image set*. Since $\{C_\alpha\}$ is locally finite, the intersection $\Sigma_b$ of $\Sigma$ with each fiber $F_b = p^{-1}(b)$ will be a finite set.

A closed subset $\Sigma$ of $E$ is defined to be the *image set* of a *system of local sections* if it locally looks likes like the image set of a system of sections. In other words, there is a covering of $B$ by opens sets $U$ so that $\Sigma \cap p^{-1}(U)$ is the image set of a system of sections on $U$. In particular this means that the projection $p : \Sigma \to B$ is locally a closed embedding. Note that the domains $C_\alpha$ of the local sections are only locally defined. (These domains are "immersed" subsets of $B$.)

If $p : E \to B$ is a smooth principal $U(1)$-bundle over a smooth manifold $B$ then a system of local sections for $E$ is called *smooth* if the image set $\Sigma$ is a smooth codimension-one closed submanifold of $E$ with boundary which is *horizontal* in the sense that $p : \Sigma \to B$ is an immersion. Note that $p(\partial\Sigma)$ is a two-sided codimension-one immersed submanifold of $B$. We may view $\Sigma$ as an immersed submanifold of $B$ together with a smooth section in $E$ which separates overlapping segments of $\Sigma$.

**Proposition 7.1.1.** *Any principal $U(1)$-bundle over a paracompact space admits a system of local sections. In the smooth case the system can be chosen to be smooth.*

*Proof.* Take a fiberwise positive generalized Morse function $f : E \to \mathbb{R}$ and let $\Sigma \subseteq E$ be the set of local minima. In the smooth case we need to perturb this manifold along its boundary so that it is not vertical. □

## 7.2 Smoothly varying incidence matrices

Suppose that we have a smooth principal $U(1)$-bundle $p : E \to B$ together with a smooth system of local sections with image set $\Sigma$. Over each $b \in B$ this gives a finite subset $\Sigma_b$ of $F_b$. We label the elements $w_0, w_1, \cdots, w_n$ in reverse cyclic order. We imagine that each $w_j$ is a pair of critical points of $x_j^+, x_j^-$ of a Morse function $f_b : F_b \to \mathbb{R}$. (Fig. 7.1)

When $b$ reaches the boundary $\partial C_\alpha$ of one of the locally defined domains $C_\alpha$ then two critical points $x_j^+, x_j^-$ represented by the point $s_\alpha(b) = w_j \in \partial\Sigma_b$ cancel at a birth-death point and disappear.

Let $\theta_j$ be the distance from $w_{j-1}$ to $w_j$ ($\theta_0$ is the distance from $w_n$ to $w_0$). Then $\theta_j > 0$ and $\sum \theta_j = 2\pi$. We choose a fixed positive real number $c < 1$ and we define the *preliminary version* of the *smooth incidence matrix* at $b \in B$ to be

$$G(b) = \begin{pmatrix} 1 & & & & -e^{ic\theta_0} \\ -e^{ic\theta_1} & 1 & & & \\ & -e^{ic\theta_2} & 1 & & \\ & & \cdots & 1 & \\ & & & -e^{ic\theta_n} & 1 \end{pmatrix}$$

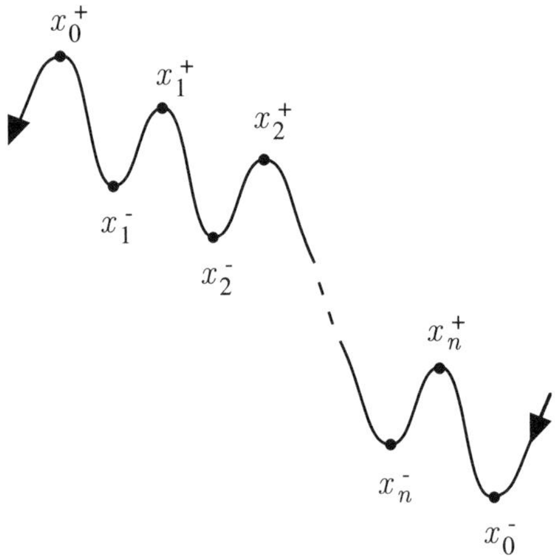


Figure 7.1: Pairs of critical points $x_j^+, x_j^-$ of a Morse function on a circle

This can also be written as

$$G(b) = I - DP$$

where $P$ is a cyclic permutation matrix and $D$ is the diagonal matrix

$$D = diag(e^{ic\theta_0}, \cdots, e^{ic\theta_n}).$$

The determinant of $G(b)$ is

$$\det G(b) = 1 - e^{2\pi ic} \neq 0$$

since $c$ is not an integer. We note that $G(b)$ is only well-defined up to conjugation by some power of the permutation matrix $P$.

The matrix $G(b)$ is a conjugate of the more traditional locally constant incidence matrix $g_b(x)$ given as follows. First fix some large positive integer $N$ and let $x = e^{2\pi ic/N}$. Then $\det G(b) = 1 - x^N$ and $e^{ic\theta_j} = x^{a_j}$ where $a_j = N\theta_j/2\pi$. Choose positive integers $c_0, c_1, \cdots, c_n$ with sum $\sum c_j = N$. (We imagine that $c_j$ is close to $a_j$ although this is not necessary.)

Let

$$g(x) = I - diag(x^{c_0}, x^{c_1}, \cdots, x^{c_n})P$$

$$D_b(x) = diag(1, x^{a_1-c_1}, x^{a_1+a_2-c_1-c_2}, \cdots)$$

Then

$$G(b) = D_b(x)g(x)D_b(x)^{-1}$$

The matrix $G(b)$ varies smoothly with $b \in B$ and $g(x)$ is independent of $b$ as long as $b$ stays away from $p\partial\Sigma$. When $b$ goes through $p\partial\Sigma$ both $G(b)$ and $g(x)$ change discontinuously. To fix this we will modify $G(b)$ and $g(x)$ is a small collar neighborhood of $\partial C_\alpha$ for each $\alpha$.

Let $\psi : \Sigma \to [0,1]$ be a smooth function. Then locally, i.e., over small open sets $U$ in $B$, $\psi$ gives smooth functions $C_\alpha \to [0,1]$ which we assume extend to smooth functions $\psi_\alpha : U \to [0,1]$ with support in $C_\alpha$. We also assume that $U$ is covered by the interiors of the closed sets $\psi^{-1}(1)$ and that $\psi^{-1}(0,1]$ is the interior of $C_\alpha$. If $w_j = s_{\alpha_j}(b)$ then let $t_j = \psi_{\alpha_j}(b)$. We assume that the indices are chosen so that $t_0 = 1$ over all of $U$. (If not then we cyclically permute the indices and/or make $U$ smaller.)

We define the *final version* of the smooth incidence matrix to be the matrix $F_b(x)$ with entries given by:

$$\begin{array}{ll} F_b(x)_{ji} = \delta_{ji} & \text{if} \quad 1 \le j \le i \le n \\ F_b(x)_{ji} = -x^{a_{i+1}+a_{i+2}+\cdots+a_j} t_i u_{i+1} \cdots u_{j-1} t_j & \text{if} \quad i < j \\ F_b(x)_{0i} = \delta_{0i} - x^{a_{i+1}+\cdots+a_n+a_0} t_i u_{i+1} \cdots u_n & \end{array}$$

where $\delta_{ji}$ is the Kronecker delta function and $u_i = 1 - t_i^2$. For example, when $n = 3$ we have:

$$F_b(x) = \begin{pmatrix} \begin{array}{c} 1 - x^{a_1+a_2+a_3+a_0} \\ t_0 u_1 u_2 u_3 \end{array} & -x^{a_2+a_3+a_0} t_1 u_2 u_3 & -x^{a_3+a_0} t_2 u_3 & -x^{a_0} t_3 \\ -x^{a_1} t_0 t_1 & 1 & 0 & 0 \\ -x^{a_1+a_2} t_0 u_1 t_2 & -x^{a_2} t_1 t_2 & 1 & 0 \\ -x^{a_1+a_2+a_3} t_0 u_1 u_2 t_3 & -x^{a_2+a_3} t_1 u_2 t_3 & -x^{a_3} t_2 t_3 & 1 \end{pmatrix}$$

In particular this means that

$$F_b(x)_{ji} = 0 \quad \text{if} \quad 1 \le j < i \le n$$

and, in general, all other entries of $F_b(x)$ are nonzero.

Let $f_t(x)$ be defined similarly with $a_i$ replaced by $c_i$ in the definition of $F_b(x)$:

$$f_t(x)_{ji} = \delta_{ji} - x^{c_{i+1}+c_{i+2}+\cdots+c_j} t_i u_{i+1} \cdots u_{j-1} t_j$$

Note that $f_t(x)$ depends only on $(x,t)$ since the $c_i$'s are constant (whereas each $a_i$ is a function of $b$). Also:

$$F_b(x) = D_b(x) f_t(x) D_b(x)^{-1}$$

where $D_b(x)$ is as before.

**Proposition 7.2.1.** *The families of incidence matrices $f_t(x), F_b(x)$ satisfy the following.*

1. *The entries of $f_t(x)$ are polynomials in $x, t_1, \cdots, t_n$.*

2. *$f_t(0) = I$.*

3. *$f_t(x)$ is well defined up to conjugation by a unitary matrix independent of $t$.*

4. *$\{f_t(x)\}$ is an admissible family of incidence matrices determining the higher FR-torsion invariant of the bundle.*

5. *The entries of $F_b(x)$ are smooth functions on any sufficiently small open set $U$ in $B$ if we stabilize the matrix $F_b(x)$ to*

$$\begin{pmatrix} F_b(x) & 0 \\ 0 & I \end{pmatrix}$$

*so that all entries are defined over all of $U$.*

6. *$F_b(x)$ is well-defined up to conjugation by a power of the cyclic permutation matrix $P$ (assuming the functions $\psi_\alpha$ are fixed).*

*Proof.* (1),(2) are obvious. (5) is also easy.

To prove (3) we note that if the integers $c_j$ are changed to $b_j$ then $f_t(x)$ is conjugated by the diagonal matrix

$$C(x) = diag(1, x^{b_1-c_1}, x^{b_1+b_2-c_1-c_2}, \cdots)$$

which is unitary since $|x| = 1$. Since the real numbers $a_i$ are canonically determined, $F_t(x)$ is more well-defined.

To prove (4) we note that $f_t(x)$ differs from $g_b(x)$ by allowable elementary operations:

$$\begin{aligned}
&\prod_{j=1}^{n} E_{0j}(t_j x^{c_{j+1}+c_{j+2}+\cdots+c_n+c_0}) f_t(x) \prod_{j>i} E_{ji}(t_i t_j x^{c_{i+1}+\cdots+c_j}) \\
&= diag(1-x^N, 1, \cdots, 1) \\
&= \prod_{j=1}^{n} E_{0j}(x^{c_{j+1}+c_{j+2}+\cdots+c_n+c_0}) g_b(x) \prod_{j>i} E_{ji}(x^{c_{i+1}+\cdots+c_j})
\end{aligned}$$

where the elementary operations $E_{ji}$ are to be placed in lexicographic order according to $(i,j)$. We indicated the generic case when $t_1, \cdots, t_n$ are all less than 1. In the general case we only take the elementary column operations $E_{ji}$ where $i < j$ and $u_{i+1}u_{i+2}\cdots u_j \neq 0$ (i.e., $t_{i+1} \neq 1, t_{i+2} \neq 1, \cdots, t_j \neq 1$) and we take elementary row operations

$$E_{kj}(t_j x^{c_{j+1}+\cdots+c_k})$$

when $1 \leq j < k \leq n+1$ (subscripts are modulo $n+1$), $t_k = 1$ and $u_j u_{j+1} \cdots u_{k-1} \neq 0$. These are admissible row and column operations because when $t_j \neq 1$ the pair of critical points at $w_j$ are cancelling and thus $x_j^+ > x_k^+$ allowing the column operations $E_{kj}$ when $u_{j+1} \cdots u_k \neq 0$ and $x_j^- > x_k^-$ allowing the row operations $E_{kj}$ when $u_j \cdots u_{k-1} \neq 0$.

Since the row and column operation matrices have polynomial entries, the correction terms $X, Y, U, V, W$ are polynomial and thus do not contribute to the polylogarithm invariant. Thus $f_t(x)$ determines the invariant. □

## 7.3 Another version of the integral invariant

The smooth family of incidence matrices $F_b(x)$ is canonical (given the functions $\psi_\alpha$) but the decomposition

$$F_b(x) = D_b(x) f_t D_b(x)^{-1}$$

depends on the choices of the integers $c_0, c_1, \cdots, c_m$ approximating $a_1, \cdots, a_m$. We let $h_r(f_t, D_b), 0 \leq r \leq 1$, be the family of positive definite Hermitian matrices given by

$$h_r(f_t, D_b) = D_b(r) f_t(rx) D_b(r)^{-2} f_t(rx)^* D_b(r)$$

This is polynomial in $r$ but not in $x$ since $f_t(rx)^*$ is polynomial in $r\bar{x} = rx^{-1}$.

**Lemma 7.3.1.** *$h_r(f_t, D_b)$ is well-defined up to conjugation by a constant unitary matrix. (By* constant *we mean independent of $t, b, r$.)*

*Proof.* If the integers $c_j$ are changed to $b_j$ then $f_t(x)$ changes to

$$C(x) f_t(x) C(x)^{-1}$$

and $D_b(x)$ changes to $D_b(x) C(x)^{-1}$ where $C(x)$ is the unitary matrix:

$$C(x) = diag(1, x^{b_1 - c_1}, x^{b_1 + b_2 - c_1 - c_2}, \cdots)$$

Then, since $C(rx) = C(r)C(x)$ and $C(x), D_b(x)$ commute (being diagonal matrices) we have:

$$h_r(Cf_tC^{-1}, D_bC^{-1}) = D_b(r)C(r)^{-1}C(rx)f_t(rx)C(rx)^{-1}C(r)D_b(r)^{-1}[same]^*$$

$$= C(x)D_b(r)f_t(rx)D_b(r)^{-1}C(x)^{-1}[C(x)D_b(r)f_t(rx)D_b(r)^{-1}C(x)^{-1}]^*$$

$$= C(x)h_r(f_t, D_b)C(x)^{-1}$$

□

**Theorem 7.3.2.** *The $2k$-cochains defined on a smooth $2k$-simplex $\sigma$ in $B$ by*

$$\int_{\sigma \times I} \mathrm{Tr}((h^{-u} dh^u)^{2k+1}), \tag{7.1}$$

$$\int_{\sigma \times I} \mathrm{Tr}((h_r(f_t, D_b)^{-1} dh_r(f_t, D_b))^{2k+1}), \tag{7.2}$$

$$\int_{\sigma \times I} \mathrm{Tr}((K^{-1}dK + (K^{-1}dK)^*)^{2k+1}) \tag{7.3}$$

*where $h = f_t(x) f_t(x)^*, K = D_b(r) f_t(rx) D_b(r)^{-1}$ differ by coboundaries.*

*Proof.* By Stokes the difference between (1) and (2) is the alternating sum over the $2k-1$ faces $\tau_i = \partial_i \sigma$ of $\sigma$:

$$\sum_{i=0}^{2k}(-1)^i \int_{(b,r,u)\in\tau_i\times I^2} \operatorname{Tr}((h_r(f_t, D_b)^{-u} dh_r(f_t, D_b))^{2k+1})$$

By Lemma 7.3.1 the integrands are well-defined.

To see that (2) = (3) note that

$$h_r(f_t, D_b) = KK^*$$

so

$$h_r(f_t, D_b)^{-1} dh_r(f_t, D_b) = (K^*)^{-1} K^{-1} dKK^* + (K^*)^{-1} dK^*$$

which is conjugate to $K^{-1}dK + (K^{-1}dK)^*$. □

**Corollary 7.3.3.** *The higher FR-torsion invariant $\tau_k(E) \in H^{2k}(B;\mathbb{R})$ is represented by the cohomology class of the $2k$-cocycle on $B$ given by the $(k+1)$-polylogarithm component of any integral in Theorem 7.3.2 divided by* $(2k+1)!2i^k$.

**Remark 7.3.4.** Our method of projecting to the $(k+1)$-polylogarithm component is explained in §7.7.

*Proof.* As we observed in the proof of Proposition 7.2.1, the correction terms are polynomial and thus do not contribute to the $(k+1)$-polylogarithm component. □

## 7.4 Explicit local sections for the canonical $S^1$-bundle over $(S^2)^k$

Let $\gamma$ denote the canonical circle bundle over $S^2 = \mathbb{C}P^1$ with total space

$$E(\gamma) = \{(L, v) : v \in L, L \in \mathbb{C}P^1\}$$

Then we will construct a system of disjoint local sections for the $k$-th exterior tensor power $\gamma \otimes \gamma \otimes \cdots \otimes \gamma$ over $(S^2)^k$. The fiber over each point $b = (b_1, \cdots, b_k) \in (S^2)^k$ will be the balanced product

$$F_b = S_{b_1} \otimes S_{b_2} \otimes \cdots \otimes S_{b_k} = S_{b_1} \times \cdots \times S_{b_k} / \sim$$

where $(ta_1, \cdots, a_k) \sim (a_1, ta_2, \cdots, a_k) \sim \cdots \sim (ta_1, \cdots, ta_k)$ for every unit complex number $t \in S^1$ and $a_i \in S_{b_i}$.

We will construct an explicit system of local sections for $\gamma$ giving finite subsets $C_L$ of the fiber $S_L$ of $E(\gamma)$ over each $L \in \mathbb{C}P^1 = S^2$. Then the system of local sections for $\gamma^{\otimes k}$ will be given by finite subsets

$$\Sigma_b \subseteq C_{b_1} \otimes C_{b_2} \otimes \cdots \otimes C_{b_k}$$

for each $b \in (S^2)^k$.

(a)

(b)

(c)

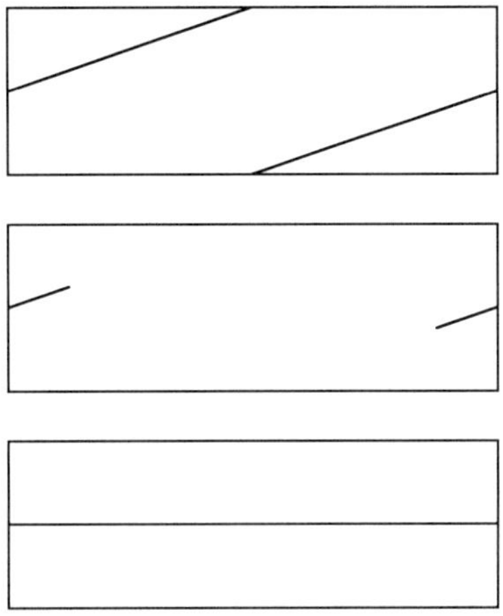

Figure 7.2: Image of $s_\infty$

**Lemma 7.4.1.** *Suppose that $m, c$ are positive integers so that*

$$c < 2^m.$$

*Then there exists a system of local section $s_\alpha$ for $\gamma$ with the property that for any two sections $s_\alpha, s_\beta$ defined in a neighborhood of the same point $b$ either*

1. *$s_\beta(b) = s_\alpha(b)e^{\pm 2\pi/2^j}$ for some $1 \leq j \leq m$ or*
2. *$s_\beta(b) = e^{\pm 2\pi i t}s_\alpha(b)$ for some $c/2^m \leq t \leq (c+1)/2^m$*

*Proof.* Elements of $\mathbb{C}P^1$ will be denoted $L = [x, y]$ where $(x, y) \in \mathbb{C}^2 - 0$ is any nonzero element of $L$. Then a local section $s_\infty$ over a neighborhood of $\infty = [1, 0]$ in $S^2 = \mathbb{C}P^1$ is given by

$$s_\infty[1, z] = N(1, z)$$

where $N(x) = x/\|x\|$. Then

$$s_\infty[re^{i\theta}, 1] = N(r, e^{-i\theta})$$

In other words, as $z = re^{i\theta}$ goes counterclockwise around $0, s_\infty(z)$ goes clockwise around the fiber. The image of this section is given in Fig. 7.2.

Here the vertical interval denotes the fiber $S^1$ normalized to length 1. The horizontal line in (c) and the diagonal lines in (a), (b) represent the image of the section $s_\infty$. We need to fill in the middle of (b) to deform (a) into (c).

Step 1. Starting at (c) we introduce $2^m$ local sections each having a distance of $c/2^m$ from $s_\infty$ ($2\pi c/2^m$ on the circle).

Step 2. Then we perform the following four moves on each of these $2^m$ locations:

1. A gap is created in $s_\infty$.

(c′)

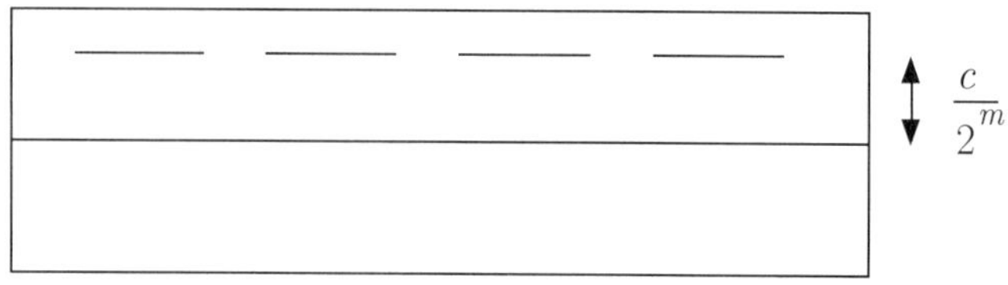


Figure 7.3: Step 1

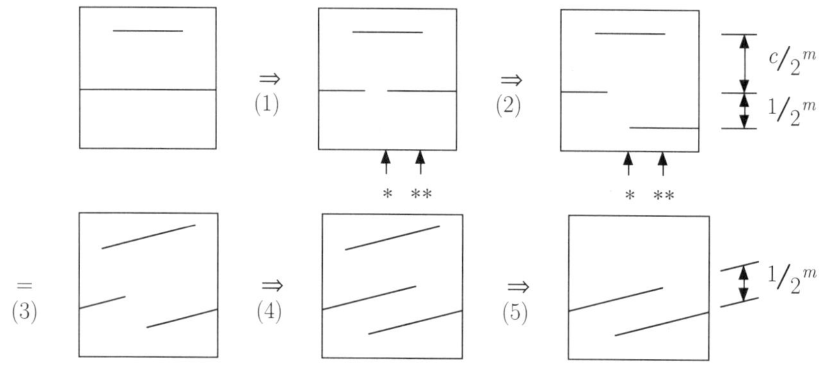


Figure 7.4: Four local moves

2. The right hand segment is pushed $1/2^m$ units down ($2\pi/2^m$ counterclockwise on the fiber).
3. We redraw the third square by tilting the lines so that the left hand edge matched the right.
4. The two segments are pushed together. ($s_\infty$ is now two local sections.)
5. The additional section is removed.

Since there are $2^m$ gaps of width $1/2^m$ the slope of the lines are all equal to 1 as in Fig. 7.5.

Step 3. Now modify this by doing the local move on every other gap indicated in Figure 7.6. The result looks just like before (Fig. 7.5) except that there are now $2^{m-1}$ gaps of width $2^{1-m}$. Repeat this process until there are only two gaps of width $1/2$.

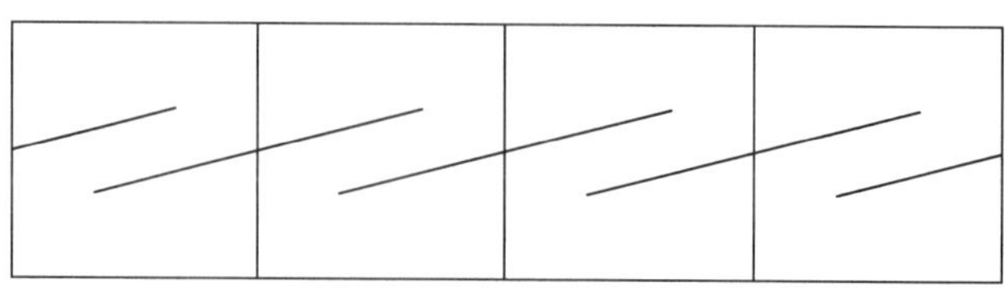

Figure 7.5: Result of Step 2

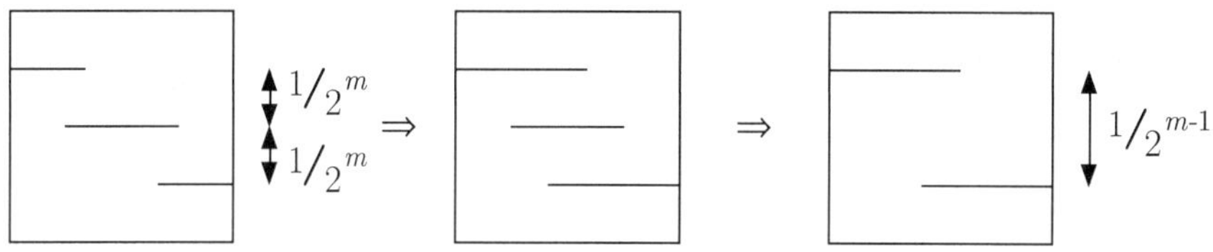


Figure 7.6: Step 3 at each gap

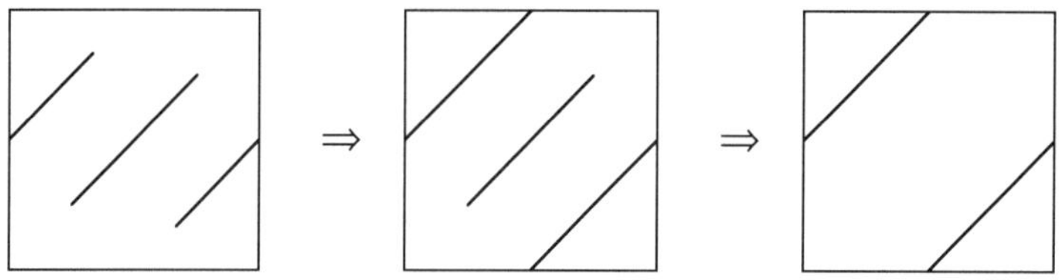

Figure 7.7: The final step (4)

Step 4. The final step is given in Figure 7.7. We note that only during the second stage of Step 2 does the vertical separation between sections change. (It changes from $c/2^m$ to $(c+1)/2^m$.) At all other times the vertical separations remain constantly equal to a negative power of 2 or $c/2^m$ or $(c+1)/2^m$. $\square$

The next step is to choose positive integers $m, c_1, c_2, \cdots, c_k$ so that if the construction above is carried out using $m, c_i$ over the $i$-th copy of the canonical circle bundle over $S^2$ then the resulting balance product

$$C_{b_1} \otimes \cdots \otimes C_{b_k} \subseteq F_{b_1} \otimes \cdots \otimes F_{b_k}$$

behaves nicely in the sense that the distance between any two distinct elements is bounded below. This is both necessary and sufficient in order for $C_{b_1} \otimes \cdots \otimes C_{b_k}$ to be the image over $b = (b_1, \cdots, b_k)$ of a system of local sections over $(S^2)^k$.

If we compare the coordinates of two points $A, B$ in $C_{b_1} \otimes \cdots \otimes C_{b_k}$ we see that the distance from $A$ to $B$ divided by $2\pi$ is equal to

$$\sum_{i=1}^{k} \epsilon_i a_i 2^{-m} - \sum_{j=1}^{k} \epsilon_j' 2^{n_j}$$

where $a_i \in [c_i, c_{i+1}]$ are real numbers, $n_j$ are positive integers $\leq m$ and the coefficients $\epsilon_i, \epsilon_j'$ are $\pm 1$ or 0. Multiply this by $2^m$ and we see that the condition we need is that $\sum \epsilon_i a_i$ be bounded away from any integer which is a linear combination of $k$ powers of 2 with coefficients $\pm 1$ and 0.

Since

$$\left| \sum \epsilon_i a_i - \sum \epsilon_i c_i \right| \leq k$$

it suffices to choose the integers $c_i$ so that $\sum \epsilon_i c_i$ differs by at least $k+1$ from any integer which is a linear combination of $k$ powers of 2 with coefficients $\pm 1$ and 0.

**Lemma 7.4.2.** *Let $n$ be an even integer which is a linear combination of $k$ powers of 2 with coefficients $\pm 1$ and 0. Then the pair "10" occurs at most $k$ times in the binary expansion of $n$.*

*Proof.* By induction on $k$. □

As an example the integer

$$4 + 4^2 + \cdots + 4^{k+1} = 4(4^{k+1} - 1)/3$$

cannot be expressed as a linear combination of $k$ powers of 2 with coefficients $\pm 1$ and 0. Thus a solution of our problem is given by taking

$$m = 2(k+1)^2$$

$$c_i = 4^{i(k+1)}(4^{k+1} - 1)/3$$

## 7.5 Reduction to isolated regions

As $b_i \in S_i^2$ moves around in $S_i^2$ there are two things that happen to the section image sets $\Sigma_{b_i} \subseteq F_{b_i}$ ignoring rotations.

1. Elements of $\Sigma_{b_i}$ appear (or disappear). The corresponding parameter $t_i = \psi_*(b_i)$ goes from 0 to 1 (or 1 to 0).

2. $\Sigma_{b_i}$ consists of two points and the distance between these two points is given by $(c_i + s_i)/2^m$ where $s_i$ goes from 0 to 1.

There are exactly $2^{m+1}$ disjoint rectangles in $S_i^2$ where both of these happen simultaneously. Local coordinates for these rectangles are given by $(t_i, s_i)$.

**Lemma 7.5.1.** *The $(k+1)$-polylogarithm component of the integrals in Theorem 7.3.2 are equal to zero except at the points $b = (b_1, \cdots, b_k) \in (S^2)^k$ where each $b_i$ lies in one of the rectangles described above.*

*Proof.* We assume first that each $b_i$ lies in one of its corresponding rectangles. We look at $K^{-1}dK$ where

$$K = D_b(r) f_t(rx) D_b(r)^{-1}$$

Since $D_b(r)$ depends only on $r$ and the local coordinates $s_1, \cdots, s_k$ we write $D_b(r) = D_s(r)$ where $s = (s_1, \cdots, s_k) \in I^k$. Thus:

$$K^{-1}dK = D_s(r) f_t(rx)^{-1} D_s(r)^{-1} dD_s(r) f_t(rx) D_s(r)^{-1}$$

$$+ D_s(r) f_t(rx)^{-1} df_t(rx)^{-1} D_s(r)^{-1} - D_s(r)^{-1} dD_s(r)$$

The entries of $f_t(rx)$ are polynomials in $t_1, \cdots, t_k, r, x$ with

$$\det f_t(rx) = 1 - r^N x^N$$

where $N = 2^m$. Consequently, the entries of $f_t(rx)^{-1}$ are rational functions of $t_1, \cdots, t_k, r, x$ with denominator $1 - r^N x^N$.

$D_s(r)$ is a diagonal matrix with diagonal entries $r^{u_i}$ where each $u_i$ is a constant plus a real linear combination of the $s_j$'s:

$$u_i = w_i + \sum_j a_{ij} s_j$$

By Lemma 7.3.1 we may assume that each $w_i = 0$. Therefore $D_s(r)^{-1} dD_s(r)$ is a diagonal matrix with diagonal entries

$$u_i r^{-1} dr + \ln r \sum_j a_{ij} ds_j \tag{7.4}$$

We will show later that the term $u_i^{-1} ds$ can be eliminated.

The $pq$-entry of $D_s(r) f_t(rx)^{-1} df_t(rx) D_s(r)^{-1}$ has terms of the form:

$$\frac{r^{u_p - u_q} \text{polynomial}(t, r, x)}{1 - r^N x^N} (dr \quad \text{or} \quad dt_i) \tag{7.5}$$

The $pq$-entry of $D_s(r) f_t(rx)^{-1} D_s(r)^{-1} dD_s(r) f_t(rx) D_s(r)^{-1}$ has terms of the form:

$$\frac{r^{u_p - u_q} \text{polynomial}(t, r, x)}{1 - r^N x^N} (u_i r^{-1} dr \quad \text{or} \quad a_{ij} \ln r ds_j) \tag{7.6}$$

The terms with $u_i r^{-1} dr$ in (7.4) and (7.6) will be eliminated from $K^{-1} dK$ using the following lemma.

**Lemma 7.5.2 (Elementary Operations).** *Suppose that we have a sequence of $n \times n$ matrices*

$$A_0, A_1, \cdots, A_{2k}.$$

*Then the quantity*

$$\sum_{\sigma \in S_{2k+1}} sgn(\sigma) \operatorname{Tr}((A_{\sigma(0)} + A^*_{\sigma(0)}) \cdots (A_{\sigma(2k)} + A^*_{\sigma(2k)})) \tag{7.7}$$

*does not change if we replace $A_j$ with $A_j + tA_i$ where $i \neq j$ and $t \in \mathbb{R}$.*

*Proof.* If $A_j + A_j^*$ is replaced with

$$A_j + A_j^* + t(A_i + A_i^*)$$

then (7.7) changes by

$$t \sum_\sigma sgn(\sigma) \operatorname{Tr}(B_{\sigma(0)} \cdots B_{\sigma(2k)}) = 0$$

where $B_p = A_p + A_p^*$ for $p \neq j$ and $B_j = B_i$ □

To apply this to the case at hand we define $A_0, \cdots, A_{2k}$ by

$$K^{-1}dK = A_0 dr + \sum A_i ds_i + \sum A_{k+j} dt_j$$

Thus $A_0 = K^{-1}\partial K/\partial r$, etc. Then

$$\mathrm{Tr}(\omega^{2k+1}) = \mathrm{Tr}((K^{-1}dK + (K^{-1}dK)^*)^{2k+1}) =$$

$$\sum_{\sigma \in S_{2k+1}} sgn(\sigma)\,\mathrm{Tr}((A_{\sigma(0)} + A^*_{\sigma(0)}) \cdots (A_{\sigma(2k)} + A^*_{\sigma(2k)}))dr dt_1 \cdots dt_k ds_1 \cdots ds_k$$

By elementary operations (Lemma 7.5.2) this does not change if we replace $A_0$ with

$$A_0 - \sum_{j=1}^{k} \frac{s_j}{r \ln r} A_j$$

since $u_i = \sum a_{ij} s_j$ the effect of this change is to eliminate the $u_i r^{-1} dr$ terms in (7.4) and (7.6). The terms of $\mathrm{Tr}(\omega^{2k+1})$ therefore have the form:

$$\frac{r^{\sum \pm u_i} (\ln r)^k \mathrm{polynomial}(t, r, x, \bar{x})}{(1 - r^N x^N)^\alpha (1 - r^N \bar{x}^N)^\beta} \tag{7.8}$$

times $dr dt_1 \cdots dt_k ds_1 \cdots ds_k$ where $\alpha + \beta = 2k$. Then integral of this over $I^{2k+1}$ gives a linear combination over the field $\mathbb{C}(x)$ of the $(k+1)$-polylogarithms of $\zeta^j x$ and $\zeta^j \bar{x}$ plus lower terms where $\zeta$ is a primitive $N$-th root of unity. (See §7.7 on polylogarithms.)

To prove the lemma suppose that one of the coordinates $b_i$ does not lie in an $(t_i, s_i)$ rectangle. Then in order for $\mathrm{Tr}(\omega^{2k+1})$ to be nonzero we need two variable $t$-coordinates for $b_i$, i.e., there must be two elements of $\Sigma_{b_i}$ which are appearing or disappearing. Then everything works as before except that at the end, instead of terms like (7.8), we get:

$$\frac{r^{\sum \pm u_i} (\ln r)^{k-1} \mathrm{polynomial}(t, r, x, \bar{x})}{(1 - r^N x^N)^\alpha (1 - r^N \bar{x}^N)^\beta} \tag{7.9}$$

times $dr dt_1 \cdots dt_{k+1} ds_1 \cdots ds_{k-1}$. The integral of this is a $\mathbb{C}(x)$-linear combination of $k$-polylogarithms of $\zeta^j x$ and lower terms.

If more than one $b_i$-coordinate fails to be in an $(t_i, s_i)$-rectangle then we get even lower terms. □

We will now reparametrize the $(t_i, s_i)$ rectangles in $S_i^2$ as indicated in Fig. 7.8. Note that this includes two of the previous rectangles which occur at $t_i = \frac{1}{k+1}$ and $t_i = \frac{k}{k+1}$.

Since $\Sigma_{b_i}$ contains two elements (at $t_i = \frac{1}{2}$).

$$\Sigma_{(t,s)} = \Sigma_b = \Sigma_{b_1} \otimes \cdots \otimes \Sigma_{b_k}$$

contains $2^k$ elements at $t = (\frac{1}{2}, \cdots, \frac{1}{2})$. If $\Sigma = \cup \Sigma_{t,s} \times \{t\}$ then $\Sigma$ is a union of $2^k$ sheets which are images of $2^k$ local sections. If we label these sheets with

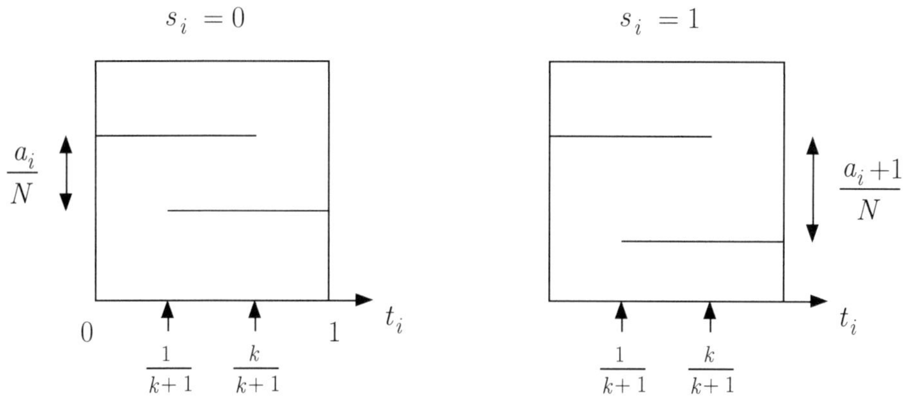


Figure 7.8: $(t_i, s_i)$ rectangle in $S_i^2$

multi-indices $\lambda = (\lambda_1, \cdots, \lambda_k)$ where $\lambda_i = 0, 1$ then the $\lambda$-sheet lies over points $t$ where

$$\min\{t_i : \lambda_i = 1\} \geq \frac{1}{k+1}$$

$$\max\{t_j : \lambda_j = 0\} \leq \frac{k}{k+1}$$

We will modify these sheets by moving them horizontally in the $t$-direction. After the movement the $\lambda$-sheet will lie over the region $R_\lambda$ in $I^k$ given by

$$R_\lambda = \{t \in I^k : \min\{t_i : \lambda_i = 1\} - \max\{t_j : \lambda_j = 0\} \geq M(t) - \epsilon\}$$

where $0 < \epsilon \leq \frac{1}{2k+2}$ is fixed and $M(t)$ is the largest barycentric coordinate of $t$.

We recall that the *barycentric coordinates* of $t$ are the nonnegative real numbers $u_0, u_1, \cdots, u_k$ uniquely determined by the property that they add up to 1 and for some permutation $\sigma \in S_k$ we have

$$u_0 + u_1 + \cdots + u_{i-1} = t_{\sigma(i)}$$

for all $i$.

In the definition of $R_\lambda$ we interpret the minimum of an empty set to be 1 and the maximum of an empty set to be 0. Since $M(t) \geq \frac{1}{k+1} \geq 2\epsilon$,

$$\min\{t_i : \lambda_i = 1\} - \max\{t_j : \lambda_j = 0\} \geq M(t) - \epsilon \geq \epsilon > 0$$

is one of the nonnegative barycentric coordinates of $t$.

In fact we need to paste in the new local sections by making this deformation in a small exterior collar of $I^{2k}$ in $(S^2)^k$.

**Proposition 7.5.3.** *For each $t \in I^k$ let $L(t)$ be the set of all $\lambda \in \{0,1\}^k \subseteq \mathbb{R}^k$ so that $t \in \mathbb{R}_\lambda$. Then*

1. *$L(t)$ is nonempty for all $t$ in $I^k$.*

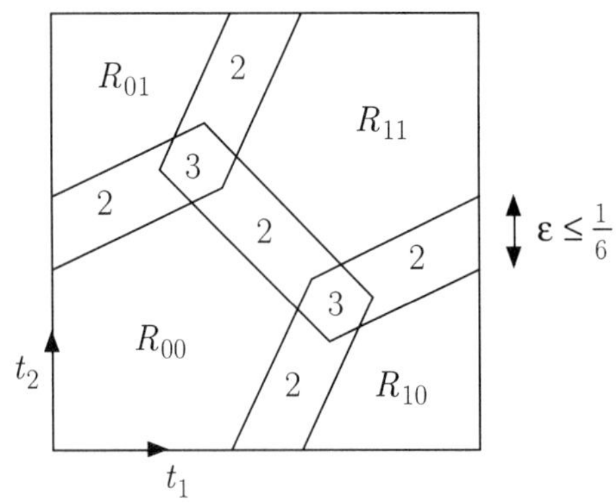


Figure 7.9: Local section domains for $k = 2$

2. *$L(t)$ has at most $k+1$ elements for all $t$.*
3. *$L(t)$ is totally ordered for all $t$. ($\lambda \leq \lambda'$ if $\lambda' - \lambda$ has no negative coordinates.)*
4. *If $t$ is within $\epsilon$ of $\partial I^k$ then $L(t)$ lies in a hyperplane given by $\lambda_i = 0$ or $1$.*
5. *The $(k+1)$-polylogarithm invariant is nontrivial only at the $(k+1)!$ points in $I^k$ where the maximum barycentric coordinate of $t$ is equal to*

$$M(t) = \frac{1+k\epsilon}{k+1}$$

*and the other barycentric coordinates are equal to*

$$M(t) - \epsilon = \frac{1-\epsilon}{k+1}$$

**Example 7.5.4.** An accurate drawing of the case $k = 2$ is given in Figure 7.9. The numbers indicate the cardinality of $L(t)$. We note that $(0,1), (1,0)$ are not comparable in the partial ordering so $R_{01}$ and $R_{10}$ do not meet. The $(k+1)! = 6$ points referred to in 7.5.3(5) are the points where the boundaries of the $R_\lambda$ intersect.

*Proof.* Choose a permutation $\sigma \in S_k$ so that

$$0 \leq t_{\sigma(1)} \leq t_{\sigma(2)} \leq \cdots \leq t_{\sigma(k)} \leq 1$$

Let $u_i = t_{\sigma(i+1)} - t_{\sigma(i)} = M(t)$ be maximal. Then

$$\lambda = e_{\sigma(i+1)} + e_{\sigma(i+2)} + \cdots + e_{\sigma(k)} \in L(t)$$

where $e_i$ is the $i$-th unit coordinate vector in $\mathbb{R}^k$. This proves (1).

The other elements of $L(t)$ are given by $e_{\sigma(j+1)} + \cdots + e_{\sigma(k)}$ where

$$u_j = t_{\sigma(j+1)} - t_{\sigma(j)} \leq M(t) - \epsilon$$

This proves (2) and (3).

If $t$ is within $\epsilon$ of $\partial I^k$ then either the first or last barycentric coordinate of $t$ is less than $\epsilon$. In the first case, $\lambda_{\sigma(1)} = 0$ for all $\lambda \in L(t)$. In the second $\lambda_{\sigma(k)} = 1$ for all $\lambda \in L(t)$. This proved (4).

If $L(t)$ lies in any hyperplane then the exponents of $r$ in the diagonal matrix $D_s(r)$ have fewer than $k$ degrees of freedom so the integrand $\mathrm{Tr}((K^{-1}dK + (K^{-1}dK)^*)^{2k+1})$ would be identically 0. $(k+1)$-polylogarithm component of the integral of $\mathrm{Tr}(\omega^{2k+1})$ will be nonzero only in the regions where $L(t)$ has $k+1$ elements. This happens in the neighborhoods of $k!$ barycenters of the $k$-simplices of $I^k$ given by the condition that the difference between the minimum and maximum barycentric coordinates of $t$ is $\leq \epsilon$.

These are the locations where the $s$ parameters in $D_s(r)$ have $k$ degrees of freedom. The $t$ parameters have $k$ degrees of freedom in $f_t(x)$ only at the points where $k$ sheets come to an end simultaneously. However the boundary of $R_\lambda$ for $\lambda = e_{\sigma(i+1)} + \cdots + e_{\sigma(k)}$ is exactly where

$$u_i = \min(u_j) = M(t) - \epsilon$$

For each permutation $\sigma \in S_k$ there are exactly $k+1$ points in the $k$-simplex corresponding to $\sigma$ where $k$ barycentric coordinates of $t$ are equal to $M(t) - \epsilon$. This proves (5). □

There is one more thing. We need to show that the new system of local sections over the regions $R_\lambda \times I^k$ in $I^k \times I^k$ can be pasted together with the old data outside of these cubes in $(S^2)^k$ so that the $(k+1)$-polylogarithm invariant vanishes everywhere else. This is accomplished by simply allowing $\epsilon$ to grow continuously to 1 in a small exterior collar of the boundary of $I^{2k}$ with the added conditions on $R_\lambda$ (vacuous for small $\epsilon$) given by

$$\{\min(t_i) \,:\, \lambda_i = 1\} \geq \frac{\epsilon}{k+1}$$

$$\{\min(t_j) \,:\, \lambda_j = 0\} \leq 1 - \frac{\epsilon}{k+1}$$

Since $M(t)$ is always $\geq 1/(k+1)$ this defines a system of local sections.

## 7.6 An explicit variable incidence matrix

In each of the $N^k = 2^{mk}$ cubes we have constructed $(k+1)!$ points where the $(k+1)$-polylogarithm invariant may be nonzero. We will see that the invariant is the same at all of these points. Furthermore the orientation is the same at all points so none of the contributions cancel. We get

$$N^k(k+1)!$$

times the $(k+1)$-polylogarithm invariant at any one point.

It is clear that the $N^k$ cubes are identical so we concentrate on the $(k+1)!$ points in each cube.

**Proposition 7.6.1.** *At each of the $(k+1)!$ points $b$ of Proposition 7.5.3(5), the orientation of $I^{2k}$ given below is the same. (It is positive.)*

1. *Let $w_0, w_1, \cdots, w_k$ be the elements of $\Sigma_b$ in reverse cyclic order around the fiber $F_b$. Suppose that $w_0$ is the element that lies in the interior of its section image component and that $w_1, \cdots, w_k$ lie on the boundaries of their components. Let $v_1, v_2, \cdots, v_k$ be the projection to the $t$-plane of the inward normal vectors to the boundaries of the sheets on which $w_1, \cdots, w_k$ lie. Then $(v_1, \cdots, v_k)$ gives the orientation of the $t$-plane $I^k$.*

2. *Let $\theta_j$ be the distance from $w_{j-1}$ to $w_j$ in the fiber so that*

$$w_{j-1} = w_j e^{i\theta_j}$$

*Then $\theta = (\theta_1, \cdots, \theta_k)$ is an affine linear function of $s = (s_1, \cdots, s_k)$ and therefore $d\theta = (d\theta_1, \cdots, d\theta_k)$ gives an orientation of the $s$-plane $I^k$.*

*Our proposition can now be stated as:*

$$sgn\ \det(\frac{\partial\theta_i}{\partial s_j}) = sgn\ \det(v_1 \cdots v_k)$$

*Proof.* When $k = 2$ the vertex $b$ lies on a corner of one of the two hexagonal regions in Fig. 7.9. These are neighborhoods of the two barycenters $(2/3, 1/3)$ and $(1/3, 1/3)$ of the two $k$-simplices in $I^k$.

More generally, $I^k$ is triangulated into a union of $k!$ simplices of dimension $k$. There is one $k$-simplex for each permutation $\sigma \in S_k$. The vertices of this simplex are

$$0, e_{\sigma(k)}, e_{\sigma(k)+\sigma(k-1)}, \cdots, e_{\sigma(k)} + \cdots + e_{\sigma(1)}$$

The orientation of this simplex is thus

$$(-1)^{[k/2]} sgn(\sigma).$$

At the barycenter of each simplex there are $(k+1)$ vertices $b$ where one sheet extends and the other $k$ sheets come to an end.

Suppose that $b$ is in the interior of the region $R_{\lambda_i}$ where

$$\lambda_i = e_{\sigma(i)} + e_{\sigma(i+1)} + \cdots + e_{\sigma(k)}.$$

Then the other regions on which $b$ lie, in clockwise order of their images in $F_b$ are

$$R_{\lambda_{i-1}}, R_{\lambda_{i-2}}, \cdots, R_{\lambda_1}, R_0, R_{\lambda_k}, R_{\lambda_{k-1}}, \cdots, R_{\lambda_{i+1}}$$

The orientation of $(v_1, \cdots, v_k)$ given by the inward normals of these regions is the same as

$$sgn\ \det(e_{\sigma(i-1)}, e_{\sigma(i-2)} + e_{\sigma(i-1)}, \cdots, e_{\sigma(1)} + \cdots + e_{\sigma(i-1)},$$

$$
\begin{aligned}
-e_{\sigma(i)} - \cdots - e_{\sigma(k)}, e_{\sigma(i)} - \cdots - e_{\sigma(k-1)}, \cdots, -e_{\sigma(i)}) \\
= sgn\ \det(e_{\sigma(i-1)}, e_{\sigma(i-2)}, \cdots, e_{\sigma(1)}, -e_{\sigma(k)}, \cdots, -e_{\sigma(i)}) \\
= (-1)^{i(k-i+1)} sgn(\sigma) \\
= (-1)^{ik} sgn(\sigma)
\end{aligned}
$$

By construction the gaps between the sheets starting at the extended sheet are a constant plus $2\pi/N$ times

$$
s_{\sigma(i+1)}, s_{\sigma(i+2)}, \cdots, s_{\sigma(k)}, -\sum s_j, s_{\sigma(1)}, \cdots, s_{\sigma(i-1)}
$$

Therefore the orientation of the $s$-plane is

$$
sgn\ \det(e_{\sigma(i+1)}, e_{\sigma(i+2)}, \cdots, e_{\sigma(k)}, -\sum e_j, e_{\sigma(1)}, \cdots, e_{\sigma(i-1)})
$$

$$
sgn\ \det(e_{\sigma(i+1)}, \cdots, e_{\sigma(k)}, -e_i, e_{\sigma(1)}, \cdots, e_{\sigma(i-1)})
$$

$$
(-1)^{i+i(k-i)} sgn(\sigma) = (-1)^{ik} sgn(\sigma).
$$

This is the same as the orientation of the $t$-plane. □

Now we choose an explicit local coordinate system at each of the equivalent $N^k(k+1)!$ points in $(S^2)^k$ discussed above.

Let $\sigma_0, \sigma_1, \cdots, \sigma_k$ be the elements of the local section image set $\Sigma$ in reverse cyclic order. Let $t_1, \cdots, t_k, s_1, \cdots, s_k$ be local coordinates for $B = (S^2)^k$ so that the point $\sigma_i$ disappears when and only when $t_i$ becomes negative and the distance from $\sigma_i$ to $\sigma_{i-1}$ is

$$
\frac{2\pi(c_i + s_i)}{N}
$$

where $c_i$ are fixed integers and $0 \le s_i \le 1$.

**Theorem 7.6.2.** *The higher FR-torsion invariant is given by the $(k+1)$-polylogarithm component of*

$$
\frac{N^k(k+1)!}{(2k+1)!2i^k} \int_{I^{2k} \times I} \mathrm{Tr}([K^{-1}dK + (K^{-1}dK)^*]^{2k+1})
$$

*where*

$$
\begin{aligned}
K &= D_s(r) f_t(rx) D_s(r)^{-1} = D_s(r) C(rx) g_t(rx) C(rx)^{-1} D_s(r)^{-1} \\
D_s(r) &= diag(1, r^{s_1}, r^{s_1+s_2}, r^{s_1+s_2+s_3}, \cdots) \\
C(rx) &= diag(1, (rx)^{c_1}, (rx)^{c_1+c_2}, \cdots)
\end{aligned}
$$

*and $g_t(rx)$ is the $(k+1) \times (k+1)$ matrix with entries:*

$$
\begin{aligned}
g_t(rx)_{ji} &= \delta_{ji} && \text{if } 1 \le j \le i \le k \\
g_t(rx)_{ji} &= -t_i u_{i+1} \cdots u_{j-1} t_j && \text{if } i < j \\
g_t(rx)_{0i} &= \delta_{0i} - y t_i u_{i+1} \cdots u_k
\end{aligned}
$$

*where* $y = r^N x^N$ *and* $u_i = 1 - t_i^2$, *i.e.*,

$$g_t(rx) = \begin{pmatrix} 1 - yu_1 \cdots u_k & -yt_1u_2\cdots u_k & -yt_2u_3\cdots u_k & \cdots & -yt_k \\ -t_1 & 1 & 0 & \cdots & 0 \\ -u_1t_2 & -t_1t_2 & 1 & \cdots & 0 \\ -u_1u_2t_3 & -t_1u_2t_3 & -t_2t_3 & \cdots & 0 \\ \cdots & \cdots & \cdots & \cdots & \cdots \\ -u_1\cdots u_{k-1}t_k & -t_1u_2\cdots u_{k-1}t_k & -t_2u_3\cdots u_{k-1}t_k & \cdots & 1 \end{pmatrix}$$

## 7.7 Polylogarithms as functions on $S^1$

We consider complex valued functions on $S^1$ which are smooth outside of a finite set. Let $\mathcal{C}(S^1)$ denote the ring or all such functions modulo those with finite support. Thus two functions $f(x), g(x)$ give the same element of $\mathcal{C}(S^1)$ if they agree outside of a finite set. Also, functions are allowed to diverge or be undefined on a finite subset of $S^1$.

We view $\mathcal{C}(S^1)$ as a vector space over the field of rational functions $\mathbb{C}(x)$. We are interested in the $2k+1$ -dimensional subspace $\mathcal{C}(k+1)$ of $\mathcal{C}(S^1)$ generated by 1 and the polylogarithms $\mathcal{L}_i(x), \mathcal{L}_i(\bar{x})$ for $i = 1, \cdots, k+1$ and in the projections $\pi_{k+1}, \bar{\pi}_{k+1} : \mathcal{C}(S^1) \to \mathbb{C}(x)$ which pick out the coefficients of $\mathcal{L}_{k+1}(x), \mathcal{L}_{k+1}(\bar{x})$.

More generally we have:

**Definition 7.7.1.** Let $\mathcal{C}(N, k+1)$ denote the $2kN+3$ -dimensional $\mathbb{C}(x)$-vector subspace of $\mathcal{C}(S^1)$ spanned by $\mathcal{L}_{k+1}(x^N), \mathcal{L}_{k+1}(\bar{x}^N), 1$ and the polylogarithms

$$\mathcal{L}_j(x\zeta), \mathcal{L}_j(\bar{x}\bar{\zeta})$$

for $j = 1, 2, \cdots, k$ and all $N$-th roots of unity $\zeta$. For $f(x) \in \mathcal{C}(N, k+1)$ let $\pi_{k+1}(f), \bar{\pi}_{k+1}(f) \in \mathbb{C}(x)$ denote the coefficients of $\mathcal{L}_{k+1}(x^N), \mathcal{L}_{k+1}(\bar{x}^N)$. We say that two elements $f, g$ of $\mathcal{C}(N, k+1)$ are *k-equivalent* and we write

$$f \approx_k g$$

if $\pi_{k+1}(f - g)$ has an $x$ in its numerator and $\bar{\pi}_{k+1}(f - g)$ has negative degree.

In order for this definition to make sense we need to verify that the generators of $\mathcal{C}(N, k+1)$ are linearly independent over $\mathbb{C}(x)$. Since

$$\mathcal{L}_{k+1}(x^N) = \frac{1}{N^k} \sum_{\zeta^N=1} \mathcal{L}_{k+1}(x\zeta)$$

it suffices to show the following.

**Lemma 7.7.2.** *The* $2kN+1$ *functions of* $x \in S^1$ *given by* 1 *and the polylogarithms*

$$\mathcal{L}_j(x\zeta) = \sum_{n=1}^{\infty} \frac{(x\zeta)^n}{n^j}$$

*and* $\mathcal{L}_j(\bar{x}\bar{\zeta})$ *for* $j = 1, 2, \cdots, k$ *and all* $N$*-th roots of unity* $\zeta$ *are linearly independent over* $\mathbb{C}(x)$.

*Proof.* First we note that $x\frac{d}{dx}$ is a well-defined operation on $\mathcal{C}(S^1)$ since

$$x\frac{d}{dx}f(x) = -i\frac{d}{d\theta}f(e^{i\theta})$$

This operation takes polylogarithms to polylogarithms:

$$x\frac{d}{dx}\mathcal{L}_k(x\zeta) = \mathcal{L}_{k-1}(x\zeta), \quad x\frac{d}{dx}\mathcal{L}_k(\bar{x}\bar{\zeta}) = -\mathcal{L}_{k-1}(\bar{x}\bar{\zeta})$$

where $\mathcal{L}_0(x\zeta), \mathcal{L}_0(\bar{x}\bar{\zeta})$ are given by

$$\mathcal{L}_0(x\zeta) = \sum_{n=1}^{\infty} x^N\zeta^N = \frac{x\zeta}{1-x\zeta}, \quad \mathcal{L}_0(\bar{x}\bar{\zeta}) = \frac{\bar{x}\bar{\zeta}}{1-\bar{x}\bar{\zeta}} = \frac{-1}{1-x\zeta}$$

We note also that

$$x\frac{d}{dx}\frac{h(x)}{(1-x^N)^\alpha} = \frac{xh'(x)(1-x^N) + \alpha N x^N h(x)}{(1-x^N)^{\alpha+1}}$$

Consequently the operation $x\frac{d}{dx}$ leaves invariant the $\mathbb{C}[x]$-module $M_k$ ($k \geq 0$) which is generated by the polylogarithms $\mathcal{L}_k(x\zeta), \mathcal{L}_k(\bar{x}\bar{\zeta}), \ldots, \mathcal{L}_1(x\zeta), \mathcal{L}_1(\bar{x}\bar{\zeta})$ and all negative powers of $1-x^N$.

Let $k \geq 1$ be minimal so that the lemma does not hold. Then there are polynomials $f_\zeta(x), g_\zeta(x) \in \mathbb{C}[x]$ not all zero so that

$$\sum_\zeta f_\zeta(x)\mathcal{L}_k(x\zeta) + g_\zeta(x)\mathcal{L}_k(\bar{x}\bar{\zeta}) \in M_{k-1} \tag{7.10}$$

Let $m \geq 0$ be the maximum degree of the polynomials $f_\zeta(x), g_\zeta(x)$. Applying the operation $x\frac{d}{dx}$ to (7.10) and subtracting $m$ times (7.10) we get

$$\sum_\zeta (xf_\zeta'(x) - mf_\zeta(x))\mathcal{L}_k(x\zeta) + (xg_\zeta'(x) - mg_\zeta(x))\mathcal{L}_k(\bar{x}\bar{\zeta}) \in M_{k-1} \tag{7.11}$$

But the coefficients in (7.11) are polynomials of degree $< m$. Therefore by induction on $m$ they are all zero. This means that $f_\zeta(x), g_\zeta(x)$ are all monomials of degree $m$:

$$f_\zeta(x) = a_\zeta x^m, \quad g_\zeta(x) = b_\zeta x^m$$

Case 1 ($k = 1$) If $k = 1$ the original equation (7.10) can be written as:

$$\sum_\zeta a_\zeta x^m \mathcal{L}_1(x\zeta) + b_\zeta x^m \mathcal{L}_1(\bar{x}\bar{\zeta}) = \frac{h(x)}{(1-x^N)^\alpha} \tag{7.12}$$

where $h(x) \in \mathbb{C}[x]$ is a polynomial not divisible by $1-x^N$. Applying $x\frac{d}{dx} - m$ to (7.12) we get

$$\sum_\zeta \frac{(a_\zeta x\zeta + b_\zeta)x^m}{1-x\zeta} = \frac{(xh'(x) - mh(x))(1-x^N) + \alpha N x^N h(x)}{(1-x^N)^{\alpha+1}}$$

which is impossible unless $\alpha = 0$ and $a_\zeta + b_\zeta = 0$ for all $\zeta$. In that case (7.12) reduces to:

$$\sum_\zeta a_\zeta x^m (\mathcal{L}_1(x\zeta) - \mathcal{L}_1(\bar{x}\bar{\zeta})) = h(x) \tag{7.13}$$

However the left hand side of (7.13) has a discontinuity of size $2\pi|a_\zeta|$ at $x = \zeta^{-1}$ since

$$\mathcal{L}_1(x\zeta) - \mathcal{L}_1(\bar{x}\bar{\zeta}) = -\ln(1 - x\zeta) + \ln(1 - \bar{x}\bar{\zeta}) = \ln(-\bar{x}\bar{\zeta})$$

so we conclude that $a_\zeta = -b_\zeta = 0$ for all $\zeta$.

Case 2 ($k \geq 2$) When $k \geq 2$ equation (7.10) can be expanded to

$$\sum_\zeta a_\zeta x^m \mathcal{L}_k(x\zeta) + b_\zeta x^m \mathcal{L}_k(\bar{x}\bar{\zeta}) \equiv \sum_\zeta p_\zeta(x)\mathcal{L}_{k-1}(x\zeta) + q_\zeta(x)\mathcal{L}_{k-1}(\bar{x}\bar{\zeta})$$

modulo $M_{k-2}$. Applying the operation $x\frac{d}{dx} - m$ we get:

$$\sum_\zeta a_\zeta x^m \mathcal{L}_{k-1}(x\zeta) - b_\zeta x^m \mathcal{L}_{k-1}(\bar{x}\bar{\zeta})$$

$$\equiv \sum_\zeta (xp'_\zeta - mp_\zeta)\mathcal{L}_{k-1}(x\zeta) + (xq'_\zeta - mq_\zeta)\mathcal{L}_{k-1}(\bar{x}\bar{\zeta})$$

which, by induction on $k$, is impossible since $xp'_\zeta - mp_\zeta, xq'_\zeta - mq_\zeta$ have no terms of degree $m$. □

We need to find the $(k+1)$-polylogarithm component of the following type of integral.

$$\int_0^1 \frac{dr}{r} \ln^k r \frac{r^n}{(1 - r^N x^N)^a (1 - r^N \bar{x}^N)^b} \tag{7.14}$$

where $n$ is a positive integer and $a, b$ are nonnegative integers with $a + b > 0$. By partial fractions we will reduce to the case where either $a$ or $b = 0$. Then we separate into cases depending on whether $n$ is divisible by $N$ or not.

**Lemma 7.7.3.** *If $n$ is a positive real number and $k$ is a nonnegative integer then*

$$\int_0^1 \frac{dr}{r} r^n \ln^k r = (-1)^k \frac{k!}{n^{k+1}}$$

**Proposition 7.7.4.** *Suppose that $a, n, N$ are positive integers and $k \geq a - 1$ is also an integer.*

(a) *Suppose that $n$ is divisible by $N$. Then*

$$\int_0^1 \frac{dr}{r} \frac{x^n r^n \ln^k r}{(1 - r^N x^N)^a} = \sum_{i=0}^{k} c_i \mathcal{L}_{k+1-i}(x^N) + f(x^N)$$

*where $f$ is a polynomial and $c_i$ are constants with*

$$c_0 = \frac{(-1)^{k+a-1}k!}{N^{k+1}}\binom{n/N-1}{a-1} \quad \textit{if} \quad n \geq aN$$

*and $c_0 = 0$ if $n < aN$.*

(b) *If $n$ is not divisible by $N$ then*

$$\int_0^1 \frac{dr}{r}\frac{x^n r^n \ln^k r}{(1-r^N x^N)^a} = \sum_{i=0}^{k}\sum_{\zeta^N=1} a_{i,\zeta}\mathcal{L}_{k+1-i}(x\zeta) + f(x)$$

*where $f(x)$ is a polynomial and $a_{i,\zeta}$ are constants so that*

$$\sum_{\zeta^N=1} a_{0,\zeta} = 0$$

*Proof.* Suppose that $q = n/N$ is a positive integer. Then:

$$\frac{1}{(1-r^N x^N)^a} = \sum_{j=0}^{\infty}\binom{a+j-1}{j} r^{jN} x^{jN}$$

$$\int_0^1 \frac{dr}{r}\ln^k r \frac{(xr)^{qN}}{(1-(rx)^N)^a} = \int_0^1 \frac{dr}{r}\ln^k r \sum_{j=0}^{\infty}\binom{a+j-1}{j} x^{(q+j)N} r^{(q+j)N}$$

$$= \frac{(-1)^k k!}{N^{k+1}}\sum_{j=0}^{\infty}\binom{a+j-1}{j}\frac{x^{(q+j)N}}{(q+j)^{k+1}}$$

But

$$(a-1)!\binom{a+j-1}{j} = (j+1)\cdots(j+a-1)$$

is a polynomial of degree $a-1 \leq k$ in $j$ so

$$\frac{(a-1)!}{(q+j)^{k+1}}\binom{a+j-1}{j} = \frac{b_0}{(q+j)^{k+1}} + \frac{b_1}{(q+j)^k} + \cdots + b_{k+1} \tag{7.15}$$

where $b_i$ are polynomials in $q$. To determine $b_0$ multiply both sides of (7.15) by $(q+j)^{k+1}$ and set $j = -q$. We get:

$$b_0 = (-1)^{a-1}(q-1)(q-2)\cdots(q-a+1)$$

$$= \begin{cases} (a-1)! & \text{if} \quad q = 0 \\ 0 & \text{if} \quad 1 \leq q \leq a-1 \\ (-1)^{a-1}(a-1)!\binom{q-1}{a-1} & \text{if} \quad q \geq a \end{cases}$$

Part (a) of the proposition follows.

To prove (b) we use Lemma 7.7.5 below to obtain

$$\int_0^1 \frac{dr}{r}\frac{x^n r^n \ln^k r}{(1-r^N x^N)^a} = \sum_{\zeta^N=1}\sum_{i=0}^{a-1} b_N(a,i)\int_0^1 \frac{dr}{r}\frac{x^n r^n \ln^k r}{(1-rx\zeta)^{a-i}}$$

$$= \sum_{i=0}^{a-1} b_N(a,i) \sum_{\zeta^N=1} \zeta^{-n}(-1)^{k+a-i-1} k! \binom{n-1}{a-i-1} \mathcal{L}_{k+1}(x\zeta) + \text{lower terms}$$

by part (a). Since $n$ is not a multiple of $N$, $\sum \zeta^{-n} = 0$. □

**Lemma 7.7.5.**

$$\frac{1}{(1-r^N x^N)^a} = \sum_{i=0}^{a-1} \sum_{\zeta^N=1} \frac{b_N(a,i)}{(1-rx\zeta)^{a-i}}$$

*where $b_N(a,i)$ are rational numbers depending only on $a, i, N$.*

*Proof.*

$$\binom{k+a-1}{k} = \frac{1}{(a-1)!}(k+1)(k+2)\cdots(k+a-1)$$

This is a polynomial of degree $a-1$ in $k$ with rational coefficients depending only on $a$. Similarly, $\binom{kN+s-1}{kN}$ is a polynomial of degree $s-1$ in $k$ with rational coefficients depending only on $s$ and $N$. Thus there exist rational numbers $b_N(a,i)$ so that

$$\binom{k+a-1}{k} = \sum_{i=0}^{a-1} b_N(a,i) \binom{kN+a-i-1}{kN}$$

Then

$$\frac{1}{(1-r^N x^N)^a} = \sum_{k=0}^{\infty} \binom{k+a-1}{k} r^{kN} x^{kN}$$

$$= \sum_{k=0}^{\infty} \sum_{i=0}^{a-1} b_N(a,i) \binom{kN+a-i-1}{kN} r^{kN} x^{kN}$$

$$= \sum_{i=0}^{a-1} \sum_{\zeta^N=1} \sum_{k=0}^{\infty} b_N(a,i) \binom{k+a-i-1}{k} r^k x^k \zeta^k$$

$$= \sum_{i=0}^{a-1} \sum_{\zeta^N=1} \frac{b(a,i)}{(1-rx\zeta)^{a-i}}$$

By partial fractions the values of $b_N(a,i)$ making this true is independent of $r, x$. Therefore we may take $r$ small enough so that all series converge absolutely so the rearrangement of the order of summation is justified. □

In order to use Proposition 7.7.4 to compute the integral (7.14) we need an identity involving negative binomials.

**Definition 7.7.6.** If $\alpha$ is a nonnegative integer then let

$$f^{\alpha}_{\beta}(x) = \sum_{i=0}^{\alpha-1} x^i \binom{i+\beta-1}{i}$$

(This is $(1-x)^{-\beta}$ expanded to degree $\alpha - 1$ in $x$.)

**Lemma 7.7.7.** *If $p + q = 1$ then for any nonnegative integers $\alpha, \beta$ not both zero we have:*

$$f^\alpha_\beta(p)q^\beta + f^\beta_\alpha(q)p^\alpha = 1$$

*Proof.* We use the *negative binomial distribution* in probability theory. Suppose that two teams $A, B$ are playing a match. Assume the games are independent and that team $A$ has probability $p$ of winning each game. Then

$$f^\alpha_\beta(p)q^\beta = \sum_{i=0}^{\alpha-1} p^i q^\beta \binom{i+\beta-1}{i}$$

is the probability that team $B$ will win $\beta$ games before team $A$ wins $\alpha$ games and $f^\beta_\alpha(q)p^\alpha$ is the probability that team $A$ will win $\alpha$ games before team $B$ wins $\beta$ games. Since these are complementary events the probabilities must add up to 1.

Since the equation hold for $p$ in the infinite set $[0, 1]$ it is a polynomial identity in $p$ with $q = 1 - p$. □

**Example 7.7.8.** Let

$$w = \frac{z}{z - \bar{z}} = \frac{z^2}{z^2 - 1}$$

Then $w + \bar{w} = 1$ and $w\bar{z} + \bar{w}z = 0$ so

$$w(1 - r^N\bar{z}) + \bar{w}(1 - r^N z) = 1$$

By Lemma 7.7.7 we then have:

$$1 = f^a_b(\bar{w}(1 - r^N z))w^b(1 - r^N\bar{z})^b + f^b_a(w(1 - r^N\bar{z}))\bar{w}^a(1 - r^N z)^a$$

So

$$\frac{1}{(1 - r^N z)^a(1 - r^N\bar{z})^b} = \frac{f^a_b(\bar{w}(1 - r^N z))w^b}{(1 - r^N z)^a} + \frac{f^b_a(w(1 - r^N\bar{z}))\bar{w}^a}{(1 - r^N\bar{z})^b}$$

$$= \sum_{i=0}^{a-1} \frac{\bar{w}^i w^b}{(1 - r^N z)^{a-i}} \binom{i+b-1}{i} + \sum_{j=0}^{b-1} \frac{w^j \bar{w}^a}{(1 - r^N\bar{z})^{b-j}} \binom{j+a-1}{j} \tag{7.16}$$

**Proposition 7.7.9.** *Suppose that $a, b, c, k$ are positive integers so that $a + b \leq k + 1$.*

(a) $$\int_0^1 \frac{dr}{r} \ln^k r \frac{r^c}{(1 - r^N z)^a(1 - r^N\bar{z})^b}$$

*is a $\mathbb{C}(x)$-linear combination of 1 and the polylogarithms $\mathcal{L}_i(x\zeta)$ for $N$-th roots of unity $\zeta$ and $i = 1, \cdots, k + 1$.*

(b) *If $c$ is divisible by $N$ then*

$$\int_0^1 \frac{dr}{r} \ln^k r \frac{r^c}{(1 - r^N z)^a(1 - r^N\bar{z})^b}$$

$$= \sum_{i=0}^{k} f_i(z)\mathcal{L}_{k+1-i}(z) + \sum_{i=0}^{k} g_i(\bar{z})\mathcal{L}_{k+1-i}(\bar{z}) + h(z)$$

*where* $z^{2b-c/N}$ *divides the numerator of* $f_0(z)$ *and* $\bar{z}^{2a-c/N}$ *divides the numerator of* $g_0(\bar{z})$ *as a polynomial in* $\bar{z}$, *i.e.,* $\deg g_0(\bar{z}) \leq \frac{c}{N} - 2a$.

*Proof.* (a) follows directly from (7.16) and Prop. 7.7.4. In case (b) we have

$$f_0(z) = z^{-c/N} w^b \sum_{i=1}^{\min(a,c/N)} \binom{a+b-i-1}{a-i}\binom{c/N-1}{i-j} \frac{\bar{w}^{a-i}(-1)^{k+i-1}k!}{N^{k+1}}$$

whose numerator is divisible by $z^{2b-c/N}$ and similarly for $g_0(\bar{z})$. □

**Corollary 7.7.10.** *Suppose that* $a, b, k$ *are positive integers and* $c, d$ *are non-negative integers with* $a \geq c \geq 0, b \geq d \geq 0$ *then*

$$\int \frac{dr}{r} \ln^k r \frac{(r^N z)^c (r^N \bar{z})^d}{(1-r^N)^a (1-r^N \bar{z})^b}$$

$$= \sum_{i=0}^{k} f_i(z)\mathcal{L}_{k+1-i}(z) + \sum_{i=0}^{k} g_i(\bar{z})\mathcal{L}_{k+1-i}(\bar{z}) + h(z)$$

*where* $f_i(z), g_i(\bar{z}), h \in \mathbb{C}(z)$ *are rational functions so that* $f_0(0) = g_0(0) = 0$, *i.e., the integral is* $k$*-equivalent to zero in the sense of Definition 7.7.1.*

*Proof.* $z^{c-d+2b-c} = z^{2b-d}$ divides $f_0(z)$ and $\bar{z}^{d-c+2a-d} = \bar{z}^{2a-c}$ divides $g_0(\bar{z})$. □

## 7.8 Simplification of the integral

We return to the integral

$$\int_{I^{2k}\times I} \mathrm{Tr}([K^{-1}dK + (K^{-1}dK)^*]^{2k+1})$$

of Theorem 7.6.2 where

$$K = D_s(r)C(rx)g_t(rx)C(rx)^{-1}D_s(r)^{-1}$$

$$D_s(r) = diag(1, r^{s_1}, r^{s_1+s_2}, r^{s_1+s_2+s_3}, \cdots)$$

$$C(rx) = diag(1, (rx)^{c_1}, (rx)^{c_1+c_2}, \cdots)$$

and the matrix $g_t(rx)$ is as given in 7.6.2.

We need the derivatives of

$$D_s(r)C(r) = diag(1, r^{s_1+c_1}, r^{s_1+s_2+c_1+c_2}, \cdots):$$

$$\frac{\partial D_s(r)C(r)}{\partial r} = diag(0, s_1 + c_1, s_1 + s_2 + c_1 + c_2, \cdots)D_s(r)C(r) \tag{7.17}$$

$$\begin{aligned}\frac{\partial D_s(r)C(r)}{\partial s_i} &= \ln r \; diag(\overbrace{0, \cdots, 0}^{i}, 1, \cdots, 1)D_s(r)C(r) \\ &= \ln r(E_i + E_{i+1} + \cdots + E_k)D_s(r)C(r)\end{aligned}$$

where $E_i$ denotes the diagonal matrix with $(i,i)$-entry equal to 1 and all other entries 0. In particular $\sum E_i = I$, the identity matrix.

Recall that $y = r^N x^N$ so $\partial y/\partial r = Ny/r$ and

$$\frac{\partial}{\partial r} g_t(rx) = \frac{N}{r}(g_t(rx) - I) \tag{7.18}$$

The derivative with respect to $t_i$ is a little tricky. First we need the inverse of $g_t(rx)$:

$$g_t(rx)^{-1} = \frac{1}{1-y}\begin{pmatrix} 1 & yt_1 & yt_2 & \cdots & yt_k \\ t_1 & 1-yu_1 & yt_1t_2 & \cdots & yt_1t_k \\ t_2 & t_1t_2 & 1-yu_2 & \cdots & yt_2t_k \\ t_3 & t_1t_3 & t_2t_3 & \cdots & yt_3t_k \\ \cdots & \cdots & \cdots & \cdots & \cdots \\ t_k & t_1t_k & t_2t_k & \cdots & 1-yu_k \end{pmatrix}$$

Since $1 - yu_i = 1 - y + yt_i^2$ this can be written as:

$$g_t(rx)^{-1}_{ij} = \begin{cases} \delta_{ij} + t_it_j(\frac{y}{1-y}) & \text{if} \quad i \le j \\ \\ t_it_j(\frac{1}{1-y}) & \text{if} \quad i > j \end{cases}$$

It is not hard to verify that this is the inverse of $g_t(rx)$ using the identity:

$$t_i^2 u_{i+1} \ldots u_j + t_{i+1}^2 u_{i+2} \ldots u_j + \cdots + t_j^2 = 1 - u_i u_{i+1} \ldots u_j$$

Since $t_i$ occurs only in the $i$-th row and $i$-th column of $g_t(rx)^{-1}$ it is very easy to see that

$$\begin{aligned}\frac{\partial g_t(rx)^{-1}}{\partial t_i} &= \frac{1}{t_i}(I - g_t(rx)^{-1})E_i + \frac{1}{t_i}E_i(I - g_t(rx)^{-1}) \\ &= \frac{1}{t_i}(-2E_i + g_t(rx)^{-1}E_i + E_i g_t(rx)^{-1})\end{aligned}$$

Consequently,

$$\begin{aligned}\frac{\partial g_t(rx)}{\partial t_i} &= -g_t(rx)\frac{\partial g_t(rx)^{-1}}{\partial t_i} g_t(rx) \\ &= \frac{1}{t_i}(2g_t(rx)E_i g_t(rx) - E_i g_t(rx) - g_t(rx)E_i) \end{aligned} \tag{7.19}$$

**Lemma 7.8.1.**

$$\mathrm{Tr}((K^{-1}dK + (K^{-1}dK)^*)^{2k+1}) = \mathrm{Tr}((A + A^*)^{2k+1})$$

*where the matrix* 1*-form* $A$ *is given by:*

$$A = A_r dr + \sum A_{t_i} dt_i + \sum A_{s_i} ds_i$$

$$\begin{aligned}\phi A_r &= \frac{m}{r}(I - g_t(rx)^{-1})E_0 \\ \phi A_{t_i} &= \frac{1}{t_i}(I - g_t(rx)^{-1})E_i \\ \phi A_{s_i} &= 2\ln r E_i(g_t(rx) - I)\end{aligned}$$

$\phi$ *being right conjugation by* $C(r)D_s(r)$, *i.e.,*

$$\phi X = D_s(r)^{-1}C(r)^{-1}XC(r)D_s(r)$$

**Remark 7.8.2.** Note that

(a) The entries of $\phi A_{s_i}/\ln r = 2E_i(g_t(rx) - I)$ are polynomials in $t_1, \cdots, t_i$ with integer coefficients.

(b)

$$(\phi A_{t_j})_{ij} = \begin{cases} \frac{t_i y}{1-y} & \text{if} \quad i \le j \\ \frac{t_i}{1-y} & \text{if} \quad i > j \end{cases}$$

(c)

$$(\phi A_r)_{i0} = \begin{cases} \frac{y}{1-y} & \text{if} \quad i = 0 \\ \frac{t_i}{1-y} & \text{if} \quad i > 0 \end{cases}$$

*Proof.* First we note that $C(rx) = C(r)C(x)$ where $C(x)$ is a unitary diagonal matrix which is constant (independent of $r, s, t$). Therefore,

$$\mathrm{Tr}((K^{-1}dK + (K^{-1}dK)^*)^{2k+1}) = \mathrm{Tr}((B + B^*)^{2k+1})$$

where

$$\begin{aligned}B &= C(x)^{-1}K^{-1}dKC(x) \\ &= [C(x)^{-1}K^{-1}C(x)]d[C(x)^{-1}KC(x)] \\ &= [D_s(r)C(r)g^{-1}C(r)^{-1}D_s(r)^{-1}]d[D_s(r)C(r)gC(r)^{-1}D_s(r)^{-1}]\end{aligned}$$

with $g = g_t(rx)$. Conjugating by $D_s(r)C(r)$ we get:

$$\begin{aligned}\phi B &= g^{-1}C(r)^{-1}D_s(r)^{-1}d[D_s(r)C(r)gC(r)^{-1}D_s(r)^{-1}]D_s(r)C(r) \\ &= g^{-1}C(r)^{-1}D_s(r)^{-1}d[D_s(r)C(r)]g \\ &\qquad + g_r(rx)^{-1}dg - C(r)^{-1}D_s(r)^{-1}d[D_s(r)C(r)]\end{aligned}$$

In the expansion

$$B = B_r dr + \sum B_{t_i} dt_i + \sum B_{s_i} ds_i$$

we then have, by (7.17), (7.18) and (7.19):

$$\phi B_{t_i} = g^{-1}\frac{\partial g}{\partial t_i} = \frac{1}{t_i}(2E_i g - g^{-1}E_i g - E_i)$$

$$\phi B_{s_i} = g^{-1}\ln r(E_i + E_{i+1} + \cdots + E_k)g \\ - \ln r(E_i + E_{i+1} + \cdots + E_k)$$

$$\phi B_r = g^{-1}\frac{\partial g}{\partial r} + [\frac{1}{r}g^{-1}diag(0, s_1 + c_1, s_1 + s_2 + c_1 + c_2, \cdots)g \\ - \frac{1}{r}diag(0, s_1 + c_1, s_1 + s_2 + c_1 + c_2, \cdots)]$$

Since

$$diag(0, s_1 + c_1, s_1 + s_2 + c_1 + c_2, \cdots) = \sum_{i=1}^{k}(s_i + c_i)(E_i + E_{i+1} + \cdots + E_k)$$

the terms in square brackets in $\phi B_r$ add up to

$$\sum \frac{s_i + c_i}{r\ln r}\phi B_{s_i}$$

Using elementary operations (7.5.2) we can therefore replace $B_r$ by

$$newB_r = B_r - \sum \frac{s_i + c_i}{r\ln r}A_{s_i}$$

Then

$$\phi newB_r = g^{-1}\frac{\partial g}{\partial r} = \frac{N}{r}g^{-1}(E_0 g - E_0)$$

If we replace $B_{s_i}$ by $B_{s_i} - B_{s_i+1}$ (doing $s_1$ then $s_2$, etc., with no change to $B_{s_k}$) we get:

$$\phi newB_{s_i} = \phi B_{s_i} - \phi B_{s_{i+1}} = \ln r g^{-1}[E_i g - gE_i] = \ln r(g^{-1}E_i g - E_i)$$

Let

$$A_r = new^2 B_r = newB_r + \frac{N}{r\ln r}\sum_{i=1}^{k} newB_{s_i}$$

Then

$$\phi A_r = \phi new^2 B_r = \frac{N}{r}(I - g^{-1})E_0$$

Using more elementary operations let

$$A_{s_i} = new^2 B_{s_i} = newB_{s_i} + t_i \ln r B_{t_i}$$

then

$$\phi A_{s_i} = \phi new^2 B_{s_i} = 2\ln r E_i(g - I)$$

**Lemma 7.8.3.** *For $i = 1, \cdots, k$ the expression*

$$t_i g^{-1}\frac{\partial g}{\partial t_i} + (g^{-1} - I)E_i$$

*is an $\mathcal{R}$-linear combination of the matrices*

$$(g^{-1} - I)E_0, \cdots, (g^{-1} - I)E_{i-1}, E_1(g - I), \cdots, E_k(g - I)$$

*where $\mathcal{R}$ is the ring of real valued functions on $S^1 \times I^{2k+1}$.*

Assume for a moment that this is true. Then by the elementary operations lemma (7.5.2) we can replace $B_{t_i}$ by $A_{t_i} = newB_{t_i}$ where

$$\phi A_{t_i} = \phi newB_{t_i} = \frac{1}{t_i}(I - g^{-1})E_i$$

by adding an $\mathcal{R}$-linear combination of $A_r, B_{t_1}, \cdots, B_{t_{i-1}}, A_{s_1}, \cdots, A_{s_k}$. This proves Lemma 7.8.1. □

*Proof of 7.8.3.* By (7.19) we have:

$$\begin{aligned} t_i g^{-1}\frac{\partial g}{\partial t_i} &+ (g^{-1} - I)E_i \\ &= (2E_i g - g^{-1}E_i g - E_i) + (g^{-1}E_i - E_i) \\ &= -(g^{-1} - I)E_i(g - I) + E_i(g - I) \end{aligned}$$

so it suffices to show that

$$(g^{-1} - I)E_i(g - I)$$

is an $\mathcal{R}$-linear combination of the matrices listed in Lemma 7.8.3. But this follows by induction on $i$:

$$(g^{-1} - I)E_1(g - I) = -t_1^2(g^{-1} - I)E_0 - t_1^2E_1(g - I)$$

and

$$\begin{aligned} (g^{-1} - I)E_i(g - I) = &\frac{1 - t_{i-1}^2}{t_{i-1}}(g^{-1} - I)E_{i-1}(g - I) \\ &- t_{i-1}t_i(g^{-1} - I)E_{i-1} - t_iE_i(g - I) \end{aligned}$$

for $i \geq 2$. □

**Lemma 7.8.4.** *Let $T = T_r dr + \sum T_{t_i} dt_i + \sum T_{s_i} ds_i$ where*

$$T_r = E_0\phi A_r = \frac{N}{r}(E_0 - E_0 g_t(rx)^{-1}E_0)$$

$$T_{t_i} = \sum_{j=0}^{i} E_j\phi A_{t_i} = \frac{1}{t_i}(E_i - \sum_{j=0}^{i} E_j g_t(rx)^{-1}E_i)$$

$T_{s_i} = \phi A_{s_i} = 2\ln r(E_i g_t(rx) - E_i)$

*Then*

(a) $\displaystyle\int_{I^{2k}\times I} \mathrm{Tr}((A+A^*)^{2k+1})$ *and*

(b) $\displaystyle\int_{I^{2k}\times I} \mathrm{Tr}(T^{2k+1} + (T^*)^{2k+1})$

*are $k$-equivalent in the sense of Definition 7.7.1. In other words, the integral* (a) *is a $\mathbb{C}(x)$-linear combination of the $2kN+3$ functions*

$$1, \mathcal{L}_1(x\zeta), \mathcal{L}_i(\bar{x}\bar{\zeta}), \cdots, \mathcal{L}_k(x\zeta), \mathcal{L}_k(\bar{x}\bar{\zeta}), \mathcal{L}_{k+1}(z), \mathcal{L}_{k+1}(\bar{z})$$

*for $\zeta^N = 1$, the integral* (b) *is a $\mathbb{C}(x)$-linear combination of the $2k+3$ functions*

$$1, \mathcal{L}_1(z), \mathcal{L}_1(\bar{z}), \cdots, \mathcal{L}_{k+1}(z), \mathcal{L}_{k+1}(\bar{z}),$$

*the coefficients of $\mathcal{L}_{k+1}(z)$ in* (a) *and* (b) *are rational functions of $x$ which agree at $x=0$ and the coefficients of $\mathcal{L}_{k+1}(\bar{z})$ in* (a) *and* (b) *are rational functions of $x$ which agree at $x=\infty$ ($\bar{x}=0$).*

*Proof.* $\mathrm{Tr}((A+A^*)^{2k+1})$ is a sum of terms of the form

$$\text{polynomial}(t_i) r^{\lambda(s)} \ln^k r \frac{dr}{r} \frac{y^\alpha}{(1-y)^{\alpha+\gamma}} \frac{\bar{y}^\beta}{(1-\bar{y})^{\beta+\delta}} dt_1 ds_1 \cdots dt_k ds_k \tag{7.20}$$

where $y = r^N x^N = r^N z, \alpha+\beta+\gamma+\delta = k+1, \alpha,\beta,\gamma,\delta \geq 0$ and $\alpha+\beta \geq 1$. ($\alpha+\gamma$ is the number of factors coming from $A$ and $\beta+\delta$ is the number of factors coming from $A^*$.)

The factors $r^{\lambda(s)}$ come from $D_s(r)C(r)$ and its inverse. Thus,

$$\lambda(s) = \sum n_i(s_i + c_i)$$

for some integers $n_i$ with $|\sum n_i| \leq k+1$. Furthermore the $s_i$ do not occur anywhere else in the integrand (7.20). When $r^{\lambda(s)} ds_1 \cdots ds_k$ is integrated we will lose one factor of $\ln r$ for each $n_i \neq 0$. The $(k+1)$-polylogarithm terms would all vanish. Therefore we may ignore all terms where $\lambda(s) \neq 0$ and assume that $r^{\lambda(s)} = 1$.

Collecting together the remaining factors involving $r$ we get the following integral

$$\int_0^1 \frac{dr}{r} \ln^k r \frac{y^i}{(1-y)^{i+\alpha}} \frac{\bar{y}^j}{(1-\bar{y})^{j+\beta}} \tag{7.21}$$

By Prop. 7.7.4 and Cor. 7.7.10 this is a $\mathbb{C}(x)$-linear combination of $1, \mathcal{L}_1(z)$, $\mathcal{L}_1(z), \cdots, \mathcal{L}_{k+1}(z), \mathcal{L}_{k+1}(z)$ which is $k$-equivalent to 0 unless either $\alpha = k+1$ and $\beta=\gamma=\delta=0$ or $\beta=k+1$ and $\alpha=\gamma=\delta=0$.

Therefore

$$\int_{I^{2k+1}} \mathrm{Tr}((A+A^*)^{2k+1}) \approx_k \int_{I^{2k+1}} \mathrm{Tr}(A^{2k+1} + (A^*)^{2k+1})$$

Since $\mathrm{Tr}(A^{2k+1}) = \mathrm{Tr}(\phi A^{2k+1})$ and $\mathrm{Tr}((A^*)^{2k+1}) = \mathrm{Tr}((\phi A^*)^{2k+1})$ we can replace $A$ by $\phi A$. Since the $(k+1)$-polylogarithm component is zero whenever $\gamma$ or $\delta$ is nonzero we can set the entries

$$(\phi A_{t_j})_{ij} = \frac{t_i}{1-y} = (\phi A_r)_{i0}$$

for $i > j$ equal to zero. The lemma follows. □

## 7.9 Partitions and necklaces

In this section we will count certain types of partitions (of $k$) which we call "descending partitions" and certain necklaces (with $2k+1$ beads) which are related to the trace of $T^{2k+1} + (T^*)^{2k+1}$. The answers will allow us to complete the computation of the higher FR-torsion invariant in the universal case.

### 7.9.1 Descending partitions

A descending partition of $k$ is a sequence of nonnegative integers $p(i)$ adding up to $k$ and satisfying certain other conditions as described in 7.9.1 below. They can also be defined in terms of a game using $k$ stones. Suppose there are $k+1$ trays numbered $0, 1, \cdots, k$ with tray 0 being empty and one stone in each of the positively numbered trays. Take each stone and move it to a lowered number tray. Then the result will be called a "descending partition" $k = \sum p(i)$ where $p(i)$ is the number of stones in the $i$-th tray. The function $p$ can be characterized as follows.

**Definition 7.9.1.** A *descending partition* of $k$ is a function

$$p : \{0, 1, \cdots, k\} \to \mathbb{N}$$

with the following properties.

(1) $\sum_{i=0}^{k} p(i) = k$

(2) $\sum_{i=0}^{n} p(i) \geq n+1$ for $n < k$.

**Proposition 7.9.2.** (a) *Let $f : \{1, 2, \cdots, k\} \to \{0, \cdots, k\}$ be a strictly decreasing function, i.e., $f(i) < i$ for all $i$. Then*

$$p(i) = |f^{-1}(i)|$$

*gives a descending partition of $k$.*

(b) *Every descending partition arises in this way.*

(c) *If $f, g : \{1, 2, \cdots, k\} \to \{0, \cdots, k\}$ are strictly decreasing functions corresponding to the same descending partition of $k$ then there is a sequence of strictly decreasing functions $f_0 = f, f_1, f_2, \cdots, f_n = g$ so that each function differs from the previous function by interchanging two values ($f_i(a) = f_{i+1}(b), f_i(b) = f_{i+1}(a)$).*

*Proof.* (a) is obvious. For (b) take any descending partition $p$. Then a corresponding strictly decreasing function $f_p$ is given by letting $f_p(i) = j \geq 0$ be the smallest integer so that

$$\sum_{t=0}^{j} p(t) \geq i$$

This function has the property $f_p(i) \leq f_p(i+1)$ for $1 \leq i < k$.

To prove (c) let $f$ be any strictly decreasing function which gives the descending partition $p$ according (a). If $f \neq f_p$ then we must have $f(i) > f(j)$ for some $i < j$. Let $\sigma(f)$ be the function obtained from $f$ by switching the values of $f(i)$, $f(j)$ for $i, j$ minimal with this property. Then $f_p = \sigma^n(f)$ for some $n \leq \binom{k}{2}$. □

**Proposition 7.9.3.** *The number of descending partitions of $k$ is given by*

$$P_k = \frac{(2k)!}{k!(k+1)!}$$

*Proof.* Let $n \geq 1$ be minimal so that

$$\sum_{i=0}^{n-1} p(i) = n$$

If $n = k$ then we say that $p$ is a *special partition* of $k$. In that case $(p(0) - 1, p(1), \cdots, p(k-1))$ gives a descending partition of $k-1$ and conversely, by adding 1 to $p(0)$ and setting $p(k) = 0$, a descending partition of $k-1$ gives a special partition of $k$.

Since $(p(0), \cdots, p(n-1), 0)$ is a special partition of $n$ and $(p(n), \cdots, p(k))$ is an arbitrary descending partition of $n-k$, we have the recursion:

$$P_k = \sum_{n=1}^{k} P_{n-1} P_{k-n} \tag{7.22}$$

with $P_0 = 1$. Thus $P_k$ is the $k$-th Catalan number

$$P_k = C_k = \frac{1}{k+1}\binom{2k}{k} = \frac{(2k)!}{k!(k+1)!}$$

For those not familiar with Catalan numbers, the recursion (7.22) can be expressed in terms of the generating function $g = \sum_{k \geq 0} P_k z^k$ as $g = 1 + zg^2$ whose solution is

$$g = \frac{1 - \sqrt{1-4z}}{2z}$$

(Taking the negative sign on $\sqrt{1-4z}$ to make $g(0)$ finite.) Then substitute

$$(1-4z)^{1/2} = \sum_{k=0}^{\infty} \binom{1/2}{k} (-4)^k z^k = 1 + \sum_{k=1}^{\infty} \frac{-2(2k-2)!}{k!(k-1)!} z^k$$

□

### 7.9.2 Necklaces

We now turn to necklaces. A *necklace* with $n$ beads and color set $B$ is defined to be an equivalence class $[b]$ of a sequence $b = (b_1, b_2, \cdots, b_n)$, $b_i \in B$, where two sequences are equivalent if they differ by a cyclic permutation. If there are $n$ distinct sequences representing the same necklace it is called *aperiodic.*

**Definition 7.9.4.** Let $N_k$ be the set of necklaces $[b_0, \cdots, b_{2k}]$ where $b_i \in \{0, \cdots, k\}$ satisfying the following conditions.

1. For each color $n > 0$ there is a unique bead with color $n$ so that the next bead has a smaller color. $(n = b_i > b_{i+1})$

2. For each color $n$ there is a unique bead with color $n$ so that the previous bead has color $\leq n$. $(b_{i-1} \leq b_i = n)$

The conditions in Definition 7.9.4 can be rephrased as follows. Between every pair of consecutive beads we insert the relation $\leq$ or $>$ whichever is true. Then the resulting $2k+1$ relations are $\ell_0, \ell_1, \cdots, \ell_k$, $g_1, g_2, \cdots, g_k$ where $\ell_i$ is a relation of the form $* \leq i$ and $g_i$ has the form $i > *$. For example one element of $N_4$ is

$$[0, 0, 2, 1, 1, 3, 4, 3, 1] \tag{7.23}$$

This gives the sequence of relations:

$$\ell_0, \ell_2, g_2, \ell_1, \ell_3, \ell_4, g_4, g_3, g_1$$

The *sign* of the necklace is $+$ or $-$ depending on whether the sequence of relations is an even or odd permutation of

$$\ell_0, \ell_1, g_1, \ell_2, g_2, \cdots, \ell_k, g_k$$

In the example above the sign is $+$.

To each necklace in $N_k$ there is an associated descending partition $p$ where $p(i)$ one less than the number of beads colored $i$. In terms of relations, $p(i)$ is the number of relations of the form $* > i$. In the example above, $(p(0), p(1), \cdots, p(k)) = (1, 2, 0, 1, 0)$. Let $N_k(p)$ be the set of all necklaces in $N_k$ associated to $p$. We need to count the number of elements of $N_k(p)$ with sign.

To see how this is related to the higher FR-torsion let $B = (b_{ij})$ be a $(k+1) \times (k+1)$ matrix 1-form on $I^{2k+1}$ where

$$b_{ij} = \begin{cases} f_{ij}(t)dt_j & \text{if} \quad i \leq j \\ f_{ij}(t)ds_i & \text{if} \quad i > j \end{cases}$$

where $f_{ij}(t)$ are positive real valued functions of $t = (t_0, t_1, \cdots, t_k)$. Then the nonzero entries of $\mathrm{Tr}(B^{2k+1})$ are given by products

$$b_{i_0 i_1} b_{i_1 i_2} \cdots b_{i_{2k} i_0} = \pm f_{i_0 i_1} f_{i_1 i_2} \cdots f_{i_{2k} i_0} \; dt_0 dt_1 ds_1 \cdots dt_k ds_k$$

where the sequence $i_0, i_1, \cdots, i_{2k}$ forms a necklace in $N_k$ and the sign $\pm$ is the sign of this necklace. (If the necklace does not lie in $N_k$ then at least one $dt_i$ or $ds_j$ is repeated.) Each of these terms occurs $2k+1$ times since there are that many sequences representing each necklace.

Suppose now that the functions $f_{ij}$ are given by

$$f_{ij}(t) = \begin{cases} 1 & \text{if } \ 0 = i \leq j \\ t_i & \text{if } \ 0 < i \leq j \\ 2u_1u_2\cdots u_{i-1}t_i & \text{if } \ i > j = 0 \\ 2t_ju_{j+1}\cdots u_{i-1}t_i & \text{if } \ i > j > 0 \end{cases}$$

where $u_i = 1 - t_i$. For example for $k = 3$ we have:

$$B = \begin{pmatrix} dt_0 & dt_1 & dt_2 & dt_3 \\ 2t_1ds_1 & t_1dt_1 & t_1dt_2 & t_1dt_3 \\ 2u_1t_2ds_2 & 2t_1t_2ds_2 & t_2dt_2 & t_2dt_3 \\ 2u_1u_2t_3ds_3 & 2t_1u_2t_3ds_3 & 2t_2t_3ds_3 & t_3dt_3 \end{pmatrix}$$

In that case the product

$$f_{i_0i_1}f_{i_1i_2}\cdots f_{i_{2k}i_0}$$

depends only on the descending partition $p$ associated to the necklace of subscripts $[i_0, i_1, \cdots, i_{2k}] \in N_k$. In fact we have:

$$f_{i_0i_1}f_{i_1i_2}\cdots f_{i_{2k}i_0} = \prod_{n=1}^{k} 2t_n^{2p(n)+1}u_n^{p(0)+p(1)+\cdots+p(n-1)-n} \tag{7.24}$$

*Proof of (7.24).* A factor of 2 occurs for each descending pair of indices $(i > j)$. $t_n$ occurs each of the $p(n)+1$ time that $n$ occurs as a first index and each of the $p(n)$ times it occurs as the second of a descending pair. $u_n$ occurs each time there is a descending pair of indices $(i, j)$ where $i > n > j$. But there are $p(0) + p(1) + \cdots + p(n-1)$ descending pairs $(i, j)$ where $j < n$ and out of these there are exactly $n$ pairs where $i \leq n$. □

Consequently,

$$\mathrm{Tr}(B^{2k+1}) = \tag{7.25}$$

$$(2k+1)\sum_p |N_k(p)| \prod_{n=1}^{k} 2t_n^{2p(n)+1}u_n^{p(0)+p(1)+\cdots+p(n-1)-n} \, dt_0dt_1ds_1\cdots dt_kds_k$$

where $|N_k(p)|$ is the number of elements of $N_k(p)$ counted with sign, i.e., the number of $(+)$ elements minus the number of $(-)$ elements.

The integral of (7.24) with respect to $t_1, \cdots, t_k$ is easy to compute:

$$\int_0^1 2t^{2a+1}(1-t^2)^b dt = \int_0^1 s^a(1-s)^b ds = \frac{a!b!}{(a+b+1)!}$$

Therefore

$$\iint_{0\le t_1,t_2,\cdots,t_k\le 1} f_{i_0i_1}\cdots f_{i_{2k}i_0}\,dt_1\cdots dt_k$$
$$= \prod_{n=1}^{k} \frac{p(n)!(p(0)+p(1)+\cdots+p(n-1)-n)!}{(p(0)+p(1)+\cdots+p(n)-n+1)!}$$

By a lucky coincidence we will see that:

$$|N_k(p)| = \prod_{n=1}^{k} \frac{(p(0)+p(1)+\cdots+p(n)-n+1)!}{p(n)!(p(0)+p(1)+\cdots+p(n-1)-n)!} \tag{7.26}$$

Therefore every summand in (7.25) has integral equal to 1 and we are left with $2k+1$ times the number of summands as given in 7.9.3.

**Theorem 7.9.5.**

$$\iint_{0\le t_1,s_1,\cdots,t_k,s_k\le 1} \mathrm{Tr}(B^{2k+1}) = (2k+1)P_k dt_0 = \frac{(2k+1)!}{k!(k+1)!}dt_0$$

This allows us to compute the integral in Lemma 7.8.4.

**Corollary 7.9.6.**

$$\int_{I^{2k+1}} \mathrm{Tr}(T^{2k+1}+(T^*)^{2k+1}) =$$
$$-\frac{(2k+1)!}{(k+1)!N^k}(\mathcal{L}_{k+1}(z)+(-1)^k\mathcal{L}_{k+1}(\bar z)) + \text{lower terms}$$

*Proof.* The relationship between the matrix 1-forms $T$ of §7.8 and $B$ above is given by:

$$T_r = \frac{-Ny}{r(1-y)}B_r,\quad T_{t_i} = \frac{-y}{1-y}B_{t_i},\quad T_{s_i} = -\ln r B_{s_i}$$

Therefore,

$$\mathrm{Tr}(T^{2k+1}) = -\frac{N\ln^k r}{r}\frac{y^{k+1}}{(1-y)^{k+1}}\mathrm{Tr}(B^{2k+1})$$

and

$$\int_{I^{2k+1}} \mathrm{Tr}(T^{2k+1}+(T^*)^{2k+1}) =$$
$$-\frac{(2k+1)!}{k!(k+1)!}\int_0^1 \frac{dr}{r}N\ln^k r\left(\frac{y^{k+1}}{(1-y)^{k+1}}+(-1)^k\frac{\bar y^{k+1}}{(1-\bar y)^{k+1}}\right)$$
$$= -\frac{(2k+1)!}{(k+1)!N^k}(\mathcal{L}_{k+1}(z)+(-1)^k\mathcal{L}_{k+1}(\bar z)) + \text{lower terms}$$

by Prop. 7.7.4. □

### 7.9.3 Counting necklaces

The rest of this section is devoted to the proof of the formula (7.26) counting with sign the number of elements of $N_k(p)$ for a fixed descending partition $p$ of a fixed integer $k$. We will proceed by induction on $n \leq k$.

For each nonnegative $n \leq k$ let $C_n = C(p(0), p(1), \cdots, p(n))$ be the set of all necklaces with $\sum_{i=0}^{n}(p(i)+1)$ beads, $p(i)+1$ colored $i$ for $i = 0, \cdots, n$, together with $2n+1$ marked relations between consecutive beads of the form:

$$\begin{array}{lll} (\ell_i) & * \leq i & \text{for } i = 0, \cdots, n \\ (g_i) & i > * & \text{for } i = 1, \cdots, n \end{array}$$

For example for $n = 2, k = 6$ an element of $C(3, 2, 1)$ has $4+3+2 = 9 \leq 2k+1$ beads with $2n+1 = 5$ relations:

$$[0 \leq 0 \leq 1, 2 > 1, 0, 1 > 0, 0 \leq 2] \tag{7.27}$$

starting with the mandatory $\ell_0$ the sequence of marked relations in this element is: $\ell_0, \ell_1, g_2, g_1, \ell_2$ making the sign positive. (Each element of $C_n$ has a sign $(\pm)$ according to whether the $2n+1$ marked relations form an even or odd permutation of $\ell_0, \ell_1, g_1, \ell_2, g_2, \cdots, \ell_n, g_n$.)

If $n < k$ then there are more numbers (beads) than relations. So the numbered beads form (linear) strings of one or more elements connected by zero or more relations. In the example above there are 4 such strings. When $n = k$ there is exactly one (circular) string with $2k+1$ beads and $2k+1$ relations.

Let $D_n$ be the subset of $C_n$ consisting of all elements having exactly one odd string and all other strings even where the parity of a string is given by the number of relations, not the number of beads. Then $D_k = C_k$. Since there are $2n+1$ relations, each element of $C_n$ has an odd number of odd strings.

Claim 1 $|D_n| = |C_n|$ for all $0 \leq n \leq k$.

*Proof.* Each necklace in $C_n - D_n$ must have at least three odd strings. A sign reversing involution on this set is thus given by interchanging the first two odd strings that occur after the string containing a fixed relation, say $g_n$ $(n > *)$. □

It now suffices to prove the following formula.

$$|D_n| = \prod_{i=1}^{n} \frac{(p(0)+p(1)+\cdots+p(i)-i+1)!}{p(i)!(p(0)+p(1)+\cdots+p(i-1)-i)!} \tag{7.28}$$

for $0 \leq n \leq k$. ($|D_0| = 1$.)

*Proof of (7.28).* For $n = 0$, $C_0 = D_0$ contains only one element: $[0 \leq 0, 0, \cdots, 0]$ which is positive. So $|C_0| = |D_0| = 1$.

Suppose that $1 \leq n \leq k$ and the formula holds for $n-1$. Then we will count the number of elements of $C_n$ corresponding to each element of $D_{n-1}$.

Any element of $C_n$ is a necklace with $p(n)+1$ beads labelled $n$. There are exactly two relations involving these highest numbered bead ($\ell_n : * \leq n$ and

$g_n : n > *$). By deleting these beads and the two relations $\ell_n, g_n$ we obtain an element of $C_{n-1}$ and this describes a mapping

$$\pi : C_n \to C_{n-1}$$

For example, the element of $C_2$ given in (7.27) is mapped to the following element of $D_1 \subseteq C_1$.

$$[0 \le 0 \le 1, 1, 0, 1 > 0, 0] \tag{7.29}$$

Claim 2 $|C_n| = |\pi^{-1} D_{n-1}|$

*Proof.* A necklace in $C_n - \pi^{-1} D_{n-1} = \pi^{-1}(C_{n-1} - D_{n-1})$. has the property that the lower numbered beads form at least 3 odd strings. We can transpose the first two such strings after the relation $g_n$ $(n > *)$. This will not affect the validity or position of the two relations $\ell_n, g_n$ so describes a sign reversing involution on the set $C_n - \pi^{-1} D_{n-1}$. □

The elements of $\pi^{-1} D_{n-1}$ fall into two disjoint subsets $A, B$ depending on whether or not the two relations $\ell_n, g_n$ involve the same highest numbered bead. For example, (7.27) is in the subset $B$.

Subset $A$ (The same bead colored "$n$" is used in both relations: $* \le n > *$.) The necklaces in this subset can be formed as follows.

1. Taking any element of $D_{n-1}$.
2. Choosing one of the $p(0) + \cdots + p(n-1) - n + 1$ strings of lower beads and inserting the pair of relations "$\le n >$" on its left.
3. Inserting the remaining $p(n)$ high beads into the $p(0) + \cdots + p(n-1) - n + 1$ gaps.

Since the pair of added relations $\ell_n . g_n$ are consecutive they do not change the parity of the necklace from $D_{n-1}$. Thus

$$|A| = |D_{n-1}|(p(0) + \cdots + p(n-1) - n + 1)\binom{p(0) + \cdots + p(n) - n}{p(n)}$$

Subset $B$ (Two separate beads labelled "$n$" are linked by the relations $\ell_n, g_n$.) Let $B_0$ be the set of all necklaces $b \in B$ so that the relation $\gamma_n : n > *$ is placed on the left of the single odd string in $\pi(b) \in D_{n-1}$.

Claim 3 $|B_0| = |B|$

*Proof.* If $b \in B - B_0$ (like example 7.27 above) then the consecutive pair of symbols $(\le, n)$ in $b$ can be interchanged with the unique odd string of lower beads. For example 7.27 we would switch $\le 2$ with $1 > 0$ to obtain:

$$[0 \le 0 \le 1, 2 > 1, 0 \le 2, 1 > 0]$$

(if $b \in B_0$ we would get "$2 > \le 2$".) This gives a sign reversing involution on $B - B_0$. □

The necklaces in $B_0$ can be constructed as follows.

1. Take any element of $D_{n-1}$.
2. Attach "$n >$" in front of the unique odd string of lower numbers.
3. Insert the remaining $p(n)$ high beads in the $p(0) + \cdots + p(n-1) - n + 1$ gaps.
4. Place a "$\geq$" symbol on the left of one of these $p(n)$ high beads.

After steps (2),(3) the strings are all even. Therefore the sign of the resulting necklace is the same regardless of where the last symbol "$\leq$" ($\ell_n$) is placed. This includes the one place where it is not allowed: to the left of $g_n : n > *$. But if we were to place $\ell_n$ to the left of $g_n$ we would get an element of subset $A$ with the same sign as its projection to $D_{n-1}$. Therefore

$$|B| = |B_0| = |D_{n-1}|\binom{p(0)+\cdots+p(n)-n}{p(n)}p(n)$$

Combining $A, B$ we have:

$$|D_n| = |C_n| = |\pi^{-1}D_{n-1}| = |A| + |B| =$$

$$|D_{n-1}|\binom{p(0) + \cdots + p(n) - n}{p(n)}(p(0) + \cdots + p(n-1) - n + 1 + p(n))$$

$$= \prod_{i=1}^{n} \frac{(p(0) + p(1) + \cdots + p(i) - i + 1)!}{p(i)!(p(0) + p(1) + \cdots + p(i-1) - i)!}$$

by induction on $n$. This proves formulas (7.28) and thus (7.26) as claimed. □

## 7.10 Statement of the Theorem

The counting of necklaces gives the final answer to the computation of the higher Franz-Reidemeister torsion invariant for the case of the "universal bundle" $\gamma^{\otimes k} : E \to (S^2)^k$.

**Theorem 7.10.1.** *The higher FR-torsion invariant of the principal $S^1$-bundle $\gamma^{\otimes k}$ over $(S^2)^k$ when evaluated on the orientation class $[(S^2)^k]$ gives*

$$\langle \tau_k(\gamma^{\otimes k}), [(S^2)^k] \rangle = -\frac{1}{2i^k}(\mathcal{L}_{k+1}(z) + (-1)^k \mathcal{L}_{k+1}(\bar{z}))$$

$$= -(-1)^{[k/2]}(\Re or \Im)\mathcal{L}_{k+1}(z) = -\Re\left(\frac{1}{i^k}\mathcal{L}_{k+1}(z)\right)$$

*Proof.* The trace integral in the definition of the higher FR-torsion is reduced to the integral in 7.9.6 by 7.3.2, 7.8.1 and 7.8.4. As already noted in Theorem 7.6.2 this integral is evaluated at $N^k(k+1)!$ points in $(S^2)^k$ and our definition of $\tau_k$ divides the integral by $(2k+1)!2i^k$. □

**Theorem 7.10.2.** *Let $E \to B$ be the $S^1$-bundle associated to a complex line bundle $\lambda$ over $B$. Let $E_n = E/(\mathbb{Z}/n)$ be the quotient of $E$ by the free action of the group of $n$-th roots of unity. Take the $U(1)$ coefficient system on $E_n$ with holonomy $z^{-1} \neq 1$ on each fiber ($z$ is the clockwise holonomy). Then*

$$\tau_k(E_n) = -\frac{n^k}{k!}\Re\left(\frac{1}{i^k}\mathcal{L}_{k+1}(z)\right) c_1(\lambda)^k$$

**Remark 7.10.3.** For an ordinary coefficient system over $E_n$ we need to assume that $z$ is an $n$-th root of unity. The extension to all $z \neq 1$ in $U(1)$ using isovariant $K$-theory was defined in Chapter II. The case $z = 1$ is settled in Theorem 5.7.13 with the same formula. (Note that Chapters V and VI come logically after this chapter.)

*Proof.* For the bundle $\gamma^{\otimes k}$ the first Chern class is the sum of the generators of $H^2((S^2)^k; \mathbb{Z})$ and therefore

$$\langle c_1(\gamma^{\otimes k})^k, [(S^2)^k]\rangle = k!$$

Comparing this with the formula in 7.10.1 we see that our theorem holds for $\gamma^{\otimes k}$. We showed in Chapter II that the higher FR-torsion $\tau_k(\lambda)$ of a principal $S^1$ bundle $\lambda$ is always a fixed multiple of ($\Re$ or $\Im$)$\mathcal{L}_{k+1}(z)c_1(\lambda)^k$. Therefore the theorem holds for all $\lambda$ when $n = 1$. The extension to all $n$ follows from the transfer formulas of Chapter II. □

# Chapter 8

# Framed Graphs

## Introduction

The space of graphs has been used by M. Culler, K. Vogtmann, A. Hatcher, R. Gersten to study the homology of the automorphism groups of free groups ([CV86], [Hat95], [HV98b], [Ger83]). Fat graphs or ribbon graphs are used by R. Penner, J. Harer , M. Kontsevich and many others to study the moduli space of Riemann surfaces ([Pen93], [Har88], [Kon92]). Our goal is to define the higher FR-torsion on these spaces of graphs and to show how they are related to the known cohomology classes such as the Miller-Morita-Mumford classes.

In order to construct the higher FR-torsion invariants we need to put "framed structures" on graphs. The idea is similar to Hatcher-Vogtmann's "Cerf theory on graphs" [HV98a] since it is the cell structure on the graph that would arise from a framed function on a tubular manifold neighborhood of the graph when embedded in Euclidean space. The theorem is:

**Theorem 8.0.1 (Framed graph theorem 8.4.21).** *Any continuous family of metric graphs admits a continuous family of framed structures which is unique up to homotopy and any two families of framings are homotopic by a framed homotopy which is fixed on the underlying family of metric graphs.*

This implies that any subcategory of the category of graphs is homotopy equivalent to the corresponding category of framed graphs.

The main property of framed graphs is that a family of framed graphs parametrized by a space $K$ gives a continuous mapping

$$K \to |\mathcal{W}h_\bullet(\mathbb{Z}, 1)|.$$

There are several ways that this can be modified to give a mapping

$$K \to |\mathcal{W}h^h_\bullet(\mathbb{Z}, 1)|$$

and produce higher FR-torsion invariants.

- When the associated homology bundle over $K$ is trivial.
- Using Poincaré duality to kill the homology bundle (in the case of fat graphs).

- Using the Volodin space to kill the homology.

Section 8.1 goes over the definition of the category of graphs and the space of metric graphs and the more or less obvious fact that these have the same homotopy type.

In Section 8.2 we define framed graphs. Basically these are graphs together with cell decompositions of their geometric realizations where the nodes of the graph are not necessarily vertices of the cell decomposition.

Section 8.3 shows that the space of framed structures on the interval $I$ and the circle $S^1$ are contractible by comparing them with (combinatorial versions of) the corresponding spaces of framed functions.

In Section 8.4 we prove the framed graph theorem which says that any continuous family of graphs admits a continuous family of framings unique up to framed homotopy.

Section 8.5 contains applications of the framed graph theorem. We construct the higher FR-torsion for various spaces of graphs and use it to get some results about the homology of the related groups. We also compute the higher FR-torsion of the Torelli group in terms of the Miller-Morita-Mumford classes.

In Section 8.6 we give a detailed proof of the fact that punctured surfaces have canonical framed functions without maxima. By the framed graph theorem this is equivalent to the well-known fact that a punctured surface has a canonical "fat graph" as its core.

## 8.1 Graphs and metric graphs

We consider unoriented graphs with loops, multiple edges and no isolated points. We use Gersten's definition [Ger83]

**Definition 8.1.1.** A *graph* is a finite set $G$ together with an involution ($e \mapsto \overline{e}$) and a retraction $t$ onto the set $G_0$ of fixed points of the involution which we call *nodes*. We also assume that $t^{-1}(v)$ contains at least one element (other than $v$) for each $v \in G_0$. We will assume that $G$ is a subset of some fixed infinite set $\Omega$ so that the collection of all graphs forms a set. We assume that $\Omega$ contains $\mathbb{R}$.

Let $\widetilde{G}_1 = G - G_0$. We call the elements of $\widetilde{G}_1$ the *oriented edges* of $G$. The *edges* of $G$ are defined to be the equivalence classes $\{e, \overline{e}\}$. Thus $G_1 = \widetilde{G}_1/\sim$ is the set of edges. The source and target maps $s, t : \widetilde{G}_1 \to G_0$ are defined by $s(e) = t(\overline{e})$.

The *degree* of each node $v \in G_0$ can be defined to be the number of oriented edges having $v$ as a source. This is equal to the number of oriented edges having $v$ as target. Nodes of degree 1 are called *leaves*.

An *oriented graph* is defined to be a graph having a preferred orientation for each edge. Thus, for an oriented graph $G$, we have well defined mappings $s, t : G_1 \to G_0$ and each node has an *in-degree* and *out-degree* which add up to its degree in the underlying unoriented graph.

A *path* in a graph $G$ is defined to be a sequence of oriented edges $e_0, e_1, \cdots, e_n$ in $\widetilde{G}_1$ so that $t(e_{i-1}) = s(e_i)$ for $1 \leq i \leq n$. The path is called a *cycle* if in addition we have $t(e_n) = s(e_0)$. If the graph $G$ is oriented and the orientation of the path or cycle agrees with the orientation of $G$ then we call it an *oriented path* or *oriented cycle*. Otherwise we call it an *unoriented path* or *unoriented cycle*.

Let $C_0(G)$ be the free abelian group generated by $G_0$ and let $C_1(G)$ be the abelian group generated by $\widetilde{G}_1$ modulo the relation that the sum of the two orientations of the same edge is zero. Then the homology groups $H_1(G), H_0(G)$ are defined to be the kernel and cokernel of the natural map

$$\partial : C_1(G) \to C_0(G)$$

given by $\partial e = t(e) - s(e)$. (So each loop goes to zero.)

The *geometric realization* of a graph $G$ is given by

$$|G| = \left(G_0 \coprod \widetilde{G}_1 \times I\right) / \sim$$

where we identify $(e, 0), (e, 1) \in \widetilde{G}_1 \times I$ with $s(e), t(e) \in G_0$ respectively and we also identify $(e, t)$ with $(\overline{e}, 1 - t)$. Note that $|G|$ can be given the structure of a metric space where the length of each edge is 1 and the distance between any two points in the same component is equal to the length of the shortest path. (Set $d(x, y) = 1$ plus the total length of $G$ if $x, y$ lie in different components.) Note also that there is a well defined isomorphism:

$$H_1(|G|; \mathbb{Z}) \cong H_1(G).$$

If two graphs have homeomorphic realizations then we will call them *topologically equivalent*. Every connected graph is topologically equivalent to a graph without nodes of degree 2 except in one case, namely, when the realization is a circle.

A *morphism of graphs* $f : G \to H$ is a set mapping which commutes with the involution ($f(\overline{e}) = \overline{f(e)}$) and target map ($f(t(e)) = t(f(e))$). A *morphism of oriented graphs* will also preserve the preferred orientation of the edges. Note that in either case the inverse image of each subgraph of $H$ is a subgraph of $G$. Also note that $f$ induces a continuous mapping $|f| : |G| \to |H|$ which is linear on edges in the obvious sense.

**Definition 8.1.2.** An *admissible graph epimorphism* is a morphism $g : G \to H$ so that

1. The inverse image of every node of $H$ is a tree which we call a *collapsing tree* in $G$.

2. Each edge of $H$ is the image of exactly one edge of $G$ which we sometime call a *long edge*.

The category of finite graphs and admissible graph epimorphisms will be denote $\mathcal{GE}$.

**Definition 8.1.3.** If $X$ is a *topological graph*, i.e., a space homeomorphic to $|G|$ for some graph $G$, then let $\mathcal{GE}(X)$ denote the *category of graph structures on* $X$. This is defined to be the category of all pairs $(G,[\phi])$ where $G$ is a graph and $[\phi]$ is an isotopy class of homeomorphisms $\phi : |G| \to X$. A morphism $(G,[\phi]) \to (G',[\phi'])$ is an admissible graphs epimorphism $g : G \to G'$ so that $[\phi' \circ g] = [\phi]$.

Note that a topological graph $X$ has a well defined subset $X_0$ consisting of points of degree not equal to 2. These are the *nodes* of $X$. (The *degree* of any point in any topological space can be defined to be the minimal number of components of a deleted neighborhood.) Let $X_1$ be the set of components of the complement of $X_0$ in $X$. These are the *edges* of $X$. Since each edge has two orientations we have a two fold covering $\widetilde{X}_1$ of $X_1$ consisting of all oriented edges of $X$.

Any automorphism (self homeomorphism) $\phi$ of a topological graph $X$ induces a $\mathbb{Z}/2$-equivariant permutation of $X_0 \coprod \widetilde{X}_1$. This permutation is the identity if and only if $\phi$ is isotopic to the identity. Therefore $X$ has only a finite number of isotopy classes of automorphisms and the forgetful functor $\mathcal{GE}(X) \to \mathcal{GE}$ is a finite covering of its image. (A *covering* $\tilde{\mathcal{C}} \to \mathcal{C}$ is any functor which induces a covering map of nerves.) We denote this image by $\mathcal{GE}(X/\cong)$. This is the full subcategory of $\mathcal{GE}$ of graphs $G$ whose realizations are homeomorphic to $X$.

It is more or less obvious that $\mathcal{GE}(X)$ is contractible if $X$ has no circle components. We go through the details in the case $X = I = [0,1]$.

**Lemma 8.1.4.** *$|\mathcal{GE}(I)|$ is contractible and the forgetful functor $\mathcal{GE}(I) \to \mathcal{GE}(I/\cong)$ gives a two fold covering map on realizations.*

*Proof.* It is obvious that an isotopy class of homeomorphisms $|G| \to I$ is determined by its values at the endpoints. Any morphism in $\mathcal{GE}(I/\cong)$ either preserves or reverses these values. Therefore the forgetful functor $\mathcal{GE}(I) \to \mathcal{GE}(I/\cong)$ is a two-fold covering functor inducing a two-fold covering map on realizations.

A null-deformation of $\mathcal{GE}(I)$ is given as follows. For each object $A = (G,[f])$ of $\mathcal{GE}(I)$ add one leaf, say $a$, at the endpoint $f^{-1}(0)$. Let $g(A)$ be the resulting object. We have a natural morphism $g(A) \to A$ given by collapsing the new leaf. If we collapse the original graph $G$ to a single node, say $b$, we get a natural morphism from $g(A)$ to a new object $h(A)$. Since $h(A)$ has only two nodes, it is unique up to homotopy. Therefore the functor $h$ is null homotopic. Since $id \simeq g \simeq h$, $\mathcal{GE}(I)$ is contractible. □

**Proposition 8.1.5.** *Suppose that $X$ is homeomorphic to the realization of a graph $G(X)$ with no nodes of degree 2, i.e., none of its components are circles. Then:*

1. *$|\mathcal{GE}(X)|$ is contractible.*
2. *$|\mathcal{GE}(X)|$ is a regular covering space of $|\mathcal{GE}(X/\cong)|$ with finite covering group $Aut(G(X))$. Thus*

$$|\mathcal{GE}(X/\cong)| \simeq BAut(G(X)).$$

3. *$\mathcal{GE}(X/\cong)$ contains as a deformation retract the full subcategory of all graphs isomorphic to $G(X)$.*

4. *$\pi_0 Aut(X) \cong Aut(G(X))$ and the connected component $Aut_0(X)$ of the identity is contractible. (Here $Aut(X)$ is the group of self-homeomorphisms of $X$ with the compact-open topology.)*

**Remark 8.1.6.** Note that $G(X)$ can be given by $G_0(X) = X_0$ (the set of points of degree $\neq 2$) and $G_1 = X_1$ (the set of components of $X - X_0$). When $X$ is allowed to have circular components we modify this by letting $G_0$ be the disjoint union of $X_0$ and the set of circular components of $X$ so that the corresponding edge is a loop at that node, i.e., we formally add a node to each circle component.

*Proof.* Since $X$ has no circular components, $\mathcal{GE}(X)$ is isomorphic to a product of copies of $\mathcal{GE}(I)$:

$$\mathcal{GE}(X) \cong \prod_{e \in X_1} \mathcal{GE}(I)$$

Thus $|\mathcal{GE}(X)|$ is contractible by Lemma 8.1.4. Statement (4) are obvious and (2) follows from (4) since the isotopy classes of homeomorphisms $|G| \to X$ are in 1-1 correspondence with $\pi_0 Aut(X)$ for any $G$ in $\mathcal{GE}(X/\cong)$. Finally, (3) follows from (1) and (2). □

Let $\mathcal{GE}^+$ be the full subcategory of $\mathcal{GE}$ of all graphs having no circle components. Then $\mathcal{GE}^+$ is the disjoint union of all $\mathcal{GE}(X/\cong)$ in 8.1.5 above so it contains as a deformation retract the full subcategory $\mathcal{GE}_0$ of all graphs without nodes of degree 2. Since all retractions onto a deformation retract are homotopic we have:

**Corollary 8.1.7.** *The functor $\mathcal{GE}^+ \to \mathcal{GE}_0$ sending $G$ to $G(|G|)$ is a homotopy equivalence on realizations.*

By Proposition 8.1.5 and an additional argument about circles, we will see that the realization of $\mathcal{GE}$ is homotopy equivalent to the space of metric graphs defined below.

**Definition 8.1.8.** A *metric graph* is defined to be a compact metric space $X$ having the following properties.

1. $X$ is a topological graph, i.e., homeomorphic to the geometric realization $|G|$ of a graph $G$.

2. For any homeomorphism $f : |G| \to X$, the components of $X - f(G_0)$ are *locally isometrically equivalent to* $\mathbb{R}$, i.e., they are covered by open sets which are isometrically equivalent to open intervals.

3. If $x, y \in X$ lie in the same component then each minimal path from $x$ to $y$ in $X$ is isometrically equivalent to a closed interval of length $d = d(x, y)$.

4. $d(x, y) = 1$ if $x, y$ lie in different components of $X$.

5. Each component of $X$ has total length equal to 1.

Since our graphs do not have isolated nodes, any metric space $X$ satisfying (1)-(4) above can be modified to satisfy (5) by dividing the distance function on each component by the total length of that component. We denote this by $rX$ and call $r$ the *rescaling operation.*

It is clear that every metric graph $X$ is homeomorphic to the realization of the graph $G(X)$ given in 8.1.6. Because of conditions (3),(4) in the definition above, $X$ is uniquely determined up to isometry by the graph $G(X)$ and the length $\lambda(e)$ of each $e \in X_1 = G_1(X)$.

If $X, Y$ are metric graphs then we note that any morphism of the underlying graphs gives a unique continuous mapping $f : X \to Y$ which is *linear* on each edge in an obvious sense.

Suppose that $G$ is a graph and $k \geq 0$. For each edge $e \in G_1$ choose $k+1$ real numbers

$$0 \leq \lambda_0(e) \leq \lambda_1(e) \leq \cdots \leq \lambda_k(e)$$

not all of which are zero so that $\lambda_0(e)$ is nonzero on at least one edge in each nontrivial cycle in $G$. For each $t \in \Delta^k$ and $e \in G_1$ let

$$\lambda_t(e) = \sum_{i=0}^{k} t_i \lambda_i(e)$$

where $t_i$ are the barycentric coordinates of $t$. Then we obtain a family of metric graphs $X = (X_t, t \in \Delta^k)$ given by:

$$X_t = r|(G, \lambda_t)| = r\left(G_0 \coprod \coprod_{e \in \widetilde{G}_1} [0, \lambda_t(e)]\right) / \sim$$

where $r$ is the rescaling operator. We will call such a family a *standard* $\Delta^k$-family of metric graphs and denote it by $X(G, \lambda)$. We define the *total space* $EX(G, \lambda)$ of this family to be the space over $\Delta^k$ given by

$$EX(G, \lambda) = r\left(G_0 \times \Delta^k \coprod \coprod_{e \in \widetilde{G}_1} \{(x, t) \in I \times \Delta^k \, : \, x \leq \lambda_t(e)\}\right) / \sim$$

**Definition 8.1.9.** We define the *space of metric graphs* to be the simplicial set $\mathcal{MG}_\bullet$ which, in degree $k$, consists of $\Delta^k$-families of metric graphs which are isometrically equivalent to the standard families constructed above and which live in some fixed universe $\Omega$.

Take any $k$-simplex

$$(G^0 \leftarrow G^1 \leftarrow \cdots \leftarrow G^k) \in \mathcal{N}_k \mathcal{GE}^{op}$$

in the nerve of $\mathcal{GE}^{op}$. For each edge $e \in G_1^k$ let

$$0 \leq \lambda_0(e) \leq \lambda_1(e) \leq \cdots \leq \lambda_k(e) = 1$$

be given by $\lambda_i(e) = 0$ if $e$ collapses to a node in $G^i$ and $\lambda_i(e) = 1$ if not. Then $\rho_k(G^0 \leftarrow \cdots \leftarrow G^k) = X(G^k, \lambda) \in \mathcal{MG}_{\bullet k}$ gives a simplicial map $\rho : \mathcal{N}_\bullet \mathcal{GE}^{op} \to \mathcal{MG}_\bullet$.

Suppose that $\mathcal{MG}_\bullet{}^+$ is the simplicial subset of $\mathcal{MG}_\bullet$ of all metric graphs without circle components (analogous to the subcategory $\mathcal{GE}^+$ of $\mathcal{GE}$). Then we have a functor

$$\gamma : \operatorname{simp} \mathcal{MG}_\bullet{}^+ \to \mathcal{GE}^{+op}$$

sending each $k$ simplex $(X_t, t \in \Delta^k)$ to $G(X_w)$ where $w$ is the last vertex of $\Delta^k$. The composition

$$\mathcal{N}_\bullet \operatorname{simp} \mathcal{MG}_\bullet{}^+ \xrightarrow{\mathcal{N}_\bullet \gamma} \mathcal{N}_\bullet \mathcal{GE}^{+op} \xrightarrow{\rho} \mathcal{MG}_\bullet{}^+$$

sends a sequence of simplices $\sigma_0 \subseteq \sigma_1 \subseteq \cdots \subseteq \sigma_k$ first to a sequence of admissible epimorphisms

$$G(X_{w_0}) \leftarrow G(X_{w_1}) \leftarrow \cdots \leftarrow G(X_{w_k})$$

where $w_i$ is the last vertex of $\sigma_i$ and this goes to a $\Delta^k$-family of metric graphs which is isomorphic to the $k$-face $a^*\sigma_k$ of $\sigma_k$ given by $a(i) = w_i$. In other words, $\rho \circ \mathcal{N}_\bullet \gamma$ is the last vertex map on $\mathcal{MG}_\bullet{}^+$ and is thus a homotopy equivalence. (([Wal85],p.359)

In the special case when the $X_{w_i}$ are all homeomorphic, the graphs $G(X_{w_i})$ are isomorphic so we get:

**Proposition 8.1.10.** *If $X_{v_i} = r|(G, \lambda_i)|$ are homeomorphic to a fixed metric graph $X$ for all vertices $v_i$ of $\Delta^k$ then the total space $EX(G, \lambda)$ is fiberwise homeomorphic to $X \times \Delta^k$ over $\Delta^k$.*

The reverse composition

$$\operatorname{simp} \mathcal{N}_\bullet \mathcal{GE}^{+op} \xrightarrow{\operatorname{simp}\ \rho} \operatorname{simp} \mathcal{MG}_\bullet{}^+ \xrightarrow{\gamma} \mathcal{GE}^{+op}$$

takes a simplex $((G^0 \leftarrow \cdots \leftarrow G^k), [k])$ in $\mathcal{N}_\bullet \mathcal{GE}^{+op}$ first to $(X(G, \lambda), [k])$ in $\operatorname{simp} \mathcal{MG}_\bullet{}^+$ and then to $G(|G^k|)$, i.e., $\gamma \circ \operatorname{simp}\ \rho$ is the last object map followed by the functor in Corollary 8.1.7. Since both $\gamma$ and $\rho$ preserve the number of nodes $\neq 2$ we get:

**Proposition 8.1.11.** *Geometric realization $\rho : N_\bullet \mathcal{GE}^{op} \to \mathcal{MG}_\bullet$ induces a homotopy equivalence*

$$N_\bullet \mathcal{GE}^{+op} \to \mathcal{MG}_\bullet{}^+.$$

*Furthermore, for any $n > 0$, we get an induced homotopy equivalence*

$$N_\bullet \mathcal{GE}^{[n]+op} \to \mathcal{MG}_\bullet{}^{[n]+}$$

*where the $^{[n]}$ refers to connected graphs with exactly $n$ nodes of degree $\neq 2$.*

For any topological graph $X$, let $\mathcal{MG}_\bullet(X/\cong)$ be the component of $\mathcal{MG}_\bullet$ of all metric graphs homeomorphic to $X$. Then by restricting to the components of $N_\bullet \mathcal{GE}^{[n]+op}$ and $\mathcal{MG}_\bullet{}^{[n]+}$ we get:

**Corollary 8.1.12.** *Geometric realization induces a homotopy equivalence:*

$$N_\bullet \mathcal{GE}(X/\cong)^{op} \to \mathcal{MG}_\bullet(X/\cong)$$

*for all topological graphs $X$ without circle components.*

We look at the remaining case $X = S^1$. Let $\mathcal{MG}_\bullet(S^1)$ be the two-fold covering space of $\mathcal{MG}_\bullet(S^1/\cong)$ given in degree $k$ by pairs $(X, o)$ where $X \in \mathcal{MG}_{\bullet k}(S^1/\cong)$ and $o$ is an orientation of the circle fibers of $X$. Then geometric realization gives a simplicial map $\tilde{\rho} : \mathcal{N}_\bullet \mathcal{GE}(S^1) \to \mathcal{MG}_\bullet(S^1)$ and we get a map of covering spaces:

$$\begin{array}{ccc} \mathcal{N}_\bullet \mathcal{GE}(S^1) & \xrightarrow{\tilde{\rho}} & \mathcal{MG}_\bullet(S^1) \\ \downarrow & & \downarrow \\ N_\bullet \mathcal{GE}(S^1/\cong) & \xrightarrow{\rho} & \mathcal{MG}_\bullet(S^1/\cong) \end{array}$$

**Lemma 8.1.13.** $|\mathcal{MG}_\bullet(S^1)| \simeq BU(1)$ *and* $|\mathcal{MG}_\bullet(S^1/\cong)| \simeq BO(2)$.

*Proof.* This is obvious since both $|\mathcal{MG}_\bullet(S^1)|$ and $BU(1)$ are classifying spaces for oriented circle bundles and both $|\mathcal{MG}_\bullet(S^1/\cong)|$ and $BO(2)$ classify unoriented circle bundles. □

Suppose that $(G, o)$ is an object in $\mathcal{GE}(S^1)$ and $G$ has $n$ nodes $v_1, \cdots, v_n$ and $n$ edges $(v_i, v_{i+1})$ (where indices are taken modulo $n$) so that $\sum(v_i, v_{i+1})$ represents $o \in H_1(G)$. Then let $p : \mathbb{R} \to |G|$ be the unique local isometry sending $i \in \mathbb{Z}$ to $v_i$ and $(i, i+1)$ to $(v_i, v_{i+1})$.

Let $\Delta_\bullet(G, o)$ be the simplicial set of all continuous mappings $f : \Delta^k \to |G|$ which lift to linear mappings $\tilde{f} : \Delta^k \to \mathbb{R}$ sending each vertex of $\Delta^k$ to an integer with image of length $\leq n$, i.e.,

$$\max_{t \in \Delta^k} \tilde{f}(t) - \min_{t \in \Delta^k} \tilde{f}(t) \leq n.$$

**Lemma 8.1.14.** $\Delta_\bullet(G, o)$ *is a deformation retract of the simplicial set of all continuous maps* $\Delta^k \to |G|$.

*Proof.* Let $\widetilde{\Delta}_\bullet(G, o)$ be the simplicial set of all linear mappings $\tilde{f} : \Delta^k \to \mathbb{R}$ sending each vertex of $\Delta^k$ to an integer and having image of length $\leq n$. Then $\widetilde{\Delta}_\bullet(G, o)$ is contractible. (To see this, restrict the range of $\tilde{f}$ to the closed interval $[a, b]$. This simplicial subset is contractible for all integers $a \leq b$ by induction on $b - a$.) Consequently, the inclusion of $\widetilde{\Delta}_\bullet(G, o)$ into the total singular set $\mathbb{R}^{\Delta^\bullet}$ of $\mathbb{R}$ is a homotopy equivalence. Factoring out the free action of $n\mathbb{Z}$ by translation on both simplicial sets we get a homotopy equivalence

$$\Delta_\bullet(G, o) \simeq |G|^{\Delta^\bullet}$$

induced by inclusion. □

**Lemma 8.1.15.** *Geometric realization gives homotopy equivalences:*

$$|\mathcal{GE}(S^1)| \simeq |\mathcal{MG}_\bullet(S^1)| \simeq BU(1)$$
$$|\mathcal{GE}(S^1/\cong)| \simeq |\mathcal{MG}_\bullet(S^1/\cong)| \simeq BO(2)$$

**Remark 8.1.16.** $\mathcal{GE}(S^1)$ is also equivalent to the category of cyclically ordered sets.

*Proof.* The universal circle bundle over $\mathcal{MG}_\bullet(S^1)$ is homotopy equivalent to the bisimplicial set $E_{\bullet\bullet}$ which in bidegree $(k,n)$ is the set of all pairs $((X,o),f)$ where $X \in \mathcal{MG}_{\bullet k}(S^1)$ with homology orientation $o$ and

$$f : \Delta^k \times \Delta^n \to E(X)$$

is a continuous map over $\Delta^k$, i.e., a continuous family of continuous maps $f_t : \Delta^n \to X_t$, $t \in \Delta^k$. (This is the bundle with fiber $(S^1)^{\Delta^\bullet}$ associated to the universal $S^1$-bundle over $\mathcal{MG}_\bullet(S^1)$.)

We will show that the pull-back $F_{\bullet\bullet}$ of $E_{\bullet\bullet}$ along

$$\rho : \mathcal{GE}(S^1) \to \mathcal{MG}_\bullet(S^1)$$

is contractible. This will imply that $\rho$ is a homotopy equivalence. The other statement follow easily.

In bidegree $(k,n)$, $F_{kn}$ is the set of all continuous $\Delta^k$-families of continuous mappings

$$f_t : \Delta^n \to X_t = |(G,\lambda_t)|, \; t \in \Delta^k$$

where $X = X(G,\lambda) = \rho(G_0 \leftarrow \cdots \leftarrow G_k)$.

Let $\Delta_\bullet\mathcal{GE}(S^1)$ be the simplicial category consisting, in degree $k$, of all pairs $((G,o),f)$ where $(G,o)$ is an object of $\mathcal{GE}(S^1)$ and $f : \Delta^k \to |G|$ is an element of $\Delta_k(G,o)$. The simplicial operators act on $f$ and morphisms $((G,o),f) \to ((G',o),f')$ are admissible epimorphisms $g : G \to G'$ preserving the orientation and so that $f' = |g| \circ f$.

We claim that $\Delta_\bullet\mathcal{GE}(S^1)$ is contractible and that its nerve is a deformation retract of $F_{\bullet\bullet}$. The second statement follows from Lemma 8.1.14 since the inclusion map

$$\mathcal{N}_k\Delta_\bullet\mathcal{GE}(S^1) \to F_{k\bullet}$$

is homotopy equivalent to a disjoint union of copies of the homotopy equivalence in Lemma 8.1.14:

$$\coprod \Delta_\bullet(G,o) \to \coprod |G|^{\Delta^\bullet}$$

where both disjoint unions are over all $(G_0 \leftarrow \cdots \leftarrow G_k) \in \mathcal{N}_k\mathcal{GE}(S^1)$ where $G_k = G$.

The contractibility of $\Delta_\bullet\mathcal{GE}(S^1)$ is easy. It is enough to show that for each $k$, $\Delta_k\mathcal{GE}(S^1)$ is contractible. For each $((G,o),f)$ in $\Delta_k\mathcal{GE}(S^1)$ we let $g((G,o),f) = ((G',o),f')$ be given by replacing the node $v = f(0)$ in $G$ by a pair $v^-, v^+$ connected by a new edge $(v^-,v^+)$ and letting $f'(0) = v^+$ (and $f'(k) = v^-$ if $f(k) = v$). This gives a functor $g$ from $\Delta_k\mathcal{GE}(S^1)$ to itself with two natural transformations $g \to id$ and $g \to *$ given by collapsing $(v^-,v^+)$ and collapsing all other edges of the graph $G'$ respectively. □

**Theorem 8.1.17.** *The simplicial map*

$$\rho : \mathcal{N}_\bullet \mathcal{GE}^{op} \to \mathcal{MG}_\bullet$$

*given by geometric realization is a homotopy equivalence.*

*Proof.* It is enough to prove this for connected graphs. We can assume that our graphs are homotopy equivalent to circles since all other cases follow from Proposition 8.1.11. In that case each graph $G$ contains a unique subgraph $\psi(G)$ with $|\psi(G)|$ homeomorphic to a circle and there is a unique admissible retraction of $G$ to $\psi(G)$. Consequently, $\mathcal{GE}(S^1/\cong)$ is a deformation retract of this component of $\mathcal{GE}$ and similarly $\mathcal{MG}_\bullet(S^1)$ is a deformation retract of the corresponding component of $\mathcal{MG}_\bullet$. Therefore the theorem follows from Lemma 8.1.15. □

**Definition 8.1.18.** Suppose that $K$ is a topological space. Then by a *continuous family of metric graphs* over $K$ we mean a space $E$ with a map $p : E \to K$ together with a continuous fiberwise metric on $E$ satisfying the following.

1. Each fiber $G_t = p^{-1}(t)$ is a metric graph.
2. Each $v \in K$ has a neighborhood $U(v)$ and a continuous family of linear graph epimorphisms

$$g_t : G_t \to G_v, \ t \in U(v).$$

Note that the pull-back of $E$ along any continuous map $f : L \to K$ is a continuous family of metric graphs over $L$. In particular, any continuous mapping $K \to |\mathcal{MG}_\bullet|$ gives a family of metric graphs over $K$ by pulling back the universal family.

As an example of a well-known general principle we have the following.

**Proposition 8.1.19.** *If $K$ is the realization of a simplicial complex then any continuous family of metric graphs $G_t, t \in K$ is homotopic to the pull-back of the universal family along a continuous mapping $f : K \to |\mathcal{MG}_\bullet|$ which is uniquely determined up to homotopy.*

*Proof.* Take a locally finite covering of $K$ by the open sets $U(v)$. Choose a triangulation $K'$ of $K$ so that each simplex $\sigma$ of $K'$ lies in an open set $U(v(\sigma))$ which we take to be minimal so that $U(v(\sigma))$ contains $U(v(\tau))$ for all faces $\tau$ of $\sigma$. This gives a functor $\sigma \mapsto G_{v(\sigma)}$ from the category of simplices of $K'$ to $\mathcal{GE}$. Take geometric realization and we get a continuous mapping

$$f : K \to |\mathcal{MG}_\bullet|$$

which is simplicial on the first barycentric subdivision $SdK'$ of $K'$.

It is easy to see that the resulting family of metric graphs $G'_t$ is homotopic to the original family $G_t$. (The homotopy can be constructed on each simplex $\sigma$ of $K'$ assuming it exists on $\partial\sigma$ using that fact that $\sigma \times I$ is the cone on $\sigma \times 0 \cup \partial\sigma \times I$ with cone point the barycenter $(b(\sigma), 1)$ over which we have $G'_{b(\sigma)} = G_{v(\sigma)}$.)

A relative construction over $K \times I$ gives the homotopy between any two choices of $f$. □

Let $\mathcal{MG}_\bullet{}^{(n)}$ be the space of all connected metric graphs with Euler characteristic $1-n$, i.e., with first homology $\mathbb{Z}^n$. Then it is obvious that

$$|\mathcal{MG}_\bullet{}^{(1)}| \simeq |\mathcal{MG}_\bullet(S^1/\cong)| \simeq BO(2)$$

(8.1.15) where the deformation retraction

$$\psi : \mathcal{MG}_\bullet{}^{(1)} \to \mathcal{MG}_\bullet(S^1/\cong)$$

is given by stripping off all external edges and $|\mathcal{MG}_\bullet{}^{(0)}|$ is easily seen to be contractible:

**Corollary 8.1.20.** *The space of trees $\mathcal{MG}_\bullet{}^{(0)}$ is contractible.*

*Proof.* Take the universal family of metric trees over $|\mathcal{MG}_\bullet{}^{(0)}|$. For all $0 \leq u \leq 1$, attach an edge of length $u$ to the center of mass of each tree (the unique point at which each component of the complement has total length $\leq \frac{1}{2}$) and rescale the metric on the graph by a factor of $1-u$. By the proposition, this gives a null homotopy of the identity map on $|\mathcal{MG}_\bullet{}^{(0)}|$. □

The remaining cases are given by Culler and Vogtmann [CV86]:

**Theorem 8.1.21.** *For all $n \geq 2$, $|MG^{(n)}| \simeq |\mathcal{GE}^{(n)}|$ has the homotopy type of the classifying space of the group $Out(F_n)$ of outer automorphisms of the free group $F_n$ on $n$ letters.*

The proof of Theorem 8.1.21 depends on the contractibility of "outer space" which can be described as follows. Choose any fixed graph $G_0$ with one node and $n$ edges. Then a *marking* on any graph $G \in \mathcal{GE}^{(n)}$ is defined to be a free homotopy class of homotopy equivalences

$$g : |G_0| \to |G|.$$

The set of all distinct markings of $G$ are in 1-1 correspondence with the outer automorphisms group $Out(F_n)$ of $F_n = \pi_1(|G_0|)$. Therefore $Out(F_n)$ acts freely on the set of marked graphs in $\mathcal{GE}^{(n)}$.

**Definition 8.1.22.** Let $OS(n)$ be category of all pairs $(G, g)$ where $G \in \mathcal{GE}^{(n)}$ with no nodes of degree 1 or 2 and $g$ is a marking on $G$. A morphism $(G, g) \to (G', g')$ in $OS(n)$ is any admissible epimorphism $f : G \to G'$ so that $g' \circ |f| \simeq g$.

Note that $Out(F_n)$ acts freely on $OS(n)$ with quotient the full subcategory of $\mathcal{GE}^{(n)}$ of graphs without nodes of degree 1 or 2. Thus Theorem 8.1.21 follows from the following theorem of Culler and Vogtmann.

**Theorem 8.1.23 ([CV86]).** *$|OS(n)|$ is contractible.*

**Remark 8.1.24.** Culler and Vogtmann took isomorphism classes of marked graphs which is equivalent to choosing one object in each isomorphism class of objects in $OS(n)$. The group $Out(F_n)$ does not act freely on this subcategory, however the action has finite stabilizers and the quotient is finite dimensional proving that $Out(F_n)$ has virtual cohomological dimension $\leq 2n-3$.

Any continuous family of metric graphs on $K$ gives a well defined locally trivial fiberwise homology bundle. When this bundle is trivial we will show below that we obtain a continuous mapping

$$K \to |\mathcal{W}h^h_\bullet(\mathbb{Z}, 1)|$$

well defined up to homotopy. One consequence of this is that the higher FR-torsion invariants for the Torelli group pull back from cohomology classes for the group $Out^h(F_n)$ of all outer automorphisms of $F_n$ which induce the identity on $H_1(F_n)$.

## 8.2 Framed graphs

A framed graph is a finite 1-dimensional cell complex in which 1-cells may be attached on top of other 1-cells according to some partial ordering of the set of cells. As in the case of graphs and metric graphs we first explain the combinatorial structure and add the continuous structure later.

A framed graphs can be viewed as a graph with two kinds of nodes:

1. *smooth* nodes which lie in the interior of 1-cells and
2. *singular* nodes which are the 0-cells of the cell decomposition.

At each smooth node $v$, the framing picks out a set of two incident edges $f(v) = \{e_1, e_2\}$ so that $v$ is the target of $e_1$ and the source of $e_2$. The singular nodes will be called *vertices*. In a morphism of framed graphs, 1-cells will be allowed to collapse only if their target nodes are vertices. A collapsing 1-cell and its target vertex will be called a *collapsing pair*.

As an example take the letter "H" considered as a framed graph with two smooth nodes and four vertices. If we let $a, b$ be the vertical 1-cells and $c$ the horizontal 1-cells then the two smooth nodes are $s(c) \in a$ and $t(c) \in b$ and the four vertices are $s(a), t(a), s(b), t(b)$ with partial ordering given by $s(a), t(a) < a$, $s(b), t(b) < b$ and $a, b < c$. Note that the underlying oriented graph has four possible framed structures depending on the choice of pairs of incident edges at the two smooth nodes.

**Definition 8.2.1.** A *framed graph* $(G, P)$ is an oriented graph $G$ with the following additional structure:

1. A partition of $G_0$ into smooth and singular nodes so that all leaves are singular. The 0-*cells*, or *vertices*, are defined to be the singular nodes.
2. A pair of incident edges $f(v) = \{e_1, e_2\}$ at each smooth node $v$ so that $v$ is the target of $e_1$ and the source of $e_2$ and so that each edge is contained in an oriented path $e_0, e_1, \cdots, e_n$ so that $f(s(e_i)) = \{e_{i-1}, e_i\}$ for $1 \le i \le n$ and so that $e_0 \notin f(s(e_0))$, $e_n \notin f(t(e_n))$. These oriented paths are defined to be the 1-*cells* of the framed graph.

3. A partial ordering of the set $P = P_0 \coprod P_1$ where $P_0$ is the set of vertices and $P_1$ is the set of 1-cells so that each 1-cell $e$ is greater in the partial ordering than the two cells which contain its endpoints.

4. A designated set of disjoint collapsing pairs $(e, v)$ which act together in the partial ordering in the sense that $z \leq v$ iff $z < e$ and $z > v$ iff $z \geq e$ for all $z \in P$.

A *morphism* of framed graphs $(f, \alpha) : (G, P) \to (H, Q)$ is an admissible epimorphism $f : G \to H$ so that

1. $f$ preserves the preferred orientation of the edges.

2. Every 1-cell in $G$ which collapses to a node in $H$ is part of a collapsing pair $(e, v)$ in $G$. Let $S \subseteq P$ be the union of these collapsing pairs.

3. Every cell in $H$ is the image of a cell in $G$ which does not lie in $S$ and $f$ induces an order preserving bijection $\alpha : P - S \to Q$.

4. Every collapsing pair in $Q$ is the image of a collapsing pair in $P - S$.

If $\alpha : P - S \to Q$ is an isomorphism of posets then we call $(f, \alpha)$ a *collapsing morphism*. If $f$ is an isomorphism of oriented graphs we call the morphism *monomial*. (Intuitively, the monomial morphisms are "permutations" since they change only the partial ordering of $P$. The collapsing morphisms incorporate both elementary collapses and "handle slides." The latter occur when edges which is not 1-cells collapse.)

**Proposition 8.2.2.** *The category of framed graphs has as a deformation retract the full subcategory of framed graphs without smooth nodes of degree* 2.

*Proof.* This follows from Lemma 8.1.4 in the same way as Proposition 8.1.5. □

We assume from now on that *all of our framed graphs are without smooth nodes of degree* 2 unless they are explicitly allowed. Let $\mathcal{FG}$ denote the category of framed graphs (without smooth nodes of degree 2). Then we will eventually prove that the forgetful functor $\mathcal{FG} \to \mathcal{GE}$ is a homotopy equivalence.

A *framed structure* on a topological graph $X$ is defined to be a framed graph $(G, P)$ together with a homeomorphism $\phi : |G| \to X$. These form the objects of a category $\mathcal{FG}(X)$. A morphism $((G, P), \phi) \to ((H, Q), \psi)$ is a pair $(f, [h])$ where $f : (G, P) \to (H, Q)$ is a morphism of framed graphs and $[h]$ is an isotopy class of fiberwise homeomorphisms over $I$:

$$h : EX(G, \lambda) = E(\rho(H \leftarrow G)) \to X \times I$$

with $h|0 = \psi$ and $h|1 = \phi$ where $EX(G, \lambda) \cong |G| \times I$ is the $\Delta^1$-family of metric graphs corresponding to $(H \leftarrow G : f) \in \mathcal{N}_1\mathcal{FG}^{op}$.

Now we add the continuous structure.

**Definition 8.2.3.** A *metric framed graph* is a triple $((G, P), \lambda, \epsilon)$ where

1. $(G, P)$ is a framed graph.

2. $(G, \lambda)$ is a metric graph.

3. $\epsilon > 0$ is a real number so that the total length of any 1-cell less than $\epsilon$ if and only if it is part of a collapsing pair.

We call $\epsilon$ the *expansion parameter*.

Suppose that $(G, P)$ is a connected framed graph and $k \geq 0$. Choose positive real numbers $0 < \epsilon_0 \leq \epsilon_1 \leq \cdots \leq \epsilon_k$ and, for each $e \in G_1$, choose:

$$0 \leq \lambda_0(e) \leq \cdots \leq \lambda_k(e) \neq 0$$

so that

1. $((G, P), r_k\lambda_k, r_k\epsilon_k)$ is a metric framed graph (for a uniquely determined rescaling factor $r_k$),

2. $\lambda_i(e) < \epsilon_i$ whenever $\lambda_{i-1}(e) < \epsilon_{i-1}$ and

3. $\lambda_i(e) > \epsilon_i$ whenever $\lambda_{i-1}(e) > \epsilon_{i-1}$.

For each $t \in \Delta^k$ let $\epsilon_t = \sum_{0 \leq i \leq k} t_i \epsilon_i$ and $\lambda_t(e) = \sum_{0 \leq i \leq k} t_i \lambda_i$ for all $e \in G_1$. Then we get a family of metric framed graphs $((G_t, P_t), r_t\lambda_t, r_t\epsilon_t)$ for all $t \in \Delta^k$ where $(G_t, P_t)$ is equal to $(G, P)$ with all collapsing pairs $(e, v)$ with $\lambda_t(e) = 0$ collapsed in $G_t$ and deleted in $P_t$ and those with $\lambda_t(e) = \epsilon_t$ no longer designated as collapsing pairs in $(G_t, P_t)$ and $r_t$ is a uniquely determined linear family of rescaling factors.

Note that the framed graph $(G_t, P_t) = (G_\sigma, P_\sigma)$ is constant on open simplices $\sigma$ of $\Delta^k$. Consequently, we get a functor $\sigma \mapsto (G_\sigma, P_\sigma)$ from the category of nondegenerate simplices in $\Delta^k$ to $\mathcal{FG}^{op}$. Using the same continuous data $(\lambda_t, \epsilon_t)$ we could alter the finite data (the names of the elements of $G_t$, $P_t$ and the partial ordering of each $P_t$) and still have a functor from the category of simplices in $\Delta^k$ to $\mathcal{FG}^{op}$. The result will be a $k$-simplex in the space of framed graphs. We make this more precise.

Suppose we have another functor $\sigma \mapsto (H_\sigma, Q_\sigma)$ from $\mathrm{simp}^{nd}\,\Delta^k$ to $\mathcal{FG}^{op}$ and a natural transformation $(f_\sigma, \alpha_\sigma) : (H_\sigma, Q_\sigma) \to (G_\sigma, P_\sigma)$ so that each $f_\sigma$ is a graph isomorphism (and, consequently, $\alpha_\sigma : Q_\sigma \to P_\sigma$ is an order preserving bijection). Suppose also that $((H_\sigma, Q_\sigma), \lambda_t, \epsilon_t)$ is a framed graph for all $t$ in the interior of $\sigma$. Then letting $(H_t, Q_t) = (H_\sigma, Q_\sigma)$ for all such $t$ we get a $k$-simplex of framed graphs.

**Definition 8.2.4.** We define the *space of framed graphs* to be the simplicial set $\mathcal{MFG}_\bullet$ which in degree $k$ consists of all $\Delta^k$-families of metric framed graphs given by the above construction. Given any metric graph $X$ we define the *space of framed structures on* $X$ to be the simplicial set $\mathcal{MFG}_\bullet(X)$ which in degree $k$ consists of a $\Delta^k$-family $((H_t, Q_t), r_t\lambda_t, r_t\epsilon_t)$, $t \in \Delta^k$, together with a fiberwise isometry with $X \times \Delta^k$.

By construction of $\mathcal{MFG}_\bullet$ we have a forgetful functor

$$\gamma : \operatorname{simp} \mathcal{MFG}_\bullet \to \mathcal{FG}^{op}$$

and a geometric realization functor

$$\rho : \mathcal{N}_\bullet \mathcal{FG}^{op} \to \mathcal{MFG}_\bullet$$

can be given as follows.

Take any element

$$((G^0, P^0) \leftarrow (G^1, P^1) \leftarrow \cdots \leftarrow (G^k, P^k))$$

of $\mathcal{N}_k \mathcal{FG}^{op}$. Let $\epsilon_0 = \epsilon_1 = \cdots = \epsilon_k = 1$. For each $e \in G_1^k$ and $0 \leq i \leq k$ let $\lambda_i(e)$ be given by:

$$\lambda_i(e) = \begin{cases} 0 & \text{if } e \text{ collapses to a node in } G^i \\ \frac{1}{2} & \text{if the image of } e \text{ in } G^i \text{ is part of a collapsing pair} \\ 1 & \text{otherwise} \end{cases}$$

For each face $\sigma$ of $\Delta^k$ let $(H_\sigma, Q_\sigma) = (G^i, P^i)$ where $i$ is the last vertex of $\sigma$.

**Proposition 8.2.5.** *The forgetful functor* $\gamma : \operatorname{simp} \mathcal{MFG}_\bullet \to \mathcal{FG}^{op}$ *and geometric realization functor* $\rho : \mathcal{N}_\bullet \mathcal{FG}^{op} \to \mathcal{MFG}_\bullet$ *are homotopy equivalences. Furthermore,* $\rho$ *gives a homotopy equivalence*

$$\mathcal{N}_\bullet \mathcal{FG}(X)^{op} \to \mathcal{MFG}_\bullet(X)$$

*for any metric graph* $X$.

*Proof.* Same as the proof of Proposition 8.1.11. □

**Definition 8.2.6.** Suppose that $K$ is a topological space. Then by a *continuous family of framed graphs* over $K$ we mean a continuous family of metric graphs $G_t$, $t \in K$ together with a continuous function $\epsilon : K \to (0, 1)$ and a continuously varying framed structure on $G_t$ so that the maps $g_t : G_t \to G_v$, $t \in U(v)$ in Definition 8.1.18 are morphisms of framed graphs. In particular, the number of nodes, the number of cells, the number of order relations among the cells and the number of collapsing pairs are all lower semicontinuous functions on $K$.

**Proposition 8.2.7.** *If* $K$ *has the homotopy type of a simplicial complex then a continuous family of framed graphs over* $K$ *is homotopic to the pull back of the universal family over* $|\mathcal{MFG}_\bullet|$ *along a map*

$$f : K \to |\mathcal{MFG}_\bullet|$$

*which is unique up to homotopy. Furthermore, given a fiberwise isometry of the total space* $E$ *with* $X \times K$ *for some fixed metric graph* $X$*, we get a lifting of* $f$ *to* $|\mathcal{MFG}_\bullet(X)|$ *which is unique up to homotopy.*

*Proof.* See Proposition 8.1.19 and its proof. □

One of the main points of framed graphs is that there is a "linearization functor" from the space of framed graphs to the simplicial bicategory of matrices with row and column operations. This is a special case of a more general map

$$\xi : A(X) \to \mathcal{W}h^h_\bullet(\mathbb{Z}[\pi_1(X)], \pi_1(X))$$

which is basically given by sending every cell complex to its cellular chain complex which is unique up to contractible choice by the theory of filtered chain complexes. (4.1.7). We will discuss later.

For any framed graph $(G, P)$, the cellular chain complex $C_*(G, P)$ is given by $C_0(G, P) = \mathbb{Z}^{P_0}$, $C_1(G, P) = \mathbb{Z}^{P_1}$ with boundary map $\partial^G : \mathbb{Z}^{P_1} \to \mathbb{Z}^{P_0}$ given by

$$\partial^G e = \sum_{v \in P_0} v\psi_v(e)$$

where the incidence numbers $\psi_v(e)$ are defined for any 1-cell $e \in P_1$ and node $v \in G_0$ by

$$\psi_v(e) = \chi_v(t(e)) - \chi_v(s(e))$$

where $\chi_v : G_0 \to \{0, 1\}$ is given recursively by:

$$\chi_v(w) = \begin{cases} 1 & \text{if } w = v \\ 0 & \text{if } w \in P_0 \text{ and } w \neq v \\ \chi_v(t(e)) & \text{if } w \text{ is an smooth node of } e \text{ and } w \neq v. \end{cases}$$

Suppose that $(f, \alpha) : (G, P) \to (H, Q)$ is a morphism of framed graphs. We deal first with the case when $\alpha : P \to Q$ is a bijection. In this case we have a chain isomorphism:

$$\begin{array}{ccc} C_1(H,Q) & \xrightarrow{g_1} & C_1(G,P) \\ \partial^H \downarrow & & \downarrow \partial^G \\ C_0(H,Q) & \xrightarrow{g_0} & C_0(G,P) \end{array}$$

given uniquely by $g_0(\alpha(v)) = v$ (and thus $g_1 = (\partial^G)^{-1} \circ g_0 \circ \partial^H$). We claim that for each $e \in P_1$, $g_1(\alpha(e))$ is equal to $e$ plus a linear combination of lower terms, i.e, that we have a 1-simplex in the space of matrices. To see this we need another description of $g_1$.

Whenever the mapping $f : G \to H$ collapses an smooth node of a 1-cell $e$ to its source, then any higher 1-cell which is incident to this node will change its boundary map by the boundary of $e$. Thus the mapping $g_1 : C_1(G, P) \to C_1(H, Q)$ is a composition of upper triangular mappings based on the incidence numbers $\psi_v(e)$, defined above, of 1-cells $e$ with smooth nodes $v$.

Let $\{(v_1, e_1), \cdots, (v_n, e_n)\}$ be the set of all pairs where $v_i \in G_0$ is an smooth node of $e_i \in P_1$ so that $\alpha(v_i) = \alpha(s(e_i)) = s(\alpha(e_i))$, i.e., the portion of $e_i$ from

$v_i$ to $s(e_i)$ collapses to one node in $H$. Then $\partial^H = \partial_n$ is obtained from $\partial^G = \partial_0$ recursively by:

$$\partial_{k+1}(e) = \partial_k(e) - \partial_k(e_k)\psi_{v_k}(e)$$

for all $0 < k < n$. In other words, $\partial_{k+1} = \partial_k \circ (I - e_k\psi_{v_k})$ and therefore,

$$g_1 = (I - e_0\psi_{v_0}) \circ \cdots \circ (I - e_{n-1}\psi_{v_{n-1}}).$$

Note that since $g_1 = (\partial^G)^{-1} \circ g_0 \circ \partial^H$ is uniquely determined, the order of the pairs $(v_i, e_i)$ is irrelevant.

Now suppose that $(f, \alpha) : (G, P) \to (H, Q)$ is an arbitrary morphism of framed graphs. Let $S$ be the set of collapsing pairs in $P$ and let $\partial^S : \mathbb{Z}^{S_1} \to \mathbb{Z}^{S_0}$ be given by $\partial^S e = t(e)$. Then we have a chain isomorphism:

$$\begin{array}{ccc} C_1(H,Q) \oplus \mathbb{Z}^{S_1} & \xrightarrow{(g_1,h_1)} & C_1(G,P) \\ \downarrow{\scriptstyle \partial^H \oplus \partial^S} & & \downarrow{\scriptstyle \partial^G} \\ C_0(H,Q) \oplus \mathbb{Z}^{S_0} & \xrightarrow{(g_0,h_0)} & C_0(G,P) \end{array}$$

given uniquely by $g_0(\alpha(v)) = v$ and

$$h_0(t(e)) = t(e) - \sum_{w \in P_0 - S_0} w\psi^H_{\alpha(w)}(\alpha(s(e)))$$

As before, $(g_1, h_1)$ is given by

$$(g_1, h_1) = (I - e_0\psi'_{v_0}) \circ \cdots \circ (I - e_{n-1}\psi'_{v_{n-1}})$$

where $\psi'_{v_i}$ is the same as $\psi_{v_i}$ except on $e_i$ where it is zero.

Since framed graphs have no 2-cells or higher, the rest of the structure is uniquely determine. Thus, for any composition of morphisms $h = f \circ g$ there is a unique chain homotopy $h^* \simeq g^* \circ f^*$ which involves targets of collapsing 1-cells sliding over other 1-cells. The higher homotopies are necessarily all zero.

*In this way we get a system of higher homotopies for each simplex in the nerve of* $\mathcal{FG}$. The rest of the construction is just some technicalities to paste these together to form a simplicial functor. The problem is that some of the arrows in our category are going the wrong way. To fix this we transform all collapsing morphisms into expanding morphisms with the aid of "ghosts."

Let $\mathcal{FG}h$ be the category of framed graphs with "ghosts." A *ghost* is by definition a collapsing pair in $P$ which does not correspond to any points in the graph $G$. Thus $(G, P^\delta)$ is a framed graph where $P^\delta = P - \{ghosts\}$. In the ghost category, when a pair $(e, v)$ is supposed to collapse, it turns into a ghost instead. And ghosts can always be created. Thus $\alpha$ is always a monomorphism in the ghost category. The idea is that then the collapsing morphisms are handle slides only (without any collapsing).

We have a forgetful functor

$$\pi : \mathcal{FG}h \to \mathcal{FG}$$

given by deleting all the ghosts. Given any $(G,P)$ in $\mathcal{FG}$, the fiber category $(G,P)\backslash\pi$ is contractible since it has an initial object, namely the one with no ghosts. Therefore, $\pi$ is a homotopy equivalence by Quillen's Theorem A.

A morphism $(f,\alpha):(G,P)\to(H,Q)$ in the ghost category factors canonically as a composition of

1. an "expansion morphism" $(id, inc):(G,P)\to(G,P\coprod S)$ where $S$ is a disjoint union of ghosts $(e_i,v_i)$,
2. a collapsing morphism $(f,id):(G,P\coprod S)\to(H,P\coprod S)$ which is the identity on the poset
3. and a monomial morphism $(id,\alpha\cup\beta):(H,P\coprod S)\to(H,Q)$ where $\beta: S\to Q$ maps $S$ bijectively onto the ghosts in $Q-\alpha(P)$.

Therefore, by the modified swallowing lemma (5.2.4), $\mathcal{FG}h$ is homotopy equivalent to a tricategory
$$|\mathcal{FG}h|\simeq|mec\mathcal{FG}h|$$
where $m,e,c$ indicate monomial, expansion and collapsing morphisms respectively. Taking the nerve in the collapsing direction and reversing the monomial direction we get a simplicial bicategory $m^{op}ec_\bullet\mathcal{FG}h$. The construction above creates a simplicial bifunctor
$$\xi: m^{op}ec_\bullet\mathcal{FG}h\to me\mathcal{W}h_\bullet^{[0,1]}(\mathbb{Z},1)\simeq\mathcal{W}h_\bullet^{[0,1]}(\mathbb{Z},1) \tag{8.1}$$
(See §5.3.) This makes sense because the collapsing direction $c$ in the ghost category gives only handle slides which make up the simplicial direction in $\mathcal{W}h_\bullet(\mathbb{Z},1)$.

Combining the above we get the desired mapping
$$|N_\bullet\mathcal{FG}|\xrightarrow{\simeq}|\mathcal{FG}h|\simeq|mec\mathcal{FG}h|\to\mathcal{W}h_\bullet^{[0,1]}(\mathbb{Z},1).$$

## 8.3 Comparison with framed functions

We will show that framed structures on the interval $I$ and the circle $S^1$ are equivalent to framed functions and therefore form contractible spaces. We will use "contractible carriers" to construct the homotopy equivalences.

Let $\mathcal{A}$ be a category and $X$ an object in another category $\mathcal{B}$. Then by a *carrier* from $\mathcal{A}$ to $X$ we mean a functor
$$C:\mathcal{A}^{op}\to\mathcal{B}/X$$
from $\mathcal{A}^{op}$ to the category of objects over $X$, i.e., for each $\alpha\in\mathcal{A}$ there is an object $C(\alpha)$ and a morphism $p_\alpha: C(\alpha)\to X$ and for each morphism $\alpha\to\beta$ in $\mathcal{A}$ there is a morphism $g_{\alpha\beta}: C(\beta)\to C(\alpha)$ so that $p_\alpha\circ g_{\alpha\beta}=p_\beta$.

If $\mathcal{B}$ is the category of topological spaces then we say that a continuous map $f:|\mathcal{A}|\to X$ is *carried* by a carrier $C$ if $f$ lifts to a family of continuous maps

$f_\sigma : \Delta^k \to C(\sigma_0)$ for all $\sigma = (\sigma_0 \to \cdots \to \sigma_k) \in \mathcal{N}_k\mathcal{A}$ so that $f_\sigma \circ a_* = g_{\sigma_0\tau_0} \circ f_\tau : \Delta^j \to C(\sigma_0)$ for all $a : [j] \to [k]$ where $\tau = a^*\sigma$.

$$\begin{array}{ccccc} \Delta^j & \xrightarrow{a_*} & \Delta^k & \longrightarrow & |\mathcal{A}| \\ {\scriptstyle f_\tau}\downarrow & & \downarrow{\scriptstyle f_\sigma} & & \downarrow{\scriptstyle f} \\ C(\tau_0) & \xrightarrow{g_{\sigma_0\tau_0}} & C(\sigma_0) & \xrightarrow{p_{\sigma_0}} & X \end{array}$$

**Lemma 8.3.1.** *Given any small category $\mathcal{A}$ and any carrier $C$ from $\mathcal{A}$ to a space $X$ which is contractible in the sense that $C(\alpha)$ is contractible for all $\alpha$ in $\mathcal{A}$, there is at least one map $f : |\mathcal{A}| \to X$ carried by $C$ and any two such maps are homotopic by a homotopy which is also carried by $C$.*

*Proof.* We give the obvious proof since we need to refer to it later.

For each $\alpha$ in $\mathcal{A}$ choose a point $f(\alpha)$ in $C(\alpha)$. For each nonidentity arrow $\alpha \to \beta$ we choose a path from $g_{\alpha\beta}(f(\beta))$ to $f(\alpha)$ in $C(\alpha)$. For any nondegenerate $k$-simplex $\sigma = (\sigma_0 \to \cdots \to \sigma_k)$ in $\mathcal{N}_k\mathcal{A}$, the map $f_{\partial\sigma} : \partial\Delta^k \to C(\sigma_0)$ which is defined by induction on $k$ extends to a map $f_\sigma : \Delta^k \to C(\sigma_0)$ since $C(\sigma_0)$ is contractible. □

**Theorem 8.3.2.** *The category of framed structures $\mathcal{FG}(I)$ on $I$ is contractible as are the nine full subcategories $\mathcal{FG}_{ij}(I)$ given by restricting the framed structure on the two endpoints $0, 1$ by restrictions $r_i, r_j$, respectively, where*

*(0) $r_0$ is no restriction. (So $\mathcal{FG}_{00}(I) = \mathcal{FG}(I)$.)*

*1. $r_1$ means the endpoint is not allowed to be a member of a collapsing pair.*

*2. $r_2$ means the endpoint is the source of a 1-cell which is not a member of a collapsing pair. (So $\mathcal{FG}_{2j}(I) \subseteq \mathcal{FG}_{1j}(I) \subseteq \mathcal{FG}_{0j}(I)$.)*

*Proof.* The category of framed structures on $I$ (or on any nonempty metric graph $X$ without circle components) has the following properties.

1. The only morphism from an object to itself is the identity.
2. There are only finitely many isomorphism classes of morphisms with any fixed source.

Consider first the case $\mathcal{FG}_{10}(I)$. The isomorphism classes of objects in this category are in one-to-one correspondence with the cells of the CW complex $CF(I)$ minus one vertex, namely the empty configuration corresponding to the one point tree which we are not allowing. Therefore we first pass to the subcategory $\mathcal{FG}_{10}^0(I)$ of $\mathcal{FG}_{10}(I)$ which has a collapsing pair $(e, v)$ at 1 pointing outward. We take morphisms which fix this pair. There is a functor

$$\mathcal{FG}_{10}(I) \to \mathcal{FG}_{10}^0(I)$$

given by adding the fixed collapsing pair $(e, v)$ at the endpoint 1. By collapsing $(e, v)$ we get a homotopy with the identity functor so it suffices to show that $\mathcal{FG}_{10}^0(I)$ is contractible.

Choosing one object from each isomorphism class of objects, we may assume that no two distinct objects of $\mathcal{FG}_{10}^0(I)$ are isomorphic. Because of (1), the set of all sequences of nonidentity morphisms in $\mathcal{FG}_{10}^0(I)$ forms a subset $\mathcal{N}_\bullet^{nd}\mathcal{FG}_{10}^0(I)$ of the nerve of $\mathcal{FG}_{10}^0(I)$ which is closed under face operators. Furthermore, the nondegenerate realization

$$\|\mathcal{FG}_{10}^0(I)\| = \coprod \mathcal{N}_k^{nd}\mathcal{FG}_{10}^0(I) \times \Delta^k /_{face\ identifications}$$

is homeomorphic to the usual realization $|\mathcal{N}_\bullet \mathcal{FG}_{10}^0(I)|$. A homotopy equivalence

$$f : \|\mathcal{FG}_{10}^0(I)\| \to CF(I)$$

to the space of linearly ordered configuration $CF(I)$ (defined in 2.10.1) is given by the contractible carrier $C$ which sends the object in $\mathcal{FG}_{10}^0(I)$ with vertex-edge sequence

$$(v_0, e_1, v_1, e_2, \cdots, e_n, v_n, e, v)$$

with degree $k$, i.e., having $k$ collapsing pairs (not including $(e, v)$), to the $k$-cell $I^k \to CF(I)$ given by the configuration $(a_1, 0, a_2, 0, \cdots, a_n, 0)$, where $a_i$ is $+$ or $-$ depending on the orientation of $e_i$, together with the data of which pairs are cancellation pairs. This is a carrier by definition of the cell structure of $CF(I)$.

The map $f$ carried by $C$ is a homotopy equivalence since $\mathcal{FG}_{10}^0(I)$ is the union of its finite terminal subcategories (i.e., subcategories $\mathcal{Z}$ having the property that no object of $\mathcal{Z}$ maps to any object outside of $\mathcal{Z}$.) For each such $\mathcal{Z}$ let $k$ be the largest degree of an object in $\mathcal{Z}$ and let $m$ be the number of objects of degree $k$. Then we will show by induction on $(k, m)$ that

$$C(\mathcal{Z}) = \coprod_{\alpha \in \mathcal{Z}} C(\alpha) / \sim$$

is embedded in $CF(I)$ and that $\{f_\alpha\}$ induces a homotopy equivalence $f_{\mathcal{Z}} : \|\mathcal{Z}\| \to C(\mathcal{Z})$ and we will conclude that

$$\|\mathcal{FG}_{10}^0(I)\| = \bigcup \|\mathcal{Z}\| \simeq \bigcup C(\mathcal{Z}) = CF(I) \simeq *.$$

If $(k, m) = (0, 1)$ then $\mathcal{Z}$ has only one object $\alpha = (*, e, v)$ with weight 0 and $C(\alpha)$ is contractible and contained in $CF(I)$ (since $C(\alpha)$ is one point).

Suppose that $\mathcal{Z}$ has $m$ objects of maximal degree $k$ and that the induction hypothesis holds for $(k', m') < (k, m)$. Choose any object $\alpha$ of $\mathcal{Z}$ of degree $k$ and let $\mathcal{Y}$ be the full subcategory of $\mathcal{Z}$ with all the other objects. Then $\mathcal{Y}$ is finite and terminal with $m - 1$ objects of degree $k$ so $f_{\mathcal{Y}} : \|\mathcal{Y}\| \to C(\mathcal{Y})$ is a homotopy equivalence and $C(\mathcal{Y}) \subseteq CF(I)$.

Let $\alpha\backslash\mathcal{Z}$ be the category of all objects of $\mathcal{Z}$ under $\alpha$. Since $\alpha\backslash\mathcal{Z}$ has an initial object, $\|\alpha\backslash\mathcal{Z}\| \simeq *$ and $C(\alpha\backslash\mathcal{Z}) = C(\alpha)$. The point is that $\|\mathcal{Z}\|$ and

$C(\mathcal{Z})$ are push-outs in the following diagrams and therefore, $f_{\mathcal{Z}} : \|\mathcal{Z}\| \to C(\mathcal{Z})$ is a homotopy equivalence.

$$\begin{array}{ccc} \|\alpha\backslash\mathcal{Y}\| & \longrightarrow & \|\alpha\backslash\mathcal{Z}\| \\ \downarrow & & \downarrow \\ \|\mathcal{Y}\| & \longrightarrow & \|\mathcal{Z}\| \end{array} \qquad \begin{array}{ccc} C(\alpha\backslash\mathcal{Y}) & \longrightarrow & C(\alpha) \\ \downarrow & & \downarrow \\ C(\mathcal{Y}) & \longrightarrow & C(\mathcal{Z}) \end{array}$$

The fact that $C(\mathcal{Z}) = C(\mathcal{Y}) \cup C(\alpha)$ follows from the fact that

1. each open cell of $CF(I)$ is the image of $intC(\alpha)$ for a unique $\alpha$ in $\mathcal{FG}$ and
2. $\partial C(\alpha)$ is the image of $C(\alpha\backslash\mathcal{Z} - \{\alpha\})$.

Thus we conclude that $\|\mathcal{FG}_{10}^0\| \simeq CF(I) \simeq *$ by Theorem 2.10.15.

To show that the other eight subspaces of $\mathcal{FG}(I)$ are contractible we show that the inclusion maps

$$\mathcal{FG}_{2j}(I) \subseteq \mathcal{FG}_{1j}(I) \subseteq \mathcal{FG}_{0j}(I)$$

are homotopy equivalences for all $j$.

Given any $(G, [\phi])$ in $\mathcal{FG}_{0j}(I)$ we add a collapsing pair to $\phi^{-1}(0)$ oriented outward and then unpair it. This gives a deformation retraction of $\mathcal{FG}_{0j}(I)$ to $\mathcal{FG}_{1j}(I)$.

Given any $(G, [\phi])$ in $\mathcal{FG}_{1j}(I)$, since $v = f^{-1}(0)$ is not part of a collapsing pair, we can add a collapsing pair to $v$ oriented inward and then unpair it. This gives us a deformation retraction of $\mathcal{FG}_{1j}(I)$ to $\mathcal{FG}_{2j}(I)$.

The analogous argument shows that $\mathcal{FG}_{i2}(I) \simeq \mathcal{FG}_{i1}(I) \simeq \mathcal{FG}_{i0}(I)$ for all $i$. □

**Theorem 8.3.3.** *The space of framed structures on $S^1$ is contractible:*

$$|\mathcal{N}_\bullet\mathcal{FG}(S^1)| \simeq |\mathcal{MFG}_\bullet(S^1)| \simeq *$$

*Proof.* This time we use the continuous model $\mathcal{MFG}(S^1)$ and compare it with the space $CF(S^1)$ of cyclically ordered configurations defined in 2.10.3. We will need to rescale by a factor of $2\pi$ since all metric framed graphs have total length equal to 1 by definition.

Given a metric framed graph $((G, P), \lambda, \epsilon)$ and an isometry $\phi : |(G, \lambda)| \to \mathbb{R}/\mathbb{Z}$ we can form a cyclically ordered configuration $(a_*, \mu, \theta)$ where

1. $a_2, a_4, \cdots, a_{2n}$ represent the vertices of $G$.
2. $a_1, a_3, \cdots, a_{2n-1}$ represent the edges of $G$, each oriented $+$ or $-$.
3. $\mu(a_i) = 1$ if $a_i$ is not part of a collapsing pair.

4. If $(e, v)$ is a collapsing pair then
$$\mu(e) = \mu(v) = \frac{\lambda(e)}{\epsilon}.$$

5. $\theta \in \mathbb{R}/2\pi\mathbb{Z}$ is $2\pi$ times the center of mass of the configuration $(a_1, a_2, \cdots, a_{2n})$ lifted up to a half open interval of length 1 in $\mathbb{R}$ with the mass of each edge concentrated in its center.

This defines a continuous mapping $C : |\mathcal{MFG}(S^1)| \to CF(S^1)$. A reverse mapping
$$M : CF(S^1) \to |\mathcal{MFG}(S^1)|$$
is given up to homotopy by a continuous family of framed structures on $S^1$ parametrized by $CF(S^1)$. For each cyclically ordered configuration $(a_*, \mu, \theta)$ we take the framed graph with $n$ vertices $a_2, a_4, \cdots, a_{2n}$ and $n$ edges $a_1, a_3, \cdots, a_{2n-1}$ with lengths
$$\lambda_i = \frac{\mu(a_{2i-1})}{\sum_{j=1}^{n} \mu(a_{2j-1})}$$
with $\epsilon = \max \lambda_i$ and adjusting the embedding of $a_1, a_2, \cdots, a_{2n}$ into $\mathbb{R}$ so that its center of mass is $\frac{\theta}{2\pi}$.

If we pull back this family of framed structures to $|\mathcal{MFG}(S^1)|$ we get a family isomorphic to the canonical family over $|\mathcal{MFG}(S^1)|$. Consequently the composition
$$M \circ C : |\mathcal{MFG}(S^1)| \to CF(S^1) \to |\mathcal{MFG}(S^1)|$$
is homotopic to the identity by Proposition 8.2.7. Since $CF(S^1)$ is contractible by Theorem 2.10.15, we conclude that $|\mathcal{MFG}(S^1)|$ is contractible. □

## 8.4 The framed graph theorem

In this section we prove the framed graph theorem. The framed graph theorem is an easy consequence of the framed tree lemma which says that any continuous family of trees admits a continuous family of framed structures which is unique up to homotopy. The framing of trees and graphs is done in two steps. First we construct a homotopy framing. Then we show that any family of homotopy framed trees admits a family of framings unique up to homotopy.

### 8.4.1 Trees, magma and homotopy framings

We want to consider the space of all trees having a fixed set of leaves $S$. This space is equivalent to the realization of the poset of all "magma" on $S$. In order to put a framed structure on such a tree we first need a homotopy framing. The set of all pairs $(M, \eta)$ where $M$ is a magma and $\eta$ is a homotopy framing of the associated tree will be a finite poset over the poset of magma.

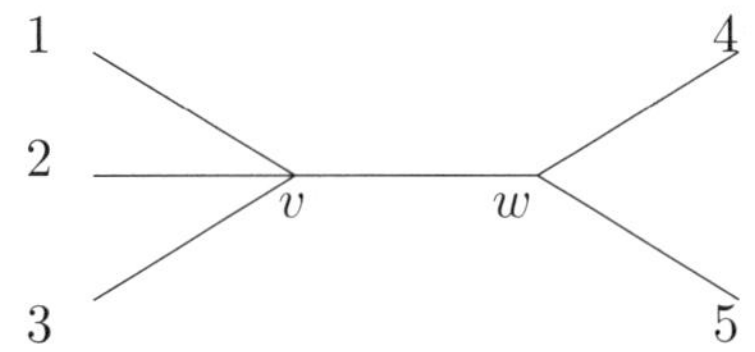


Figure 8.1: $M(T) = S \coprod \{\{1, 2, 3\}, \{4, 5\}\}$

Suppose that $T$ is a tree, i.e., a contractible graph, with a fixed set of leaves $S$. Then the *magma* $M(T)$ of $T$ is defined to be the set of all subsets $J$ of $S$ which have the property that the subtree $T_J$ spanned by $J$ is incident to only one edge not in $T_J$. The complement $J^c = S - J$ of $J$ in $S$ also has this property with respect to the same edge $e$. In other words $T - e = T_J \coprod T_{J^c}$. Therefore, each edge of $T$ gives two complementary elements of $M(T)$. In particular, every singleton and its complement are in $M(T)$. The magma $M(T)$ determines $T$ up to an isomorphism fixing $S$.

**Definition 8.4.1.** A *magma* on a finite set $S$ is defined to be a collection $M$ of nonempty subsets $J$ of $S$ having the following properties.

1. $J \in M$ if and only if $J^c = S - J \in M$.

2. $\{i\} \in M$ for all $i \in S$.

3. If $I, J \in M$ then either $I \subseteq J$, $J \subseteq I$, $I \subseteq J^c$ or $J^c \subseteq I$.

Let $\mathrm{Mag}(S)$ denote the poset of all magma on $S$ ordered by inclusion.

Given any magma $M$ on $S$ the corresponding tree $T(M)$ with leaf set $S$ is given as follows.

1. $\widetilde{T}_1(M) = M$ with involution $-J = J^c$.

2. $T_0(M) - S$ is the set of all minimal partitions of $S$ into a disjoint union of at least three elements of $M$.

For example in the tree $T$ in Figure 8.1 with leaf set $S = \{1, 2, 3, 4, 5\}$ the magma is $M(T) = S \coprod \{\{1, 2, 3\}, \{4, 5\}\}$. The vertices $v, w$ correspond to the minimal partitions:

$$\text{(v)}: \quad S = \{1\} \coprod \{2\} \coprod \{3\} \coprod \{4, 5\}$$

$$\text{(w)}: \quad S = \{1, 2, 3\} \coprod \{4\} \coprod \{5\}$$

The smooth edge $(v, w)$ corresponds to the element $\{4, 5\} \in M(T)$ and $(w, v)$ corresponds to $\{1, 2, 3\} = \{4, 5\}^c$.

Take the geometric realization $|\mathrm{Mag}(S)| = |\mathcal{N}_\bullet \mathrm{Mag}(S)|$ of $\mathrm{Mag}(S)$. As is true for the realization of any finite poset, this is homeomorphic to the nondegenerate realization $\|\mathrm{Mag}(S)\|$ of the nondegenerate nerve of $\mathrm{Mag}(S)$:

$$|\mathrm{Mag}(S)| = \coprod_{M_0 \leq \cdots \leq M_k} \Delta^k / \sim \cong \coprod_{M_0 < \cdots < M_k} \Delta^k /_{face\ ident} = \|\mathrm{Mag}(S)\|$$

This space $|\mathrm{Mag}(S)| \cong \|\mathrm{Mag}(S)\|$ can also be viewed as the realization of a finite simplicial complex whose vertex set is $\mathrm{Mag}(S)$ where a set of $k+1$ vertices spans a $k$-simplex if and only if they are linearly ordered. Let $C(M) = |\{M' \leq M\}|$. This is a cone on $\partial C(M) = |\{M' < M\}|$ so it is contractible. The relatively open subset

$$U(M) = C(M) - \partial C(M) = C(M) - \bigcup_{M' < M} C(M')$$

is also contractible since it is the open cone. As a set $|\mathrm{Mag}(S)|$ is a disjoint union of the $U(M)$'s.

Given magma $M \leq M'$ on $S$, there is a unique admissible graph epimorphism $T(M') \to T(M)$ and, conversely, an admissible graph epimorphism $T \to T'$ exists which is the identity on the set of leaves $S$, if and only if $M(T)$ contains $M(T')$. Consequently the poset of magma on $S$ is equivalent to the category of trees with a fixed set of leaves.

Pulling back the canonical family of metric graphs over the realization of $\mathcal{GE}$ (induced from $\mathcal{MG}_\bullet$ by geometric realization $\rho : \mathcal{N}_\bullet\mathcal{GE} \to \mathcal{MG}_\bullet$) we get a canonical family of metric trees $T_x$ over $|\mathrm{Mag}(S)|$. These trees have the property that $M(T_x) = M$ for each $x \in U(M)$.

Given a framed structure $\phi : |(G, \lambda)| \to X$ on any topological graph $X$, each node $v \in X_1$ will either be (the image of) an smooth node or a vertex. If $v$ is an smooth node of $\phi$ then the definition of framed graph gives a pair of incident edges at $v$ indicating that these two edges are part of the same 1-cell. If $v$ is a vertex of the framing then it may be paired with one of its incident edges. The list of all possibilities gives a finite set, ordered by inclusion, which we call the poset of "homotopy framings" of $X$.

**Definition 8.4.2.** By a *homotopy framing* of a combinatorial or topological graph $X$ we mean a function $\eta$ which assigns to each node $v$ of $X$ a set $\eta(v)$ of at most 2 incident edges. (This could include the same edge twice if it is a loop at $v$.) The elements of $\eta(v)$ will be called the *labels* of $v$. A framed structure $\phi$ on $X$ will be called *subordinate* to a homotopy framing $\eta$ if:

1. The homotopy framing $\eta(v)$ agrees with the framed structure given in Definition 8.2.1(2) at all smooth nodes.

2. $e \in \eta(v)$ whenever $(e, v)$ is a collapsing pair.

Note that a homotopy framing on a tree is a discrete structure on the set of leaves. There is only a finite number of homotopy framed trees with any fixed set of leaves. This set will be partially ordered by "degeneration" which indicates that the set of framed trees subordinate to one homotopy framed tree converges to the framed trees subordinate to the other homotopy framed tree.

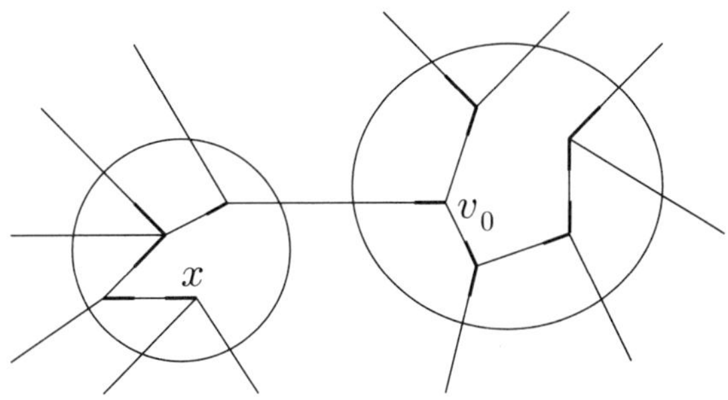


Figure 8.2: Collapsing trees (circled) with labels (dark)

**Definition 8.4.3.** Suppose $\eta, \eta'$ are homotopy framings of trees $T, T'$ represented by magma $M, M'$ respectively. Then we say that $(M', \eta')$ is a *degeneration* of $(M, \eta)$ if the following conditions are satisfied.

1. $M' \subseteq M$ with corresponding graph epimorphism $g : T \to T'$. We use the notation $T_w$ for the "collapsing tree" $g^{-1}(w)$ for each node $w$ of $T'$.

2. Each collapsing tree $T_w$ has at least one node $v_0$ so that each edge in the minimal path from $v_0$ to any other node in $T_w$ is a label of its target node. (See Figure 8.2.)

3. If $e' \in \eta'(w)$ then there exists $e \in \eta(v_0)$ so that $g(e) = e'$ (and $v_0$ is as given in (2)).

These definitions can be visualized in terms of city traffic in a circular city. The nodes are disjoint simply connected regions in the city. The internal edges are roads connecting these regions and the external edges are roads going out of the city. If the edge $e$ is a label of the vertex $v$ ($e \in \eta(v)$) then cars are allowed to travel on the road $e$ in the direction of $v$. (The end of $e$ attached to each $v$ is darkened in Figure 8.2.) If $e$ is a label of both endpoints then $e$ is a two-way street.

In a degeneration, sets of regions of the city connected by roads are collected to form larger but still disjoint regions which we will call *districts*. (These are the collapsing trees, encircled in Figure 8.2.) Within each district there must be at least one smaller region (a "hub") from which one can travel to all other regions in the district. (These are $v_0$ and $x$ in Figure 8.2.) A necessary condition for interdistrict travel is that the target is a hub and travel in that direction was previously allowed. There are at most two roads into any district having this property. In Figure 8.2, cars are allowed to travel from any node in the left hand district to any node in the right hand district.

With this interpretation the following is obvious.

**Proposition 8.4.4.** *Degeneration is a transitive relation on the set of homotopy framed trees.*

Let HFT$(S)$ denote the poset of homotopy framed trees with leaf set $S$. Then we want to prove:

1. $|\mathrm{HFT}(S)|$ is contractible
2. $|\mathrm{HFT}(S)|$ is homotopy equivalent to the category of framed trees with leaf set $S$.

We will show the second statement first.

Since HFT$(S)$ is a finite poset, its realization is the realization of a simplicial complex. Consequently, given a continuous mapping of any triangulable space $K$, such as a sphere, into $|\mathrm{HFT}(S)|$ can be approximated by a simplicial map from a triangulation of $K$ into $|\mathrm{HFT}(S)|$.

## 8.4.2 Homotopy framed trees admit unique framings

This subsection shows that a simplicial family of homotopy framed trees admits a continuous family of framings which is unique up to deformation.

**Definition 8.4.5.** Suppose that $S$ is a fixed set of leaves and $K$ is a *triangulated space*, i.e., the geometric realization of a simplicial complex. Then a family of homotopy framed trees $(T_t, \eta_t), t \in K$ with leaf set $S$ will be called *simplicial* if for every magma $M \in M(S)$ and every homotopy framing $\eta$ on $M$ the subset of $K$ given by

$$K(M, \eta) = \{t \in K : (M(T_t), \eta_t) \leq (M, \eta)\}$$

is a closed subcomplex of $K$. This is equivalent to saying that the pair $(M(T_t), \eta_t)$ is fixed on each open simplex of $K$ and the assignment of this pair to each simplex gives a morphism of posets $simp(K) \to \mathrm{HFT}(S)$. The family of homotopy framed trees $(T_t, \eta_t)$ is *continuous* if $K(M, \eta)$ is a closed subset of $K$ for all $(M, \eta) \in \mathrm{HFT}(S)$.

**Proposition 8.4.6.** *Suppose we have a simplicial family of homotopy framed trees $(T_t, \eta_t)$ over a triangulated space $K$ and a continuous family of framings $\phi_t$ for $T_t$ subordinate to $\eta_t$ for $t$ in a subcomplex $C$ of $K$. Then the subordinate framing extends to all of $K$.*

**Remark 8.4.7.** Taking $C$ to be empty, the proposition says that a family of framings $\phi_t$ subordinate to $\eta_t$ exists. Taking $K = X \times I$ and $C = X \times 0 \cup X \times 1$ the proposition says that $\phi_t$ is unique up to homotopy.

*Proof.* It suffices to construct the framing $\phi_t$ over $K$ one simplex at a time. Thus we may assume that $K = K(M_0, f_0)$ is a $k$-simplex and that $C = \partial\Delta^k$ contains $K(M_1, f_1)$ for all $(M_1, f_1) < (M_0, f_0)$.

At this point we can completely forget about the family of trees $T_t$ for $t \in K - C$. It suffices to extend the given family of framed trees $(T_t, \phi_t)$ for $t \in C$ to any family of framed trees $(T'_t, \phi'_t)$, $t \in K$ with the property $M(T'_t) = M_0$. The reason is that any two trees with the same magma are related by a unique linear isomorphism $g_t : T'_t \cong T_t$ which varies continuously with the trees. Then $(T_t, g_t(\phi'_t))$ is the extension of $\phi_t$ over the original family of homotopy framed trees $(T_t, \phi_t)$. Thus it suffices to prove the following. $\square$

**Lemma 8.4.8.** *Let $(T_0, \eta_0)$ be a fixed homotopy framed tree with magma $M_0$ on the leaf set $S$.*

a) *The space of all framed trees $(T, \phi)$ with $M(T) = M_0$ and $\phi$ subordinate to $\eta_0$ is contractible.*

b) *Let $(T_t, \phi_t), t \in K$, be any continuous family of framed trees subordinate to $(T_0, \eta_0)$ and its degenerations. Let $\epsilon(t) > 0$ be a continuous function of $t \in K$ so that no internal node of $T_t$ is within $\epsilon(t)$ of any of its leaves. Then there is a continuous deformation $(T_{t,u}, \phi_{t,u}), (t, u) \in K \times I$, of $(T_t, \phi_t)$ so that*

1. *$M(T_{t,u}) = M_0$ for all $u > 0$.*
2. *$\phi_{t,u}$ is subordinate to $\eta_0$ for all $t \in X$ when $u > 0$.*
3. *The framing $\phi_{t,u}$ agrees with $\phi_t$ within a distance of $\frac{\epsilon(t)}{2}$ from each leaf of $T_t$ for all $u > 0$.*

*Proof.* First consider the case when $S$ has two elements. There are two possibilities for the homotopy framing $\eta_0$ of $T_0$ at each endpoint of $I$: either $\eta_0$ is empty or it contains the edge $I$. This gives four possibilities for $(T_0, \eta_0)$. The corresponding spaces of framed trees equivalent to four of the nine cases considered in Theorem 8.3.2 ($|\mathcal{FG}_{ij}(I)|$ for $i, j \leq 1$) and thus are contractible.

We want to proceed by induction on the number of nodes of the tree $T_0$. The induction step will be carried out by breaking the tree into pieces and/or stripping off some of its leaves. However, unlike the entire tree, the leaves of these subtrees are not fixed and the entire subtree is allowed to collapse to a point. Thus, in order for the induction to work, we need to prove the lemma for trees having nodes of degree 2 (where edges were removed) and with "collapsible leaves," i.e., those leaves which are not leaves of the original tree. Unlike nodes of degree $\geq 3$ which have an intrinsic multiplicity (the degree minus 2), nodes of degree 2 can coalesce without leaving any indication of how many there used to be. Therefore, we also add multiplicities to our nodes which add up when they coalesce.

Therefore, for the purpose of this proof, we will consider metric trees with designated "collapsible" and "noncollapsible" leaves. Our family of trees $T_t$ will have the property that no internal node is within $\epsilon(t)$ of any noncollapsible leaf. The collapsible leaves are allowed to converge to internal nodes, possibly producing a one point tree. The nodes of degree $\geq 2$ are assigned positive multiplicities at least equal to the degree minus 2. We add the multiplicities of coalescing nodes. (Thus over each component of $K$, the total multiplicity of $T_t$ will be fixed.) We will prove the lemma for these trees with "magma" replaced by "isomorphism class of trees with fixed noncollapsible leaves" and with condition (b3) required only for noncollapsible leaves. The proof will be by induction of the total multiplicity $m$ of $T_0$. We may assume without loss of generality that the nodes of $T_0$ have minimal multiplicity.

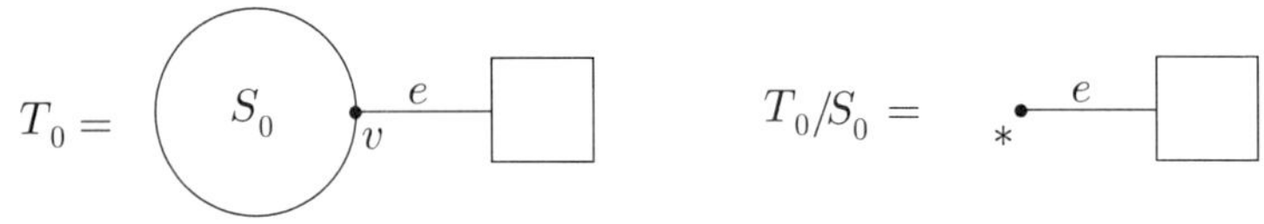


Figure 8.3: Case II: Collapse $S_0$ to a point $*$

Now that the statement has changed we need to start over. Take the minimal nontrivial case $m = 2$. Then $T_0$ is a line segment with no internal nodes. There are two cases: either at least one of the leaves is noncollapsible, in which case the tree cannot be collapsed, or both leaves are collapsible, in which case the tree could collapse (assuming that $\eta_0$ is nonempty at one endpoint of $T_0$). In both cases, statement (a) follows from Theorem 8.3.2 as explained at the beginning of the proof. As for (b), this is trivial when the tree is not collapsible. When it is collapsible, there must be one endpoint of $T_0$, say $v_0$, where $\eta_0(v_0)$ is nonempty. Then we can let $(T_{tu}, \phi_{tu})$ be equal to $(T_t, \phi_t)$ with one edge of length $u$ attached to $v_0$ and oriented outward toward the new (collapsible) leaf.

Now suppose that $m \geq 3$. Then $T_0$ has at least one internal node and there are two cases depending on the homotopy framing $\eta_0$ at the internal nodes.

Case I. Suppose that $T_0$ has no nodes of degree $\geq 3$ and that $\eta_0(v)$ contains both incident edges for each node $v$ of degree 2.

In this case, $T_0$ is again a line segment. The homotopy framing at the internal nodes allows any framing so there is no implied relationship between the placement of the internal nodes and any framing of the line subordinate to $\eta_0$. Therefore, by Proposition 8.2.2 we can ignore the internal nodes and apply Theorem 8.3.2 as before. This proves part (a). To prove part (b) we simply move the internal nodes into general position. For example, if one leaf of $T_0$ is collapsible and the other isn't then the $m-2$ internal nodes of $T_t$ have positions

$$0 \leq p_1(t) \leq p_2(t) \leq \cdots \leq p_{m-2}(t) \leq L(t) - \epsilon(t).$$

Let $(T_{t,u}, \phi_{t,u})$ be equal to $(T_t, \phi_t)$ with the internal nodes repositioned at

$$p_j(t,u) = p_j(t) + u\left(\frac{j\epsilon(t)}{2(m-2)}\right).$$

Case II. $T_0$ contains an internal node $v$ whose label set $\eta_0(v)$ does not contain one of its incident edges $e$.

Let $S_0$ be the component of $T_0 - int(e)$ containing the node $v$. We define the multiplicity of $v$ in $S_0$ to be the same a its multiplicity in $T_0$. Let $S_t$ be the image of $S_0$ in $T_t$ for all $t \in K$. Then $S_t$ is a subcomplex of $(T_t, \phi_t)$ for all $t$ and has an induced framing $\phi_t|S_t$.

Let $R_0 = T_0/S_0$ be $T_0$ with $S_0$ collapsed to one point $*$. (See Figure 8.3.) This will be a collapsible leaf with empty label set and multiplicity one. Let $R_t = T_t/S_t$ be $T_t$ with $S_t$ collapsed to one point $*$ which we assign the unique multiplicity $\geq 1$ making the total multiplicity of $R_t$ equal to that of $R_0$. Since $S_t$ is a subcomplex of $T_t$, $R_t$ has a coinduced framing $\overline{\phi}_t$ for all $t \in K$.

Since $v$ is an internal node of $T_0$, the new trees $S_0, R_0$ have smaller total multiplicity than $T_0$. Therefore, by induction, we can construct the required framed deformations (or null deformations) $S_{t,u}$ and $R_{t,u}$ of $S_t$, $R_t$. The point $*$ will be a vertex in the framing of $R_{t,u}$ for all $u$ and therefore, we may attach $R_{t,u}$ to $S_{t,u}$ by identifying $*$ with the node $v$ in $S_{t,u}$. The resulting point will be a vertex or not in $T_{t,u}$ depending on whether $v$ is a vertex of not in $S_{t,u}$. This gives us the deformation $(T_{t,u}, \phi_{t,u})$ required in part (b) or the null deformation required in part (a). □

**Definition 8.4.9.** Let $\mathcal{FT}(S)$ denote the category of all framed trees $(T, \phi)$ with leaf set $S$ and morphism fixing $S$. For any $(M, \eta) \in \mathrm{HFT}(S)$, let $\mathcal{FT}(M, \eta)$ denote the full category of all framed trees $(T, \phi)$ whose corresponding homotopy framed trees are degenerations of $(M, \eta)$.

Letting $K = |\mathcal{FT}(M, \eta)|$ in the above lemma we get the following.

**Proposition 8.4.10.** *$|\mathcal{FT}(M, \eta)|$ is contractible for all homotopy framed magma $(M, \eta)$.*

By Quillen's Theorem A we conclude that:

**Corollary 8.4.11.** $|\mathcal{FT}(S)| \simeq |\mathrm{HFT}(S)|$.

### 8.4.3 Existence and uniqueness of homotopy framings of trees

In this subsection we want to show that any simplicial family of trees over a simplicial complex $K$, i.e. an order preserving map

$$T : K \to \mathrm{Mag}(S),$$

lifts to a family of homotopy framings on a subdivision of $K$

$$\widetilde{T} : K' \to \mathrm{HFT}(S)$$

and which agrees with an already given homotopy framing on a given subcomplex $C$ of $K$. As usual this will show that homotopy framings exist and are unique up to homotopy.

First we clarify the statement. If $K'$ is a subdivision of $K$ then each open simplex in the geometric realization $|K'|$ is a subset of a unique open simplex in $|K|$ so we get an order preserving mapping $\lambda : K' \to K$. The composition

$$T \circ \lambda : K' \to K \to \mathrm{Mag}(S)$$

represents the same continuous family of trees (up to linear homeomorphism) as $T$.

If we apply the nondegenerate nerve functor $\mathcal{N}^{nd}_\bullet$ to the order preserving map $T : K \to \mathrm{Mag}(S)$ we get a map of ordered simplicial complexes

$$T' : K' \to \mathcal{N}^{nd}_\bullet \mathrm{Mag}(S)$$

where $K' = \mathcal{N}_\bullet^{nd} K$ can be interpreted as the first barycentric subdivisions of $K$ since the $k$-simplices of $K'$ are increasing sequences

$$\sigma = (\sigma_0 \subset \sigma_1 \subset \cdots \subset \sigma_k)$$

of simplices in $K$. In this case the simplicial map $\lambda : K' \to K$ is the last (maximal) vertex map given by $\lambda(\sigma) = \sigma_k$. The map $\lambda$ is, evidently, a natural transformation $\lambda : \mathcal{N}_\bullet^{nd} \to id$ of functors on the category of posets, i.e., $\lambda \circ T' = T \circ \lambda$ in this case.

Given a lifting $f : C \to \mathrm{HFT}(S)$ of the restriction of $T$ to a simplicial subcomplex $C$ of $K$ our job is to show the following.

**Proposition 8.4.12.** *There exists a subdivision $K'$ of $K$ relative to $C$ (i.e., with $C$ not subdivided) so that $T \circ \lambda : K' \to \mathrm{Mag}(S)$ lifts to $\mathrm{HFT}(S)$ and the lifting agrees with the given lifting on $C \subseteq K'$.*

This is a purely combinatorial property of the posets $\mathrm{Mag}(S)$ and $\mathrm{HFT}(S)$ and the morphism $p : \mathrm{HFT}(S) \to \mathrm{Mag}(S)$ and is equivalent to the following lemma.

**Lemma 8.4.13.** *Let $M$ be any magma on $S$.*

a) *The poset $p^{-1}(M)$ is contractible, i.e., $|p^{-1}(M)| \simeq *$.*

b) *If $(M_0, \eta_0) \in \mathrm{HFT}(S)$ and $M_0 \subseteq M$ then the subposet*

$$P = \{(M, \eta) \in p^{-1}(M) : (M_0, \eta_0) \leq (M, \eta)\}$$

*of $p^{-1}(M)$ is also contractible.*

We will first show that the proposition is equivalent to the lemma. First, it is obvious that the proposition implies the lemma. For (a), we let $C = \mathcal{N}_\bullet^{nd} p^{-1}(M)$ and let $K$ be the cone on $C$ with $T(\sigma) = M$ for all $\sigma \in K$. Then any lifting of $|T|$ to $|\mathrm{HFT}(S)|$ gives a contraction of $|p^{-1}(M)|$.

For (b) we let $C = v_0 * \mathcal{N}_\bullet^{nd}(P)$ be the cone on the nondegenerate nerve of $P$ with $f : C \to \mathrm{HFT}(S)$ mapping the cone point $v_0$ of $C$ to $(M_0, \eta_0)$ and all other simplices to their maximal vertex which lies in $P$. Let $K = v_1 * C$ be the cone on $C$ with $T : K \to \mathrm{Mag}(S)$ mapping the cone point $v_0$ of $C$ to $M_0$ and all other simplices to $M$. Given any subdivision $K'$ of $K$ the link $L$ of $v_0$ in $K'$ has realization $|L|$ homeomorphic to $|v_1 * \mathcal{N}_\bullet^{nd}(P)|$ where $v_1 * \mathcal{N}_\bullet^{nd}(P)$ is the cone on $\mathcal{N}_\bullet^{nd}(P)$ using the cone point $v_1$ of $K$. The homeomorphism is given by radial projection from $v_0$ in

$$|K'| \cong |K| = |v_1 * v_0 * \mathcal{N}_\bullet^{nd}(P)| \cong |v_0 * v_1 * \mathcal{N}_\bullet^{nd}(P)|.$$

It is the standard mistake to assume that this radial projection is simplicial. It is not. However it does send linear simplices homeomorphically (nonlinearly) onto linear simplices. A lifting of $T' : K' \to \mathrm{Mag}(S)$ to $\mathrm{HFT}(S)$ when restricted to $L$ will therefore give a contraction of $|P| \cong |\mathcal{N}_\bullet^{nd}(P)|$.

*Proof of Proposition 8.4.12.* We will prove that the lemma implies the proposition. We first consider the following special case.

Suppose that $K' = \Delta^k$ with vertices $v_0 < v_1 < \cdots < v_k$ and $C' = \partial\Delta^k$. Let $T' : K' \to \mathcal{N}_\bullet^{nd}\mathrm{Mag}(S)$ be a simplicial map with

$$T'(v_0) \le T'(v_1) \le \cdots \le T'(v_k).$$

Let $j < k$ be maximal so that $T'(v_j) \ne T'(v_k)$. (Let $j = -1$ if no such $j$ exists.) Suppose that $C''$ is a subdivision of $C' = \partial K'$ in which the $j$-simplex $(v_0, v_1, \cdots, v_j)$ has not been subdivided and let

$$f'' : C'' \to \mathcal{N}_\bullet^{nd}\mathrm{HFT}(S)$$

be a lifting of $T' \circ \lambda : C'' \to C' \subseteq K' \to \mathcal{N}_\bullet^{nd}\mathrm{Mag}(S)$. Then we claim that there exists a subdivision $K''$ of $K'$ extending the subdivision $C''$ of $C'$ and a lifting of

$$T' \circ \lambda : K'' \to K' \to \mathcal{N}_\bullet^{nd}\mathrm{Mag}(S)$$

to $\mathcal{N}_\bullet^{nd}\mathrm{HFT}(S)$ which extends $f''$.

It is easy to see that this special case implies the general case. We work one simplex at a time on $K' = \mathcal{N}_\bullet^{nd}K$. When we come to the simplex $\tau$ of $K' - C'$ we have a homotopy framing over a subdivision of the boundary of $\tau$ which we want to extend to some subdivision of the interior of $\tau$ without subdividing the boundary further. We first subdivide $\tau$ as the cone on $\partial\tau$ with the barycenter of $\tau$ as cone point. Each simplex $\sigma$ of this cone ($\sigma \subsetneq \partial\tau$) satisfies the conditions of the special case so we can inductively subdivide each such $\sigma$ relative to $\sigma \cap \partial\tau$ and extend the homotopy framing. The composition

$$K'' \to \mathcal{N}_\bullet^{nd}\mathrm{HFT}(S) \xrightarrow{\lambda} \mathrm{HFT}(S)$$

of the resulting lifting with the maximal vertex map $\lambda$ is the required lifting.

Returning to the special case suppose first that $j = -1$, i.e., $T'(v_i) = M$ for all vertices $v_i$ in $\Delta^k$. Then the statement follows from part (a) of the lemma. In more detail, we first subdivide $\Delta^k$ as the cone $b * \partial\Delta^k$ where $b \in \Delta^k$ is the barycenter. Since $|p^{-1}(M)|$ is contractible, we can find an extension $\overline{f}$ of $|f'| : \partial\Delta^k \to |p^{-1}(M)|$ to $\Delta^k$ which, near $\partial\Delta^k$ is equal to $|f'|$ composed with radial projection away from the cone point $b$. We can take an iterated relative barycentric subdivision of $(b * \partial\Delta^k, \partial\Delta^k)$ and find a simplicial approximation of $\overline{f}$ which agrees with $f$ on $\partial\Delta^k$.

Now suppose that $j \ge 0$ and the special case holds for numbers less than $j$. Let $\tau = (v_0, \cdots, v_j)$ with $f''(v_0) \le f''(v_0) \le \cdots \le f''(v_j) = (M_0, \eta_0)$ and $pf''(v) = M > M_0$ for all other vertices $v$ in $C''$. Let $L$ be the link of $\tau$ in the subdivided boundary $C''$ of $K' = \Delta^k$. Then $f''$ maps $L$ into the n.d. nerve $\mathcal{N}_\bullet^{nd}P$ of the poset $P$ of the lemma. Since $P$ is contractible, we can subdivide the cone $b * L$ of $L$ (using the barycenter $b$ of $K'$ as cone point) without subdividing $L$ and extend $f''$ to a simplicial map $\overline{f} : (b * L)' \to \mathcal{N}_\bullet^{nd}P$ where $(b * L)'$ is the subdivided cone of $L$.

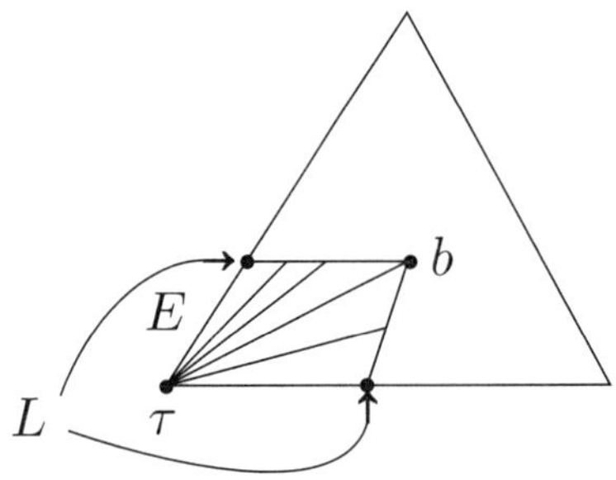


Figure 8.4: $E$ is the closure of all simplices of $C''$ containing $\tau$

Since $\overline{f}$ sends the vertices of $(b*L)'$ to elements of HFT$(S)$ which are greater than $(M,\eta_0)$, we get a well-defined simplicial map from the join of $(b*L)'$ and $\tau$ to $\mathcal{N}_\bullet^{nd}$HFT$(S)$. This join is a subdivision of the cone $b*E$ on the closure $E$ of all simplices of $C''$ containing $\tau$. (See Figure 8.4.) By construction, $\overline{f}|E = f''|E$ and by induction on $j$, we can extend $f'' \cup \overline{f}$ to the rest of $K'$ by an appropriate subdivision of the cone $b * C''$ of $C''$. (Each simplex in the cone $b * C''$ not contained in $b * E$ satisfies the conditions of the special case for some $j' < j$.) This completes the proof of the proposition assuming the lemma. □

*Proof of Lemma 8.4.13.* Part (a) is trivial since $p^{-1}(M)$ has a unique minimal element given by taking $\eta(v)$ to be empty for all nodes $v$.

For part (b) we will show that for any poset $X$ and order preserving map $\eta : X \to P$ (sending $x$ to $(M, \eta(x)) \in P$) there exists a sequence of order preserving maps $\eta = \eta_1, \eta_2, \cdots, \eta_n : X \to P$ so that $\eta_1(x) \geq \eta_2(x) \leq \eta_3(x) \geq \cdots$ for all $x \in X$ and so that $(M, \eta_n(X))$ is a single point in $P$.

First choose trees $T, T_0$ representing the magma $M, M_0$ and let $g : T \to T_0$ be the unique linear epimorphism which fixes the leaves. Let $T(v)$ denote the collapsing subtree $T(v) = g^{-1}(v) \subseteq T$ for each node $v$ of $T_0$. The inverse image of the edge $vw$ of $T_0$ is an edge of $T$ connecting the collapsing subtrees $T(v), T(w)$. We call these "long edges" of $T$. The edges in the collapsing subtrees are "short edges."

There are three types of collapsing subtrees $T(v)$ which we refer to as "type 0, 1 or 2" depending on the size of $\eta_0(v)$. If $\eta_0(v)$ has 1 or 2 elements $e_i$ then the corresponding long edges are incident to nodes $u_i$ in $T(v)$ and there is a unique minimal admissible homotopy framing $\eta_2$ for $T(v)$ given as follows. (See Figure 8.5.) For each node $u$ in $T(v)$ this minimal label set $\eta_2(u)$ consists of

1. any long edge $e$ incident to $u$ so that $g(e) \in \eta_0(v)$ and

2. the short edges incident to $u$ which lead to the nodes $u_i$.

For each $x \in X$ we let $\eta_2(x)$ be equal to this unique minimal homotopy framing on all $T(v)$ where $\eta_0(v)$ is nonempty. If $u \in T(v)$ where $\eta_0(v)$ is empty ($T(v)$ has "type 0") then we let $\eta_2(x)(u)$ be the set of short edges in $\eta(x)(u)$.

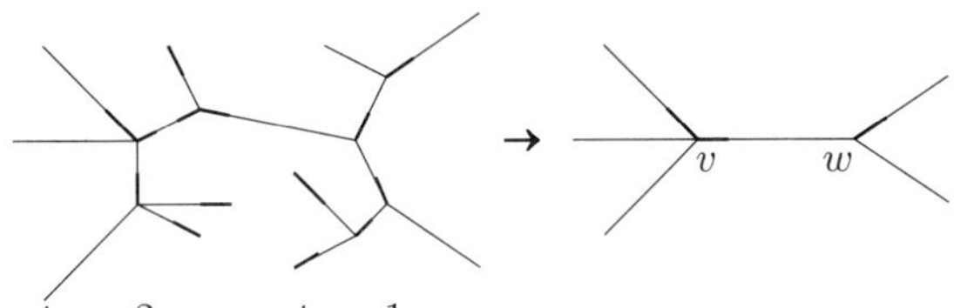


Figure 8.5: $\eta_2$ on type 1 and type 2 subtrees

Then $(M, \eta_2(x)) \geq (M_0, \eta_0)$ and $\eta_2(x) \leq \eta_2(y)$ whenever $x \leq y$. (Note that $\eta_2(x), \eta_2(y)$ differ only on type 0 collapsing subtrees.)

We will replace the homotopy framing on each type 0 subtree $T(v)$ with the *canonical homotopy framing* $\eta_\infty$ which is defined recursively as follows.

1. For every leaf $u$ of $T(v)$, $\eta_\infty(u)$ contains the unique edge $e_u$ of $T(v)$ incident to $u$
2. The restriction of $\eta_\infty$ to the subtree $ST(v)$ of $T(v)$ obtained by deleting all leaves and incident edges is canonical.
3. The edges $e_u$ are not labels of nodes in $ST(v)$.

If we iterate the leaf stripping operator $S$ we get a decreasing sequence of subtrees of $S^iT(v)$ of $T(v)$. This will terminate after $\left[\frac{n-2}{2}\right]$ steps where $n$ is the number of elements of $S$ since $T(v)$ has at most $n-2$ elements an every nontrivial tree has at least two leaves.

For each $x \in X$, let $\eta_3(x)$ be equal to $\eta_2(x)$ except at the leaves $u$ of type 0 subtrees $T(v)$ where $\eta_3(x)(u)$ is canonical, i.e., consists of the unique edge $e_u$ of $T(v)$ incident to $u$.

Let $\eta_4(x)$ be equal to $\eta_3(x)$ except that the edges $e_u$ above are not labels of any node of $ST(v)$. Then $\eta_3, \eta_4$ give order preserving maps $X \to P$ and

$$\eta_2(x) \leq \eta_3(x) \geq \eta_4(x)$$

for all $x \in X$. Also $\eta_4(x)$ is canonical outside of $S^1T(v)$ for all $x \in X$.

If we repeat this process on the leaves of each $ST(v)$ we get $\eta_5, \eta_6$. Continuing in this way we get $\eta_{2i-1}, \eta_{2i}$ where $\eta_{2i}$ is canonical outside of $S^{i-1}T(v)$. Consequently $\eta_{2i}(x)$ is canonical for all $x \in X$ and therefore a constant function when $i \geq \left[\frac{n}{2}\right]$. $\square$

One immediate consequence of Proposition 8.4.12 is that the projection

$$p : \mathrm{HFT}(S) \to \mathrm{Mag}(S)$$

induces a homotopy equivalence on realizations. Since $\mathrm{Mag}(S)$ is contractible (having a unique minimal element), and $|\mathcal{FT}(S)| \simeq |\mathrm{HFT}(S)|$ by Corollary 8.4.11, we get:

**Corollary 8.4.14.** *$\mathcal{FT}(S)$ and $\mathrm{HFT}(S)$ have contractible realizations.*

If we combine Propositions 8.4.6 and 8.4.12 we get the following.

**Theorem 8.4.15 (Framed tree lemma).** *Any simplicial family of metric trees with fixed leaves admits a continuous family of framings and any two families of framings are homotopic by a framed homotopy with is fixed on the underlying family of metric trees.*

**Remark 8.4.16.** The family of metric trees $X_t, t \in K$ only needs to be *continuously simplicial* in the sense that the set

$$K(M) = \{s \in K : M(X_t) \leq M\}$$

is a subcomplex of $K$ for all magma $M$ since, in that case, the corresponding family of magma $M(X_t)$ will be simplicial in the sense of 8.4.5 and Proposition 8.4.12 will apply.

## 8.4.4 Framing families of graphs

Now we are ready to show that any continuously simplicial family of metric graphs (8.4.16) admits a continuous family of framings unique up to homotopy. This will follow easily from the results of the previous two subsections.

As in the case of trees, the first step is to show:

**Proposition 8.4.17.** *Any simplicial family of graphs admits a simplicial family of homotopy framings which is unique up to simplicial homotopy.*

As we explained in the last subsection this statement is equivalent to the following.

**Lemma 8.4.18.** *Let $G$ be a fixes graph.*

a) *The poset of all homotopy framings of $G$ is contractible.*

b) *Let $g : G \to G_0$ be an admissible epimorphism and let $\eta_0$ be any homotopy framing of $G_0$. Then the poset $P$ of all homotopy framings of $G$ degenerating to $\eta_0$ is contractible.*

*Proof.* This proof is the same as the proof of Lemma 8.4.13. Part (a) is trivial since every graph has a unique minimal homotopy framing given by an empty label set $\eta(v)$ for all vertices $v$.

For part (b) we take any poset $X$ and any order preserving map $\eta : X \to P$. Then there is a sequence of poset morphisms $\eta = \eta_1 \geq \eta_2 \leq \eta_3 \geq \cdots$ with $\eta_{n+1}$ a constant mapping where $n$ is the maximum number of nodes of a collapsing tree $T(v) = g^{-1}(v)$.

As before, $\eta_2$ fixes the homotopy framing along all long edges and on all collapsing trees of type 1, 2. For $i = 1, 2, \cdots, \left[\frac{n+1}{2}\right]$ the value of $\eta_{2i}(x)$ is canonical and thus independent of $x$ outside of the sets $S^{i-1}T(v)$ for collapsing trees $T(v)$ of type 0. Since $S^{[n/2]}T(v)$ has at most one node, $\eta_{2[n/2]+2}(x)$ is canonical everywhere and thus independent of $x$. □

The next step is to show:

**Proposition 8.4.19.** *Every simplicial family of homotopy framed graphs admits a continuous family of framings which is unique up to framed homotopy.*

As before, this is equivalent to the following lemma.

**Lemma 8.4.20.** *Let $(G_0, \eta_0)$ be a fixed homotopy framed graph.*

a) *The space of framings of $G_0$ subordinate to $\eta_0$ is contractible.*

b) *Given any continuous family of framed graphs $(G_t, \phi_t), t \in K$ and any family of admissible epimorphisms $g_t : G_0 \to G_t$ with respect to which $\phi_t$ is subordinate to $\eta_0$ for all $t$, there exists a deformation $(G_{t,u}, \phi_{t,u}), (t, u) \in K \times I$, of $(G_t, \phi_t)$ and $g_{t,u} : G_0 \to G_{t,u}$ of $g_t$ so that*

1. *$g_{t,u}$ is a graph isomorphism for all $u > 0$*

2. *$\phi_{t,u}$ is subordinate to $\eta_0$ for all $u > 0$.*

*Proof.* Since we can work with each component of $G_0$ separately we may assume that $G_0$ is connected.

Let $m$ be the number of nodes of $G_0$ and let $r$ be the rank of $H_1(G_0)$. If $r = 0$ then $G_0$ is a tree and the lemma follows from Lemma 8.4.8. (Take the case where all leaves are collapsible.) So we may assume that $(r, m) \geq (1, 1)$ in lexicographic order and we proceed by induction on $(r, m)$. As in the case of trees there are two cases.

<u>Case I.</u> Suppose that $G_0$ has no nodes of degree $\geq 3$ and that $\eta_0(v)$ contains both incident edges for all nodes $v$ of degree 2.

In this case, $G_0$ must be a circle. The homotopy framing at the nodes of $G_0$ allows any framing so there is no implied relationship between the placement of the nodes and the subordinate framing. Thus (a) follows from Theorem 8.3.3. To prove (b) we just move the internal nodes into general position.

<u>Case II.</u> $G_0$ contains a node $v$ of degree $\geq 2$ whose label set $\eta_0(v)$ does not contain one of its incident edges, say $e$.

Suppose first that $G_0 - int(e)$ is disconnected, i.e., $e$ is a separating edge. Then we have the same situation as in Figure 8.3. We let $S_0$ be the component containing $v$, let $S_t$ be the image of $S_0$ in $G_t$. Let $R_0 = G_0/S_0$ and $R_t = G_t/S_t$. Then $S_t$ is a subcomplex of $(G_t, \phi_t)$ for all $t$ and therefore $R_t$ inherits a framing. We can find appropriate deformations of these by induction on $(r, m)$ and paste these deformations together.

Now suppose that $G_0 - int(e)$ is connected. In that case $e$ represents a cohomology class in $H^1(G_0)$ from which we can form a $\mathbb{Z}$-covering $\widetilde{G}_0$ containing a vertex $v^i$ with incident edge $e^i$ for each $i \in \mathbb{Z}$ as indicated in Figure 8.6. Since $g_t : G_0 \to G_t$ is a homotopy equivalence, we get a corresponding $\mathbb{Z}$-covering $\widetilde{G}_t$ of $G_t$ and an induced map of covering spaces

$$\tilde{g}_t : \widetilde{G}_0 \to \widetilde{G}_t.$$

Let $S^i$ be the component of $\widetilde{G}_0 - int(e^i)$ containing $v^i$ and let $S^i_t = \tilde{g}_t(S^i)$. Then $S^i_t$ is a subcomplex of $(G_t, \phi_t)$ and thus has an induced framing for all $t$. If

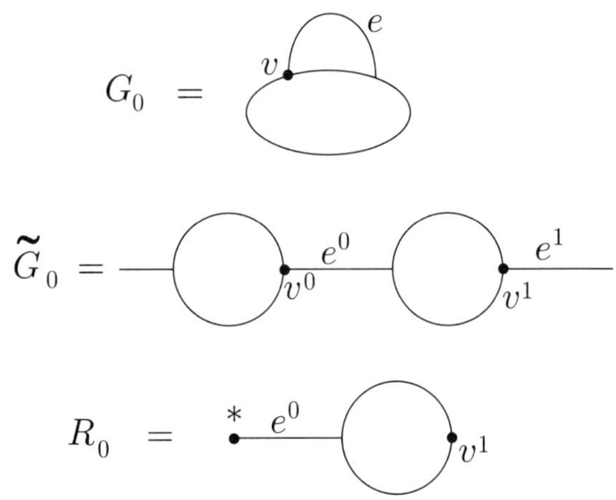


Figure 8.6: Case II when $e$ is a separating edge

we let $R_0 = S^1/S^0$ and $R_t = S_t^i/S_t^o$ then $R_t$ has a coinduced framing $\overline{\phi}_t$. Since $H^1(R_0)$ is smaller in rank than $H^1(G_0)$, the lemma holds for $R_0$. Therefore,

1. the space of framings of $R_0$ subordinate to the induced homotopy framing $\overline{f}_0$ is contractible and
2. the family of framings $(R_t, \overline{\phi}_t)$ has a deformation $(R_{t,u}, \overline{\phi}_{t,u})$ so that $R_{t,u}$ is equivalent to $R_0$ for all $u > 0$.

If we coalesce the collapsed point $*$ in $R_0$ (and $R_t$) with the point $v^1$ in the same graph, we get the desired deformation. □

Putting together Propositions 8.4.17 and 8.4.19 we get the following.

**Theorem 8.4.21 (Framed graph theorem).** *Any simplicial family of metric graphs admits a continuous family of framings and any two families of framings are homotopic by a framed homotopy which is fixed on the underlying family of metric graphs.*

## 8.5 Applications of the framed graph theorem

### 8.5.1 Framing of graphs and the homology of $Out(F_n)$

As we explained in detail in §8.2, the framed graph theorem gives a mapping

$$\xi : \mathcal{GE} \simeq \mathcal{MG}_\bullet \to \mathcal{W}h_\bullet(\mathbb{Z}, 1)$$

unique up to homotopy. (Each simplex in the nerve of $\mathcal{FG}$ gives a simplex in $\mathcal{W}h_\bullet(\mathbb{Z}, 1)$. The morphisms need to be partially reversed as in (8.1).) Since the first homology functor $H_1(-;\mathbb{Z})$ is not constant on the category $\mathcal{GE}$ this mapping does not lift to $\mathcal{W}h^h_\bullet(\mathbb{Z},1)$ and since $H_1(-,\mathbb{Q})$ is not upper triangular we cannot define the higher FR-torsion. However, the existence of the mapping $\xi$ has other unexpected consequences.

First we note that for any Waldhausen category $(\mathcal{K}, co\mathcal{K}, w\mathcal{K})$ we have a natural mapping

$$j : |w\mathcal{K}| \to \Omega|w\mathcal{S}_\bullet\mathcal{K}|.$$

Using the equivalence between filtered chain complexes and multiple mapping cylinders (Theorem 5.3.4) and Theorem 5.6.17 we get a mapping

$$\mathcal{GE} \to \mathcal{W}h_\bullet(\mathbb{Z},1) \simeq FC(*,\xi_\mathbb{Z}) \simeq we_\bullet\mathcal{K}(*,\xi_\mathbb{Z}) \to \Omega|w\mathcal{S}_\bullet e_\bullet\mathcal{K}(*,\xi_\mathbb{Z})| \simeq QS^0$$

where $Q = \Omega^\infty S^\infty$. Restricting to $|\mathcal{GE}^{(n)}| \simeq BOut(F_n)$ for $n \geq 2$ and composing with the mapping of classifying spaces corresponding to the obvious homomorphism

$$S_n \to Aut(F_n) \to Out(F_n)$$

where $S_n$ is the symmetric group on $n$ letters, we get:

$$BS_n \to BAut(F_n) \to BOut(F_n) \to QS^0.$$

Since the mapping of $\mathcal{K}(BG_\bullet, \xi_R)$ to $Q(|BG_\bullet|_+)$ is given by the forgetful functor $(E,P) \mapsto P$ (the poset over $BG_\bullet$) we see that this composition is the same as the mapping

$$BS_n \to BS_\infty \to BS_\infty^+ = QS^0$$

and we recover the following result of Allen Hatcher.

**Theorem 8.5.1.** *[Hat95] For $n$ much larger than $k$, $H_k(S_\infty)$ is a direct summand of both $H_k(Aut(F_n))$ and $H_k(Out(F_n))$.*

Hatcher also proved the following.

**Theorem 8.5.2.** *[Hat95] $H_k(Aut(F_n)) \to H_k(Out(F_n))$ is an isomorphism for $n > k^2/4 + 5k/2$.*

Hatcher used Waldhausen's space $A(*)$. Since we are using the linearized version of $A(*)$, namely $BGL(\mathbb{Z})^+$, we get the following consequence.

**Theorem 8.5.3.** *The composition*

$$BOut(F_n) \to BGL_n(\mathbb{Z}) \to BGL(\mathbb{Z})^+$$

*factors through $QS^0$.*

**Remark 8.5.4.** This implies that $H_k(Out(F_n),\mathbb{Q}) \to H_k(GL_n(\mathbb{Z}),\mathbb{Q})$ is trivial in the stable range $n \geq 2k+1$ (by Suslin [Sus82]).

*Proof.* When a Waldhausen category $\mathcal{C}$ has two kinds of weak equivalences $w\mathcal{C}, h\mathcal{C}$ we get a commuting square:

$$\begin{array}{ccc}
|w\mathcal{C}| & \to & \Omega|wS_\bullet\mathcal{C}| \\
\downarrow & & \downarrow \\
|h\mathcal{C}| & \to & \Omega|hS_\bullet\mathcal{C}|
\end{array}$$

Applying this to the bisimplicial Waldhausen category $e_\bullet\mathcal{K}(X_\bullet, \xi)$ where $X_\bullet$ is contractible and $\xi$ is the trivial $\mathbb{Z}$-bundle over $X_\bullet$ we get the following.

$$\begin{array}{ccccc}
|\mathcal{W}h_\bullet(\mathbb{Z},1)| & \xrightarrow{\simeq} & |we_\bullet\mathcal{K}(*,\xi_\mathbb{Z})| & \xrightarrow{j_0} & \Omega|wS_\bullet e_\bullet\mathcal{K}(*,\xi_\mathbb{Z})| \simeq QS^0 \\
\uparrow{\scriptstyle \xi} & \nearrow_{C} & \downarrow & & \downarrow \\
|\mathcal{FG}^{(n)}| & & |he_\bullet\mathcal{K}(*,\xi_\mathbb{Z})| & \longrightarrow & \Omega|hS_\bullet e_\bullet\mathcal{K}(*,\xi_\mathbb{Z})| \\
\downarrow{\scriptstyle \simeq} & & \downarrow{\scriptstyle \simeq} & & \downarrow{\scriptstyle \simeq} \\
BOut(F_n) \simeq |\mathcal{GE}^{(n)}| & & |hh_\bullet\mathcal{K}^{sf}(\mathbb{Z})| & \xrightarrow{j} & \Omega|hS_\bullet h_\bullet\mathcal{K}^{sf}(\mathbb{Z})| \\
\downarrow{\scriptstyle H_1} & & & \nearrow_{k} & \downarrow{\scriptstyle \simeq} \\
\coprod_{m\geq 0} BGL_m(\mathbb{Z}) \simeq |i\mathcal{K}_0^{sf}(\mathbb{Z})| & \xrightarrow{\Sigma} & |h\mathcal{K}^{sf}(\mathbb{Z})| & & \mathbb{Z}\times BGL(\mathbb{Z})^+
\end{array}$$

Here $\mathcal{FG}^{(n)}$ is the component of the category of framed graphs corresponding to $\mathcal{GE}^{(n)}$ and $\mathcal{FG}^{(n)} \simeq \mathcal{GE}^{(n)}$ by the framed graph theorem. The mapping $\xi : \mathcal{FG}^{(n)} \to \mathcal{W}h_\bullet(\mathbb{Z},1)$ is given by sending each framed graph to its based free cellular chain complex. Going across and down to $hh_\bullet\mathcal{K}^{fg}(\mathbb{Z})$ forgets the basis (and enlarges the category of allowable morphisms). By Remark 5.6.21 (see also the proof of Proposition 8.5.6), if we compose with $j$ we get the alternating sum of the homology functors.

The first homology functor $H_1 : \mathcal{GE}^{(n)} \to i\mathcal{K}_0^{sf}(\mathbb{Z})$ has image in the component $\mathcal{G}\ell(n,\mathbb{Z}) \simeq BGL_n(\mathbb{Z})$. The 0-th homology $H_0$ is a constant functor with image $[\mathbb{Z}] \in \mathcal{G}\ell(1,\mathbb{Z}) \simeq BGL_1(\mathbb{Z})$. Consequently, this diagram commutes up to homotopy provided that $\Sigma$ is suspension and $k = j + [\mathbb{Z}]$ so that $k\circ\Sigma\circ H_1$ has image in the component of $\Omega|hS_\bullet h_\bullet\mathcal{K}^{sf}(\mathbb{Z})|$ corresponding to $(1-n)\times BGL(\mathbb{Z})^+$. $\square$

In order to define the higher FR-torsion we need to trivialize the homology bundle. One way to do this is to take the fiber of the first homology functor

$$H_1 : \mathcal{GE} \to \mathcal{G}\ell(\mathbb{Z})$$

where $\mathcal{G}\ell(\mathbb{Z})$ is the category of isomorphisms of f.g. free $\mathbb{Z}$-modules. We note that

$$\mathcal{G}\ell(\mathbb{Z}) = \coprod_{n\geq 0} \mathcal{G}\ell(n,\mathbb{Z})$$

where $\mathcal{G}\ell(n,\mathbb{Z})$ is the full subcategory of $\mathcal{G}\ell(\mathbb{Z})$ of all free groups of rank $n$.

Let $\mathcal{GE}^h$ be the category of all pairs $(G,\theta)$ where $G$ is a graph and $\theta = (\theta_1,\cdots,\theta_n)$ is a basis for $H_1(G)$. We call $\theta$ a *homological marking* of $G$. Let $\mathcal{MG}_\bullet{}^h$ be the simplicial set of all homologically marked metric graphs. Let $\mathcal{GE}^{(n)h}$, $\mathcal{MG}_\bullet{}^{(n)h}$ be the subcategory/simplicial subset of graphs with Euler characteristic $1-n$.

**Proposition 8.5.5.** *$\mathcal{MG}_\bullet{}^{(n)h} \simeq \mathcal{GE}^{(n)h}$ is the homotopy fiber of the homology functor $H_1 : \mathcal{GE}^{(n)} \to \mathcal{G}\ell(n,\mathbb{Z})$ and consequently,*

$$|\mathcal{MG}_\bullet{}^{(n)h}| \simeq |\mathcal{GE}^{(n)h}| \simeq BOut^h(F_n)$$

*where $Out^h(F_n)$ is the kernel in the exact sequence*

$$1 \to Out^h(F_n) \to Out(F_n) \to GL_n(\mathbb{Z}) \to 1.$$

*Proof.* This follows from Quillen's Theorem B. It also follows from the obvious fact that $\mathcal{MG}_\bullet{}^{(n)h}$ is the quotient of the space of marked graphs (outer space) by the free action of $Out^h(F_n)$. The fact that $Out(F_n)$ maps onto $GL_n(\mathbb{Z})$ is well-known and follows from the fact that the Nielsen moves generate $GL_n(\mathbb{Z})$ [Nie24]. □

By the framed graph theorem and the canonical process of killing a trivial projective homology bundle we get a mapping

$$|\mathcal{GE}^{(n)h}| \to |\mathcal{W}h^h_\bullet(\mathbb{Z},1)|$$

well defined up to homotopy for all $n \geq 0$. For $n \geq 3$ ($Out^h(F_n)$ being trivial for $n \leq 2$) these define higher FR-torsion classes

$$\tau_{2k}(\mathcal{GE}^{(n)h}) = \tau_{2k}(Out^h(F_n)) \in H^{4k}(Out^h(F_n);\mathbb{R}).$$

**Proposition 8.5.6.** *The mapping $|\mathcal{FG}^{(n)h}| \xrightarrow{\simeq} |\mathcal{GE}^{(n)h}| \to |\mathcal{W}h^h_\bullet(\mathbb{Z},1)|$ defining the higher FR-torsion of $Out^h(F_n)$ is the map induced by the homotopy commuting square:*

$$\begin{array}{ccccc}
BOut(F_n) \simeq & |\mathcal{FG}^{(n)}| & \xrightarrow{f} & \Omega|w\mathcal{S}_\bullet e_\bullet \mathcal{K}(*,\xi_\mathbb{Z})| & \simeq QS^0 \\
 & \big\downarrow H_1 & & \big\downarrow p & \\
BGL_n(\mathbb{Z}) \simeq & |\mathcal{G}\ell(n,\mathbb{Z})| & \xrightarrow{g} & \Omega|h\mathcal{S}_\bullet h_\bullet \mathcal{K}^{sf}(\mathbb{Z})| & \simeq \mathbb{Z}\times BGL(\mathbb{Z})^+
\end{array}$$

*where $f = j_0 \circ C \oplus \mathbb{Z}^{n-1}$ and $g$ is the mapping of $BGL_n(\mathbb{Z})$ into $0 \times BGL(\mathbb{Z})^+$ given by $g = j \circ \Sigma \oplus \mathbb{Z}^n \simeq [\mathbb{Z}^n] - j$ (i.e., the negative of the usual mapping).*

*Proof.* In order to get an explicit map of fibers we need an explicit homotopy $p \circ f \simeq g \circ H_1$. For this we use Waldhausen's additivity theorem (in the form of Proposition 1.3.2(3) of [Wal85]) which says that, for any admissible sequence of exact functors $f \rightarrowtail g \twoheadrightarrow h$, the induced mappings in K-theory are related by a (homotopy) natural homotopy $|g| \simeq |f| \vee |h|$, i.e., the homotopy is natural up to homotopy. We use the additivity theorem in the case when $f, g, h$ are functors $\mathcal{C} \to \mathcal{D}$ where only $\mathcal{D}$ is a Waldhausen category but so that $f, g, h$ factor as a composition of exact functors $F \rightarrowtail G \twoheadrightarrow H$ from another Waldhausen category $\mathcal{E}$ to $\mathcal{D}$ (and a fixed functor $\mathcal{C} \to \mathcal{E}$). We use the short-hand notation $\mathcal{A}_\bullet = we_\bullet\mathcal{K}(*, \xi_\mathbb{Z})$, $\mathcal{B}_\bullet = hh_\bullet\mathcal{K}^{sf}(\mathbb{Z})$, $\mathcal{S}_\bullet\mathcal{A}_\bullet = w\mathcal{S}_\bullet e_\bullet \mathcal{K}(*, \xi_\mathbb{Z})$ and $\mathcal{S}_\bullet\mathcal{B}_\bullet = h\mathcal{S}_\bullet h_\bullet\mathcal{K}^{sf}(\mathbb{Z})$.

Let $C : \mathcal{FG}^{(n)} \to \mathcal{A}_\bullet$ be the functor which takes each framed graph $G$ to its based free cellular chain complex $C(G) = (C_1(G) \xrightarrow{\partial} C_0(G))$ where $C_0(G)$ is the free abelian group generated by the vertices of $G$. Adding the augmentation map $\epsilon : C_0(G) \to \mathbb{Z}$ (sending each vertex of $G$ to $1 \in \mathbb{Z}$), we get the augmented chain complex

$$\overline{C}(G) = (C_1(G) \xrightarrow{\partial} C_0(G) \xrightarrow{\epsilon} \mathbb{Z})$$

where $\mathbb{Z}$ is in degree $-1$. This gives another functor $\overline{C} : \mathcal{FG}^{(n)} \to \Sigma^{-2}\mathcal{A}_\bullet$. (Here $\Sigma^{-2}$ means we add terms of degree $-1$ and $-2$. Since double suspension is homotopic to the identity map on $\mathcal{S}_\bullet\mathcal{A}_\bullet$ and $\mathcal{S}_\bullet\mathcal{B}_\bullet$ we are not changing the homotopy type of these spaces. We could also deal with this problem by suspending everything twice in this proof.)

The functor $\overline{C}$ fits into a short exact sequence of functors

$$0 \to \Sigma^{-1}\mathbb{Z} \to \overline{C} \to C \to 0$$

where $\Sigma^{-1}\mathbb{Z}$ is $\mathbb{Z}$ in degree $-1$. By Waldhausen's additivity theorem we get a homotopy of functors (from $\mathcal{FG}^{(n)}$ to $\Omega|\mathcal{S}_\bullet\mathcal{A}_\bullet|$)

$$h_0 : j_0\overline{C} \simeq j_0 C \oplus \Sigma^{-1}\mathbb{Z}.$$

Choosing a fixed path $\gamma$ from $\mathbb{Z} \oplus \Sigma^{-1}\mathbb{Z}$ to 0 in $\Omega|\mathcal{S}_\bullet\mathcal{A}_\bullet|$ this gives a homotopy

$$\gamma h_0 : j_0\overline{C} \oplus \mathbb{Z}^n \simeq j_0 C \oplus \Sigma^{-1}\mathbb{Z} \oplus \mathbb{Z}^n \simeq j_0 C \oplus \mathbb{Z}^{n-1} = f.$$

Since $H_1(G) = \ker \partial$ there is a natural inclusion map $h_1(G) : \Sigma H_1(G) \to \overline{C}(G)$ which is an $h$-equivalence giving a homotopy $h_1 : \Sigma H_1 \simeq \overline{C}$ in $\mathcal{B}_\bullet$. Combining $jh_1$ and $\gamma h_0$ gives the explicit homotopy

$$h : gH_1 = j\Sigma H_1 \oplus \mathbb{Z}^n \simeq pj_0\overline{C} \oplus \mathbb{Z}^n \simeq pj_0 C \oplus \mathbb{Z}^{n-1} = pf.$$

Now we go to the homotopy fibers of $H_1$ and $p$. An object of the homotopy fiber $F(H_1)$ of $H_1$ is a framed graph $G \in \mathcal{FG}^{(n)}$ together with a path from $H_1(G)$ to the base point $\mathbb{Z}^n$ of $\mathcal{G}\ell(n, \mathbb{Z})$. This is given by a homological marking $\theta$ for $G$.

The induced map on homotopy fibers $F(H_1) \to F(p)$ sends $(G, \theta)$ to its image $f(G) = j_0 C(G) \oplus \mathbb{Z}^{n-1}$ in $\Omega|\mathcal{S}_\bullet\mathcal{A}_\bullet|$ together with the path in $\Omega|\mathcal{S}_\bullet\mathcal{B}_\bullet|$

from $g(\mathbb{Z}^n)$ to $pf(G) = pj_0C(G) \oplus \mathbb{Z}^{n-1}$ given by the paths $g(\theta) = (\Sigma\theta, id)$ and $h(G)$. Dropping the letters $j$ and $p$ this path is given as follows.

$$\Sigma\mathbb{Z}^n \oplus \mathbb{Z}^n \xrightarrow{(\Sigma\theta,id)} \Sigma H_1 \oplus \mathbb{Z}^n \xrightarrow{h_1} \overline{C} \oplus \mathbb{Z}^n \xrightarrow{h_0} C \oplus \Sigma^{-1}\mathbb{Z} \oplus \mathbb{Z}^n \xrightarrow{\gamma} C \oplus \mathbb{Z}^{n-1} \tag{8.2}$$

Note that the first two maps are natural transformations but the last two are homotopies.

We need to compare this with the functor

$$\xi^h : \mathcal{FG}^{(n)h} \to \Omega|w\mathcal{S}_\bullet e_\bullet \mathcal{K}^h(*, \xi_\mathbb{Z})|$$

which can be given by sending $(G, \theta)$ to the acyclic based chain complex $\overline{E}(G)$ given by

$$\overline{E}(G) = (\mathbb{Z}^n \xrightarrow{\theta} C_1(G) \to C_0(G) \xrightarrow{\epsilon} \mathbb{Z}).$$

Note that the $h$-equivalence $0 \xrightarrow{\simeq} \overline{E}(G)$ which puts this in the fiber of $p$ can be factored as:

$$0 \xrightarrow{\simeq} (\mathbb{Z}^n \xrightarrow{id} \mathbb{Z}^n \to 0) \xrightarrow{(id,\theta,0)} (\mathbb{Z}^n \xrightarrow{\theta} H_1(G) \to 0) \subseteq \overline{E}(G)$$

Choosing a fixed path from $\mathbb{Z}^n$ to $\Sigma^2\mathbb{Z}^n$ in $\Omega|\mathcal{S}_\bullet\mathcal{A}_\bullet|$ we can replace each $\mathbb{Z}^n$ in (8.2) above with $\Sigma^2\mathbb{Z}^n$ to obtain:

$$\Sigma\mathbb{Z}^n \oplus \Sigma^2\mathbb{Z}^n \xrightarrow{(\Sigma\theta,id)} \Sigma H_1 \oplus \Sigma^2\mathbb{Z}^n \xrightarrow{h_1} \overline{C} \oplus \Sigma^2\mathbb{Z}^n \xrightarrow{h_0} C \oplus \Sigma^{-1}\mathbb{Z} \oplus \Sigma^2\mathbb{Z}^n \simeq C \oplus \mathbb{Z}^{n-1}$$

Using the homotopy naturality of the additivity homotopy we get a homotopy of homotopies from this homotopy to the homotopy:

$$0 \to (\mathbb{Z}^n \xrightarrow{id} \mathbb{Z}^n \to 0) \to (\mathbb{Z}^n \xrightarrow{\theta} H_1(G) \to 0) \to \overline{E}(G) \simeq E \oplus \Sigma^{-1}\mathbb{Z}$$

where $E$ is $\overline{E}$ with the final $\mathbb{Z}$ removed. Therefore, the homotopy

$$\overline{E} \simeq E \oplus \Sigma^{-1}\mathbb{Z} \simeq C \oplus \Sigma^2\mathbb{Z}^n \oplus \Sigma^{-1}\mathbb{Z} \simeq C \oplus \mathbb{Z}^{n-1}$$

in $\Omega|\mathcal{S}_\bullet\mathcal{A}_\bullet|$ together with its image in $\Omega|\mathcal{S}_\bullet\mathcal{B}_\bullet|$ gives a homotopy from the functor $\xi^h$ to the induced map of homotopy fibers $F(H_1) \to F(p)$. □

When $n = 2g$ is even there is a larger subgroup of the outer automorphism group on which the higher FR-torsion is defined. This is the inverse image in $Out(F_{2g})$ of the *symplectic group*

$$Sp_{2g}(\mathbb{Z}) = \{f \in GL_{2g}(\mathbb{Z}) \, : \, f^t j f = j\}$$

where

$$j = \begin{pmatrix} 0 & I_g \\ -I_g & 0 \end{pmatrix}.$$

Another way to say this is that $Sp_{2g}(\mathbb{Z})$ is the group of automorphisms of $\mathbb{Z}^{2g}$ which preserve the *symplectic form* $\omega_0$ given by $\omega_0(x, y) = x^t jy$. Thus,

$$Out^{sp}(F_{2g}) := \phi^{-1}(Sp_{2g}(\mathbb{Z})) = \{f \in Out(F_{2g}) : \omega^{\phi(f)} = \omega_0\}$$

where $\omega^h$ is given by $\omega^h(x, y) = \omega_0(h(x), h(y))$ and $\phi$ is the epimorphism $Out(F_{2g}) \to GL_{2g}(\mathbb{Z})$.

We get a short exact sequence

$$0 \to Out^h(F_{2g}) \to Out^{sp}(F_{2g}) \to Sp_{2g}(\mathbb{Z}) \to 0$$

which we claim is equivalent to a fibration sequence of $K(\pi, 1)$'s:

$$\mathcal{GE}^{(2g)h} \to \mathcal{GE}^{(2g)sp} \to \mathcal{S}p(2g, \mathbb{Z})$$

where $\mathcal{S}p(2g, \mathbb{Z})$ and $\mathcal{GE}^{(2g)sp}$ are the categories defined below.

First we define $\mathcal{S}p(2g, \mathbb{Z})$ to be the category of all pairs $(A, \omega_A)$ where $A$ is a free abelian group of rank $2g$ and $\omega_A$ is a symplectic form on $A$, i.e., one equivalent to $\omega_0$ under some choice of basis for $A$. A morphism $f : (A, \omega_A) \to (B, \omega_B)$ is defined to be an isomorphism $f : A \to B$ so that $\omega_B(f(x), f(y)) = \omega_A(x, y)$ for all $x, y \in A$. Then the objects of $\mathcal{S}p(2g, \mathbb{Z})$ are all isomorphic so the full subcategory with the single object $(\mathbb{Z}^{2g}, \omega_0)$ is a deformation retract. Since $Aut(\mathbb{Z}^{2g}, \omega_0) = Sp_{2g}(\mathbb{Z})$, the standard classifying space $BSp_{2g}(\mathbb{Z})$ is contained in $|\mathcal{S}p(2g, \mathbb{Z})|$ as a deformation retract.

Let $\mathcal{GE}^{(2g)sp}$ be the category of all pairs $(G, \omega_G)$ where $G \in \mathcal{GE}^{(2g)}$ and $\omega_G$ is a symplectic form on $H_1(G) \cong \mathbb{Z}^{2g}$ with morphisms preserving the symplectic structure. Let $\mathcal{FG}^{(2g)sp}$ denote the corresponding category of framed graphs.

**Proposition 8.5.7.** *$\mathcal{GE}^{(2g)sp} \simeq \mathcal{FG}^{(2g)sp} \simeq BOut^{(2g)sp}(F_{2g})$. Furthermore, we have a map of homotopy fiber sequences:*

$$\begin{array}{ccccc}
\mathcal{GE}^{(2g)h} & \longrightarrow & \mathcal{GE}^{(2g)sp} & \xrightarrow{H_1} & \mathcal{S}p(2g, \mathbb{Z}) \\
\downarrow & & \downarrow & & \downarrow \\
\mathcal{GE}^{(2g)h} & \longrightarrow & \mathcal{GE}^{(2g)} & \xrightarrow{H_1} & \mathcal{G}\ell(2g, \mathbb{Z})
\end{array}$$

*Proof.* The group $Out^{sp}(F_{2g})$ acts freely on outer space and the quotient can be identified with $\mathcal{GE}^{(2g)sp}$. Thus it must be the classifying space of this group. The framed graph theorem implies that $\mathcal{GE}^{(2g)sp} \simeq \mathcal{FG}^{(2g)sp}$.

The right hand square in the diagram is a homotopy fiber square. To see this note that the two functors labelled $H_1$ have the same fiber category, namely, $\mathcal{GE}^{(2g)h}$ over the objects $(\mathbb{Z}^{2g}, \omega_0)$ and $\mathbb{Z}^{2g}$. □

If $(G, \omega_G) \in \mathcal{FG}^{(2g)sp}$ then the symplectic form $\omega_G$ determines an isomorphism

$$\widehat{\omega}_G : H_1(G) \xrightarrow{\cong} \mathrm{Hom}(H_1(G), \mathbb{Z})$$

by $\widehat{\omega}_G(x)(y) = \omega_G(x,y)$. Let $\omega_G^* = \widehat{\omega}_G^{-1}$ denote the inverse of this isomorphism. We use this to put together the augmented chain complex

$$\overline{C}(G) : C_1(G) \xrightarrow{d} C_0(G) \xrightarrow{\epsilon} \mathbb{Z}$$

with its dual to give an acyclic complex

$$D(G) : \mathbb{Z} \xrightarrow{\epsilon^*} C^0(G;\mathbb{Z}) \xrightarrow{\partial^*} C^1(G;\mathbb{Z}) \xrightarrow{\omega_G^*} C_1(G) \xrightarrow{d} C_0(G) \xrightarrow{\epsilon} \mathbb{Z} \tag{8.3}$$

which we call $D(G)$. This complex is a based free complex using the dual bases for $C^0(G;\mathbb{Z}) = \mathrm{Hom}(C_0(G),\mathbb{Z})$ and $C^1(G;\mathbb{Z}) = \mathrm{Hom}(C_1(G),\mathbb{Z})$.

The complex $D(G)$ is also natural in $(G,\omega_G)$ since morphisms (in the ghost category $\mathcal{FG}h$) are chain isomorphisms and elementary expansions. A chain isomorphism $f : \overline{C}(G) \to \overline{C}(H)$ induces a dual isomorphism $f^* : \overline{C}^*(H) \to \overline{C}^*(G)$ whose inverse can be used to give a chain map $D(G) \to D(H)$.

$$\begin{array}{ccc} \overline{C}^*(G) & \xrightarrow{(f^*)^{-1}} & \overline{C}^*(H) \\ {\scriptstyle \omega_G^*}\downarrow & & \downarrow{\scriptstyle \omega_H^*} \\ \overline{C}(G) & \xrightarrow{f} & \overline{C}(H) \end{array}$$

The assumption that $f_*$ preserves the symplectic structure on $H_1$ tells us that $f^*\widehat{\omega}_H f = \widehat{\omega}_G$ so $f\omega_G^* = \omega_H^*(f^*)^{-1}$ and the diagram above commutes. Elementary expansions $\overline{C}(G) \to \overline{C}(G) \oplus E$ go to elementary expansions $D(G) \to D(G) \oplus E \oplus E^*$. Consequently, we get a simplicial bifunctor

$$D : m^{op}ec_\bullet\mathcal{FG}h^{(2g)sp} \to me\mathcal{W}h^h_\bullet(\mathbb{Z},1)$$

defining a mapping

$$\xi^{sp} : |\mathcal{FG}^{(2g)sp}| \to |\mathcal{W}h^h_\bullet(\mathbb{Z},1)|.$$

By Proposition 8.5.7 we get:

**Proposition 8.5.8.** *$\xi^{sp}$ determines higher FR-torsion classes*

$$\tau_{2k}(\mathcal{GE}^{(2g)sp}) = \tau_{2k}(Out^{sp}(F_{2g}) \in H^{4k}(Out^{sp}(F_{2g});\mathbb{R}).$$

**Theorem 8.5.9.** *The higher FR-torsion of $Out^{sp}(F_{2g})$ restricts to double the torsion of $Out^h(F_{2g})$:*

$$\tau_{2k}(Out^{sp}(F_{2g}))|Out^h(F_{2g}) = 2\tau_{2k}(Out^h(F_{2g})).$$

*Proof.* The functor $D$ is an extension of the functor $\overline{C}$ by $\Sigma^2\overline{C}^*$ where $\overline{C}^* = ct_1(\overline{C})$. Since $ct_1$ preserves the higher torsion the theorem follows from the Splitting Lemma (5.7.8) and Theorem 2.4.13. ∎

There is another way to kill the homology of graphs and define higher FR-torsion. This is to attach "homological 2-cells" to kill $H_1(G)$. More precisely, we let $\mathcal{Y}(n,\mathbb{Z})$ be the pull-back in the following diagram.

$$\begin{array}{ccc} \mathcal{Y}(n,\mathbb{Z}) & \to & \mathcal{X}(n,\mathbb{Z}) \\ \downarrow & & \downarrow F \\ \mathcal{GE}^{(n)} & \to & \mathcal{G}\ell(n,\mathbb{Z}) \end{array}$$

Here $\mathcal{X}(n,\mathbb{Z})$ is the following category version of the $(\frac{n-1}{2})$-acyclic space $X_n(\mathbb{Z})$ of Theorem 1.1.4.

**Definition 8.5.10.** For any ring $R$ let $\mathcal{X}(n,R)$ be the simplicial category which in degree $k$ consists of all pairs $(M,P,g)$ where $M = M_R$ is a free $R$-module, $P$ is a poset with $n$ elements and $g = (g_0,\cdots,g_k)$ is a $(k+1)$-tuple of isomorphisms

$$g_i : R^P \xrightarrow{\approx} M$$

having the property that $g_i^{-1}g_j \in Aut(R^P)$ is $P$-upper triangular for all $i,j$ from 0 to $k$. The morphisms are pairs $(f,\alpha) : (M,P,g) \to (N,Q,h)$ where $f : M \to N$ is an isomorphism and $\alpha : P \to Q$ is a closed bijection so that $f \circ g_i = h_i \circ \alpha_* : R^P \to N$ for all $i$. The simplicial operators act on the last coordinate $g$ in the usual way by deleting and repeating entries.

Since $\mathcal{X}(n,R)$ contains an equivalent simplicial full subcategory of objects where $M = R^n$, we get:

**Lemma 8.5.11.** *$\mathcal{X}(n,R)$ is homotopy equivalent to $X_n(R) {=} EGL_n(\mathbb{Z}) \times_{GL_n(\mathbb{Z})} V_n(\mathbb{Z})$ and is thus $(n-1)/2$-acyclic. Furthermore the homotopy fiber of the forgetful functor*

$$F : \mathcal{X}(n,\mathbb{Z}) \to \mathcal{G}\ell(n,\mathbb{Z})$$

*over $\mathbb{Z}^n$ is naturally equivalent to the Volodin space $V_n(\mathbb{Z})$.*

By Proposition 3.4.7 the isomorphism $\mathbb{Z}^n \xrightarrow{g} H_1(G)$ can be realized at the chain level by a morphism of $G$-expansion functors, say $\phi$, which is unique up to fiber homotopy. Taking the fiberwise mapping cone we get a $G$-expansion functor $C(\phi)$ on $\mathcal{Y}(n,\mathbb{Z})$ which has only homology in degree 0: $H_0(C(\phi);\mathbb{Z}) = \mathbb{Z}$. (Details of this construction will be given below.) Consequently, the canonical cone of $C(\phi)$ gives a mapping

$$C(C(\phi)) : \mathcal{Y}(n,\mathbb{Z}) \to \mathcal{W}h^h_\bullet(\mathbb{Z},1)$$

well-defined up to homotopy.

This fits into the following homotopy commuting square.

$$\begin{array}{ccccc} V_n(\mathbb{Z}) & \longrightarrow & \mathcal{Y}(n,\mathbb{Z}) & \longrightarrow & \mathcal{GE}^{(n)} \\ \downarrow s & & \downarrow C(C(\phi)) & & \downarrow \\ \Omega BGL(\mathbb{Z})^+ & \longrightarrow & \mathcal{W}h^h_\bullet(\mathbb{Z},1) & \longrightarrow & Q^{(n)}S^0 \end{array}$$

where the rows are fibration sequences and $Q^{(n)}S^0$ is the degree $n$ component of $QS^0$. Since $s$ in the above diagram is $(n-2)$-connected by Suslin [Sus82], we see that:

**Proposition 8.5.12.** *The higher FR-torsion class* $\tau_{2k}(\mathcal{Y}(n,\mathbb{Z})) \in H^{4k}(\mathcal{Y}(n,\mathbb{Z}); \mathbb{R})$ *is nontrivial for* $n \geq 4k+2$.

**Remark 8.5.13.** We also note that $\tau_{2k}(\mathcal{Y}(n,\mathbb{Z}))$ restricts to $\tau_{2k}(\mathcal{GE}^{(n)h}) = \tau_{2k}(Out^h(F_n))$ under the natural mapping $\mathcal{GE}^{(n)h} \to \mathcal{Y}(n,\mathbb{Z})$.

We give the details of the construction of the morphism $\phi$. Let $\xi$ be the expansion functor on $\mathcal{Y}(n,\mathbb{Z})$ given by the composition

$$\mathcal{Y}(n,\mathbb{Z}) \to \mathcal{GE}^{(n)} \to \mathcal{W}h_\bullet(\mathbb{Z},1).$$

Let $\eta$ be the pull-back to $\mathcal{Y}(n,\mathbb{Z})$ of the monomial functor $\eta_\mathbb{Z}$ on $\mathcal{X}(n,\mathbb{Z})$ given as follows (over any ring $R$).

At each vertex $(M,P,g) \in \mathcal{X}_0(n,\mathcal{R})$ let $\eta_R(M,P,g) = (0,P)$ where all elements of the poset $P$ are assigned a degree of 0 and 0 is the boundary map in the chain complex:

$$\cdots \to 0 \to 0 \to R^P.$$

A morphism $(f,\alpha)$ induces a monomial morphism $(id,\alpha)$. On each $k$-simplex $(M,P,g) \in \mathcal{X}_k(n,R)$ let $\eta_R(M,P,g) = (e_g,P)$ where

$$e_g(j,i) = (g^j)^{-1}g^i - id_{R^n}$$

for all $0 \leq i,j \leq k$ and $e_g = 0$ (necessarily) in all other degrees.

We note that the first coordinate gives the homology functor of $\eta_R$:

$$H_*(\eta_R)(M,P,g) = H_0(\eta_R)(M,P,g) = M.$$

Since $\mathcal{Y}(n,\mathbb{Z})$ is the pull-back, we are given an isomorphism

$$H_0(\eta) \cong H_1(\xi).$$

By Proposition 3.4.7 there exists a unique fiber homotopy class of morphisms

$$\phi : \Sigma(\eta) \to \xi$$

which realizes this isomorphism.

## 8.5.2 Higher FR-torsion of the Torelli group

Let $T_g^s$ be the group of isotopy classes of orientation preserving self-diffeomorphisms $\phi$ of a compact oriented surface $S$ with genus $g \geq 1$ and $s \geq 1$ marked points so that $\phi$ induces the identity on the integral homology of $S$ and fixes each of the marked points. Then $T_g^s$ is called the *Torelli group*.

It is well-know that there is a canonical smooth bundle

$$S \to ET \to BT_g^s$$

having $s$ disjoint sections $s_1, \cdots, s_s$ giving the $s$ marked points in each fiber. We let $\epsilon_1, \cdots, \epsilon_s$ be the $SO(2)$-bundles over $BT_g^s$ given by pulling back the vertical tangent bundle of $ET$ along the sections $s_i$.

Since the homology bundle of $ET$ is free and trivial by definition, there is a well-defined mapping

$$\xi : BT_g^s \to \mathcal{W}h_\bullet^h(\mathbb{Z}, 1).$$

Consequently, we have well defined higher FR-torsion classes

$$\tau_{2k}(BT_g^s) = \tau_{2k}(\xi) \in H^{4k}(BT_g^s; \mathbb{R}).$$

We will compare these with the Miller-Morita-Mumford classes $\kappa_{2k}$ which we now recall and the powers of the Euler classes $e(\epsilon_i) \in H^2(BT_g^s)$.

Let $e(\gamma) \in H^2(ET)$ be the Euler class of the vertical tangent bundle $\gamma$ of $ET$. Then $e(\gamma)^{2k+1} \in H^{4k+2}(ET)$ pushes down ([BH58]) to a cohomology class $\kappa_{2k} \in H^{4k}(BT_g^s)$ called the Miller-Morita-Mumford class ([Mil86], [Mor87]). This is related to Morse theory as follows.

Let $f : ET \to \mathbb{R}$ be an oriented fiberwise GMF. Then the fiberwise gradient of $f$ is a vector field transverse to the zero section of the vertical tangent bundle and therefore the fiberwise singular set $\Sigma(f)$ gives a cycle in $ET$ which is dual to $e(\gamma)$.

To see how this relates to higher FR-torsion suppose for a moment that there were a fiberwise framed function on $E$ whose critical points of index 0 are the $s$ marked points. Then the first framing vectors at the other critical points would give a trivialization of the restriction of $T_\gamma(ET)$ to $\Sigma(f)$ outside of the marked points and consequently the self intersection of $\Sigma(f)$ would be empty away from the marked points. This would imply that $\kappa_k = \sum e(\epsilon_i)^k$. Consequently the difference between $\kappa_k$ and the sum of the classes $e(\epsilon_i)^k$ is the obstruction to the existence of a fiberwise framed function whose critical points of index 0 are the marked points.

Suppose that we have the following where $ET_0$ is the complement of an open tubular neighborhood of the $s$ sections $s_i : B \to ET$.

**Lemma 8.5.14.** *There exists a fiberwise framed function $f : ET_0 \to \mathbb{R}$ having critical points with indices* 0 *and* 1.

If we now add back the tubular neighborhoods of the marked points then we can extend $f$ to an oriented fiberwise GMF having maxima exactly at the marked points and framed everywhere else. This gives a functor $\xi_f : BT_g^s \to \mathcal{W}h_\bullet(\mathbb{Z}, 1)$ with trivial homology bundle and therefore we have higher FR-torsion invariants $\tau_{2k}(f) \in H^{4k}(B; \mathbb{R})$. By the framing principle 6.1.1 these are related to the higher FR-torsion of $T_g^s$ by

$$\tau_{2k}(T_g^s) = \tau_{2k}(f) + (-1)^k \zeta(2k+1) \sum_{i=1}^{s} \frac{e(\epsilon_i)^{2k}}{(2k)!} \tag{8.4}$$

Turn the function upside down. Then $-f : ET \to \mathbb{R}$ is an oriented fiberwise GMF with the marked points being the only critical points of index 0. The framing of $f$ gives a framing of the positive eigenspace of the second derivative of $-f$ at the index 1 critical points which can be rotated counterclockwise to give a framing of $-f$ at these points. Consequently by the discussion above, the push-downs of the powers of the Euler classes of the vertical tangent bundle $\gamma$ along the index 2 critical points of $-f$ is given by the Miller-Morita-Mumford classes minus the classes $e(\epsilon_i)^{2k}$. Consequently, by the framing principle, we have:

$$\tau_{2k}(T_g^s) = \tau_{2k}(-f) + (-1)^k \zeta(2k+1) \left( \frac{\kappa_{2k}}{(2k)!} - \sum_{i=1}^{s} \frac{e(\epsilon_i)^{2k}}{(2k)!} \right) \tag{8.5}$$

However, the chain complex of $-f$ is the conjugate transpose ($ct_2$) of the chain complex of $f$ by Theorem 5.5.1. So, by Corollary 5.8.3 we have:

$$\tau_{2k}(-f) = -\tau_{2k}(f).$$

Therefore, when we add together Equations (8.4) and (8.5) we get:

**Theorem 8.5.15.** *If $g, k, s \geq 1$ then*

$$\tau_{2k}(T_g^s) = (-1)^k \zeta(2k+1) \frac{\kappa_{2k}}{2(2k)!}$$

*in $H^{4k}(T_g^s; \mathbb{R})$.*

**Remark 8.5.16.** a) It was John Klein [Kle93] who first showed that the higher FR-torsion for the Torelli group was defined. He conjectured that $\tau_{2k}(T_g)$ was a nonzero multiple of $\kappa_{2k}$.

b) Quoting Bökstedt's Theorem (6.4.1 above), R. Hain, R. Penner and the author showed that the higher FR-torsion of the comparison map for the mapping class group must be nonzero. Since we did not have the complexified Bökstedt Theorem (7.10.2 above) we could not get an exact value for the higher FR-torsion.

If $S$ is a closed surface of genus $g$ with $s = 1$ marked point let $S^\delta$ denote the complement of the marked point. Then $S^\delta$ is homotopy equivalent to a wedge of $2g$ circles. Any self-diffeomorphism of $S$ fixing the marked point will induce a self-homotopy equivalence of $S^\delta$. Since a self-homotopy equivalence of any path-connected space induces an automorphism of its fundamental group which is well-defined up to inner automorphism, we get a homomorphism

$$h : T_g^1 \to Out(F_{2g}).$$

The image of this homomorphism is contained in $Out^h(F_n)$ since the elements of the Torelli group fix $H_1(S) \cong H_1(F_{2g})$.

By naturality of the higher FR-torsion (and Lemma 8.5.14), the higher FR-torsion class of $Out^h(F_{2g})$ pulls back to the higher FR-torsion of the punctured surface bundle $ET_0 \to BT_g^1$ which is given by $\tau_{2k}(f)$ in the above proof since $f|ET_0$ was framed. (The Splitting Lemma 5.7.8 tells us that the torsion of the GMF $f$ is the same as that of $f|ET_0$.) Consequently, by (8.4) and (8.5.15) above (with $s = 1$), we get the following.

**Corollary 8.5.17.** *For all $g, k \geq 1$ we have*

$$h^* \tau_{2k}(Out^h(F_{2g})) = (-1)^k \zeta(2k+1) \left( \frac{\kappa_{2k}}{2(2k)!} - \frac{e(\epsilon_1)^{2k}}{(2k)!} \right)$$

*in $H^{4k}(T_g^1; \mathbb{R})$.*

The *mapping class group $M_g^s$* is the group of isotopy classes of orientation preserving self-diffeomorphism of an oriented surface $S$ of genus $g \geq 1$ which fix pointwise a set of $s \geq 1$ marked points. Since the components of $Diff(S)$ are contractible so are the components of the subgroup of diffeomorphisms which fix the marked points (by induction on $s$). Thus there is a canonical smooth bundle:

$$S \to EM \to BM_g^s.$$

We also have an exact sequence of groups:

$$1 \to T_g^s \to M_g^s \to Sp_{2g}(\mathbb{Z}) \to 1.$$

When $s = 1$ this sequence fits into the commuting diagram with exact rows:

$$\begin{array}{ccccccccc}
1 & \longrightarrow & T_g^1 & \longrightarrow & M_g^1 & \longrightarrow & Sp_{2g}(\mathbb{Z}) & \longrightarrow & 1 \\
 & & \downarrow h & & \downarrow & & \downarrow & & \\
1 & \longrightarrow & Out^h(F_{2g}) & \longrightarrow & Out(F_{2g}) & \longrightarrow & GL_{2g}(\mathbb{Z}) & \longrightarrow & 1
\end{array}$$

where $h$ is the homomorphism of Corollary 8.5.17 above.

Assume for now that the following analogue of Lemma 8.5.14 holds.

**Lemma 8.5.18.** *There exists an oriented fiberwise GMF $f : EM \to \mathbb{R}$ with maxima at the sets of marked points.*

The higher FR-torsion is not defined. However we can still use the framing principle to compute the torsion of the comparison maps $F_f$ and $F_{-f}$ showing how much $f$ and $-f$ are not framable. By arguments analogous to the ones above we get:

$$\tau_{2k}(F_f) = (-1)^k \zeta(2k+1) \sum_{i=1}^{s} \frac{e(\epsilon_i)^{2k}}{(2k)!} \tag{8.6}$$

and

$$\tau_{2k}(F_{-f}) = (-1)^k \zeta(2k+1) \left( \frac{\kappa_{2k}}{(2k)!} - \sum_{i=1}^{s} \frac{e(\epsilon_i)^{2k}}{(2k)!} \right) \tag{8.7}$$

Putting these together we get the following.

**Theorem 8.5.19 (Hain-I-Penner).** *Let $g$ be any fiberwise framed function on $EM$. Then the torsion of the comparison map for $-g$ is given by*

$$\tau_{2k}(F_{-g}) = (-1)^k \zeta(2k+1)\frac{\kappa_{2k}}{(2k)!}.$$

**Remark 8.5.20.** a) The same formula applies for the Torelli group by restriction.

b) We do not need Lemma 8.5.18 to make the statement meaningful since we can just take a framed function on an even fiberwise suspension of $EM$ as we usually do and the framed function theorem 4.6.3 implies that $\tau_{2k}(F_{-g})$ is a well defined element of $H^{4k}(M_g^s;\mathbb{R})$.

*Proof.* We will explain how (8.6) and (8.7) imply the theorem. In the notation of Chapter VI, we have:

$$\tau_{2k}(F_f) = \tau_{2k}(\xi_+, \xi_f)$$
$$\tau_{2k}(F_{-f}) = \tau_{2k}(\xi_+, \xi_{-f}).$$

By the reverse framing principle (6.1.3) we have

$$\tau_{2k}(\overline{F}_{-f}) = \tau_{2k}(\xi_-, \xi_{-f}) = (-1)^{n-1}\tau_{2k}(F_f)$$

where $n = 2$. By Proposition 5.7.10 we have:

$$\tau_{2k}(F_{-g}) = \tau_{2k}(\xi_+, \xi_-) = \tau_{2k}(F_{-f}) - \tau_{2k}(\overline{F}_{-f})$$

so

$$\tau_{2k}(F_{-g}) = \tau_{2k}(F_{-f}) + (-1)^n \tau_{2k}(F_f). \tag{8.8}$$

Putting $n = 2$ we see that $\tau_{2k}(F_{-g})$ is given by adding (8.6) and (8.7). □

We note that (8.8) holds more generally under the following conditions.

1. The bundle $E \to B$ is vertically oriented.
2. $f, -f : E \to I$ are both oriented fiberwise GMF's.
3. $f, -f$ both have the property that their birth-death sets are framable.

## 8.6 Fat graphs give framed functions

In this section we prove the two technical lemmas 8.5.14 and 8.5.18 which we needed to compute the higher FR-torsion for the Torelli group and the mapping class group. This is basically already known, perhaps in a different language. What is know is that the moduli space of Riemann surfaces is homotopy equivalent to the space of "fat graphs." We note that unlike the usual approach where the Torelli group and mapping class group act on a finite dimensional contractible space (Teichmüller space or outer space) with finite stabilizers, we take an infinite dimensional contractible space on which these group act freely. We use arguments which we believe can be generalized to manifolds of arbitrary dimension (not just surfaces).

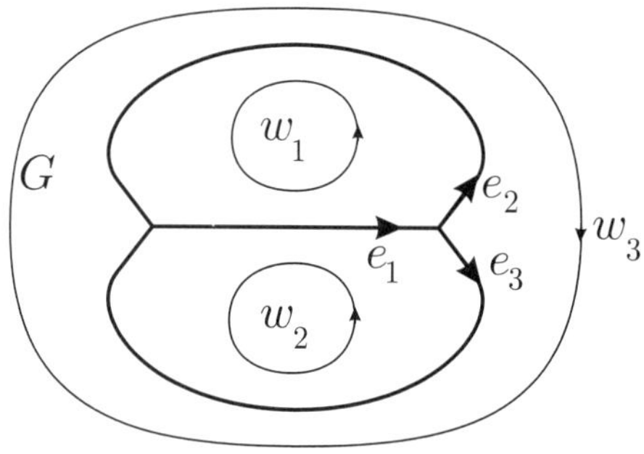


Figure 8.7: $w_1 = e_1e_2$, $w_2 = \overline{e}_3\overline{e}_1$, $w_3 = \overline{e}_2e_3$

### 8.6.1 Fat graphs

**Definition 8.6.1.** [Pen93] A *fat graph* is a graph $G$ without leaves together with a cyclic ordering on the edges incident to each node.

A fat graph determines an oriented surface with boundary given as follows. Immerse the fat graph $G$ in the plane so that at each node the counterclockwise order of the incident edges is equal to the given cyclic order. The surface $S(G)$ is the immersed tubular neighborhood of $G$ in $\mathbb{R}^2$.

The boundary components of $S(G)$ are circles which retract to oriented cycles $w_1, \cdots, w_s$ on $G$ which are *reduced* ($\overline{e}$ does not follow $e$ in any $w_i$) and have the property that each edge $e$ of $G$ occurs exactly once with each orientation in the union of these cycles. (See Figure 8.7.) We recall that oriented cycles are only well defined up to cyclic order.

Following Culler-Vogtmann [CV86] we note that a fat graph structure on $G$ is uniquely determined by any sequence of oriented cycles $w_1, \cdots, w_s$ having this property. For example, in Figure 8.7 the fat graph structure on $G$ is given by the two oriented cycles $e_1\overline{e}_3\overline{e}_2, \overline{e}_1e_2e_3$. The rule is that $e_j$ follows $e_i$ in the fat graph if $\overline{e}_i$ follows $e_j$ in one of the $w$'s. The associated surface $S(G, w) = S(G, w_1, \cdots, w_s)$ is the mapping cylinder of the map $\coprod_s S^1 \to |G|$ given by the cycles $w_1, \cdots, w_s$. We note that reduced oriented cycles are in 1-1 correspondence with the conjugacy classes of nontrivial elements of $\pi_1 G$.

**Definition 8.6.2.** A connected graph $G$ without leaves together with an ordered list of reduced oriented cycles $w_1, \cdots, w_s$ having the above property (each edge $e$ of $G$ occurs exactly once with each orientation in the union of these cycles) will be called a *marked fat graph.* A morphism of marked fat graphs $(G, w) \to (G', w')$ is an admissible epimorphism $G \to G'$ which sends $w_i$ to $w_i'$ for each $i$. We denote the category of marked fat graphs by $\mathcal{F}at$.

Let $\mathcal{F}at_g^s$ denote the full subcategory of $\mathcal{F}at$ consisting of marked fat graphs corresponding to surfaces of genus $g$ with $s$ punctures. Then $s$ is any positive integer except in the case $g = 0$ where $s \geq 2$. It follows from Lemma 8.1.13 that $|\mathcal{F}at_0^2| \simeq BU(1)$. The union of the other components is homotopy equivalent to the full subcategory $\mathcal{F}at^+$ of $\mathcal{F}at$ consisting of all fat graphs having no nodes of degree 2. (See 8.1.7.)

**Theorem 8.6.3.** *The geometric realization* $|\mathcal{F}at^+| = |\mathcal{N}_\bullet\mathcal{F}at^+|$ *of* $\mathcal{F}at^+$ *is homotopy equivalent to the disjoint union*

$$|\mathcal{F}at^+| \simeq \coprod_{s\geq \max(1,3-2g)} BM_g^s$$

*of classifying spaces of mapping class groups* $M_g^s$ *where* $s \geq 1$ *for* $g \geq 1$ *and* $s \geq 3$ *for* $g = 0$.

**Remark 8.6.4.** Since we have two different descriptions of the same thing we also have two different proofs, one by Penner [Pen93] and the other by Culler-Vogtmann [CV86]. We follow the second more combinatorial approach. There is also the approach using "ribbon graphs" and "quadratic differentials" [Str84].

*Proof.* For every pair $(g, s)$ we choose one connected oriented surface $S(g, s)$ with $s$ boundary components and Euler characteristic $\chi = 2 - 2g - s$ (so that if we cap off the $s$ boundary components we get a closed surface of genus $g$). Then $\pi_1 S(g, s) = F_n$ is the free group on $n = 2g + s - 1$ letters and the boundary components give a sequence of $s$ conjugacy classes $W = (w_1, \cdots, w_s)$ of elements in $F_n$.

Let $OS(n)$ be the "outer space" of all connected marked graphs $(G, \phi)$ with no nodes of degree 1 or 2 (Definition 8.1.22) where $\phi : F_n \cong \pi_1 G$ is a conjugacy class of isomorphisms. Let $OS(n, W)$ be the full subcategory of $OS(n)$ of all $(G, \phi)$ so that $(G, \phi(W))$ is a marked fat graph. Then Culler and Vogtmann showed that $OS(n, W)$ is a deformation retract of $OS(n)$ and thus also contractible. (See Theorem 8.1.23 and the accompanying remark.)

Let $Out(F_n, W)$ be the group of all outer automorphisms of $F_n$ which fix each of the conjugacy classes $w_i$. Then $Out(F_n, W)$ acts freely on $OS(n, W)$ and the quotient is the component $\mathcal{F}at_g^{s+} = \mathcal{F}at_g^s \cap \mathcal{F}at^+$ of the category of marked fat graphs without nodes of degree 2. Consequently,

$$|\mathcal{F}at_g^{s+}| \simeq BOut(F_n, W).$$

However, it is know that

$$Out(F_n, W) \cong M_g^s \tag{8.9}$$

when $n = 2g + s - 1 \geq 2$ (by Nielsen [Nie24]). □

A proof of Nielsen's theorem (8.9) can be given as follows. Since the surface $S(g, s)$ is homotopy equivalent to a wedge of $n$ circles, it follows that $Out(F_n, W)$ is the group of homotopy classes of self homotopy equivalences of the $(s + 1)$ tuple of spaces

$$(S(g, s), \partial_1 S, \partial_2 S, \cdots, \partial_s S)$$

where $\partial_i S$ is the component of $\partial S(g, s)$ corresponding to $w_i$. Each such homotopy equivalence is homotopic to a diffeomorphism. To see this we cut up the surface $S(g,s)$ by embedded arcs $C_j$ so that each piece is a 2-cell. Given any smooth homotopy equivalence $f : S(g,s) \to S(g,s)$ which is a diffeomorphism on $\partial S(g,s)$ we can make $f$ transverse to all of these arcs and the inverse image of each arc $C_j$ will have exactly one component which is not a circle. Call this $C_j'$. If $C_1, \cdots, C_k$ (together with parts of $\partial S(g,s)$) form the boundary of a 2-cell $e$ then $C_1', \cdots, C_k'$ (with pieces of $\partial S(g,s)$) must also bound a 2-cell $e'$ so $f$ is homotopic to a diffeomorphism $h$ which sends $C_j'$ to $C_j$ and $e'$ to $e$.

This shows that $M_g^s$ maps onto $Out(F_n, W)$. To see that the kernel is trivial we need to show that each self-diffeomorphism $f$ of the punctured surface $S(g,s)$ which is homotopic to the identity is isotopic to the identity. By rotating $\partial S(g,s)$ we may assume that $f$ fixes $\partial S(g,s)$ and that the homotopy $H : f \simeq id$ also fixes $\partial S(g,s)$. Now consider the intersection of $f(C_j)$ with $C_j$ for each $j$. By transversality this is a finite set including the endpoints. If we cut the surface $S(g,s)$ along the curve $C_j$ we get something homotopy equivalent to a wedge of $n-1$ circles. Consequently, each $C_j$ is dual to a 1-cell representing a free generator $x_j$ of $F_n$. The segments of $f(C_j)$ bounded by the intersection points with $C_j$ represent a sequence of words:

$$y_1 x_j y_2 x_j^{-1} y_3 x_j y_4 x_j^{-1} \cdots$$

where each $y_i$ is a word in the generators $x_k$ (and $x_k^{-1}$) for $k \neq j$ and the other letters are alternately $x_j$ and $x_j^{-1}$ corresponding to the fact that the crossings $f(C_j) \cap C_j$ alternate between going from left to right and from right to left. However, this word must be trivial in $F_n$ since $f$ is homotopic to the identity. Consequently, one of the $y_i$ must be trivial. This means that the embedded curve corresponding to this $y_i$, together with a portion of $C_j$ bounds a disk. This disk can be used to isotope $f$ to decrease the number of intersections of $f(C_j)$ with $C_j$ without increasing this number for other $j$. Therefore this number can be reduced to the minimum number (two) for each $j$ and we may assume that $f$ is the identity on each $C_j$. Then $f$ is isotopic to the identity using the well-known fact that $Diff(D^2 \text{ rel } \partial D^2)$ is contractible. (See the appendix to [Cer68].)

The homology functor

$$H_1 : \mathcal{F}at_g^s \to \mathcal{G}\ell(2g, \mathbb{Z}) \tag{8.10}$$

is given by sending each fat graph $(G, \phi(W))$ to the quotient group

$$H_1(G, \phi(W)) = \frac{H_1(G)}{\phi_*(W)}$$

where $\phi_* : H_1 F_n = \mathbb{Z}^n \to H_1 G$ is the isomorphism on $H_1$ induced by $\phi : F_n \xrightarrow{\cong} \pi_1 G$. This quotient group represents the first homology group of the closed surface given by attaching $s$ 2-disks to the fat graph along the boundary components $\phi(w_1), \cdots, \phi(w_s)$.

The fiber category of the functor $H_1$ in (8.10) is given by adding *relative homological markings*

$$h : \mathbb{Z}^{2g} \xrightarrow{\cong} H_1(G, \phi(W))$$

to the already marked fat graphs $(G, \phi(W))$. This gives us a category of *doubly marked graphs* $(G, \phi(W), h)$. Since both categories are $K(\pi, 1)$'s (for $(g, s) \neq (0, 2)$) the nerve of the fiber category classifies the kernel of the epimorphism

$$M_g^s \to GL_{2g}(\mathbb{Z})$$

which is the Torelli group $T_g^s$.

By (8.9) we know that $T_g^s$ is isomorphic to the subgroup of $Out(F_n, W)$ consisting of those outer automorphisms which are the identity on $H_1(G)/\phi_*(W)$. This acts freely on $OS(n)$ and the quotient is the space of doubly marked graphs $(G, \phi(W), h)$ without nodes of degree 2. These correspond to surfaces of genus $g$ with $s$ punctures.

**Corollary 8.6.5.** *For $s \geq \max(1, 3 - 2g)$ a classifying space for the Torelli group $T_g^s$ is given by the nerve of the category of all doubly marked fat graphs $(G, \phi(W), h)$ corresponding to surfaces of genus $g$ with $s$ punctures.*

The tautological bundle $ET_0 \to BT_g^s$ with fiber the fat graph $G$ has upper triangular free homology bundle. In other words, $H_1G \cong \mathbb{Z}^n$, considered as a $\mathbb{Z}[T_g^s]$-module, has a filtration giving an exact sequence of $\mathbb{Z}[T_g^s]$-modules:

$$0 \to \mathbb{Z} \xrightarrow{\Delta} \mathbb{Z}^s \xrightarrow{\phi_*} H_1G \xrightarrow{h^{-1}} \mathbb{Z}^{2g} \to 0$$

where $\Delta : \mathbb{Z} \to \mathbb{Z}^s$ is the diagonal map and the action of $T_g^s$ on $\mathbb{Z}$, $\mathbb{Z}^s$ and $\mathbb{Z}^{2g}$ is trivial. Consequently we have well-defined higher FR-torsion invariants:

$$\tau_{2k}(ET_0) \in H^{4k}(T_g^s; \mathbb{R}).$$

By the Splitting Lemma this is the same as $\tau_{2k}(f)$ in (8.4). Consequently we have:

$$\tau_{2k}(ET_0) = \tau_{2k}(T_g^s) - (-1)^k \zeta(2k+1) \sum_{i=1}^{s} \frac{e(\epsilon_i)^{2k}}{(2k)!}.$$

By the framed graph theorem we do not change the homotopy type when we add framed structures to these graphs. Thus we get the following.

**Corollary 8.6.6.** *The nerve of the category of framed marked fat graphs is homotopy equivalent to the disjoint union of $\mathbb{C}P^\infty = BU(1)$ and the classifying spaces $BM_g^s$ for $s \geq \max(1, 3 - 2g)$.*

## 8.6.2 Integrating vector fields

Instead of constructing a framed function on an existing surface the plan is to construct the surface together with a framed function on it simultaneously from a framed marked fat graph. The framed function will be constructed by integrating a vector field using the following elementary lemma.

Suppose that $S$ is a compact smooth surface and $v$ is a smooth vector field on $S$ with the following properties.

a. $v$ points outward along $\partial S$.

b. The set $v^{-1}(0) = \{x \in S : v(x) = 0\}$ of zeroes of $v$ is finite.

c. There are local coordinates $x_1, x_2$ centered at each $x \in v^{-1}(0)$ so that

$$v = g(x_1)\frac{\partial}{\partial x_1} \pm 2x_2\frac{\partial}{\partial x_2}$$

in a neighborhood of $x$ where $g$ is a smooth function of one variable so that $g(0) = 0$ but $g'(0), g''(0)$ are not both zero.

d. There is a partial ordering on $v^{-1}(0)$ so that every trajectory of $v$ either goes from $x$ to $y$ where $x, y \in v^{-1}(0)$ with $x < y$ or goes from some point in $v^{-1}(0)$ to $\partial S$.

Then we will integrate $v$ to obtain a smooth function $f : S \to \mathbb{R}$ so that $v$ is a "gradient-like vector field" for $f$. Note that when we say a trajectory of $v$ "goes from $x$ to $y$" where $x, y \in v^{-1}(0)$ we mean that the trajectory converges to $x$ in negative time and converges to $y$ in positive time.

**Lemma 8.6.7.** *Under the above conditions there exists a GMF $f : S \to I$ so that*

1. $f^{-1}(1) = \partial S$.
2. $\Sigma(f) = v^{-1}(0)$.
3. $f(x) < f(y)$ *if* $x < y$ *in* $v^{-1}(0)$.
4. $v$ *is the gradient of* $f$ *with respect to the local coordinates* $x_1, x_2$ *in a neighborhood* $V(x)$ *of each* $x \in \Sigma(f)$.
5. *The directional derivative* $v(f)$ *is positive outside the set* $\Sigma(f) = v^{-1}(0)$.

*Furthermore,* $f$ *can be chosen to agree with any given function* $r : v^{-1}(0) \to (0, 1)$ *satisfying the condition that* $r(x) < r(y)$ *whenever* $x < y$ *in* $v^{-1}(0)$.

**Remark 8.6.8.** Since the space of all functions $f$ satisfying the above conditions is obviously convex, $f$ is unique up to contractible choice if it exists.

*Proof.* First we choose a Riemannian metric on $S$ so that the local coordinates in condition c give a local isometry with $\mathbb{R}^2$. We assume that $V(x)$ is an $\epsilon$-disk neighborhood of $x$ where $\epsilon > 0$ is chosen so that the coefficient $g(x_1)$ of $\partial/\partial x_1$ in $v$ has $g'(x_1) \neq 0$ on $V(x)$ if $g'(0) \neq 0$ and $g''(x_1) \neq 0$ on $V(x)$ if $g'(0) = 0$.

Next we note that condition 1 is superfluous since any smooth function $f : S \to [0, c]$ where $0 < c \leq 1$ satisfying the other four conditions can be modified in an arbitrarily small neighborhood of $\partial S$ to satisfy condition 1 as well. In fact we can choose $f$ to agree with any prescribed function $g : \partial S \to [c, \infty)$ on $\partial S$. To see this let

$$\psi : \partial S \times [-\epsilon, 0] \xrightarrow{\cong} T$$

be the parameterization of a collar neighborhood $T$ of $\partial S$ in $S$ given by the smooth flow generated by $v$. Then we can let $f' : S \to [0, \infty)$ be given by

$$f'(z) = \begin{cases} f(z) + \phi(t)[g(y) - f(y)] & \text{if } z = \psi(y,t), \\ f(z) & \text{if } x \notin T. \end{cases}$$

where $\phi : [-\infty, 0] \to \mathbb{R}$ is a smooth function so that $\phi(t) = 0$ for $t \le -\epsilon$, $\phi(0) = 1$ and $\frac{d}{dt}\phi(t) \ge 0$ for all $t$. Then $f'$ satisfies conditions 2-5 and is equal to $g$ on $\partial S$.

The construction of $f$ will be by induction on the number of zeroes of $v$. If this number is zero then $S$ is empty and there is nothing to prove. So suppose that $v$ has at least one zero.

Let $x$ be a maximal element of $v^{-1}(0)$. If $x$ has index 2 then the prescribed value of $f$ on the $\epsilon$-disk $V(x)$ is given by

$$f = c - x_2^2 + \int_0^{x_1} g(t)\, dt \tag{8.11}$$

where $g(t) < 0$ for $|t| \le \epsilon$. Let $S' = S - \text{int}V(x)$. Then by induction on the size of $v^{-1}(0)$, there is a smooth function $f : S' \to [0, c]$ given by (8.11) in a neighborhood of $\partial V(x) \subseteq \partial S'$ and Equation 8.11 can be used to extend $f$ to all of $S$.

If the maximal element $x$ is a transverse zero of $v$ with index 0 then $S$ must be a 2-disk and the lemma is easy. So suppose that $x$ has index 1 or that $x$ is degenerate of index 0.

Let $U_+(x)$ be the set of all points $z$ in $\partial V(x)$ so that the trajectory of $v$ through $z$ converges to $x$ in negative time, i.e., $U_+(x)$ is the unstable manifold of $x$ in $\partial V(x)$. Let $U_-(x) \subseteq \partial V(x)$ be the stable manifold of $x$. Then $U_+(x), U_-(x)$ are either both two point sets or one is one point and the other is a closed arc going half way around $\partial V(x)$.

Let $W_+(x), W_-(x)$ be small closed neighborhoods of $U_+(x), U_-(x)$ in $\partial V(x)$ with the property that $U_\pm(x) \simeq W_\pm(x)$, $v$ points outward (away from $x$) at each point in $W_+(x)$ and inward along each point in $W_-(x)$ and the smooth flow of $v$ carries $\partial W_-(x)$ to $\partial W_+(x)$.

Let $\psi_t$ be the smooth flow generated by $v$. By definition of the partial ordering of $v^{-1}(0)$ (condition d) $\psi$ carries $W_+(x)$ to a closed subset $D$ in $\partial S$ and $\psi$ also carries each point $y \in W_-(x) - U_-(x)$ to $D \subseteq \partial S$, say $\psi_{\alpha(y)}(y) \in \partial S$. Let $\phi : W_-(x) \to [0, \infty)$ be any smooth function so that $\phi = 0$ on $V_-(x)$, $\phi = \alpha$ near $\partial W_-(x)$ and $0 \le \phi \le \alpha$ at all other points in $W_-(x)$. Then the "graph" of $\phi$ given by $G = \{\psi_{\phi(y)}(y) \,:\, y \in W_-(x)\}$ together with the portion $\partial S - D$ of the boundary of $S$ form the boundary of a new surface $S'$ so that $v$ points outward along $\partial S'$ and $v$ has one fewer zero in $S'$ than in $S$. (See Figure 8.8.)

By induction there is a smooth function on $S'$ which is equal to the required function on $V(x) \cap S'$. This function extends uniquely to $S' \cup V(x)$ and easily extends to the rest of $S$ using the smooth flow $\psi$. $\square$

Lemma 8.6.7 generalizes to higher dimensions without much change in its statement or proof.

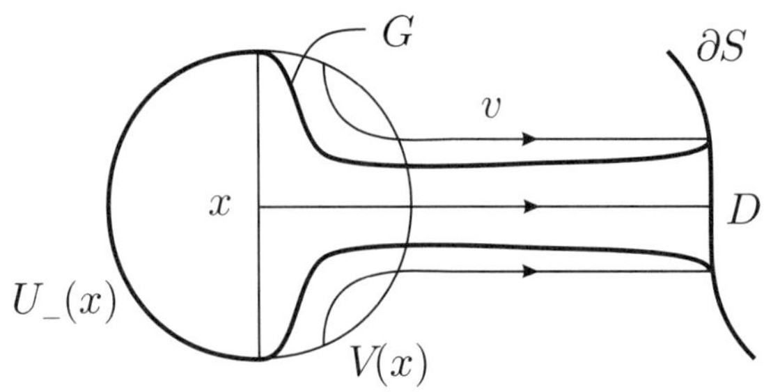


Figure 8.8: $G \cup (\partial S - D)$ is the boundary of a surface $S'$ not containing $x$

**Proposition 8.6.9.** *Suppose that $M$ is a compact smooth manifold and $v$ is a smooth vector field on $M$ with the following properties.*

*a. $v$ points outward along $\partial M$.*

*b. The set $v^{-1}(0) = \{x \in M : v(x) = 0\}$ of zeroes of $v$ is finite.*

*c. There are local coordinates $x_1, \cdots, x_n$ centered at each $x \in v^{-1}(0)$ so that*

$$v = g(x_1)\frac{\partial}{\partial x_1} + \sum_{i=2}^{n} \pm 2x_i \frac{\partial}{\partial x_i}$$

*in a neighborhood of $x$ where $g$ is a smooth function of one variable so that $g(0) = 0$ but $g'(0), g''(0)$ are not both zero.*

*d. There is a function $r : v^{-1}(0) \to (0,1)$ so that every trajectory of $v$ either goes from $x$ to $y$ where $x, y \in v^{-1}(0)$ with $r(x) < r(y)$ or goes from some point in $v^{-1}(0)$ to $\partial S$.*

*Then there exists a GMF $f : M \to I$ so that*

*1. $f^{-1}(1) = \partial M$.*

*2. $\Sigma(f) = v^{-1}(0)$.*

*3. $f(x) = r(x)$ for all $x \in v^{-1}(0)$.*

*4. $v$ is the gradient of $f$ with respect to the local coordinates $x_i$ in a neighborhood $V(x)$ of each $x \in \Sigma(f)$.*

*5. The directional derivative $v(f)$ is positive outside the set $\Sigma(f)$.*

*Furthermore the space of all such $f$ is contractible.*

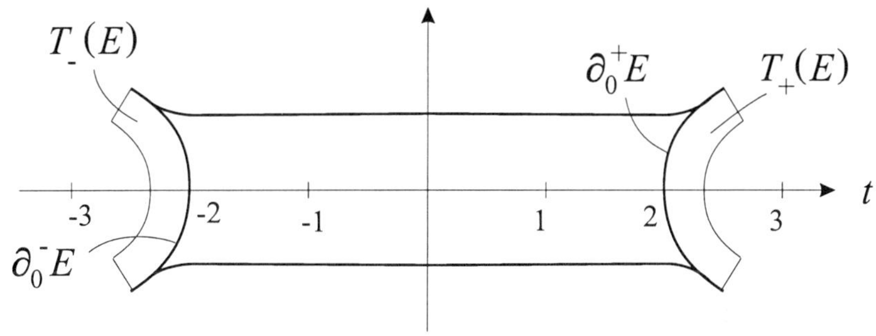


Figure 8.9: Building block $E$ (dark) with $\epsilon$-tabs $T_{\pm}(E)$

### 8.6.3 Proof of Lemmas 8.6.5 and 8.6.6

Given any framed marked fat graph $(G, P, w)$ we will construct a surface $S(G, P, w)$ together with a smooth tangent vector field $v$ by pasting together the building blocks $D, E$ given below. This is a rather long-winded version of a construction of Maxim Kontsevich [Kon92] for what he called $\mathcal{M}_{g,n}^{comb}$.

The first building block is the 2-disk $D$ of radius 1 with the vector field

$$v(x_1, x_2) = \nabla(x_1^2 + x_2^2) = (2x_1, 2x_2).$$

This has a zero at the origin with index 0.

The second building block $E$ is shown in Figure 8.9 where $\epsilon > 0$ is a fixed small real number less than 1/8. $E$ is given by the equation

$$E = \{(x_1, x_2) \in \mathbb{R}^2 \ : \ |x_1| \le \tfrac{5}{2}, |x_2| \le \lambda(x_1) \text{ and } (|x_1| - 3)^2 + x_2^2 \ge 1\}$$

where $\lambda : [-\frac{5}{2}, \frac{5}{2}] \to \mathbb{R}$ is an even function with the following properties.

- $\lambda(t) = \frac{1}{2}$ if $|t| \le 2$,
- $0 < \lambda'(t) < 1$ if $t > 2$,
- $\lambda(t) = \sqrt{1 - (|t| - 3)^2}$ for $|t| \ge \frac{12}{5}$ and
- $\lambda(t)^2 + (|t| - 3)^2 > 1$ if $|t| < \frac{12}{5}$.

Note that the boundary of $E$ consists of the graphs of the functions $\lambda$ and $-\lambda$ which we denote by $\partial_1^+ E, \partial_1^- E$ and the two arcs of length $2\pi/3$ which we denote by $\partial_0^- E, \partial_0^+ E$ with union $\partial_0 E$. The regions $T_{\pm}(E)$ in Figure 8.9 are not part of $E$ and will be explained later when needed. The number $\frac{12}{5}$ in the formula for $\lambda$ can be replaced by any element of the open interval $(3 - \frac{\sqrt{2}}{2}, \frac{5}{2})$.

The vector field on $E$ is given by

$$v(x_1, x_2) = (g(x_1), 2x_2) \tag{8.12}$$

where $g : [-4, 4] \to \mathbb{R}$ is a smooth odd function (Figure 8.10) so that

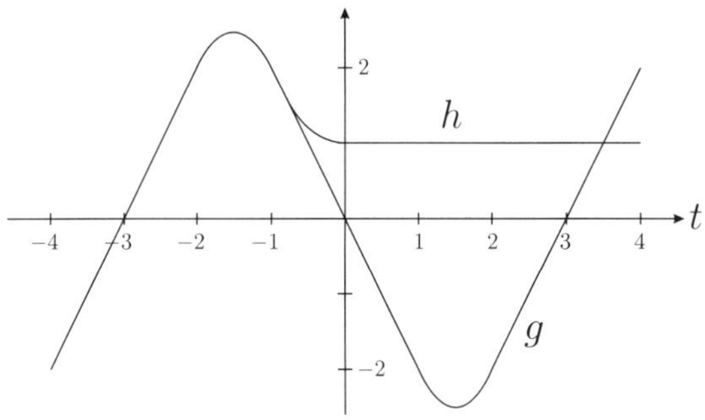


Figure 8.10: $g : [-4, 4] \to [-\frac{5}{2}, \frac{5}{2}]$ is an odd function, $h(t) = 1$ for $t \geq 0$

- $g(t) = -2t$ for $|t| \leq 1$,
- $g(t) = 2t - 6$ for $t \geq 2$,
- $g''(t) > 0$ for $1 < t < 2$ and
- $g(\frac{3}{2}) = -\frac{5}{2}$ is the minimum value of $g$.

These building blocks are designed to fit together. Copies of the disk $D$ can be placed in the two "sockets" at each end of $E$ and the result will be a smooth submanifold of $\mathbb{R}^2$ with a smooth vector field $v$ given by the same formula (8.12) on a larger domain.

Suppose that we have several copies $D(x)$, $E(y)$ of these building blocks where $x, y$ are elements of a finite graded poset $P = P_0 \coprod P_1$ (with $x \in P_0$, $y \in P_1$). Then we can put them together in the following way. At each stage we will have a smooth oriented surface $S$ built out of those pieces $D(x), E(y)$ where $x, y \in A$ and $A$ is a closed subset of $P$. There will be a smooth vector field $v$ on $S$ which restricts to the given vector field on each $D(x)$ and $E(y)$. The vector field will point outward along $\partial S$ so there will be a product structure near $\partial S$ given by the smooth flow of $v$. This gives a smooth embedding

$$\psi_S : \partial S \times [-\epsilon, 0] \to S$$

where $\psi_S(-, 0)$ is the inclusion map and $\frac{\partial}{\partial t}\psi_S(x, t) = v(\psi_S(x, t))$. In order to add the next piece, say $E(y)$, we need a smooth embedding

$$\phi_y : \partial_0 E(y) \to \partial S$$

which is orientation reversing so that the orientations of $E(y)$ and $S$ will match in the topological surface $S \cup_\phi E(y)$. The smooth structure of $S \cup_\phi E(y)$ is given as follows.

Let $U_\epsilon(y)$ be the union of $E(y) \subseteq \mathbb{R}^2$ with the sets $T_\pm(y)$ where

$$T_+(y) = \{(3 + r\cos\theta, r\sin\theta) \in \mathbb{R}^2 \, : \, (r, \theta) \in [(1-\epsilon)^2, 1] \times [-\pi/3, \pi/3]\}$$

and $T_-(y) = \{(-3 - r\cos\theta, r\sin\theta)\}$ is the reflection of $T_+(y)$ through the $x_2$-axis. We will refer to $T_\pm(y)$ as "$\epsilon$-tabs." Then let

$$S \cup_\phi E(y) = S \coprod U_\epsilon(y)/\sim$$

where for each $(t, \theta) \in [-\epsilon, 0] \times [-\pi/3, \pi/3]$ we identify the point $(-3, 0) + (1 + t)^2(\cos\theta, \sin\theta)$ in $T_-(y)$ with $\psi_S(\phi_y(\cos\theta - 3, \sin\theta), t)$ in $S$ and similarly for $T_+(y)$. Since the vector field $v$ is equal to $\partial/\partial t$ in both pieces, $\psi_S$ preserves $v$ and $v$ defines a smooth vector field on $S \cup_\phi E(y)$.

Given any framed marked fat graph $(G, P, w)$, let $P = P_0 \coprod P_1$ be the associated poset of cells. Let $S(G, P, w)$ be the smooth surface

$$S(G, P, w) = \coprod_{x \in P_0} D(x) \coprod \coprod_{y \in P_1} E(y)/\sim$$

where $D(x), E(y)$ are copies of $D, E$ and the identifications are given as follows.

A 1-cell $y \in P_1$ is a path in $G$ with internal nodes $v_1(y), v_2(y), \cdots, v_k(y)$. We take the rectangles

$$R_i = [x_i - \delta, , x_i + \delta] \times [-1/2, 1/2] \subseteq E(y) \subseteq \mathbb{R}^2$$

for $i = 1, \cdots, k(y)$ where

$$x_i = -2 + \frac{4i}{k(y) + 1}, \qquad \delta = \frac{1}{k(y) + 1}.$$

If the source or target of $z \in P_1$ is attached to $y$ at the node $v_i(y)$ then we attach $\partial_0^- E(z)$ or $\partial_0^+ E(y)$ to $E(y)$ on the "attachment area" $R_i \cap \partial E(y)$. The cyclic ordering of the edges incident to $v_i(y)$ determines on which side and in which order the $E(z)$'s are attached to $R_i \cap \partial E(y)$.

At the singular nodes $x \in P_0$ we take the incident 1-cells $y_1, \cdots, y_n$ and attach $E(y_1), \cdots, E(y_n)$ in the same cyclic order to $D(x)$. In the case of a collapsing pair $(y, x)$ we require the attaching map $\phi_y : \partial_0^+ E(y) \to \partial D(x)$ to be an isometry.

This determines the smooth surface $S(G, P, w)$ together with a smooth tangent vector field which is gradient-like near the center of each $D(x)$ and $E(y)$ and points outward along $\partial S(G, P, w)$. We use Lemma 8.6.7 to integrate the vector field to get the framed function on $S(G, P, w)$ having $(G, P, w)$ as the associated cell complex.

To construct the bundle over the classifying space of the category of framed marked fat graphs we need to go smoothly from $S(G, P, w)$ to $S(G', P', w')$ whenever there is a morphism $(G, P) \to (G', P')$ taking $W$ to $W'$. This is done by sliding the attaching map in the required way and by collapsing the collapsing pairs $D(x^-) \cup E(x^+)$ using one standard model. (See Figure 8.11.)

The *standard deformation* can be given as follows. First choose one smooth function $h : [-4, 4] \to \mathbb{R}$ so that

- $h(x) = g(x)$ for $x \leq -1$,

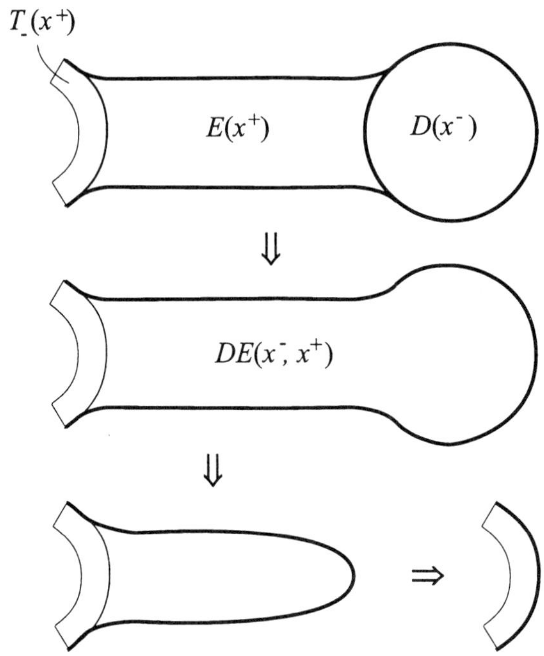


Figure 8.11: Choose a standard deformation of $D(x^-) \cup E(x^+)$ into the $\epsilon$-tab $T_-(x^+)$

- $h(x) = 1/2$ for $x \geq 0$ and
- $h'(x) \leq 0$ for $-1 < x < 0$.

(See Figure 8.10.) Then the first step in the standard deformation is given by deforming the vector field $v$ by

$$v_t(x_1, x_2) = ((1-t)g(x_1) + th(x_1), 2x_2).$$

This is obviously a gradient vector field for each $t$ since its curl is zero. Since the slope of this vector field is at least 1 in the portion of $\partial(D(x^-) \cup E(x^+)) \subseteq \mathbb{R}^2$ where $2 \leq x_1 \leq 3$ and $x_2 \geq 0$ and the slope of the function $\lambda$ whose graph forms $\partial(D(x^-) \cup E(x^+))$ is less than 1, the vector field $v_t$ points outward along the solid edge in the second part of Figure 8.11 for all $0 \leq t \leq 1$. In the second part of the standard deformation we follow the smooth flow of the new nowhere zero vector field to deform $D(x^-) \cup E(x^+)$ into the $\epsilon$-tab $T_-(x^-)$. The resulting family of (singular) surfaces will be called $DE_t(x^-, x^+)$ where $DE_0(x^-, x^+) = D(x^-) \cup E(x^+)$ and $DE_1(x^-, x^+)$ is the arc $\partial_0^- E(x^+)$ of length $2\pi/3$.

Finally we extend this construction to the nondegenerate realization of the nerve of the category of framed marked fat graphs. A $k$-simplex in the nerve of this category is a sequence of morphisms

$$(G^0, P^0, w^0) \to \cdots \to (G^k, P^k, w^k).$$

We need to associate to this a $k$-parameter family of smooth surfaces $S_t(G, P, w)$, $t \in \Delta^k$. By induction on $k$ we already have this family of surfaces for $t \in \partial\Delta^k$ and by induction on the size of $P^0$ we have an extension of this to $S_t(G, Q)$ for $t \in \Delta^k$ where $Q$ is either the complement of a maximal element $y \in P^0$ or the complement of a maximal collapsing pair $(x^+, x^-)$ in $P^0$.

In the first case the surfaces $S_t(G, P, w)$, $t \in \partial\Delta^k$ are given by

$$S_t(G, P, w) = S_t(G, Q) \cup_{\phi_t} E(y)$$

where $\phi_t : \partial_0 E(y) \to \partial S_t(G, Q)$, $t \in \partial\Delta^k$ is a smooth family of embeddings whose image lies in $S_t(G, A)$ where $A$ is the set of all $z < y$ in $P^0$. Since the admissible attachment areas for $\partial_0^\pm E(y)$ are contractible for all $t \in \Delta^k$, we can extend the family of embeddings $\phi_t$ to $\Delta^k$ in a unique way up to isotopy.

In the second case, the collapsing pair $(x^+, x^-)$ will exist only in $P^i$ for $i$ less than some $j$. In that case the sum $u = \sum_{i=j}^k t_i$ will determine how far the union $D(x^-) \cup E(x^+)$ is collapsed. We construct $S_t(G, P, w)$ by

$$S_t(G, P, w) = S_t(G, Q) \cup_{\phi_t} DE_u(x^-, x^+)$$

where $\phi_t : \partial_0 DE_u(x^-, x^+) = \partial_0^- E(x^+) \to \partial S_t(G, Q)$, $t \in \Delta^k$ is a smooth family of embeddings in the admissible attachment region which agrees with the given embedding over $\partial\Delta^k$.

The fiberwise framed function is constructed as follows. By Lemma 8.6.7 we have for each $t$ a framed function for which $v$ is a gradient-like vector field.

We can smoothly extend this to a neighborhood of $t$ by taking suitable values on $v^{-1}(0)$ and arbitrary values elsewhere. Take a finite covering of $\Delta^k$ by these neighborhoods and paste them together using a partition of unity. Since the required conditions are convex they will be satisfied by the resulting smooth family of functions. The framing of these functions comes from the framed structure of $G$.

This completes the construction of the smooth surface bundle together with a fiberwise framed function without maxima over the nondegenerate realization of the nerve of the category of framed marked fat graphs. If we ignore the framed structure we have a model for the canonical smooth surface bundle over the space of marked fat graphs which must be fiber diffeomorphic to the canonical one given by Theorem 8.6.3. Consequently, this proves that the canonical surface bundle over $BM_g^s$ has a fiberwise framed function without maxima for all $s \geq \max(1, 3-2g)$ as claimed in Lemma 8.5.18 and by restriction to the Torelli group we get Lemma 8.5.14.

# Bibliography


[Ada66] J. F. Adams, *On the groups $J(X)$. IV*, Topology **5** (1966), 21–71.

[Ada78] J. F. Adams, *Infinite Loop Spaces*, Annals of Math Studies, vol. 90, Princeton University Press, Princeton, 1978.

[Bas68] H. Bass, *Algebraic $K$-Theory*, Benjamin, New York, 1968.

[Beĭ85] A. A. Beĭlinson, *Higher regulators and values of $L$-functions*, J. Soviet Math. **30** (1985), 2036–2070, Translated from: Sovrem. Probl. Math. **24**, Moscow, Viniti (1984), 181-238.

[BG00] Jean-Michel Bismut and Sebastian Goette, *Families torsion and Morse functions*, to appear in Astérisque, 2000.

[BH58] A Borel and F. Hirzebruch, *Characteristic classes and homogeneous spaces, I*, Amer. J. of Math. **80** (1958), 458–538.

[BL77] D. Burghelea and R. Lashof, *Stability of concordances and the suspension homomorphism*, Ann. of Math. (2) **105** (1977), no. 3, 449–472.

[BL95] Jean-Michel Bismut and John Lott, *Flat vector bundles, direct images and higher real analytic torsion*, J. Amer. Math. Soc. **8** (1995), no. 2, 291–363.

[Bök84] M. Bökstedt, *The rational homotopy type of $\Omega Wh^{Diff}(*)$*, LNM 1051, Springer-Verlag, 1984, pp. 25–37.

[Bor74a] A. Borel, *Cohomologie de $SL_n$ et valuers de fonctions zêta*, Ann. Scuola Normale Superiore **7** (1974), 613–636.

[Bor74b] A. Borel, *Stable real cohomology of arithmetic groups*, Ann. Sci. ENS **7** (1974), 235–272.

[Bot59] R. Bott, *The stable homotopy of the classical groups*, Ann. of Math. **70** (1959), 313–337.

[Bot69] R. Bott, *Lectures on $K(X)$*, Math. Lecture Notes, Benjamin, New York, 1969.

[Bun] U. Bunke, *Higher analytic torsion of sphere bundles and continuous cohomology of $Diff(S^{2n-1})$*, preprint.

[BZ92] J.-M. Bismut and W. Zhang, *An extension of a theorem of Cheeger and Müller*, Asterisque, vol. 205, Société Mathématique de France, Paris, 1992.

[Cer68] Jean Cerf, *Sur les difféomorphismes de la sphère de dimension trois* ($\Gamma_4 = 0$), Springer-Verlag, Berlin, 1968.

[Cer70] Jean Cerf, *La stratification naturelle des espaces de fonctions différentiables réelles et le théorème de la pseudo-isotopie*, Inst. Hautes Études Sci. Publ. Math. (1970), no. 39, 5–173.

[Che79] J. Cheeger, *Analytic torsion and the heat equation*, Ann. of Math. **109** (1979), 259–322.

[CV86] Marc Culler and Karen Vogtmann, *Moduli of graphs and automorphisms of free groups*, Invent. Math. **84** (1986), no. 1, 91–119.

[Dup76] J. Dupont, *Simplicial deRham cohomology and characteristic classes of flat bundles*, Topology **15** (1976), 233–245.

[DWW] W. Dwyer, M. Weiss, and B. Williams, *A parametrized index theorem for the algebraic $K$-theory euler class*, preprint.

[Esn89] Hélène Esnault, *On the Loday symbol in the Deligne-Beĭlinson cohomology*, $K$-Theory **3** (1989), no. 1, 1–28.

[FH78] F. T. Farrell and W. C. Hsiang, *On the rational homotopy groups of the diffeomorphism groups of discs, spheres and aspherical manifolds*, Algebraic and Geometric Topology (Proc. Sympos. Pure Math., Stanford Univ., Stanford, Calif., 1976), Part 1 (Providence, R.I.), Amer. Math. Soc., Providence, R.I., 1978, pp. 325–337.

[Fla74] H. Flanders, *Analytic solutions of matrix equations*, Linear and Multilinear Algebra **2** (1974), 241–243.

[Ger83] S. M. Gersten, *On fixed points of automorphisms of finitely generated free groups*, Bull. Amer. Math. Soc. (N.S.) **8** (1983), no. 3, 451–454.

[Har88] John L. Harer, *The cohomology of the moduli space of curves*, Theory of Moduli (Montecatini Terme, 1985) (Berlin), Springer, Berlin, 1988, pp. 138–221.

[Hat75] Allen Hatcher, *Higher simple homotopy theory*, Ann. of Math. **102** (1975), 101–137.

[Hat95] Allen Hatcher, *Homological stability for automorphism groups of free groups*, Comment. Math. Helv. **70** (1995), no. 1, 39–62.

[Hir76] M. Hirsch, *Differential Topology*, Graduate Texts in Math., vol. 33, Springer-Verlag, New York, 1976.

[Hsi79] W. C. Hsiang, *On $\pi_i(Diff(M^n))$*, Geometric Topology (New York), Acad. Press, New York, 1979, pp. 351–365.

[Hus66] D. Husemoller, *Fiber Bundles*, McGraw-Hill, 1966.

[HV98a] Allen Hatcher and Karen Vogtmann, *Cerf theory for graphs*, J. London Math. Soc. (2) **58** (1998), no. 3, 633–655.

[HV98b] Allen Hatcher and Karen Vogtmann, *Rational homology of $Aut(F_n)$*, Math. Res. Lett. **5** (1998), no. 6, 759–780.

[HW73] Allen Hatcher and John Wagoner, *Pseudo-isotopies of compact manifolds*, Société Mathématique de France, Paris, 1973, With English and French prefaces, Astérisque, No. 6.

[Igu79] Kiyoshi Igusa, *The $Wh_3(\pi)$-obstruction for pseudoisotopy*, Ph.D. thesis, Princeton University, 1979.

[Igu84] Kiyoshi Igusa, *What happens to Hatcher and Wagoner's formulas for $\pi_0 C(M)$ when the first Postnikov invariant of $M$ is nontrivial?*, Algebraic $K$-theory, Number Theory, Geometry and Analysis (Bielefeld, 1982) (Berlin), Springer, Berlin, 1984, pp. 104–172.

[Igu87] Kiyoshi Igusa, *The space of framed functions*, Trans. Amer. Math. Soc. **301** (1987), no. 2, 431–477.

[Igu88] Kiyoshi Igusa, *The stability theorem for smooth pseudoisotopies*, $K$-Theory **2** (1988), no. 1–2, vi+355.

[Igu91] Kiyoshi Igusa, *Parametrized Morse theory and its applications*, Proceedings of the International Congress of Mathematicians, Vol. I, II (Kyoto, 1990) (Tokyo), Math. Soc. Japan, 1991, pp. 643–651.

[Igu93] Kiyoshi Igusa, *The Borel regulator map on pictures. I. A dilogarithm formula*, $K$-Theory **7** (1993), no. 3, 201–224.

[IK93] Kiyoshi Igusa and John Klein, *The Borel regulator map on pictures. II. An example from Morse theory*, $K$-Theory **7** (1993), no. 3, 225–267.

[IK95] Kiyoshi Igusa and John Klein, *Higher Franz-Reidemeister torsion. I. The algebra of filtered chain complexes*, preprint, 1995.

[JLDZ00] R. Hain J. L. Dupont and S. Zucker, *Regulators and characteristic classes of flat bundles*, The Arithmetic and Geometry of Algebraic Cycles (Providence R.I.), CRM Proceedings and Lecture Notes, vol. 24, Amer. Math. Soc., 2000.

[Kle89] John Klein, *The cell complex construction and higher R-torsion for bundles with framed Morse function*, Ph.D. thesis, Brandeis University, 1989.

[Kle93] John Klein, *Higher Franz-Reidemeister torsion and the Torelli group*, Mapping Class Groups and Moduli Spaces, Contemp. Math., vol. 150, American Math. Soc., 1993, pp. 195–204.

[KM63] Michel A. Kervaire and John W. Milnor, *Groups of homotopy spheres. I*, Ann. of Math. (2) **77** (1963), 504–537.

[Kon92] Maxim Kontsevich, *Intersection theory on the moduli space of curves and the matrix Airy function*, Comm. Math. Phys. **147** (1992), no. 1, 1–23.

[KP72] D.S. Kahn and S.B. Priddy, *Applications of the transfer to stable homotopy theory*, Bull. Amer. Math. Soc. **78** (1972), 981–987.

[KT74] F. Kamber and P. Tondeur, *Characteristic invariants of foliated bundles*, Manuscripta Math. **11** (1974), 51–89.

[Lod97] J.-L. Loday, *Cyclic Homology*, Grundlehren der mathematischen Wissenschaften, vol. 301, Springer-Verlag, New York, 1997.

[Mat69] H. Matsumoto, *Sur les sous-groupes arithmétique des groups semi-simples déployés*, Ann. Sci. Éc. Norm. Sup. **2** (1969), 1–62.

[May67] J.P. May, *Simplicial Objects in Algebraic Topology*, Van Nostrand Math. Studies, vol. 11, Van Nostrand, Princeton, N.J., 1967.

[May78] J.P. May, *$A_\infty$-ring spaces and algebraic $K$-theory*, Geometric Applications of Homotopy Theory II (Berlin-Heidelberg-New York), Lecture Notes in Math., vol. 658, Springer, 1978, pp. 240–315.

[Mil63] J. Milnor, *Morse Theory*, Annals of Math. Studies, vol. 51, Princeton University Press, Princeton, 1963.

[Mil65] J. Milnor, *Lectures on the h-Cobordism Theorem*, Princeton Math. Notes, Princeton University Press, Princeton, 1965.

[Mil66] J. Milnor, *Whitehead torsion*, Bull. Amer. Math. Soc. **72** (1966), 358–426.

[Mil71] J. Milnor, *Introduction to Algebraic $K$-Theory*, Annals of Math. Studies, vol. 72, Princeton University Press, Princeton, 1971.

[Mil86] Edward Y. Miller, *The homology of the mapping class group*, J. Differential Geom. **24** (1986), no. 1, 1–14.

[Mor87] S. Morita, *Characteristic classes of surface bundles*, Invent. Math. **90** (1987), 551–577.

[Mül78] W. Müller, *Analytic torsion and R-torsion of Riemannian manifolds*, Adv. in Math. **28** (1978), 233–305.

[NH95] I. Najfeld and T. Havel, *Derivatives of the matrix exponential and their computation*, Advances in Applied Math. **16** (1995), 321–375.

[Nie24] J. Nielsen, *Die isomorphismengruppe der frien Gruppen*, Math. Ann. **91** (1924), 169–209.

[Pen93] R. C. Penner, *Universal constructions in Teichmüller theory*, Adv. Math. **98** (1993), no. 2, 143–215.

[Qui71] Daniel Quillen, *The Adams conjecture*, Topology **10** (1971), 67–80.

[Qui73] D. Quillen, *Higher algebraic $K$-theory I*, Algebraic $K$-Theory I (Berlin-Heidelberg-New York), Lecture Notes in Math., vol. 341, Springer, Berlin-Heidelberg-New York, 1973, pp. 85–147.

[Ram88] D. Ramakrishnan, *Regulators, algebraic cycles, values of $L$-functions*, Algebraic $K$-Theory and Algebraic Number Theory, Contemp. Math., vol. 83, American Math. Soc., 1988, pp. 183–310.

[Rap88] M. Rapoport, *Comparison of the regulators of Beĭlinson and of Borel*, Beĭlinson's Conjectures on Special Values of $L$-Functions, Academic press, 1988, pp. 169–192.

[RS71] D.B. Ray and I.M. Singer, *R-torsion and the Laplacian for Riemannian manifolds*, Adv. in Math. **7** (1971), 145–210.

[Seg73] Graeme Segal, *Configuration spaces and iterated loop spaces*, Invent. Math. **21** (1973), 213–221.

[Seg74] Graeme Segal, *Categories and cohomology theories*, Topology **13** (1974), 293–312.

[Str84] Kurt Strebel, *Quadratic Differentials*, Springer-Verlag, Berlin, 1984.

[Sul74] Dennis Sullivan, *Genetics of homotopy theory and the Adams conjecture*, Ann. of Math. (2) **100** (1974), 1–79.

[Sus81] A. A. Suslin, *On the equivalence of $K$-theories*, Comm. Algebra **9** (1981), no. 15, 1559–1566.

[Sus82] A. A. Suslin, *Stability in algebraic $K$-theory*, Algebraic $K$-Theory, Part I (Oberwolfach, 1980) (Berlin), Springer, Berlin, 1982, pp. 304–333.

[Swe69] M.E. Sweedler, *Hopf Algebras*, W.A. Benjamin, Inc., New York, 1969.

[vdK80] Wilberd van der Kallen, *Homology stability for linear groups*, Invent. Math. **60** (1980), no. 3, 269–295.

[Vol71] I. A. Volodin, *Algebraic $K$-theory as an extraordinary homology theory of the category of associative rings with unit*, Math. USSR Izv. 5 (1971), 859–887.

[Wag78] J. B. Wagoner, *Diffeomorphisms, $K_2$, and analytic torsion*, Algebraic and Geometric Topology (Proc. Sympos. Pure Math., Stanford Univ., Stanford, Calif., 1976), Part 1 (Providence, R.I.), Amer. Math. Soc., Providence, R.I., 1978, pp. 23–33.

[Wal78] Friedhelm Waldhausen, *Algebraic K-theory of generalized free products*, Ann. of Math. (2) **108** (1978), 135–256.

[Wal82] Friedhelm Waldhausen, *Algebraic K-theory of spaces, a manifold approach*, Current Trends in Algebraic Topology, Part 1 (London, Ont., 1981) (Providence, R.I.), Amer. Math. Soc., Providence, R.I., 1982, pp. 141–184.

[Wal85] Friedhelm Waldhausen, *Algebraic K-theory of spaces*, Algebraic and Geometric Topology (New Brunswick, N.J., 1983) (Berlin), Springer, Berlin, 1985, pp. 318–419.

[Wal87a] Friedhelm Waldhausen, *Algebraic K-theory of spaces, concordance and stable homotopy theory*, Algebraic Topology and Algebraic K-theory (William Browder, ed.), Annals of Math. Studies, vol. 113, 1987, pp. 392–417.

[Wal87b] Friedhelm Waldhausen, *An outline of how manifolds relate to algebraic K-theory*, Homotopy Theory (Durham, 1985) (Cambridge), Cambridge Univ. Press, Cambridge, 1987, pp. 239–247.

# Index